ANALYTICAL CHEMISTRY

ANALYTICAL CHEMISTRY

Qin Yang
Tao Lin
Yong He

Kruger Brentt
Publishers
2023

Kruger Brentt Publishers UK. LTD.
Company Number 9728962

Regd. Office: 68 St Margarets Road, Edgware, Middlesex HA8 9UU

ISBN: 9781787150386

For information on all our publications visit our website at http://krugerbrentt.com/

PREFACE

Analytical chemistry is concerned with the chemical characterization of matter, both qualitative and quantitative. It is important in nearly every aspect of our lives because chemicals make up everything we use. This text deals with the principles and techniques of quantitative analysis, that is, how to determine how much of a specific substance is contained in a sample. There is constant development and change in the techniques and methods of analytical chemistry. Better instrument design and a fuller understanding of the mechanics of analytical processes enable steady improvements to be made in sensitivity, precision, and accuracy. These same changes contribute to more economic analysis as they frequently lead to the elimination of time-consuming separation steps. The ultimate development in this direction is a non-destructive method, which not only saves time but leaves the sample unchanged for further examination or processing.

Analytical chemistry is a dynamic discipline. New chemicals and increasingly sophisticated instruments and computational capabilities are constantly coming into use to improve the ways in which chemical analyses are done. Some of these improvements involve the determination of ever smaller quantities of substances; others greatly shorten the time required for analysis; and some enable analysts to tell with much greater specificity the identities of a large number of compounds in a complex sample.

The present book entitled "*Analytical Chemistry*" contains sixteen chapters covering all related disciplines. These chapters include Introduction to analytical chemistry, calibrations, standardizations, and blank corrections, errors in chemical analyses, preparing solutions of analytical chemistry, standard addition and internal standards, importance of analytical methodology, standardizing analytical methods, the assessment of analytical data, gravimetric methods, introduction to electroanalytical techniques, titrimetric methods, spectroscopic methods: general, spectroscopy based on absorption, UV/VIS and IR spectroscopy, atomic absorption and emission spectroscopy and statistical treatment of data. Related terminology is given at the end for ready reference. It is an essential resource for students, city planners, and all who are concerned with the nature of contemporary urban development problems.

We are grateful to all those persons as well as various books, manuals, periodicals, magazines, journals etc. that helped in the preparation of this book. In spite of the best efforts, it is possible that some errors may have occurred into the compilation and editing of the book. Further queries, constructive suggestions and criticisms for the improvement of the book are always welcome and shall be thankfully acknowledged.

Qin Yang

Tao Lin

Yong He

CONTENTS

1

INTRODUCTION TO ANALYTICAL CHEMISTRY

1.1 INTRODUCTION

- Analytical chemistry can be defined as the science of making and interpreting chemical measurements. Chemistry is a quantitative science, and the ability to make measurements by various means is critical to obtaining a better understanding of nature. Chemical measurements permeate all aspects of our society, from blood tests to food analysis, environmental monitoring, and forensic analysis. It is important to understand the principles and practices underlying these measurements, their capabilities and limitations.
- Analytical chemistry broadly defines the field of chemical measurements for both qualitative and quantitative purposes. It encompasses both wet chemical and instrumental techniques and applications in wide ranging fields. Chemical measurements play an important role in science and society and an understanding of their capabilities and limitations is important.

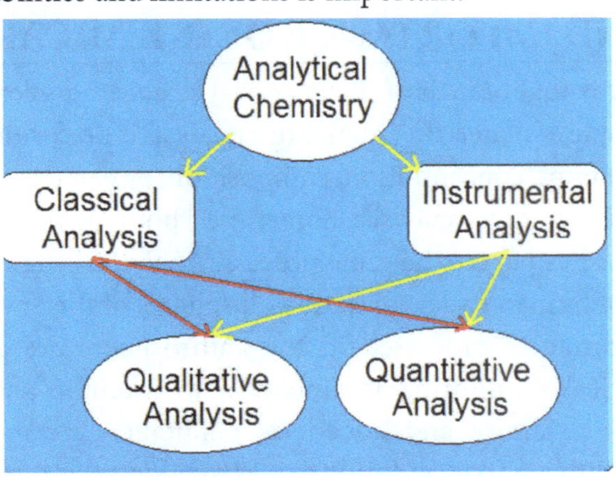

1.2 HISTORICAL PERSPECTIVE

- Antoine Lavoisier (1743-1794) was a French nobleman who is most widely regarded as the "father of modern chemistry". He achieved this distinction in large part because of the careful and systematic measurements he made in chemical experiments that allowed him to characterize the stoichiometry of reactions and predict the formulas of compounds. He was the first to recognize the importance of analytical measurements in advancing chemical theory and as a consequence was able to write the first modern chemical textbook.
- Early analytical measurements were quite limited by today's standards. Many qualitative tests for various chemical species based on chemical reactions were developed. These included tests for inorganic ions (*e.g.* Pb^{2+}, Ni^{2+}, Cu^{2+}) through the formation of precipitates or coloured solutions, and organic functionalities such as double bonds and aldehydes.
- Quantitative measurements were limited to mass and other physical properties such as pressure, as well as volumetric techniques such as titrations. Some of these measurements are still used today, and while restricted in their capabilities, were quite sophisticated. The era of modern analytical chemistry is most often considered to have begun with the introduction of the first electronic pH meter, designed by Arnold Beck-man in 1934 to measure the acidity of lemon juice. Although this device, based on vacuum tubes, was cumbersome in comparison with modern pH meters that can be contained in a device the size of a ballpoint pen, it heralded the dawn of modern analytical instrumentation. The pH meter was followed shortly thereafter by infrared and ultraviolet/visible spectrometers, as well as other instruments that made laboratory measurements simpler and more reliable. This was followed by an era of computerization and automation that has led to the very powerful instrumentation we see in laboratories today, instruments such as optical spectrometers, chromatographs, and mass spectrometers.

1.3 NATURE AND IMPORTANCE OF CHEMICAL ANALYSIS

Analytical chemistry is that branch of the chemical sciences employed to determine the composition of a sample of material. A qualitative analysis is performed to determine *what* is in a sample. The amount, concentration, composition, or percent of a substance present is determined by quantitative analysis. Sometimes both qualitative and quantitative analyses are performed as part of the same process. Analytical chemistry is important in practically all areas of human endeavor and in all spheres of the environment. Industrial raw materials and products processed in the anthrosphere are assayed by chemical analysis, and analytical monitoring is employed to monitor and control industrial processes. Hardness, alkalinity, and trace-level pollutants are measured in water by chemical analyses. Nitrogen oxides, sulfur oxides, oxidants, and organic pollutants are determined in air by chemical analysis. In the geosphere , fertilizer constituents in soil and commercially valuable minerals in ores are measured by chemical analysis. In the biosphere, xenobiotic materials and their metabolites are monitored by chemical analysis.

Analytical chemistry is a dynamic discipline. New chemicals and increasingly sophisticated instruments and computational capabilities are constantly coming into use to improve the ways in which chemical analyses are done. Some of these improvements involve the determination of ever smaller quantities of substances; others greatly shorten the time required for analysis; and some enable analysts to tell with much greater specificity the identities of a large number of compounds in a complex sample. Many of the calculations of analytical chemistry involve quantities of materials that take part in chemical reactions. Therefore, stoichiometric calculations may be very important in analytical chemistry.

1.4 CLASSIFICATION OF ANALYTICAL METHODS

The variety of analytical methods available to solve chemical problems today is quite extensive and there are often many techniques that can be used to determine the identity or concentration of a particular analyte. The selection of the most appropriate method to solve a specific problem is governed by a number of parameters associated with the analyte and the environment in which it exists. To aid in this selection, analytical methods can be classified in a variety of ways, some of which are described below.

1.4.1 Classification by Type of Result

One way to classify analytical problems is in terms the question that is to be answered. In *qualitative analysis*, the fundamental question relates to what is present rather than how much. This question can be formulated in a number of ways. One may simply be looking to determine the presence or absence of a specific analyte (*e.g.* Are there pesticides present in my drinking water?), or may be looking to identify a component of a mixture (.*e.g.* What is the contaminant in the milk?). Alternatively, one may seek the composition of an unknown mixture (*e.g.* What was the white powder found at the crime scene?) on a molecular or elemental level.

Quantitative analysis refers to those cases where the amount or concentration of a particular analyte is to be determined (*e.g.* What is the concentration of lead in my drinking water?, What is my blood sugar level?). In most introductory classes on analytical chemistry, including this one, the emphasis is placed on quantitative analysis. This is not only because of the importance of developing quantitative skills that are useful in all areas of chemistry, but also because many techniques for qualitative analysis (*e.g.* IR and NMR spectroscopy) are closely connected to particular areas of application and are better taught in the context of those subjects.

The distinction between qualitative and quantitative analysis is often blurred. It is necessary, for example, to know the identity of an analyte before it can be measured.

Some would argue that the qualitative detection of the presence of an analyte in a sample is simply an indication that its concentration is above some minimum detectable level, and is therefore quantitative. Finally, even quantitative measurements are often ultimately used to make some qualitative decision (*e.g.* Is the water safe to drink? What is the patient's illness? Does the hair at the crime scene match that of the suspect.) Nevertheless, these classifications are often useful in practice.

1.4.2 Classification by Technique

In a broad classification by technique, analytical methods are sometimes distinguished as "wet" chemical methods or instrumental methods. The former classification typically refers to methods where the analytical determination is based on traditional measurements of volume (*i.e.* titrations) or mass (gravimetric methods). These techniques are generally based on chemical reactions. For example, if one wanted to determine the concentration of a hydrochloric acid solution, a titration where it is neutralized with a strong base such as sodium hydroxide could be used. If one wanted to determine the amount of sulfate ion (SO_2^{-4}) present in a solution, it could be reacted with excess barium ions (Ba^{2+}) in another solution to form a solid barium sulfate precipitate ($BaSO^4$) that could be dried and weighed.

In contrast, instrumental methods are those in which the final measurement to be related to analyte concentration is made on an electronic instrument, such as a spectrometer or pH meter. (Students often refer to these as "machines", but "instrument" is more accurate since the primary purpose of the devices is not mechanical work). Again, the distinction between wet chemical and instrumental methods is blurred because modern balances are all electronic and most instrumental methods involve gravimetric and volumetric procedures to prepare samples for analysis.

While both of these strategies are treated in this class, the emphasis is on instrumental methods since these are more widely used for most analytical procedures today. There is often a perception that, because instrumental methods are a more recent development that they give better results than wet chemical methods. In fact, the opposite is generally true and methods such as titrations generally give superior accuracy when they can be employed.

1.4.3 Classification by Analyte Concentration

Often an important parameter to be considered in a chemical analysis is the anticipated concentration of the analyte. This can be broadly classified in the following ways:

Major component	: 1% to 100%
Minor component	: 0.01% to 1%
Trace component	: 1 x 10-5 % to 0.01% (100 ppb to 100 ppm)
Ultratrace component	: <100 ppb

The concentrations given above are in terms of mass fractions of the analyte, and "ppm" and "ppb" refer to "parts per million" and "parts per billion", respectively. (Concentration units are discussed in more detail later.) These categories are only approximate guidelines.

In general, the lower the concentration of the analyte, the more difficult it is to determine in a sample, so much effort in analytical research is directed at lowering the level at which a particular analyte can be detected. Most of the emphasis in this class is directed at minor or trace components of a mixture.

1.4.4 Other Classifications

Analytical methods can also be classified in many other ways as well. For example, instrumental techniques can be further divided by the physical means that are used to determine the analyte. Thus, spectroscopic methods exploit the interaction of light and matter, while electrochemical methods utilize various electrical measurements (voltage, current, etc.) to determine concentration. Classification according to sample type is also common. For example, if the analysis relates to biological samples, it is referred to as a bioanalytical method.

1.5 SCOPE OF ANALYTICAL CHEMISTRY

Analytical measurements permeate virtually all aspects of our society because the need for chemical information is critical to our general well-being and understanding of nature. If you have ever had a blood test or had your well water tested, then analytical chemistry has been involved. Below are some application areas and samples of questions that may be asked.

Environment	Are pesticides present in a soil sample?
	What is the concentration of Pb in drinking water?
	What is the acidity of lake?
	Who is responsible for an oil spill?
Health/Medicine:	What is a patient's cholesterol level?
	Is an athelete on steroids?
	Is a woman pregnant?
Law/Forensics	What was a suspect's blood alcohol level?
	Was a victim drugged or poisoned?
	What type of accelerant was used for arson?

There are many other examples as well. For example, in industry, analytical measurements are widely used to monitor production processes to end product quality to ensure optimum performance. In food science, analytical techniques are used confirm product quality (*e.g.* in breweries) and the absence of harmful contaminants. In geology, the analysis of ore samples is critical to assess the quality of minerals that can be obtained from a site. Even in areas such as art and archeology, analytical chemistry plays a major role in determining the authenticity of paintings or artifacts.

The answers provided by analytical chemistry can often have profound implications to individuals and society. Athletes have been stripped of their medals based on blood tests, artworks have been rendered worthless through an analysis of paint samples, and major projects have been initiated on the basis of geological or environmental analysis. It is therefore important to understand the nature of analytical measurements, their capabilities, and their limitations.

1.6 TYPES OF ANALYSIS

Analytical chemistry can be broken down into two general areas of analysis :

Qualitative analysis and Quantitative Analysis and each one of these two types can be either classical or instrumental as we will see.

1.6.1 Qualitative Classical Analysis

This analysis detects (identify) the type of all or some of the substances present in the sample (elements or ions or compounds) . In other words it gives an answer to the question : What substances are present in a sample ? these substances can be detected either by a chemical reaction For example, when you add silver nitrate solution to the sample solution a white precipitate formation indicates the presence of chloride ion in the sample.

There are also many reagents that give distinctive colors with some of the substances and can be used in the detection of these substances . The flame also can be used for the detection of some common metals.

1.6.2 Qualitative Instrumental Analysis

Currently there are many instruments that are separate and distinguish substances (organic or inorganic) in the sample, such as gas chromatography – mass spectrometry (GC – MS) , High performance liquid chromatograophy - mass spectrometry (HPLC – MS) , infrared spectra (ir) and induced coupled plasma – mass spectrometry (ICP – MS) or ICP – AES (Atomic Emission Spectrometry) .

Classical Chemical Analysis

Which depends on the chemical reaction, such as volumetric analysis and gravimetric analysis, as we shall see later in this book . And it uses simple equipment such as burettes, balances , flame , furnace . And is used to estimate high concentrations (more than 0.001 M) .

Instrumental Analysis :

This type of analysis uses instruments and depends on the physical and physico-chemical properties of the substance being analyzed (analyte) such as absorption or emission of electromagnetic radiation (spectroscopic methods of analysis) or electrical properties of the substance being analyzed such as voltage or current intensity or electrical conductivity ...... etc. (electrochemical methods of analysis) and finally the methods of separation such chromatography .

All these methods are used to estimate low concentrations of the analyte (less than 0.001 M) , but most of the instrumental analysis methods require expensive instruments .

1.7 MAIN STEPS IN CHEMICAL ANALYSIS

1.7.1 Goal Setting

We must set the target of the analysis and how to access it at the lowest cost and fastest time while maintaining good precision (the duplicate results are near to each other) and good accuracy (the mean of the results of the analysis is very close to the real value) .

1.7.2 Sampling

The number of samples taken depends on the homogeneity of this thing example homogeneous air (few samples) and a liquid such as water is less homogeneous , so larger number of samples should be taken while the solid - state which is non - homogeneous such as soil or rock very larger number of samples must be taken and so on .

1.7.3 Drying the Sample

The solid samples usually contain varying amounts of adsorbed water which is a factor cannot be controlled therefore we better get rid of it before weighing the sample by drying in the oven and if the analyte is thermally unstable we can use the desiccator for drying.

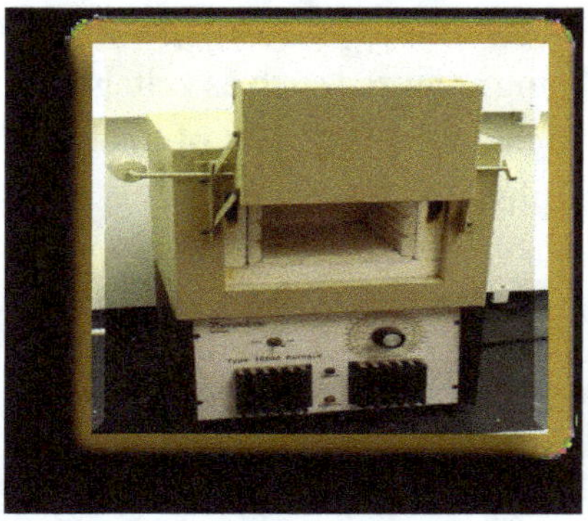

1.7.4 Sample Dissolution

The chemical reactions are very slow in the solid state compared to in the solution , therefore the sample must be dissolve before analysis. To dissolve the sample we try the water first then dilute acids solutions , hot concentrated acids solutions , mixtures of acids solutions and If the sample does not dissolve we use the fusion where an acidic sample is melted with a base material such as Na_2CO_3 and a basic sample is melted with an acidic substance such as $K_2S_2O_7$ and if the sample is an oxidant it can be melted with a reductant material and Conversely , where products of the fusion can be dissolved in a water or dilute acid solutions .

Organic compounds are soluble in organic solvents but we should choose the solvent which does not interfere in the measurement process . In the case of estimating inorganic elements in organic samples we must get rid of the organic matter by dry burning (dry ashing) in the oven or by wet burning (wet ashing) using hot oxidizing concentrated acids such as HNO_3 or $HClO_4$ where the organic matter is converted to water vapor and carbon dioxide and therefore we prevent the intervention of organic matter in the analysis of inorganic elements.

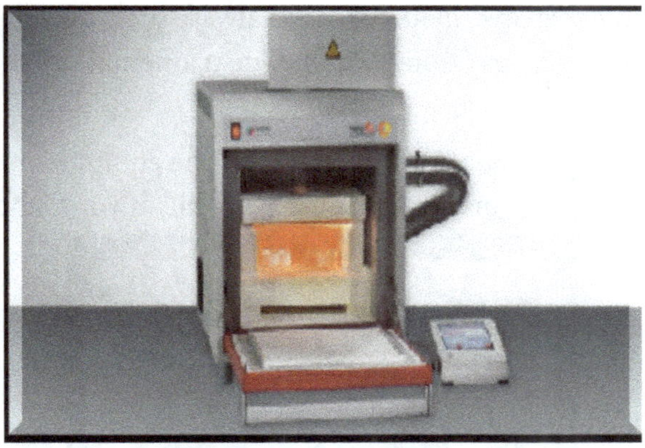

1.7.5 Storage of Sample Solution

During the storage , contamination or loss or disintegration must be avoided . The sample solution must also be stored in a container made from an inert material so that the sample does not interact with it and not be adsorbed on the inner walls of the container . The container must be closed tightly to avoid the affect of the air on the chemical composition of the sample. The sun and heat must be also taken into account because they may had an impact on the sample composition.

1.7.6 Interferences

An interference is a substance that interfere in the analyte determination for example , cobalt ion Co^{2+} can be detected by adding thiocyanate SCN- reagent which reacts with the cobalt to form a red complex, but also iron ion Fe^{3+} forms red complex with SCN- .Therefore Fe^{3+} will interfere in the detection of Co^{2+} and must be eliminated by any means before the analysis of Co^{2+} e.g. by separating iron from cobalt by extraction , precipitation or any other means.

Eliminating Interferences

Specific method of analysis or specific Reagent is a special analysis method for a single substance and can not respond to any other substance on other words , there is no interference . Unfortunately these specific reagents or methods of analysis are practically non-existent . All reagent or methods of analysis are either selective (respond to a limited number of substances) and the possibility of the existence of these substances with each other in the sample may be minimal or non-selective (respond to a large number of substances) and the possibility of interferences is too high .

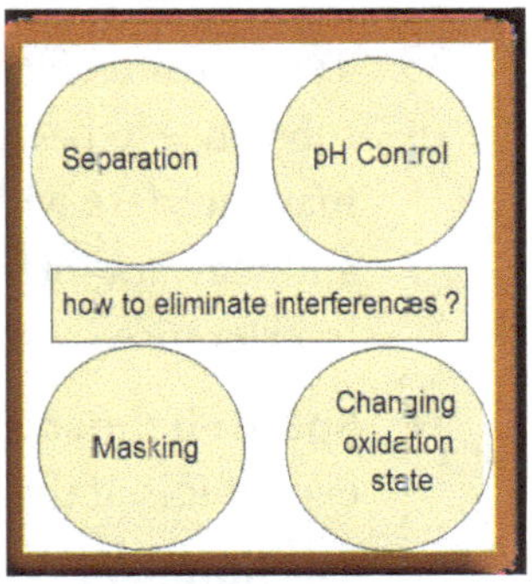

1.7.7 Selecting a Method of Analysis

There are many methods of analysis to estimate a particular substance in a sample, and to choose one of them depends on several factors such as :

- **Cost:** Some inexpensive methods of analysis, such as volumetric analysis and gravimetric analysis but most of the instrumental methods of analysis are expensive.
- **Availability of equipments and materials:** The equipments and the materials of the chosen method of analysis must be available in your laboratory.

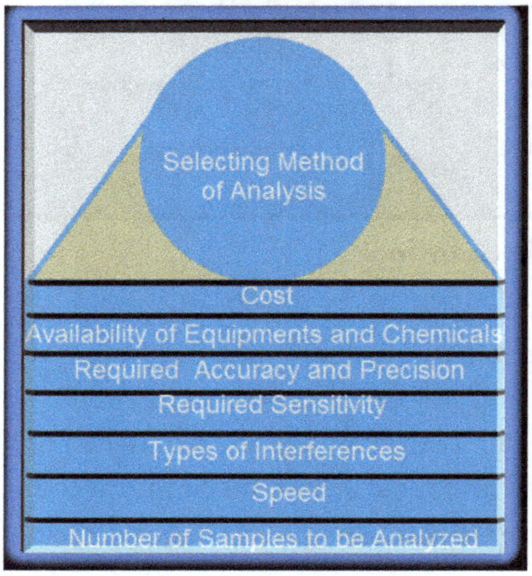

- **Accuracy and precision needed:** Usually the accurate and precise methods of analysis are expensive . Therefore when for example estimating the salt in sea water there is no need to choose an expensive method because a simple mistake will not result in a disaster but if we analyze mercury in sea water we must choose an accurate and precise method even if it is expensive because mercury is poisonous and any error could result in disaster .
- **Sensitivity :** The selected method must be able to determine the analyte in the sample , for example if the concentration of the analyte in the sample around 0.00001 M the chosen method must be able to determine 0.00001 M in other words its sensitive should be less than 0.00001 M .
- **Type of interferences :** We must know the type of impurities existing with the substance to be analyzed in the sample because these impurities or some of which may interfere with the substance to be determined in certain method of analysis and may not interfere in another method . So we choose the method in which the interferences are the least (more selective).
- **Speed :** If the analysis requires the results as soon as we must choose a fast method even if it is at the expense of accuracy and precision .
- **The number of samples to be analyzed :** If the number of samples to be analyzed is high we must choose an automatic and fast method of analysis . And to choose a suitable method for analyzing a certain substance in a certain sample go back to the following approved journals in analytical chemistry: Analytical Abstract

- Analytical Chemistry –Analytica Chimica Acta - Analytical Letters –Analyst - Talanta and many others .

1.7.8 Measurement

Measure the concentration of analyte in several aliquots. The purpose of replicate measurements (repeated measurements) is to assess the variability (uncertainty) in the analysis and to guard against a gross error in the analysis of a single aliquot. The uncertainty of a measurement is as important as the measurement itself, because it tells us how reliable a measurement is. If necessary, use different analytical methods on similar samples to make sure that each method gives the same result and that the choice of analytical method is not biasing the result. You may also wish to construct and analyse several different bulk samples to see what variations arise from your sampling procedure.

1.7.9 Calculating and Evaluating the Results

By knowing the weight of the sample taken for analysis and the amount of analyte in this sample from the results of the analysis, we can calculate the percentage of the analyte in the sample . The results of any chemical analysis must be statistically evaluated otherwise they are not useful.

2

CALIBRATIONS, STANDARDIZATIONS, AND BLANKCORRECTIONS

2.1 INTRODUCTION

We introduced a relationship between the measured signal,S_{meas}, and the absolute amount of analyte

$$Smeas = kn\text{A} + \text{Sreag} \quad \dots\dots\dots 1$$

or the relative amount of analyte in a sample

$$Smeas = kC\text{A} + Sreag \quad \dots\dots\dots 2$$

where nA is the moles of analyte, CA is the analyte's concentration, k is the method's sensitivity, and Sreag is the contribution to Smeas from constant errors introduced by the reagents used in the analysis. To obtain an accurate value for nA or CA it is necessary to avoid determinate errors affecting Smeas, k, and Sreag. This is accomplished by a combination of calibrations, standardizations, and reagent blanks.

2.2 CALIBRATING SIGNALS

Signals are measured using equipment or instruments that must be properly calibrated if Smeas is to be free of determinate errors. Calibration is accomplished against a standard, adjusting Smeas until it agrees with the standard's known signal. Several common examples of calibration are discussed here.

When the signal is a measurement of mass, Smeas is determined with an analytical balance. Before a balance can be used, it must be calibrated against a reference weight meeting standards established by either the National Institute for Standards and Technology or the American Society for Testing and Materials. With an electronic balance the sample's mass is determined by the current required to generate an upward electromagnetic force counteracting the sample's downward gravitational force. The balance's calibration procedure invokes an internally programmed calibration routine specifying the reference weight to be used. The reference weight is placed on the balance's weighing pan, and the relationship between the displacement of the weighing pan and the counteracting current is automatically adjusted.

Calibrating a balance, however, does not eliminate all sources of determinate error. Due to the buoyancy of air, an object's weight in air is always lighter than its weight in vacuum. If there is a difference between the density of the object being weighed and the density of the weights used to calibrate the balance, then a correction to the object's weight must be made. An object's true weight in vacuo, Wv, is related to its weight in air, Wa, by the equation

$$W_v = W_a \times \left[1 + \left(\frac{1}{D_o} - \frac{1}{D_w}\right) \times 0.0012\right]$$

where Do is the object's density, Dw is the density of the calibration weight, and 0.0012 is the density of air under normal laboratory conditions (all densities are in units of g/cm3). Clearly the greater the difference between Do and Dw the more serious the error in the object's measured weight.

The buoyancy correction for a solid is small, and frequently ignored. It may be significant, however, for liquids and gases of low density. This is particularly important when calibrating glassware. For example, a volumetric pipette is calibrated by carefully filling the pipette with water to its calibration mark, dispensing the water into a tared beaker and determining the mass of water transferred. After correcting for the buoyancy of air, the density of water is used to calculate the volume of water dispensed by the pipette.

Balances and volumetric glassware are examples of laboratory equipment. Laboratory instrumentation also must be calibrated using a standard providing a known response. For example, a spectrophotometer's accuracy can be evaluated by measuring the absorbance of a carefully prepared solution of 60.06 ppm $K_2Cr_2O_7$ in 0.0050 M H_2SO_4, using 0.0050 M H_2SO_4 as a reagent blank. The spectrophotometer is considered calibrated if the resulting absorbance at a wavelength of 350.0 nm is 0.640 ± 0.010 absorbance units. Be sure to read and carefully follow the calibration instructions provided with any instrument you use.

2.3 STANDARDIZING METHODS

The American Chemical Society's Committee on Environmental Improvement defines standardization as the process of determining the relationship between the measured signal and the amount of analyte. A method is considered standardized when the value of k in equation 1 or 2 is known.

In principle, it should be possible to derive the value of k for any method by considering the chemical and physical processes responsible for the signal. Unfortunately, such calculations are often of limited utility due either to an insufficiently developed theoretical model of the physical processes or to non-ideal chemical behavior. In such situations the value of k must be determined experimentally by analyzing one or more standard solutions containing known amounts of analyte. In this section we consider several approaches for determining the value of k. For simplicity we will assume that Sreag has been accounted for by a proper reagent blank, allowing us to replace Smeas in equations 1 and.2 with the signal for the species being measured.

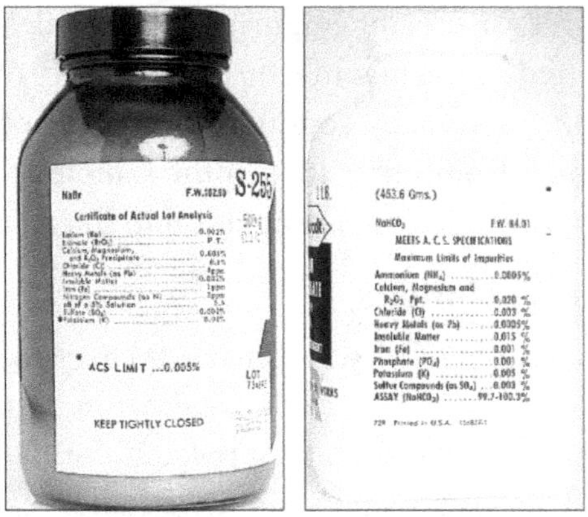

(a) (b)

2.3.1 Reagents Used as Standards

The accuracy of standardization depends on the quality of the reagents and glass- ware used to prepare standards. For example, in an acid–base titration, the amount of analyte is related to the absolute amount of titrant used in the analysis by the stoichiometry of the chemical reaction between the analyte and the titrant. The amount of titrant used is the product of the signal (which is the volume of titrant) and the titrant's concentration. Thus, the accuracy of a titrimetric analysis can be no better than the accuracy to which the titrant's concentration is known.

2.3.2 Primary Reagents

Reagents used as standards are divided into primary reagents and secondary reagents. A primary reagent can be used to prepare a standard con- taining an accurately known amount of analyte. For example, an accurately weighed sample of 0.1250 g $K_2Cr_2O_7$ contains exactly 4.249 x 10^{-4} mol of $K_2Cr_2O_7$. If this same sample is placed in a 250-mL volumetric flask and diluted to volume, the con- centration of the resulting solution is exactly 1.700 x 10^{-3} M. A primary reagent must have a known stoichiometry, a known purity (or assay), and be stable during long-term storage both in solid and solution form. Because of the difficulty in establishing the degree of hydration, even after drying, hydrated materials usually are not considered primary reagents. Reagents not meeting these criteria are called secondary reagents. The purity of a secondary reagent in solid form or the concentration of a standard prepared from a secondary reagent must be determined relative to a primary reagent. Lists of acceptable primary reagents are available.

2.3.3 Other Reagents

Preparing a standard often requires additional substances that are not primary or secondary reagents. When a standard is prepared in solution, for ex- ample, a suitable solvent and solution matrix must be used. Each of these solvents and reagents is a potential source of additional analyte that, if unaccounted for, leads to a determinate error. If available, reagent grade chemicals conforming to standards set by the American Chemical Society should be used. The packaging label included with a reagent grade chemical (Figure 1) lists either the maximum allowed limit for specific impurities or provides the actual assayed values for the impurities as reported by the manufacturer. The purity of a reagent grade chemical can be improved by purification or by conducting a more accurate assay. As discussed later, contributions to Smeas from impurities in the sample matrix can be compensated for by including an appropriate blank determination in the analytical procedure.

2.3.4 Preparing Standard Solutions

Solutions of primary standards generally are pre- pared in class A volumetric glassware to minimize determinate errors. Even so, the relative error in preparing a primary standard is typically ±0.1%. The relative error can be improved if the glassware is first calibrated as described in Example 1. It also is possible to prepare standards gravimetrically by taking a known mass of standard, dissolving it in a solvent, and weighing the resulting solution. Relative errors of ±0.01% can typically be achieved in this fashion.

It is often necessary to prepare a series of standard solutions, each with a different concentration of analyte. Such solutions may be prepared in two ways. If the range of concentrations is limited to only one or two orders of magnitude, the solutions are best prepared by transferring a known mass or volume of the pure standard to a volumetric flask and diluting to volume. When working with larger concentration ranges, particularly those extending over more than three orders of magnitude, standards are best prepared by a serial dilution from a single stock solution. In a serial dilution a volume of a concentrated stock solution, which is the first standard, is diluted to prepare a second standard. A portion of the second standard is then diluted to prepare a third standard, and the process is repeated until all necessary standards have been prepared. Serial dilutions must be prepared with extra care because a determinate error in the preparation of any single standard is passed on to all succeeding standards.

2.3.4.1 Single-Point versus Multiple-Point Standardizations*

The simplest way to determine the value of k in equation 5.2 is by a single-point standardization. A single standard containing a known concentration of analyte, CS, is prepared and its signal, Sstand, is measured. The value of k is calculated as

$$k = \frac{S_{\text{stand}}}{C_S} \qquad 3$$

A single-point standardization is the least desirable way to standardize a method. When using a single standard, all experimental errors, both determinate and indeterminate, are carried over into the calculated value for Any uncertainty in the value of k increases the uncertainty in the analyte's concentration. In addition, equation 3 establishes the standardization relationship for only a single concentration of analyte. Extending equation 3 to samples containing concentrations of analyte different from that in the standard assumes that the value of k is constant, an assumption that is often not true. Figure 2 shows how assuming a constant value of k may lead to a determinate error. Despite these limitations, single-point standardizations are routinely used in many laboratories when the analyte's range of expected concentrations is limited. Under these con ditions it is often safe to assume that k is constant (although this assumption should be verified experimentally). This is the case, for example, in clinical laboratories where many automated analyzers use only a single standard.

The preferred approach to standardizing a method is to prepare a series of standards, each containing the analyte at a different concentration. Standards are chosen such that they bracket the expected range for the analyte's concentration. Thus, a multiple-point standardization should use at least three standards, although more are preferable. A plot of Sstand versus CS is known as a calibration curve. The exact standardization, or calibration relationship, is determined by an appropriate curve-fitting algorithm.* Several approaches to standard- ization are discussed in the following sections.

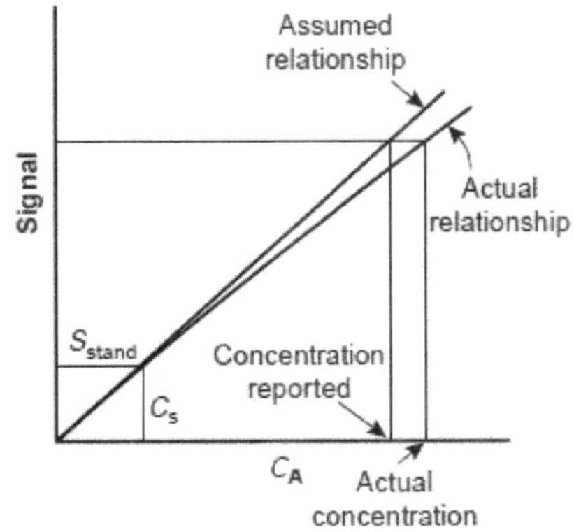

Fig. : Example showing how an impoper use of a single-point standardization can lead to a determinate error in the reported concentration of analyte.

2.3.4.2 External Standards

The most commonly employed standardization method uses one or more external standards containing known concentrations of analyte. These standards are identified as external standards because they are prepared and analyzed separately from the samples.

A quantitative determination using a single external standard was described at the beginning of this section, with k given by equation 3. Once standardized, the concentration of analyte, CA, is given as:

$$C_A = \frac{S_{samp}}{k} \quad 4$$

A multiple-point external standardization is accomplished by constructing a calibration curve, two examples of which are shown in Figure.3. Since this is the most frequently employed method of standardization, the resulting relation- ship often is called a normal calibration curve. When the calibration curve is a linear (Figure 3a), the slope of the line gives the value of k. This is the most desirable situation since the method's sensitivity remains constant throughout the standard's concentration range. When the calibration curve is nonlinear, the method's sensitivity is a function of the analyte's concentration. In Figure 3b, for example, the value of k is greatest when the analyte's concentration is small and decreases continuously as the amount of analyte is increased. The value of k at any point along the calibration curve is given by the slope at that point. In either case, the calibration curve provides a means for relating S_{samp} to the analyte's concentration.

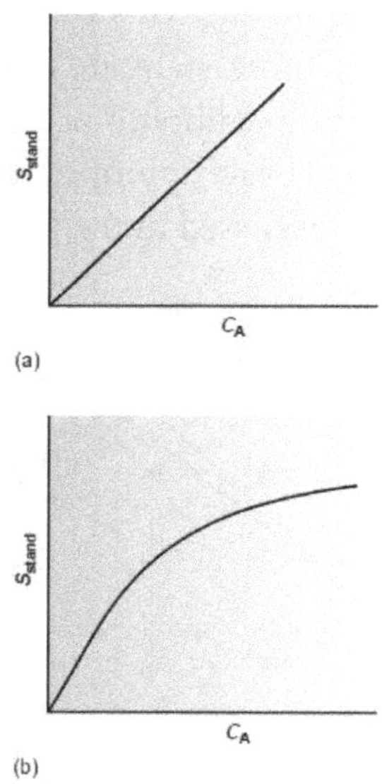

Figure 5.3
Examples of (a) straight-line and (b) curved normal calibration curves.

Fig. : Examples of (a) straight-line and (b) curved normal calibration curves

An external standardization allows a related series of samples to be analyzed using a single calibration curve. This is an important advantage in laboratories where many samples are to be analyzed or when the need for a rapid throughput of samples is critical. Not surprisingly, many of the most commonly encountered quantitative analytical methods are based on an external standardization.

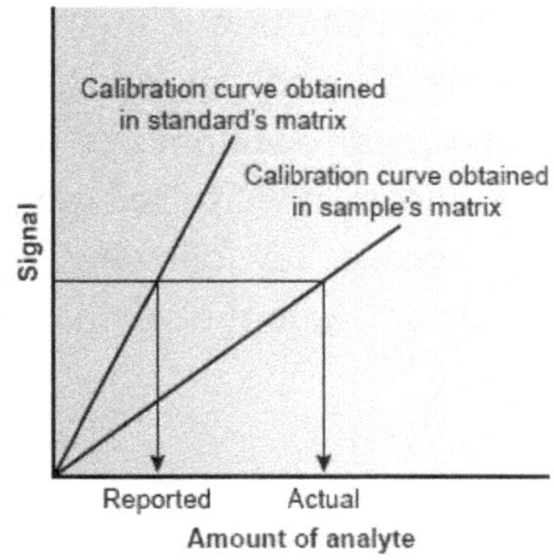

Fig. : Effect of the sample's matrix on a normal calibration curve

There is a serious limitation, however, to an external standardization. The relationship between Sstand and CS in equation 3 is determined when the analyte is present in the external standard's matrix. In using an external standardization, we assume that any difference between the matrix of the standards and the sample's matrix has no effect on the value of k. A proportional determinate error is introduced when differences between the two matrices cannot be ignored. This is shown in Figure 5.4, where the relationship between the signal and the amount of analyte is shown for both the sample's matrix and the standard's matrix. In this example, using a normal calibration curve results in a negative determinate error. When matrix problems are expected, an effort is made to match the matrix of the standards to that of the sample. This is known as matrix matching. When the sample's matrix is unknown, the matrix effect must be shown to be negligible, or an alternative method of standardization must be used. Both approaches are discussed in the following sections.

2.3.5 Standard Additions

The complication of matching the matrix of the standards to that of the sample can be avoided by conducting the standardization in the sample. This is known as the method of standard additions.

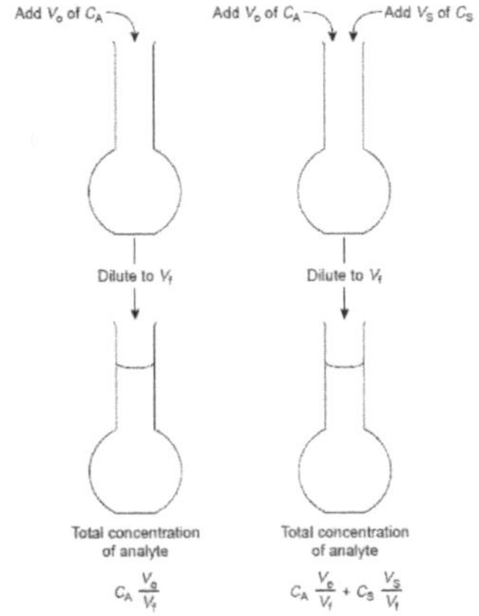

Fig. : Illustration showing the method of standard additions in which separate aliquots of sample are diluted to the same final volume. One aliquot of sample is spiked with a known volume of a standard solution of analyte before diluting to the final volume

The simplest version of a standard addition is shown in Figure 5. A volume, Vo, of sample is diluted to a final volume, Vf, and the signal, Ssamp is measured. A second identical aliquot of sample is Sspiked with a volume, Vs, of a standard solution for which the analyte's concentration, CS, is known. The Sspiked sample is diluted to the same final volume and its signal, Sspike, is recorded. The following two equations relate Ssamp and Sspike to the concentration of analyte, CA, in the original sample

$$S_{samp} = kC_A \frac{V_o}{V_f}$$

$$S_{spike} = k\left(C_A \frac{V_o}{V_f} + C_S \frac{V_s}{V_f}\right)$$

where the ratios Vo/Vf and Vs/Vf account for the dilution. As long as Vs is small relative to Vo, the effect of adding the standard to the sample's matrix is insignificant, and the matrices of the sample and the Sspiked sample may be considered identical. Under these conditions the value of k is the same in equations 5 and 6. Solving both equations for k and equating gives Equation 7 can be solved for the concentration of analyte in the original sample.

$$\frac{S_{samp}}{C_A(V_o/V_f)} = \frac{S_{spike}}{C_A(V_o/V_f) + C_S(V_s/V_f)}$$

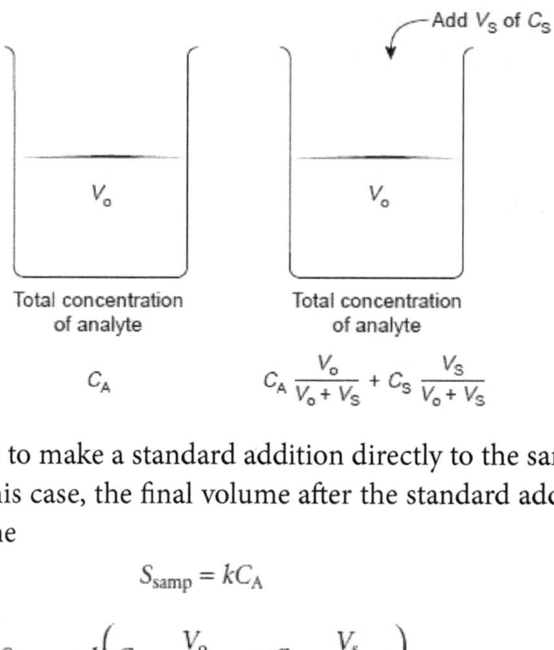

It also is possible to make a standard addition directly to the sample after measuring S_{samp} (Figure 6). In this case, the final volume after the standard addition is Vo + Vs and equations 5–7 become

$$S_{samp} = kC_A$$

$$S_{spike} = k\left(C_A \frac{V_o}{V_o + V_s} + C_S \frac{V_s}{V_o + V_s}\right)$$

$$\frac{S_{samp}}{C_A} = \frac{S_{spike}}{C_A[V_o/(V_o + V_s)] + C_S[V_s/(V_o + V_s)]}$$

The single-point standard additions outlined in Examples 4 and 5 are easily adapted to a multiple-point standard addition by preparing a series of Sspiked samples containing increasing amounts of the standard. A calibration curve is prepared by plotting Sspike versus an appropriate measure of the amount of added standard. Figure 7 shows two examples of a standard addition calibration curve based on equation 6. In Figure 7(a) Sspike is plotted versus the volume of the standard solution Sspikes, Vs. When k is constant, the calibration curve is linear, and it is easy to show that the x-intercept's absolute value is $C_A V_o / C_S$.

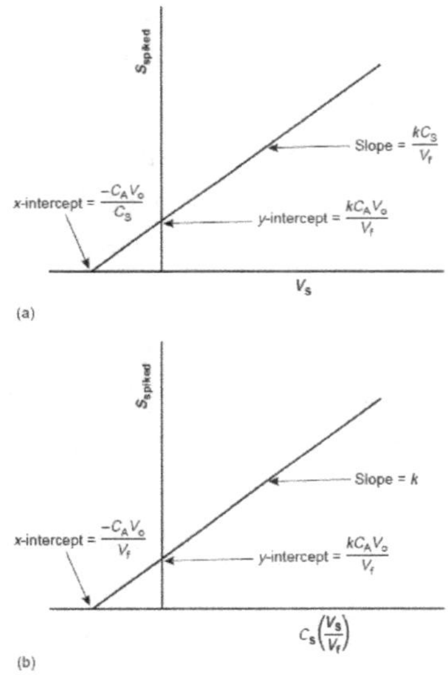

Since both Vo and CS are known, the x-intercept can be used to calculate the analyte's concentration.

Figure 7(b) shows the relevant relationships when Sspike is plotted versus the concentrations of the Sspiked standards after dilution. Standard addition calibration curves based on equation 8 are also possible.

Since a standard additions calibration curve is constructed in the sample, it cannot be extended to the analysis of another sample. Each sample, therefore, re- quires its own standard additions calibration curve. This is a serious drawback to the routine application of the method of standard additions, particularly in laboratories that must handle many samples or that require a quick turnaround time. For example, suppose you need to analyze ten samples using a three-point calibration curve. For a normal calibration curve using external standards, only 13 solutions need to be analyzed (3 standards and 10 samples). Using the method of standard additions, however, requires the analysis of 30 solutions, since each of the 10 samples must be analyzed three times (once before spiking and two times after adding successive Sspikes).

The method of standard additions can be used to check the validity of an external standardization when matrix matching is not feasible. To do this, a normal calibration curve of stand versus CS is constructed, and the value of k is determined from its slope. A standard additions calibration curve is then constructed using equation 6, plotting the data as shown in Figure 7(b). The slope of this standard additions calibration curve gives an independent determination of k. If the two val- use of k are identical, then any difference between the sample's matrix and that of the external standards can be ignored. When the values of k are different, a proportional determinate error is introduced if the normal calibration curve is used.

2.3.6 Internal Standards

The successful application of an external standardization or the method of standard additions, depends on the analyst's ability to handle samples and standards repro- ducibly. When a procedure cannot be controlled to the extent that all samples and standards are treated equally, the accuracy and precision of the standardization may suffer. For example, if an analyte is present in a volatile solvent, its concentration will increase if some solvent is lost to evaporation. Suppose that you have a sample and a standard with identical concentrations of analyte and identical signals. If both experience the same loss of solvent their concentrations of analyte and signals will continue to be identical. In effect, we can ignore changes in concentration due to evaporation provided that the samples and standards experience an equivalent loss of solvent. If an identical standard and sample experience different losses of solvent, however, their concentrations and signals will no longer be equal. In this case, an external standardization or standard addition results in a determinate error.

A standardization is still possible if the analyte's signal is referenced to a signal generated by another species that has been added at a fixed concentration to all samples and standards. The added species, which must be different from the ana- lyte, is called an internal standard.

Since the analyte and internal standard in any sample or standard receive the same treatment, the ratio of their signals will be unaffected by any lack of repro- ducibility in the procedure. If a solution contains an analyte of concentration CA, and an internal standard of concentration, CIS, then the signals due to the analyte, SA, and the internal standard, SIS, are

$$S_A = k_A C_A$$

$$S_{IS} = k_{IS} C_{IS}$$

where k_A and k_{IS} are the sensitivities for the analyte and internal standard, respectively. Taking the ratio of the two signals gives

$$\frac{S_A}{S_{IS}} = \frac{k_A}{k_{IS}} \times \frac{C_A}{C_{IS}} = K \times \frac{C_A}{C_{IS}} \qquad 5.10$$

Because equation 5.10 is defined in terms of a ratio, K, of the analyte's sensitivity and the internal standard's sensitivity, it is not necessary to independently deter- mine values for either kA or kIS.

In a single-point internal standardization, a single standard is prepared, and K is determined by solving equation 5.10

$$K = \left(\frac{C_{IS}}{C_A}\right)\left(\frac{S_A}{S_{IS}}\right)_{stand} \qquad 5.11$$

$$C_A = \left(\frac{C_{IS}}{K}\right)\left(\frac{S_A}{S_{IS}}\right)_{samp}$$

A single-point internal standardization has the same limitations as a single- point normal calibration. To construct an internal standard calibration curve, it is necessary to prepare several standards containing different concentrations of analyte. These standards are usually prepared such that the internal standard's concentration is constant. Under these conditions a calibration curve of (SA/SIS)stand versus CA is linear with a slope of K/CIS.

When the internal standard's concentration cannot be held constant the data must be plotted as (SA/SIS)stand versus CA/CIS, giving a linear calibration curve with a slope of K.

2.4 LINEAR REGRESSION AND CALIBRATION CURVES

In a single-point external standardization, we first determine the value of k by measuring the signal for a single standard containing a known concentration of analyte. This value of k and the signal for the sample are then used to calculate the concentration of analyte in the sample (see Example 2). With only a single determination of k, a quantitative analysis using a single-point external standardization is straightforward. This is also true for a single-point standard addition (see Examples 4 and 5) and a single-point internal standardization (see Example 8).

A multiple-point standardization presents a more difficult problem. Consider the data in Table 5.1 for a multiple-point external standardization. What is the best estimate of the relationship between Smeas and CS? It is tempting to treat this data as five separate single-point standardizations, determining k for each standard and reporting the mean value. Despite its simplicity, single-point standardizations, determining k for each standard and reporting the mean value. Despite its simplicity, this is not an appropriate way to treat a multiple-point standardization.

In a single-point standardization, we assume that the reagent blank (the first row in Table 1) corrects for all constant sources of determinate error. If this is not the case, then the value of k determined by a single- point standardization will have a determinate error.

Table 2.1: Data for Hypothetical Multiple Point External Standardization

C_s	S_{meas}
0.000	0.00
0.100	12.36
0.200	24.83
0.300	35.91
0.400	48.79
0.500	60.42

Table 2 demonstrates how an uncorrected constant error affects our determination of k. The first three columns show the concentration of analyte, the true measured signal (no constant error) and the true value of k for five standards. As expected, the value of k is the same for each standard. In the fourth column a constant determinate error of +0.50 has been added to the measured signals. The corresponding val- ues of k are shown in the last column. Note that a different value of k is obtained for each standard and that all values are greater than the true value. As we noted, this is a significant limitation to any single-point standardization.

How do we find the best estimate for the relationship be- tween the measured signal and the concentration of analyte in a multiple-point standardization? Figure 8 shows the data in Table 1 plotted as a normal calibration curve. Although the data appear to fall along a straight line, the actual calibration curve is not intuitively obvious. The process of mathematically determining the best equation for the calibration curve is called regression.

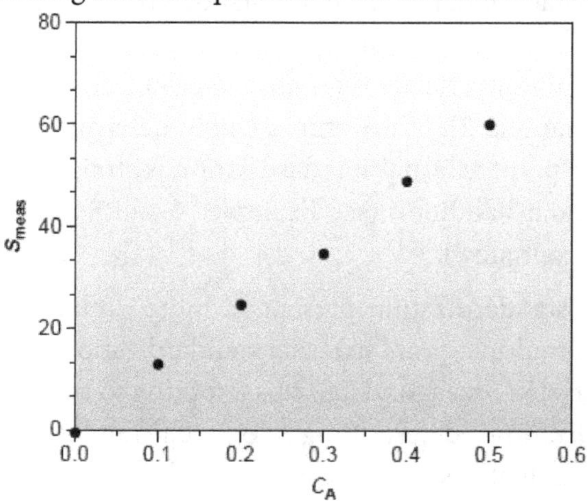

Fig. : Normal calibration plot of hypothetical data

2.4.1 Linear Regression of Straight-Line Calibration Curves

A calibration curve shows us the relationship between the measured signal and the analyte's concentration in a series of standards. The most useful calibration curve is a straight line since the method's sensitivity is the same for all concentrations of an- alyte. The equation for a linear calibration curve is

$$y = \beta 0 + \beta 1x \qquad .....12$$

where y is the signal and x is the amount of analyte. The constants β0 and β1 are the true y-intercept and the true slope, respectively. The goal of linear regression is to determine the best estimates for the slope, b1, and y-intercept, b0. This is accomplished by minimizing the residual error between the experimental val- ues, yi, and those values, yˆi, predicted by equation 12 (Figure 9). For obvious reasons, a regression analysis is also called a least-squares treatment. Several ap- proaches to the linear regression of equation 5.12 are discussed in the following sections.

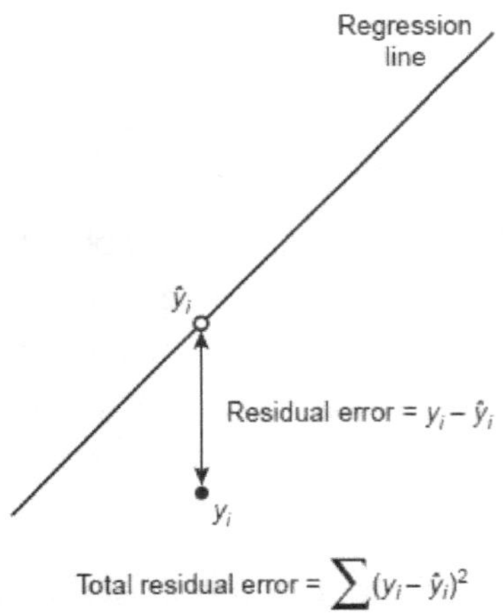

Fig. : Residual error in linear regression, where the filled circle shows the experimental value y_i and the open circle shows the predicted y_{i-}

2.4.2 Unweighted Linear Regression with Errors in y

The most commonly used form of linear regression is based on three assumptions: (1) that any difference between the experimental data and the calculated regression line is due to indeterminate errors affecting the values of y, (2) that these indeterminate errors are normally distributed, and (3) that the indeterminate errors in y do not depend on the value of x. Because we assume that indeterminate errors are the same for all standards, each standard contributes equally in estimating the slope and y-intercept. For this reason the result is considered an unweighted linear regression.

The second assumption is generally true because of the central limit theorem. The validity of the two remaining assumptions is less certain and should be evaluated before accepting the results of a linear regression. In particular, the first assumption is always suspect since there will certainly be some indeterminate errors affecting the values of x. In preparing a calibration curve, however, it is not unusual for the relative standard deviation of the measured signal (y) to be significantly larger than that for the concentration of analyte in the standards (x). In such circumstances, the first assumption is usually reasonable.

2.4.3 Finding the Estimated Slope and y-Intercept

The derivation of equations for calculating the estimated slope and y-intercept can be found in standard statistical texts 7 and is not developed here. The resulting equation for the slope is given as

$$b_1 = \frac{n\sum x_i y_i - \sum x_i \sum y_i}{n\sum x_i^2 - (\sum x_i)^2}$$

and the equation for the *y*-intercept is

$$b_0 = \frac{\sum y_i - b_1 \sum x_i}{n}$$

Although equations 13 and 14 appear formidable, it is only necessary to evaluate four summation terms. In addition, many calculators, spreadsheets, and other computer software packages are capable of performing a linear regression analysis based on this model. To save time and to avoid tedious calculations, learn how to use one of these tools. For illustrative purposes, the necessary calculations are shown in detail in the following example.

Table 2.2: Effect of a constant Determinate Error on the Value of k Calculated Using a Single-Point Standrardization

CA	Smeas	k (true)	Smeas (with constant error)	k (apparent)
1.00	1.00	1.00	1.50	1.50
2.00	2.00	1.00	2.50	1.25
3.00	3.00	1.00	3.50	1.17
4.00	4.00	1.00	4.50	1.13
5.00	5.00	1.00	5.50	1.10
	mean K(true)=	1.00	mean k (apparent)=	1.23

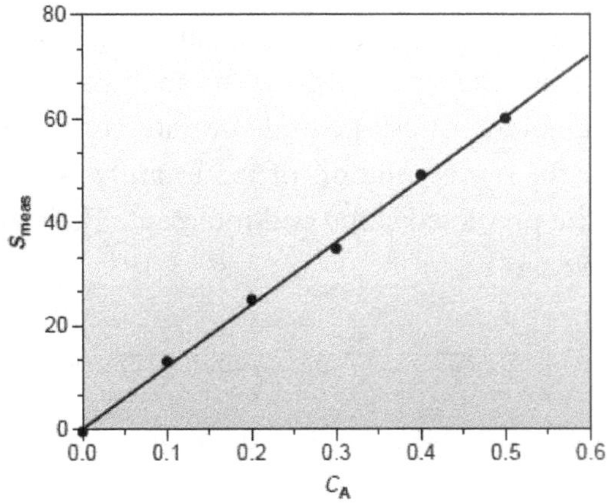

Fig. : Normal calibration curve for the hypothetical data in Table showing the regression line.

2.4.4 Uncertainty in the Regression Analysis

As shown in Figure 10, the regression line need not pass through the data points (this is the consequence of indeterminate errors affecting the signal). The cumulative deviation of the data from the regression line is used to calculate the uncertainty in the regression due to indeterminate error.

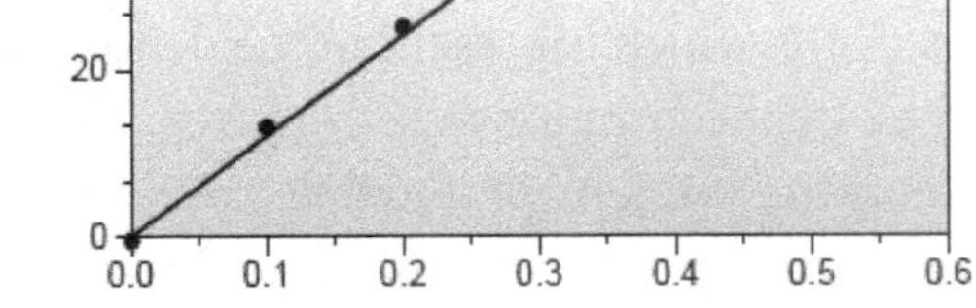

Fig. : Normal calibration curve for the hypothetical data in Table showing the regression line.

This is called the standard deviation about the regression, sr, and is given as

$$s_r = \sqrt{\frac{\sum_{i=1}^{n}(y_i - \hat{y}_i)^2}{n-2}}$$

$$\hat{y}_i = b_0 + b_1 x_i$$

There is an obvious similarity between equation 5.15, except that the sum of squares term for sr is determined relative to y^i instead of y^, and the denominator is n – 2 instead of n – 1; n – 2 indicates that the linear regression analysis has only n – 2 degrees of freedom since two pa- rameters, the slope and the intercept, are used to calculate the values of y^i. A more useful representation of uncertainty is to consider the effect of indeterminate errors on the predicted slope and intercept. The standard deviation of the slope and intercept are given as :

$$s_{b_1} = \sqrt{\frac{ns_r^2}{n\sum x_i^2 - (\sum x_i)^2}} = \sqrt{\frac{s_r^2}{\sum (x_i - \bar{x})^2}}$$

$$s_{b_0} = \sqrt{\frac{s_r^2 \sum x_i^2}{n \sum x_i^2 - (\sum x_i)^2}} = \sqrt{\frac{s_r^2 \sum x_i^2}{n \sum (x_i - \bar{x})^2}}$$

$$\beta_1 = b_1 \pm ts_{b_1}$$

$$\beta_o = b_o \pm ts_{b_0}$$

where t is selected for a significance level of α and for n – 2 degrees of freedom. Note that the terms tsb1 and tsb0 do not contain a factor of (Root of n)–1 because the confidence interval is based on a single regression line. Again, many calculators, spread-sheets, and computer software packages can handle the calculation of sb0 and sb1 and the corresponding confidence intervals for β0 and β1. Example 11 illustrates the calculations.

To minimize the uncertainty in the predicted slope and y-intercept, calibration curves are best prepared by selecting standards that are evenly spaced over a wide range of concentrations or amounts of analyte. The reason for this can be rationalized by examining equations 16 and 5.17. For example, both sb0 and sb1 can be minimized by increasing the value of the term Σ(xi – x ‘)2, which is present in the denominators of both equations. Thus, increasing the range of concentrations used in preparing standards decreases the uncertainty in the slope and the y-intercept. Furthermore, to minimize the uncertainty in the y-intercept, it also is necessary to de- crease the value of the term Σx12 in equation 5.17. This is accomplished by spreading the calibration standards evenly over their range.

2.4.5 Using the Regression Equation

Once the regression equation is known, we can use it to determine the concentration of analyte in a sample. When using a normal calibration curve with external standards or an internal standards calibration curve, we YX measure an average signal for our sample, Y'x, and use it to calculate the value of X

$$X = \frac{\overline{Y}_X - b_0}{b_1} \tag{5.20}$$

The standard deviation for the calculated value of X is given by the following equation

$$s_X = \frac{s_r}{b_1}\left\{\frac{1}{m} + \frac{1}{n} + \frac{(\overline{Y}_X - \overline{y})^2}{b_1^2 \sum (x_i - \overline{x})^2}\right\}^{1/2} \tag{5.21}$$

where m is the number of replicate samples used to establish Y'X, n is the number of calibration standards, y^ is the average signal for the standards, and xi and–x' are the individual and mean concentrations of the standards. Once sX is known the confi- dence interval for the analyte's concentration can be calculated as

$$\mu_X = X \pm ts_X$$

where µX is the expected value of X in the absence of determinate errors, and the value of t is determined by the desired level of confidence and for n – 2 degrees of freedom. The following example illustrates the use of these equations for an analysis using a normal calibration curve with external standards.

In a standard addition the analyte's concentration is determined by extrapolating the calibration curve to find the x-intercept. In this case the value of X is

$$X = x\text{-intercept} = \frac{-b_0}{b_1}$$

and the standard deviation in X is

$$s_X = \frac{s_r}{b_1}\left\{\frac{1}{n} + \frac{(\overline{y})^2}{b_1^2 \sum (x_i - \overline{x})^2}\right\}^{1/2}$$

where n is the number of standards used in preparing the standard additions cali- bration curve (including the sample with no added standard), and –y is the average signal for the n standards. Because the analyte's concentration is determined by ex- trapolation, rather than by interpolation, sX for the method of standard additions generally is larger than for a normal calibration curve.

A linear regression analysis should not be accepted without evaluating the validity of the model on which the calculations were based. Perhaps the simplest way to evaluate a regression analysis is to calculate and plot the residual error for each value of x. The residual error for a single calibration standard, ri, is given as

$$r_i = y_i - \hat{y}_i$$

If the regression model is valid, then the residual errors should be randomly distrib- uted about an average residual error of 0, with no apparent trend toward either smaller or larger residual errors (Figure 11a). Trends such as those shown in Fig- ures 5.11b and 5.11c provide evidence that at least one of the assumptions on which the regression model is based are incorrect. For example, the trend toward larger residual errors in Figure 5.11b suggests that the indeterminate errors affecting y are not independent of the value of x. In Figure 11c the residual errors are not ran- domly distributed, suggesting that the data cannot be modeled with a straight-line relationship. Regression methods for these two cases are discussed in the following sections.

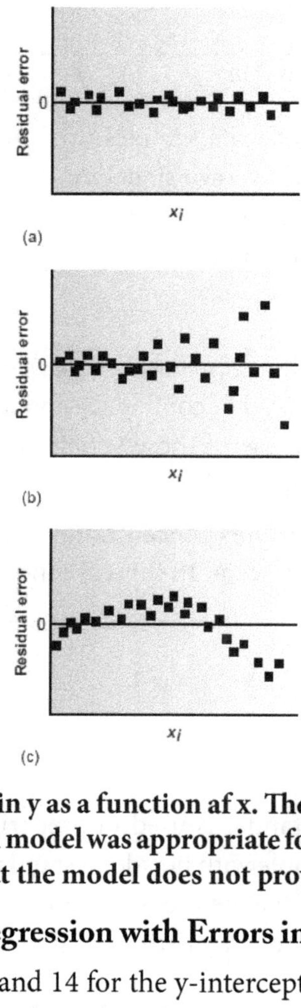

Fig. : Plot of the residual error in y as a function af x. The distribution of the residuals in (a) indicates that the regression model was appropriate for the data, and the distributions in (b) and (c) indicate that the model does not provide a good fit for the data

2.4.5.1 Weighted Linear Regression with Errors in y

Equations 13 for the slope, b1, and 14 for the y-intercept, b0, assume that indeterminate errors equally affect each value of y. When this assumption is false, as shown in Figure 11b, the variance associated with each value of y must be included when estimating β0 and β1. In this case the predicted slope and intercept are

$$b_1 = \frac{n\sum w_i x_i y_i - \sum w_i x_i \sum w_i y_i}{n\sum w_i x_i^2 - (\sum w_i x_i)^2}$$

$$b_0 = \frac{\sum w_i y_i - b_1 \sum w_i x_i}{n}$$

where wi is a weighting factor accounting for the variance in measuring yi. Values of wi are calculated using equation 5.24

$$w_i = \frac{ns_i^{-2}}{\sum s_i^{-2}}$$

where si is the standard deviation associated with yi. The use of a weighting factor ensures that the contribution of each pair of xy values to the regression line is proportional to the precision with which yi is measured.

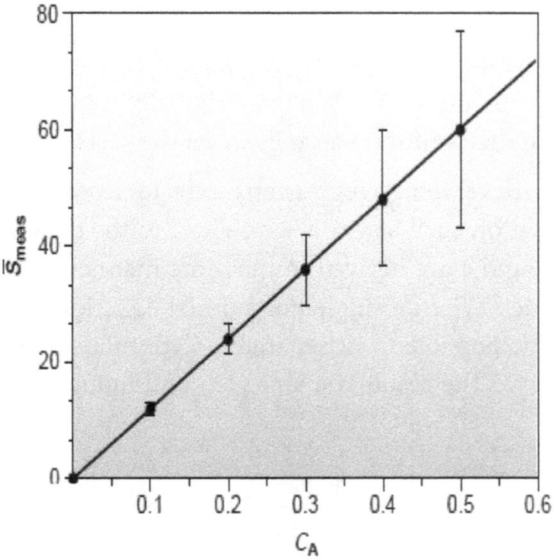

Weighted normal calibration curve for the data. The lines through the data points show the standard deviation of the signal for standard. These lines have been scaled by a factor of 50 so that they can be seen on the same scale as the calibration.

Equations for calculating confidence intervals for the slope, the y-intercept, and the concentration of analyte when using a weighted linear regression are not as easily defined as for an unweighted linear regression. The confidence interval for the concentration of an analyte, however, will be at its optimum valuewhen the analyte's signal is near the weighted centroid, –y, of the calibration curve

$$\bar{y} = \frac{1}{n}\sum w_i y_i$$

2.4.5.2 Weighted Linear Regression with Errors in Both *x* and *y*

If we remove the assumption that the indeterminate errors affecting a calibration curve are found only in the signal (y), then indeterminate errors affecting the preparation of standards containing known amounts of analyte (x) must be factored into the regression model. The solution for the resulting regression line is computationally more involved than that for either the unweighted or weighted regression lines, and is not presented in this text.

2.4.6 Curvilinear and Multivariate Regression

Regression models based on a straight line, despite their apparent complexity, use the simplest functional relationship between two variables. In many cases, calibra- tion curves show a pronounced curvature at high concentrations of analyte (see Figure 5.3b). One approach to constructing a calibration curve when curvature exists is to seek a transformation function that will make the data linear. Logarithms, expo- nentials, reciprocals, square roots, and trigonometric functions have all been used in this capacity. A plot of y versus log x is a typical example. Such transformations are not without complications. Perhaps the most obvious is that data that originally has a uniform variance for the y values will not maintain that uniform variance when the variable is transformed.

A more rigorous approach to developing a regression model for a nonlinear calibration curve is to fit a polynomial equation such as y = a + bx + cx2 to the data. Equations for calculating the parameters a, b, and c are derived in the same manner as that described earlier for the straight-line model. When a single polynomial equation cannot be fitted to the calibration data, it may be possible to fit separate polynomial equations to short segments of the calibration curve. The result is a sin- gle continuous calibration curve known as a spline function.

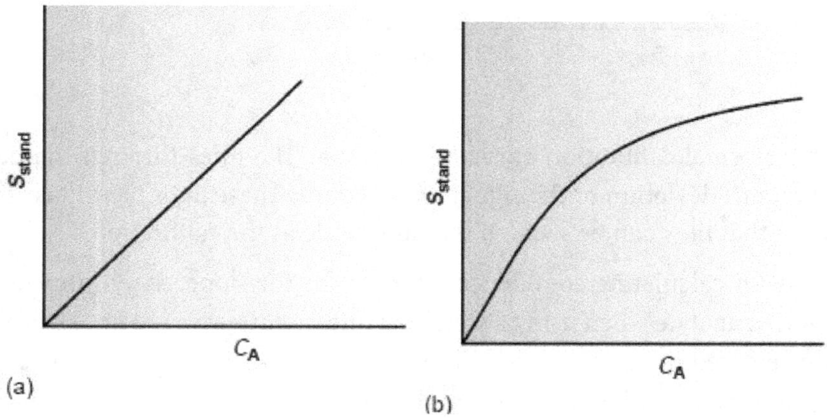

Fig. : Examples of (a) straight-line and (b) curved normal calibration curves.

The regression models considered earlier apply only to functions containing a single independent variable. Analytical methods, however, are frequently subject to determinate sources of error due to interferents that contribute to the measured signal. In the presence of a single interferent, equations 5.1 and 5.2 become

$$S_{meas} = k_A n_A + k_I n_I + S_{reag}$$

$$S_{meas} = k_A C_A + k_I C_I + S_{reag}$$

where kI is the interferent's sensitivity, nI is the moles of interferent, and CI is the interferent's concentration. Multivariate calibration curves can be prepared using standards that contain known amounts of analyte and interferent.

2.5 BLANK CORRECTIONS

In discussing ways to standardize a method, we assumed that an appropriate reagent blank had been used to correct Smeas for signals originating from sources other than the analyte. At that time we did not ask an important question— "What constitutes an appropriate reagent blank?" Surprisingly, the answer is not intuitively obvious. In one study, analytical chemists were asked to evaluate a data set consisting of a normal calibration curve, three samples of different size but drawn from the same source, and an analyte-free sample (Table 2.3). At least four different approaches for correcting the signals were used by the participants: (1) ignore the correction entirely, which clearly is incorrect; (2) use the y-intercept of the calibration curve as a calibration blank, CB; (3) use the analyte-free sample as a reagent blank, RB; and (4) use both the calibration and reagent blanks. Equa ions for calculating the concentration of analyte using each approach are shown in Table 5.4, along with the resulting concentration for the analyte in each of the three samples.

Table: 2.3 Hypothetical Data Used to Study Procedures for Method Blanks

W_s [a]	S_{stand}	Sample Number	W_x [b]	S_{samp}
1.6667	0.2500	1	62.4746	0.8000
5.0000	0.5000	2	82.7915	1.0000
8.3333	0.7500	3	103.1085	1.2000
8.5507	0.8413			
11.6667	1.0000		analyte-free[c]	0.1000
18.1600	1.4870			
19.9333	1.6200			

Calibration equation: $S_{stand} = 0.0750 \times W_s + 0.1250$

Source: Modified from cardone, M.J.Anal.Chem. 1986, 58, 433-438.

[a]W_s = weight of nalyte used to prepare standard solution by diluting to a fixed volume, V.

[b]W_x = of sample used to prepare sample solution by diluting to a fixed volume, V.

[c]Analyte-free sample prepared in the same fashion as samples, but without the analyte being present.

Table.2.4 Equations and resulting concentrations for different approaches to correcting for the method blank

Approach for Correcting Method Blank	Equation[a]	Concentration of Analyte in Sample 1	Sample 2	Sample 3
Ignore blank corrections	$C_A = \frac{W_a}{W_x} = \frac{S_{samp}}{kW_x}$	0.1707	0.1610	0.1552
Use calibration blank	$C_A = \frac{W_a}{W_x} = \frac{S_{samp} - CB}{kW_x}$	0.1441	0.1409	0.1390
Use reagent blank	$C_A = \frac{W_a}{W_x} = \frac{S_{samp} - RB}{kW_x}$	0.1494	0.1449	0.1422
Use both calibration and reagent blank	$C_A = \frac{W_a}{W_x} = \frac{S_{samp} - CB - RB}{kW_x}$	0.1227	0.1248	0.1261
Use total Youden blank	$C_A = \frac{W_a}{W_x} = \frac{S_{samp} - TYB}{kW_x}$	0.1313	0.1313	0.1313

aCA=concentration of analyte; W_a *=weight of analyte;* W_x*=weight og sample; k=slope of calibration curve=0.075 (see Table). Abbreviations: CB=calibration blank=0.125(see Table); RB=reagent blank=0.100 (see Table); TYB=total Youden blank=0.185(see text)*

That all four methods give a different result for the concentration of analyte underscores the importance of choosing a proper blank but does not tell us which of the methods is correct. In fact, the variation within each method for the reported concentration of analyte indicates that none of these four methods has adequately corrected for the blank. Since the three samples were drawn from the same source, they must have the same true concentration of analyte. Since all four methods predict concentrations of analyte that are dependent on the size of the sample, we can conclude that none of these blank corrections has accounted for an underlying constant source of determinate error.

To correct for all constant method errors, a blank must account for signals due to the reagents and solvent used in the analysis and any bias due to interactions between the analyte and the sample matrix. Both the calibration blank and the reagent blank correct for signals due to the reagents and solvents. Any difference in their values is due to the number and composition of samples contributing to the determination of the blank.

Unfortunately, neither the calibration blank nor the reagent blank can correct for bias due to analyte–matrix interactions because the analyte is missing in the reagent blank, and the sample's matrix is missing from the calibration blank. The true method blank must include both the matrix and the analyte and, consequently, can only be determined using the sample itself. One approach is to measure the signal for samples of different size and determine the regression line from a plot of signal versus the amount of sample. The resulting y-intercept gives the signal for the condition of no sample and is known as the total Youden blank.1 This is the true blank correction. The regression line for the sample data in Table 2.3 is giving a true blank correction of 0.185. Using this value to correct the signals gives identical values for the concentration of analyte in all three samples (see Table 2.4,).

$$S_{samp} = 0.009844 \text{ x } W_x + 0.185$$

The total Youden blank is not encountered frequently in analytical work, because most chemists rely on a calibration blank when using calibration curves and rely on reagent blanks when using a single-point standardization. As long as any constant bias due to analyte–matrix interactions can be ignored, which is often the case, the accuracy of the method will not suffer. It is always a good idea, however, to check for constant sources of error, by analyzing samples of different sizes, before relying on either a calibration or reagent blank.

3

ERRORS IN CHEMICAL ANALYSES

3.1 INTRODUCTION: AN OVERVIEW

The term error has two slightly different meanings.

- Error refers to the difference between a measured value and the "true" or "known" value.
- Error often denotes the estimated uncertainty in a measurement or experiment.

" We can only hope to minimize errors and estimate their size with acceptable accuracy"

Every measurement is influenced by many uncertainties, which combine to produce a scatter of results.

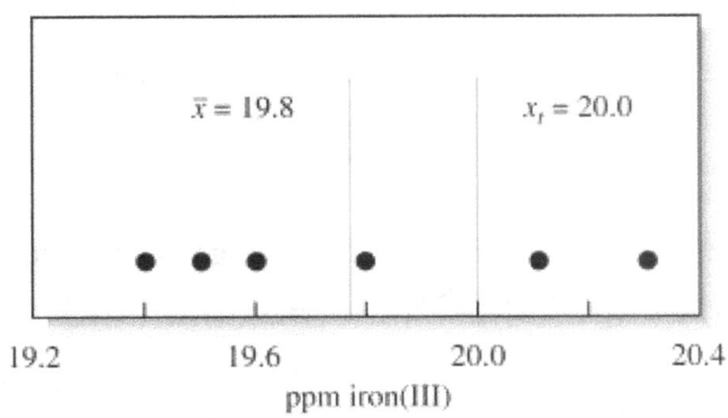

Fig. 1: Results from six replicate determinations of iron in aqueous samples of a standard solution containing 20.0 ppm iron(III). Note that the results range from a low of 19.4 ppm to a high of 20.3 ppm of iron. The average, or mean value, x , of the data is 19.78 ppm, which rounds to 19.8 ppm

Because measurement uncertainties can never be completely eliminated, *measurement data can only give us an estimate of the "true" value.*

However, the probable magnitude of the error in a measurement can often be evaluated. It is then possible to define limits within which the true value of a measured quantity lies with a given level of probability.

Before beginning an analysis ask, "What maximum error can be tolerated in the result?"

The answer to this question often determines the method chosen and the time required to complete the analysis.

3.2 SOME IMPORTANT TERMS

To improve the reliability and to obtain information about the variability of results, two to five portions (replicates) of a sample are usually carried through an entire analytical procedure.

Replicates are samples of about the same size that are carried through an analysis in *exactly the same way.*

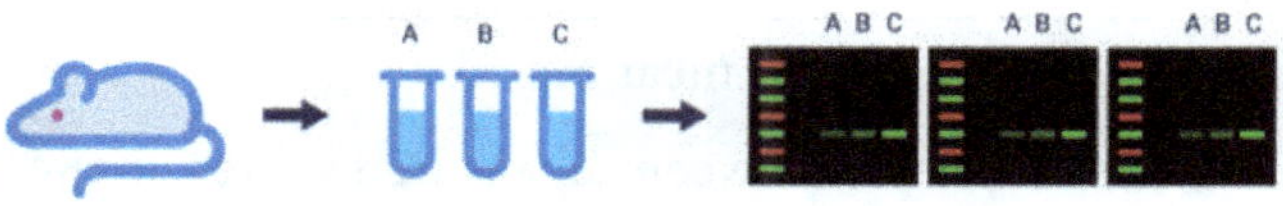

- Individual results from a set of measurements are seldom the same
- Usually, the "best" estimate is considered to be the central value for the set.
- The central value of a set should be more reliable than any of the individual results.
- Usually, the mean or the median is used as the central value for a set of replicate measurements.

An analysis of the variation in the data allows us to estimate the uncertainty associated with the central value.

3.2.1 The Mean and the Median

The mean, also called the arithmetic mean or the average, is obtained by dividing the sum of replicate measurements by the number of measurements in the set:

3.2.2 The Symbol

- xi means to add all of the values xi for the replicates; xi represents the individual values of x making up the set of N replicate measurements.
- The median is the middle value in a set of data that has been arranged in numerical order.
- The median is used advantageously when a set of data contain an outlier. An outlier is a result that differs significantly from others in the set.
- An outlier can have a significant effect on the mean of the set but has no effect on the median

Calculate the mean and median for the data shown in figure.

Solution:- $$\text{mean} = x = \frac{19.4+19.5+19.8+20.1+20.3}{6} = 19.8 \text{ ppm Fe}$$

Because the set contains as even number of measurements the median is the average of the central pair:

$$\text{median}\frac{19.6+19.8}{6}19.7 \text{ ppm Fe}$$

3.2.3 Precision

- Precision describes the agreement among several results obtained in the same way. Describes the reproducibility of measurements.
- Precision is readily determined by simply repeating the measurement on replicate samples.
- Precision of a set of replicate data may be expressed as standard deviation, variance, and coefficient of variation.
- *di*, deviation from mean, is how much *xi*, the individual result, deviates from the *mean*. x x d

3.2.4 Accuracy

- Accuracy indicates the closeness of the measurement to the true or accepted value and is expressed by the error.
- Accuracy measures agreement between a result and the accepted value.
- Accuracy is often more difficult to determine because the true value is usually unknown. An accepted value must be used instead.
- Accuracy is expressed in terms of either absolute or relative error.

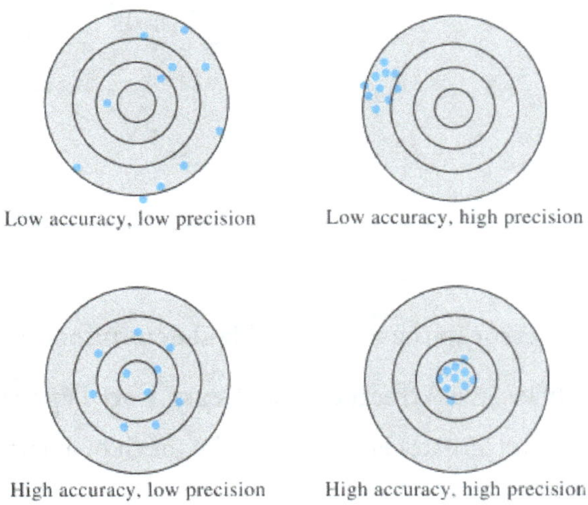

Fig. 2: Note that we can have very precise results (upper right) with a mean that is not

accurate and an accurate mean (lower left) with data points that are imprecise.

3.2.5 Absolute Error

* The absolute error of a measurement is the difference between the measured value and the true value. If the measurement result is low, the sign is negative; if the measurement result is high, the sign is positive.]

$$E = x_i - x_i$$

3.2.6 Relative Error

The relative error of a measurement is the absolute error divided by the true value.

Relative error may be expressed in percent, parts per thousand, or parts per million, depending on the magnitude of the result.

3.3 TYPES OF ERRORS IN EXPERIMENTAL DATA

- Results can be precise without being accurate and accurate without being precise.
- Each dot represents the error associated with a single determination.
- Each vertical line labeled (xi - xt) is the absolute average deviation of the set from the true value.

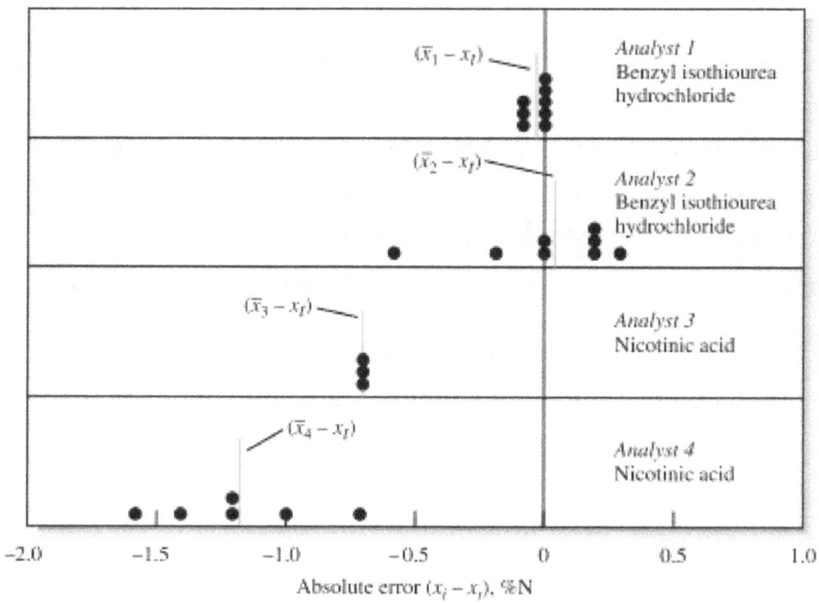

Fig. 3: Absolute error in the micro-Kjeldahl determination of nitrogen.

3.3.1 Chemical Analyses are Affected by at Least Two Types of Errors:

- Systematic (or determinate) error, causes the mean of a data set to differ from the accepted value.
- Random (or indeterminate) error, causes data to be scattered more or less

symmetrically around a mean value.

3.3.2 A Third Type of Error is Gross Error.

- These differ from indeterminate and determinate errors.
- They usually occur only occasionally, are often large, and may cause a result to be either high or low.
- They are often the product of human errors.
- Gross errors lead to outliers, results that appear to differ markedly from all other data in a set of replicate measurements.

3.4 SYSTEMATIC ERRORS

- Systematic errors
- Have a definite value,
- An assignable cause, and
- Are of the same magnitude for replicate measurements made in the same way.
- They lead to bias in measurement results.

3.4.1 There are Three Types of Systematic Errors:

- Instrumental errors.
- Method errors
- Personal errors

Instrumental Errors

- Are caused by nonideal instrument behavior, by faulty calibrations, or by use under inappropriate conditions
- Pipets, burets, and volumetric flasks may hold or deliver volumes slightly different from those indicated by their graduations.
- Calibration eliminates most systematic errors of this type.
- Electronic instruments can be influenced by noise, temperature, pH and are also subject to systematic errors.
- Errors of these types usually are detectable and correctable.

Method Errors

- The nonideal chemical or physical behavior of the reagents and reactions on which an analysis is based often introduce systematic method errors.
- Such sources of nonideality include the slowness of some reactions, the incompleteness of others, the instability of some species, the lack of specificity of

most reagents, and the possible occurrence of side reactions that interfere with the measurement process.

- Errors inherent in a method are often difficult to detect and hence, these errors are usually the most difficult to identify and correct.

Personal Errors

- Result from the carelessness, inattention, or personal limitations of the experimenter.
- Many measurements require personal judgments.
- Examples include estimating the position of a pointer between two scale divisions, the color of a solution at the end point in a titration, or the level of a liquid with respect to a graduation in a pipet or buret.
- Judgments of this type are often subject to systematic, unidirectional errors.
- A universal source of personal error is prejudice, or bias.
- Number bias is another source of personal error that varies considerably from person to person.
- The most frequent number bias encountered in estimating the position of a needle on a scale involves a preference for the digits 0 and 5.
- Also common is a prejudice favoring small digits over large and even numbers over odd.
- Digital and computer displays on ph meters, laboratory balances, and other electronic instruments eliminate number bias because no judgment is involved in taking a reading.

3.5 THE EFFECT OF SYSTEMATIC ERRORS ON ANALYTICAL RESULTS

Systematic errors may be either ***constant or proportional.***

3.5.1 Constant Errors

- The magnitude of a constant error stays essentially the same as the size of the quantity measured is varied.
- With constant errors, the absolute error is constant with sample size, but the relative error varies when the sample size is changed \

Suppose that 0.5c mg of precipitate is lost as a result of being washed with 200 mL of wash liquid. If the precipitate weighs 500 mg. the relative error due to solubility loss is $(0.50/500) \times 100\% = 0.1\%$ Loss of the same quantity from 50 mg of precipitate results in a relative error of low.

One way of reducing the effect of constant error is to increase the sample size until the error is acceptable.

- The excess of reagent needed to bring about a color change during a titration is another example of constant error.
- This volume, usually small, remains the same regardless of the total volume of reagent required for the titration. Again, the relative error from this source becomes more serious as the total volume decreases.
- One way of reducing the effect of constant error is to increase the sample size until the error is acceptable.

3.5.2 Proportional Errors

- Proportional errors decrease or increase in proportion to the size of the sample.
- A common cause of proportional errors is the presence of interfering contaminants in the sample.
- For example, a widely used method for the determination of copper is based on the reaction of copper(II) ion with potassium iodide to give iodine. The quantity of iodine is then measured and is proportional to the amount of copper. Iron(III), if present, also liberates iodine from potassium iodide. Unless steps are taken to prevent this interference, high results are observed for the percentage of copper because the iodine produced will be a measure of the copper(II) and iron(III) in the sample.
- The size of this error is fixed by the fraction of iron contamination, which is independent of the size of sample taken. If the sample size is doubled, for example, the amount of iodine liberated by both the copper and the iron contaminant is also doubled. Thus, the magnitude of the reported percentage of copper is independent of sample size.

3.6 DETECTION AND ELIMINATION OF SYSTEMATIC (INSTRUMENTAL AND PERSONAL) ERRORS

- Periodic calibration of equipment is always desirable because the response of most instruments changes with time as a result of component aging, corrosion, or mistreatment.
- Most personal errors can be minimized by careful, disciplined laboratory work.
- It is a good habit to check instrument readings, notebook entries, and calculations systematically.
- Errors due to limitations of the experimenter can usually be avoided by carefully choosing the analytical method or using an automated procedure.

Detection of Systematic (Method) Errors

- Bias in an analytical method is particularly difficult to detect.

- The best way to estimate the bias of an analytical method is by analyzing Standard reference materials (SRMs).

3.7 ANALYSIS OF STANDARD SAMPLES

- The overall composition of a synthetic standard material must closely approximate the composition of the samples to be analyzed.
- Treat care must be taken to ensure that the concentration of analyte is known exactly.
- A synthetic standard may not reveal unexpected interferences so that the accuracy of determinations may not be known.

3.7.1 Independent Analysis

- If standard samples are not available, a second independent and reliable analytical method can be used in parallel.
- The independent method should differ as much as possible from the one under study.
- This practice minimizes the possibility that some common factor in the sample has the same effect on both methods.
- Again, a statistical test must be used to determine whether any difference is a result of random errors in the two methods or due to bias in the method under study.

3.7.2 Blank Determinations

- A blank contains the reagents and solvents used in a determination, but no analyte.
- Often, many of the sample constituents are added to simulate the analyte environment, which is called the sample matrix.
- In a blank determination, all steps of the analysis are performed on the blank material. The results are then applied as a correction to the sample measurements.
- Blank determinations reveal errors due to interfering contaminants from the reagents and vessels employed in the analysis.
- Blanks are also used to correct titration data for the volume of reagent needed to cause an indicator to change color.

Variation in Sample Size

As the size of a measurement increases, the effect of a constant error decreases. Thus constant errors can often be detected by varying the sample size.

4

PREPARING SOLUTIONS OF ANALYTICAL CHEMISTRY

4.1 ANALYSIS, DETERMINATION, AND MEASUREMENT

The first important distinction we will make is among the terms "analysis," "deter mination," and " measurement." An analysis provides chemical or physical infor- mation about a sample. The components of interest in the sample are called analytes, and the remainder of the sample is the matrix. In an analysis we determine the identity, concentration, or properties of the analytes. To make this determina- tion we measure one or more of the analyte's chemical or physical properties.

An example helps clarify the differences among an analysis, a determination, and a measurement. In 1974, the federal government enacted the Safe Drinking Water Act to ensure the safety of public drinking water supplies. To comply with this act municipalities regularly monitor their drinking water supply for potentially harmful substances. One such substance is coliform bacteria. Municipal water departments collect and analyze samples from their water supply. To determine the concentration of coliform bacteria, a portion of water is passed through a membrane filter. The filter is placed in a dish containing a nutrient broth and incubated. At the end of the incubation period the number of coliform bacterial colonies in the dish is measured by counting (Figure 1). Thus, municipal water departments analyze samples of water to determine the concentration of coliform bacteria by measuring the number of bacterial colonies that form during a specified period of incubation.

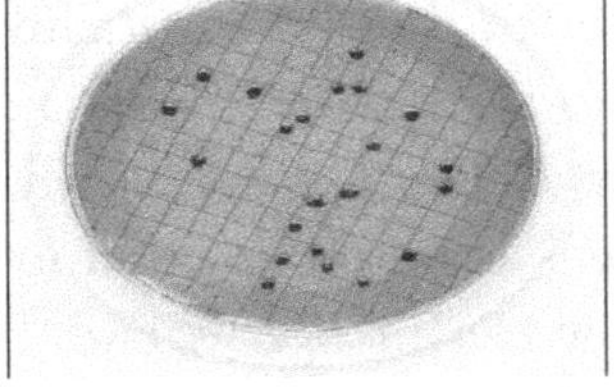

Fig. 1: Membrane filter showing colonies of coliform bacteria. The number of colonies are counted and reported as colonies /100 ml of sample pour

4.2 TECHNIQUES, METHODS, PROCEDURES, AND PROTOCOLS

Suppose you are asked to develop a way to determine the concentration of lead in drinking water. How would you approach this problem? To answer this question it helps to distinguish among four levels of analytical methodology: techniques, meth- ods, procedures, and protocols.

A technique is any chemical or physical principle that can be used to study an analyte. Many techniques have been used to determine lead levels For example, in graphite furnace atomic absorption spectroscopy lead is atomized, and the ability of the free atoms to absorb light is measured; thus, both a chemical principle (atom- ization) and a physical principle (absorption of light) are used in this technique.

A method is the application of a technique for the determination of a specific analyte in a specific matrix. As shown in Figure 2, the graphite furnace atomic ab- sorption spectroscopic method for determining lead levels in water is different from that for the determination of lead in soil or blood. Choosing a method for deter- mining lead in water depends on how the information is to be used and the estab lished design criteria (Figure 3). For some analytical problems the best method might use graphite furnace atomic absorption spectroscopy, whereas other problems might be more easily solved by using another technique, such as anodic strip- ping voltammetry or potentiometry with a lead ion-selective electrode.

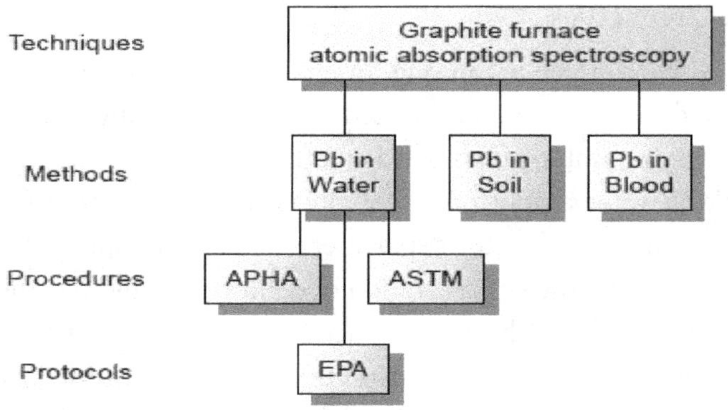

Fig. 2: Chart showing hierarchical relationship among a technique, methods using that technique, and procedures and protocols for one method. (Abbreviations: APHA = American Public Health Association, ASTM = American Society for Testing Materials, EPA = Environmental Protection Agency)

A procedure is a set of written directions detailing how to apply a method to a particular sample, including information on proper sampling, handling of interferents, and validating results. A method does not necessarily lead to a single procedure, as different analysts or agencies will adapt the method to their specific needs. As shown in Figure 2, the American Public Health Agency and the American Society for Testing Materials publish separate procedures for the determination of lead levels in water.

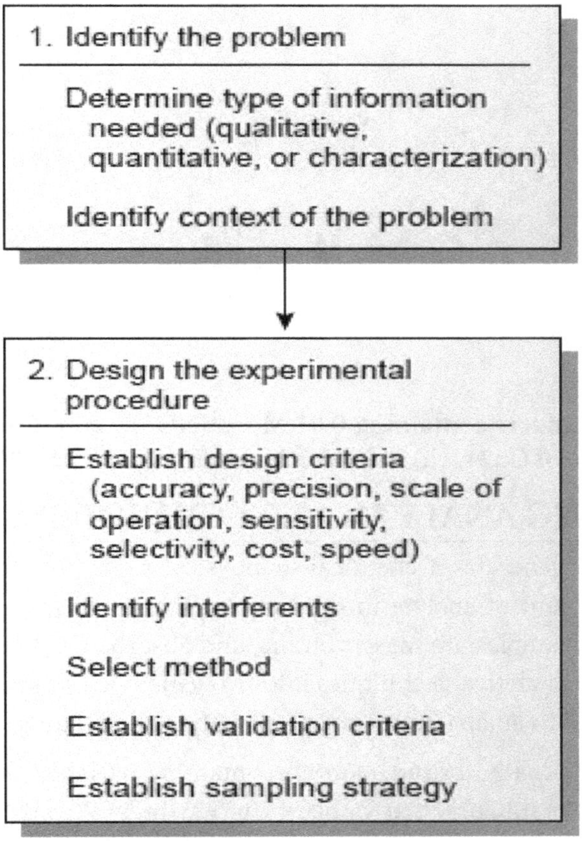

Figure.3 Subsection of the analytical approach to problem solving, of relevance to the selection of a method and the design of an analytical procedure.

Finally, a protocol is a set of stringent written guidelines detailing the procedure that must be followed if the agency specifying the protocol is to accept the results of the analysis. Protocols are commonly encountered when analytical chemistry is used to support or define public policy. For purposes of determining lead levels in water under the Safe Drinking Water Act, labs follow a protocol specified by the Environmental Protection Agency.

There is an obvious order to these four facets of analytical methodology. Ideally, a protocol uses a previously validated procedure. Before developing and validating a procedure, a method of analysis must be selected. This requires, in turn, an initial screening of available techniques to determine those that have the potential for monitoring the analyte. We begin by considering a useful way to classify analytical techniques.

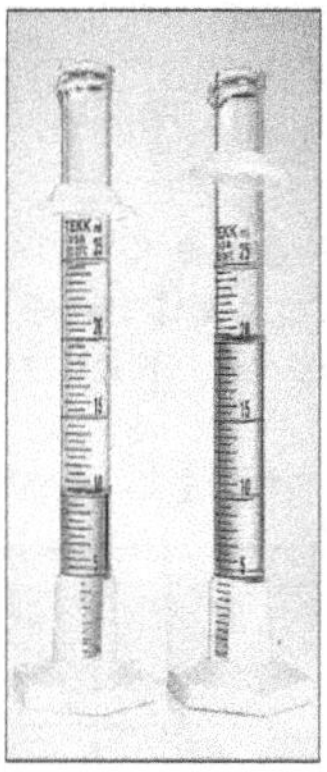

(a) (b)

Fig. 4: Graduated cylinders containing 0.01 M Cu (NO3)2 . (a) Cylinder 1 contains 10 ml, or 0.0001 mol, of Cu2+. (b) cylinder 2 contains 20 ml, or 0.0002 mol, of Cu2+.

4.3 CLASSIFYING ANALYTICAL TECHNIQUES

Analyzing a sample generates a chemical or physical signal whose magnitude is proportional to the amount of analyte in the sample. The signal may be anything we can measure; common examples are mass, volume, and absorbance. For our pur- poses it is convenient to divide analytical techniques into two general classes based on whether this signal is proportional to an absolute amount of analyte or a relative amount of analyte.

Consider two graduated cylinders, each containing 0.01 M $Cu(NO_3)_2$ (Figure 3.4). Cylinder 1 contains 10 mL, or 0.0001 mol, of Cu^{2+}; cylinder 2 contains 20 mL, or 0.0002 mol, of Cu^{2+}. If a technique responds to the absolute amount of analyte in the sample, then the signal due to the analyte, SA, can be expressed as

$$S_A = kn_A \qquad 1$$

where nA is the moles or grams of analyte in the sample, and k is a proportionality constant. Since cylinder 2 contains twice as many moles of Cu^{2+} as cylinder 1, an- alyzing the contents of cylinder 2 gives a signal that is twice that of cylinder 1.

A second class of analytical techniques are those that respond to the relative amount of analyte; thus

$$S_A = kC_A \qquad 2$$

where CA is the concentration of analyte in the sample. Since the solutions in both cylinders have the same concentration of Cu^{2+}, their analysis yields identical signals.

Techniques responding to the absolute amount of analyte are called total analysis techniques. Historically, most early analytical methods used total analysis techniques, hence they are often referred to as "classical" techniques. Mass, volume, and charge are the most common signals for total analysis techniques, and the cor- responding techniques are gravimetry , titrimetry , and coulometry . With a few exceptions, the signal in a total analysis tech- nique results from one or more chemical reactions involving the analyte. These re- actions may involve any combination of precipitation, acid–base, complexation, or redox chemistry. The stoichiometry of each reaction, however, must be known tosolve equation 1 for the moles of analyte.

Techniques, such as spectroscopy , potentiometry , and voltammetry , in which the signal is proportional to the relative amount of analyte in a sample are called concentration techniques. Since most concentration techniques rely on measuring an optical or electrical signal, they also are known as "instrumental" techniques. For a concentration technique, the rela- tionship between the signal and the analyte is a theoretical function that depends on experimental conditions and the instrumentation used to measure the signal. For this reason the value of k in equation 3.2 must be determined experimentally.

4.4 SELECTING AN ANALYTICAL METHOD

A method is the application of a technique to a specific analyte in a specific matrix. Methods for determining the concentration of lead in drinking water can be devel- oped using any of the techniques mentioned in the previous section. Insoluble lead salts such as PbSO4 and PbCrO4 can form the basis for a gravimetric method. Lead forms several soluble complexes that can be used in a complexation titrimetric method or, if the complexes are highly absorbing, in a spectrophotometric method. Lead in the gaseous free-atom state can be measured by an atomic ab- sorption spectroscopic method. Finally, the availability of multiple oxidation states (Pb, Pb^{2+}, Pb^{4+}) makes coulometric, potentiometric, and voltammetric methods feasible.

The requirements of the analysis determine the best method. In choosing a method, consideration is given to some or all the following design criteria: accuracy, precision, sensitivity, selectivity, robustness, ruggedness, scale of operation, analysis time, availability of equipment, and cost. Each of these criteria is considered in more detail in the following sections.

4.4.1 Accuracy

Accuracy is a measure of how closely the result of an experiment agrees with the ex- pected result. The difference between the obtained result and the expected result is usually divided by the expected result and reported as a percent relative error

$$\%\ \text{Error} = \frac{\text{obtained result} - \text{expected result}}{\text{expected result}} \times 100$$

Analytical methods may be divided into three groups based on the magnitude of their relative errors. When an experimental result is within 1% of the correct result, the analytical method is highly ac- curate. Methods resulting in relative errors between 1% and 5% are moderately accurate, but methods of low accuracy produce rel- ative errors greater than 5%.

The magnitude of a method's relative error depends on how accurately the signal is measured, how accurately the value of k in equations1 or.2 is known, and the ease of handling the sample without loss or contamination. In general, total analysis methods produce results of high accuracy, and concentration methods range from high to low accuracy.

4.4.2 Precision

When a sample is analyzed several times, the individual results are rarely the same. Instead, the results are randomly scattered. Precision is a measure of this variability. The closer the agreement between individual analyses, the more precise the results. For example, in determining the concentration of K+ in serum, the results shown in Figure 5(a) are more precise than those in Figure 5(b). It is important to realize that precision does not imply accuracy. That the data in Figure 5(a) are more precise does not mean that the first set of results is more accurate. In fact, both sets of results may be very inaccurate.

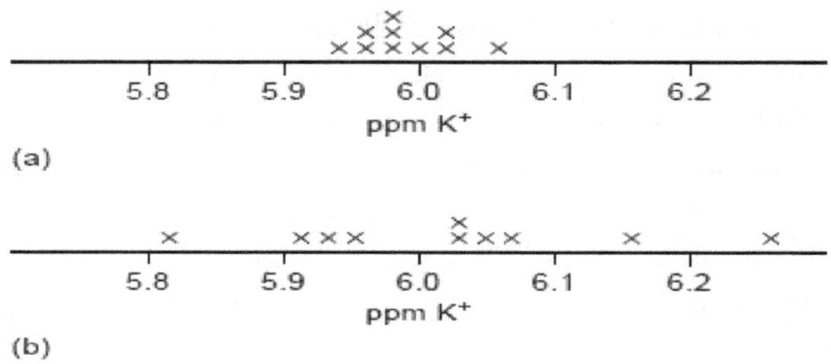

Two determinations of the concentration of K+ in serum showing the effect fo precision the data in (a) are less scattered and therefore more pracise than the data in (b)

As with accuracy, precision depends on those factors affecting the relationship between the signal and the analyte (equations 1 and 2). Of particular impor tance are the uncertainty in measuring the signal and the ease of handling samples reproducibly. In most cases the signal for a total analysis method can be measured with a higher precision than the corresponding signal for a concentration method.

4.4.3 Sensitivity

The ability to demonstrate that two samples have different amounts of analyte is an essential part of many analyses. A method's sensitivity is a measure of its ability to establish that such differences are significant. Sensitivity is often confused with a method's detection limit. The detection limit is the smallest amount of analyte that can be determined with confidence. The detection limit, therefore, is a statistical parameter.

Sensitivity is the change in signal per unit change in the amount of analyte and is equivalent to the proportionality constant, k, in equations 3.1 and 3.2. If ΔSA is the smallest increment in signal that can be measured, then the smallest difference in the amount of analyte that can be detected is

$$\Delta n_A = \frac{\Delta S_A}{k} \quad \text{(total analysis method)}$$

$$\Delta C_A = \frac{\Delta S_A}{k} \quad \text{(concentration method)}$$

Suppose that for a particular total analysis method the signal is a measurement of mass using a balance whose smallest increment is ±0.0001 g. If the method's sensitivity is 0.200, then the method can conceivably detect a difference of as little as

$$\Delta n_A = \frac{\pm 0.0001 \text{ g}}{0.200} = \pm 0.0005 \text{ g}$$

in the absolute amount of analyte in two samples. For methods with the same ΔSA, the method with the greatest sensitivity is best able to discriminate among smaller amounts of analyte. Selectivity

An analytical method is selective if its signal is a function of only the amount of an- alyte present in the sample. In the presence of an interferent, equations 1 and 2 can be expanded to include a term corresponding to the interferent's contribution to the signal, SI,

$$_{amp} = S_A + S_I = k_A n_A + k_I n_I \quad \text{(total analysis method)}$$

$$_{mp} = S_A + S_I = k_A C_A + k_I C_I \quad \text{(concentration method)}$$

where Ssamp is the total signal due to constituents in the sample; kA and kI are the sensitivities for the analyte and the interferent, respectively; and nI and CI are the moles (or grams) and concentration of the interferent in the sample.

The selectivity of the method for the interferent relative to the analyte is de- fined by a selectivity coefficient, KA,I

$$K_{A,I} = \frac{k_I}{k_A} \quad \text{......5}$$

which may be positive or negative depending on whether the interferent's effect on the signal is opposite that of the analyte.* A selectivity coefficient greater than +1 or less than –1 indicates that the method is more selective for the interferent than for the analyte. Solving equation 3.5 for k_I

kI = KA, I x kA 3.6

substituting into equations 3.3 and 3.4, and simplifying gives

$$S_{samp} = k_A(n_A + K_{A,I} \times n_I) \qquad \text{(total analysis method)}$$

$$S_{samp} = k_A(C_A + K_{A,I} \times C_I) \qquad \text{(concentration method)}$$

The selectivity coefficient is easy to calculate if kA and kI can be independently determined. It is also possible to calculate KA,I by measuring Ssamp in the presence and absence of known amounts of analyte and interferent.

Not surprisingly, methods whose signals depend on chemical reactivity are often less selective and, therefore, more susceptible to interferences. Problems with selec- tivity become even greater when the analyte is present at a very low concentration.

4.4.4 Robustness and Ruggedness

For a method to be useful it must provide reliable results. Unfortunately, methods are subject to a variety of chemical and physical interferences that contribute uncer- tainty to the analysis. When a method is relatively free from chemical interferences, it can be applied to the determination of analytes in a wide variety of sample matri- ces. Such methods are considered robust.

Random variations in experimental conditions also introduce uncertainty. If a method's sensitivity is highly dependent on experimental conditions, such as tem- perature, acidity, or reaction time, then slight changes in those conditions may lead to significantly different results. A rugged method is relatively insensitive to changes in experimental conditions.

4.4.5 Scale of Operation

Another way to narrow the choice of methods is to consider the scale on which the analysis must be conducted. Three limitations of particular importance are the amount of sample available for the analysis, the concentration of analyte in the sample, and the absolute amount of analyte needed to obtain a measurable signal. The first and second limitations define the scale of operations shown in Figure 3.6; the last limitation positions a method within the scale of operations.

The scale of operations in Figure 3.6 shows the analyte's concentration in weight percent on the y- axis and the sample's size on the x-axis. For convenience, we divide analytes into major (>1% w/w), minor (0.01% w/w – 1% w/w), trace (10–7% w/w – 0.01% w/w) and ultratrace (<10–7% w/w) components, and we divide samples into macro (>0.1 g), meso (10 mg – 100 mg), micro (0.1 mg – 10 mg) and ultramicro (<0.1 mg) sample sizes. Note that both the x-axis and the y-axis use a logarithmic scale. The analyte's concentration and the amount of sample used provide a characteristic description for an analysis. For example, samples in a macro–major analysis weigh more than 0.1 g and contain more than 1% analyte.

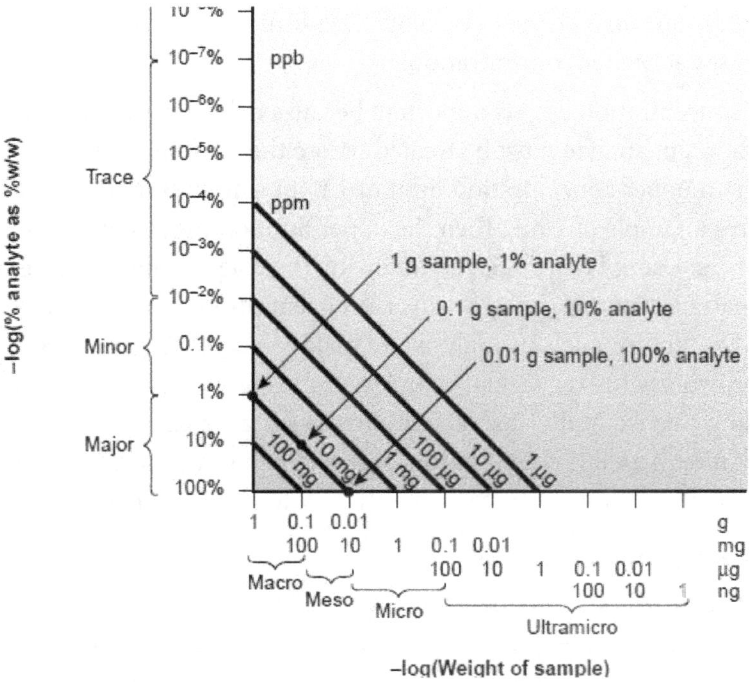

Fig 6: Scal of Peration for analytical methods adapted from reference 7a and 7b

Diagonal lines connecting the two axes show combinations of sample size and concentration of analyte containing the same absolute amount of analyte. As shown in Figure 6, for example, a 1-g sample containing 1% analyte has the same amount of analyte (0.010 g) as a 100-mg sample containing 10% analyte or a 10-mg sample containing 100% analyte.

Since total analysis methods respond to the absolute amount of analyte in a sample, the diagonal lines provide an easy way to define their limitations. Consider, for example, a hypothetical total analysis method for which the minimum de- tectable signal requires 100 mg of analyte. Using Figure 6, the diagonal line repre- senting 100 mg suggests that this method is best suited for macro samples and major analytes. Applying the method to a minor analyte with a concentration of 0.1% w/w requires a sample of at least 100 g. Working with a sample of this size is rarely practical, however, due to the complications of carrying such a large amount of material through the analysis. Alternatively, the minimum amount of required analyte can be decreased by improving the limitations associated with measuring the signal. For example, if the signal is a measurement of mass, a decrease in the minimum amount of analyte can be accomplished by switching from a con- ventional analytical balance, which weighs samples to ±0.1 mg, to a semimicro (±0.01 mg) or microbalance (±0.001 mg).

Concentration methods frequently have both lower and upper limits for the amount of analyte that can be determined. The lower limit is dictated by the small- est concentration of analyte producing a useful signal and typically is in the parts per million or parts per billion concentration range. Upper concentration limits exist when the sensitivity of the analysis decreases at higher concentrations.

An upper concentration level is important because it determines how a sam- ple with a high concentration of analyte must be treated before the analysis. Con- sider, for example, a method with an upper concentration limit of 1 ppm (micro- grams per milliliter). If the method requires a sample of 1 mL, then the upper limit on the amount of analyte that can be handled is 1 μg. Using Figure 3.6, and following the diagonal line for 1 μg of analyte, we find that the analysis of an ana- lyte present at a concentration of 10% w/w requires a sample of only 10 μg! Ex- tending such an analysis to a major analyte, therefore, requires the ability to ob- tain and work with very small samples or the ability to dilute the original sample accurately. Using this example, analyzing a sample for an analyte whose concen- tration is 10% w/w requires a 10,000-fold dilution. Not surprisingly, concentra- tion methods are most commonly used for minor, trace, and ultratrace analytes, in macro and meso samples.

4.4.6 Equipment, Time, and Cost

Finally, analytical methods can be compared in terms of their need for equipment, the time required to complete an analysis, and the cost per sample. Methods relying on instrumentation are equipment-intensive and may require significant operator training. For example, the graphite furnace atomic absorption spectroscopic method for determining lead levels in water requires a significant capital investment in the instrument and an experienced operator to obtain reliable results. Other methods, such as titrimetry, require only simple equipment and reagents and can be learned quickly.

The time needed to complete an analysis for a single sample is often fairly simi- lar from method to method. This is somewhat misleading, however, because much of this time is spent preparing the solutions and equipment needed for the analysis. Once the solutions and equipment are in place, the number of samples that can be analyzed per hour differs substantially from method to method. This is a significant factor in selecting a method for laboratories that handle a high volume of samples.

The cost of an analysis is determined by many factors, including the cost of necessary equipment and reagents, the cost of hiring analysts, and the number of samples that can be processed per hour. In general, methods relying on instruments cost more per sample than other methods.

4.4.7 Making the Final Choice

Unfortunately, the design criteria discussed earlier are not mutually independent.8 Working with smaller amounts of analyte or sample, or improving selectivity, often comes at the expense of precision. Attempts to minimize cost and analysis time may decrease accuracy. Selecting a specific method requires a careful balance among these design criteria. Usually, the most important design criterion is accuracy, and the best method is that capable of producing the most accurate results. When the need for results is urgent, as is often the case in clinical labs, analysis time may be- come the critical factor.

The best method is often dictated by the sample's properties. Analyzing a sam- ple with a complex matrix may require a method with excellent selectivity to avoid interferences. Samples in which the analyte is present at a trace or ultratrace con- centration usually must be analyzed by a concentration method. If the quantity of sample is limited, then the method must not require large amounts of sample.

Determining the concentration of lead in drinking water requires a method that can detect lead at the parts per billion concentrations. Selectivity is also im- portant because other metal ions are present at significantly higher concentrations. Graphite furnace atomic absorption spectroscopy is a commonly used method for determining lead levels in drinking water because it meets these specifications. The same method is also used in determining lead levels in blood, where its ability to detect low concentrations of lead using a few microliters of sample are important considerations.

4.4.8 Developing the Procedure

After selecting a method, it is necessary to develop a procedure that will accomplish the goals of the analysis. In developing the procedure, attention is given to compen- sating for interferences, selecting and calibrating equipment, standardizing the method, acquiring a representative sample, and validating the method.

4.4.9 Compensating for Interferences

The accuracy of a method depends on its selectivity for the analyte. Even the best methods, however, may not be free from interferents that contribute to the mea- sured signal. Potential interferents may be present in the sample itself or the reagents used during the analysis. In this section we will briefly look at how to mini- mize these two sources of interference.

In the absence of an interferent, the total signal measured during an analysis, Smeas, is a sum of the signal due to the analyte, and the signal due to the rea- gents, S_{reag}

$$S_{meas} = S_A + S_{reag} = kn_A + S_{reag} \qquad \text{(total analysis method)}$$

$$S_{meas} = S_A + S_{reag} = kC_A + S_{reag} \qquad \text{(concentration method)}$$

Without an independent determination of Sreag, equation 9 or 10 cannot be solved for the moles or concentration of analyte. The contribution of Sreag is deter- mined by measuring the signal for a reagent or method blank that does not contain the sample. Consider, for example, a procedure in which a 0.1-g sample is dissolved in 100 mL of solvent. After dissolving the sample, several reagents are added, and the signal is measured. The reagent blank is prepared by omitting the sample and adding the reagents to 100 mL of solvent. When the sample is a liquid, or is in solu- tion, an equivalent volume of an inert solvent is substituted for the sample. Once Sreag is known, it is easy to correct Smeas for the reagent's contribution to the overall signal.

Compensating for an interference in the sample's matrix is more difficult. If the identity and concentration of the interferent are known, then it can be added to the reagent blank. In most analyses, however, the identity or concentration of matrix interferents is not known, and their contribution to Smeas is not included in Sreag. Instead, the signal from the interferent is included as an additional term.

$$S_{meas} = k_A n_A + k_I n_I + S_{reag} \qquad \text{(total analysis method)}$$

$$S_{meas} = k_A C_A + k_I C_I + S_{reag} \qquad \text{(concentration method)}$$

Solving either equation 3.11 or 3.12 for the amount of analyte can be accomplished by separating the analyte and interferent before the analysis, thus eliminating the term for the interferent.

Alternatively, equations 11 or 12 can be solved for the amounts of both the analyte and the interferent. To do so, however, we must obtain two independent values for Smeas. Using a concentration method as an example, gives two equations

$$S_{meas,1} = k_{A,1} C_A + k_{I,1} C_I + S_{reag,1}$$

$$S_{meas,2} = k_{A,2} C_A + k_{I,2} C_I + S_{reag,2}$$

that can be solved simultaneously for CA and CI. This treatment is general. The composition of a solution with a total of n analytes and interferents can be deter- mined by measuring n independent signals, and solving n independent simultane- ous equations of the general form of equation 11 or 12.

A sample was analyzed for the concentration of two analytes, A and B, under two sets of conditions. Under condition 1, the calibration sensitivities are

$k_{A,1}=76 ppm^{-1}$ $k_{B,1}=186 ppm^{-1}$

and for condition

$k_{A,2}=33 ppm^{-1}$ $k_{B,2}=243 ppm^{-1}$

The signals under the two sets of conditions are

$S_{meas,1}=33.4$ $S_{meas,1}=29.7$

Determine the concentration of A and B. You may assume that S_{reg} is zero under both conditions.

Solution

Using equation 3.12, we write the follwing simultaneous equations

Using equation 3.12, we write the following simultaneous equations

33.4 = (76 ppm^{-1})CA + (186 ppm^{-1})Cg

29.7 = (33 ppm^{-1})CA + (243 ppm^{-1})Cg

Multiplying the first equation by the ratio 33/76 gives the two equations as

14.5 = (33 ppm^{-1})CA + (80.8 ppm^{-1})C3

29.7 = (33 ppm^{-1})CA + (243 ppm^{-1})CB Subtracting the first equation from the second gives

15.2 = (162.2 ppm-')CB

Solving for CB gives the concentration of B as 0.094 ppm. Substituting this concentration back into either of the two original equations gives the concentration of A, CA, as 0.21 ppm.

4.5 CALIBRATION AND STANDARDIZATION

Analytical chemists make a distinction between calibration and standardization.9 Calibration ensures that the equipment or instrument used to measure the signal is operating correctly by using a standard known to produce an exact signal. Balances, for example, are calibrated using a standard weight whose mass can be traced to the internationally accepted platinum–iridium prototype kilogram.

Standardization is the process of experimentally determining the relation- ship between the signal and the amount of analyte (the value of k in equations 1 and 2). For a total analysis method, standardization is usually defined by the stoichiometry of the chemical reactions responsible for the signal. For a concentration method, however, the relationship between the signal and the ana lyte's concentration is a theoretical function that cannot be calculated without experimental measurements. To standardize a method, the value of k is determined by measuring the signal for one or more standards, each containing a known concentration of analyte. When several standards with different concen- trations of analyte are used, the result is best viewed visually by plotting Smeas versus the concentration of analyte in the standards. Such a plot is known as a calibration curve.

4.5.1 Sampling

Selecting an appropriate method helps ensure that an analysis is accurate. It does not guarantee, however, that the result of the analysis will be sufficient to solve the problem under investigation or that a proposed answer will be correct. These latter concerns are addressed by carefully collecting the samples to be analyzed.

A proper sampling strategy ensures that samples are representative of the material from which they are taken. Biased or nonrepresentative sampling and contamination of samples during or after their collection are two sources of sampling error that can lead to significant errors. It is important to realize that sampling errors are completely independent of analysis errors. As a result, sampling errors cannot be corrected by evaluating a reagent blank.

4.5.2 Validation

Before a procedure can provide useful analytical information, it is necessary to demonstrate that it is capable of providing acceptable results. Validation is an evaluation of whether the precision and accuracy obtained by following the procedure are appropriate for the problem. In addition, validation ensures that the written procedure has sufficient detail so that different analysts or laboratories following the same procedure obtain comparable results. Ideally, validation uses a standard sample whose composition closely matches the samples for which the procedure was developed. The comparison of replicate analyses can be used to evaluate the procedure's precision and accuracy. Intralaboratory and interlaboratory differences in the procedure also can be evaluated. In the absence of appropriate standards, accuracy can be evaluated by comparing results obtained with a new method to those obtained using a method of known accuracy.

4.6 PROTOCOLS

Earlier we noted that a protocol is a set of stringent written guidelines, specifying an exact procedure that must be followed if results are to be accepted by the agency specifying the protocol. Besides all the considerations taken into account when de- signing the procedure, a protocol also contains very explicit instructions regarding internal and external quality assurance and quality control (QA/QC) procedures.10 Internal QA/QC includes steps taken to ensure that the analytical work in a given laboratory is both accurate and precise. External QA/QC usually involves a process in which the laboratory is certified by an external agency.

As an example, we will briefly outline some of the requirements in the Envi- ronmental Protection Agency's Contract Laboratory Program (CLP) protocol for the analysis of trace metals in aqueous samples by graphite furnace atomic ab- sorption spectrophotometry. The CLP protocol (Figure 3.7) calls for daily stan- dardization with a reagent blank and three standards, one of which is at the labo- ratory's contract required detection limit. The resulting calibration curve is then verified by analyzing initial calibration verification (ICV) and initial calibration blank (ICB) samples. The reported concentration of the ICV sample must fall within ±10% of the expected concentration. If the concentration falls outside this limit, the analysis must be stopped and the problem identified and corrected be- fore continuing.

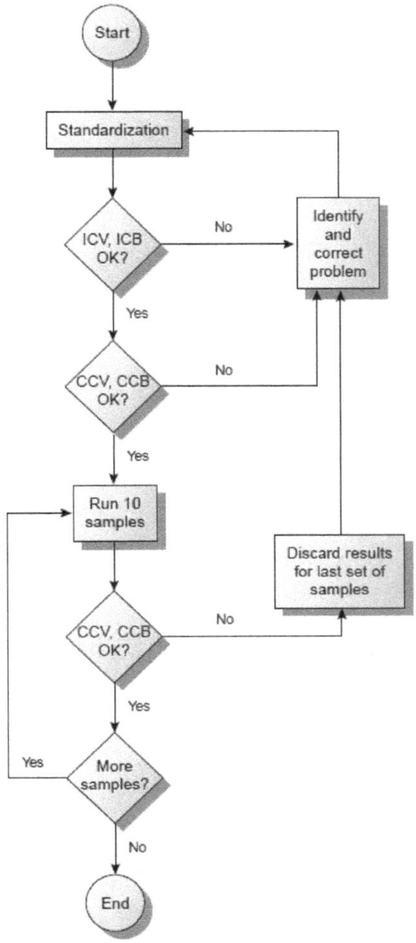

After a successful analysis of the ICV and ICB samples, standardization is reverified by analyzing a continuing calibration verification (CCV) sample and a contin- uing calibration blank (CCB). Results for the CCV also must be within ±10% of the expected concentration. Again, if the concentration of the CCV falls outside the established limits, the analysis must be stopped, the problem identified and corrected, and the system standardized as described earlier. The CCV and the CCB are analyzed before the first and after the last sample, and after every set of ten samples. Whenever the CCV or the CCB is unacceptable, the results for the most recent set of ten samples are discarded, the system is standardized, and the samples are reanalyzed. By following this protocol, every result is bound by successful checks on the standardization. Although not shown in Figure 3.7, the CLP also contains detailed instructions regarding the analysis of duplicate or split samples and the use of spike testing for accuracy.

5

STANDARD ADDITION AND INTERNAL STANDARDS

5.1 INTRODUCTION

The instrumental analysis of "real world" samples often involves complex mixtures such as blood, urine, or food that contain many chemical components. While standard calibration methods such as those previously discussed work well for laboratory standards and simple unknowns, the presence other constituents in real samples often leads to inaccuracies. In the discussion that follows, some common methods to correct for these problems in the calibration process are described.

5.2 MATRIX EFFECTS

The matrix of a sample is simply the background of all the chemical constituents in which the analyte exists. For instance, if we wish to determine the concentration of calcium in blood plasma, the matrix would consist of all of the other components present (proteins, sugars, salts, etc.). A matrix effect is a change in the instrumental response for the analyte caused by the presence of other components in the sample. For example, in the determination of magnesium in sea water, it is found that the instrumental response for magnesium ion (as magnesium chloride) in distilled water is not the same as for magnesium ion in sea water. In this case, sea water is the matrix in which the analyte is present and the change in response is mainly due to the presence of high concentrations of sodium, chloride, and sulfate ions. Thus, if the sensitivity (slope of the calibration curve) for magnesium is measured with standards prepared from magnesium chloride in distilled water, accurate results for magnesium might be obtained in measurements on rain water, but inaccurate results will be obtained on sea water samples, where the calibration curve is no longer valid.

As indicated in Figure 1, the sample matrix may affect the background (the intercept of the calibration curve), the sensitivity (the slope of the curve), or both. Matrix effects that change the background are often the most difficult to correct. Several measures can often be used to overcome matrix effects that change the sensitivity. Depending on the instrumental method, matrix effects may arise from changes in the chemical environment (e.g. the presence of other ions or compounds that change the form of the analyte) or physical properties (e.g. viscosity). In many methods of analysis, matrix effects can cause large errors and therefore techniques for dealing with them must be considered. Some of the more useful methods are listed below.

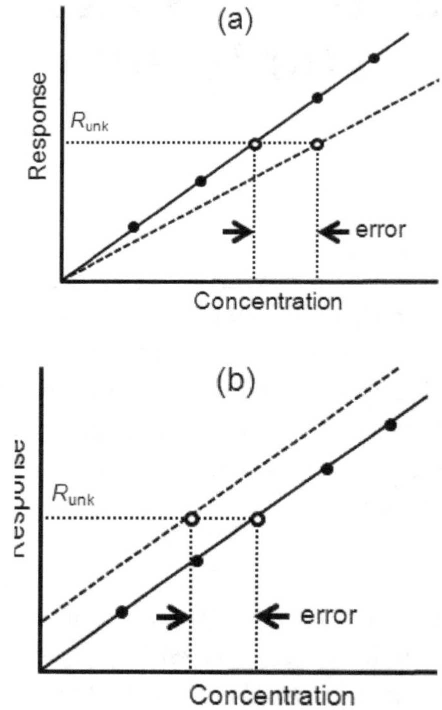

Fig. 1: Matrix effects on calibration curves: (a) change in sensitivity (slope), and (b) change in background (intercept). The solid line indicates the calibration curve constructed from laboratory samples, the dashed line indicates the (unknown) calibration curve in the presence of the matrix, and the dotted line indicates the error in the unknown concentration.

- ***Matrix matching:*** If the matrix is known and reasonably constant, chemical components that simulate the matrix can be added to all of the standards used to prepare the calibration curve. All measurements will then be made under the same conditions (with the same matrix). This method is effective for situations where the components of the matrix are simple and well known (*e.g.* sea water), but not so if the matrix is more complex (*e.g.* blood plasma) or likely to vary from sample to sample.

- ***Dilution:*** This is an alternate form matrix matching in which the sample is sufficiently diluted to remove the effects of the matrix on the sensitivity. Unfortunately, the dilution required is often so large that it may reduce the ability to detect the analyte.
- ***Matrix buffering:*** This is another form of matrix matching in which very high concentrations of a matrix component are added to both the standards and the sample to "mask" the influence of the original sample matrix. For example, enough sodium chloride could be added a seawater sample to mask the effects of the ions already present. Unfortunately, this may also complicate the analysis in other ways.
- ***Matrix destruction:*** In certain cases, it may be possible to destroy the sample matrix (but not the analyte). For example, in the determination of metals in organic substances (*e.g.* hair, food), the organic matrix may be destroyed by dissolving it in acid and heating (called "ashing"). This is not always possible, however, and may result in the loss of analyte.
- ***Matrix separation:*** For many complex samples, it is possible to separate the analyte(s) from the other components of the sample and make the measurements in the absence of a matrix. This may, for example, involve precipitation or extraction of the analyte (or the matrix components).. These techniques are widely used, but complicate the analytical procedure.
- ***Standard addition:*** This is a relatively simple procedure in which the sample itself is used to prepare the standards. In this way the sample provides the matrix and ensures that the matrix effects are the same for all measurements.
- ***Internal standard:*** In this method, a substance similar to the analyte, but not present in the sample, is added to each sample and standard and the ratio of the responses of the two substances is used for quantitative purposes.

The method used to reduce or eliminate the matrix effect (when necessary) depends on the nature of the analyte and matrix, as well as the type of analytical technique used. In this chapter, the focus will be on standard addition and internal standards, which are relatively simple techniques that are widely used.

5.3 CALIBRATION WITH STANDARD ADDITION METHODS

In the method of standard addition, the sample is made to act as both the standard and the unknown. As the name "standard addition" implies, a standard with a known amount of the analyte is added to a sample with an unknown amount of the same analyte. This is done in such a manner that the matrix is similar for all solutions. Therefore the slope of the calibration curve should not change and the problem of matrix effects should be eliminated. For this method to work, the instrument response must obey a specific relationship to the concentration of analyte, usually the linear relationship that we have already seen. It is also critical with the methods presented here that the *background contribution can be considered to be zero* (*i.e.* the calibration curve has a zero intercept).

Although not a strict requirement, the "standard" and the "unknown" which are mixed together are normally in the form of solutions, and these will be designated to have analyte concentrations of Cs and Cu prior to mixing. Although the analyte comes from different sources, as shown in Figure 2, the instrument measuring the response makes no distinction between the two, so the effect of the combined concentrations (in the presence of the matrix) is recorded.

In performing calibration by standard addition, there are two unknown quantities: the sensitivity (S, the slope of the calibration curve) and the concentration of the analyte in the unknown, Cu (recall that the background, B, is assumed to be zero). For this reason, we need response measurements for at least two different solutions to solve for the unknown concentration (two equations, two unknowns). Therefore, there is no one-point calibration method for standard addition, only two-point and multipoint methods. Both of these will be discussed in this topic.

As in traditional calibration, multipoint standard addition methods are preferred in practice since they produce more precise results and are able to detect nonlinearity in the response. However, two-point standard addition methods are faster to carry out and do not require the use of regression. (For the latter reason, they are also more likely to be asked by analytical chemistry professors on tests.) Another disadvantage of multipoint methods is that, if the matrix changes significantly between samples, a new calibration curve may be required each time, which is very time-consuming.

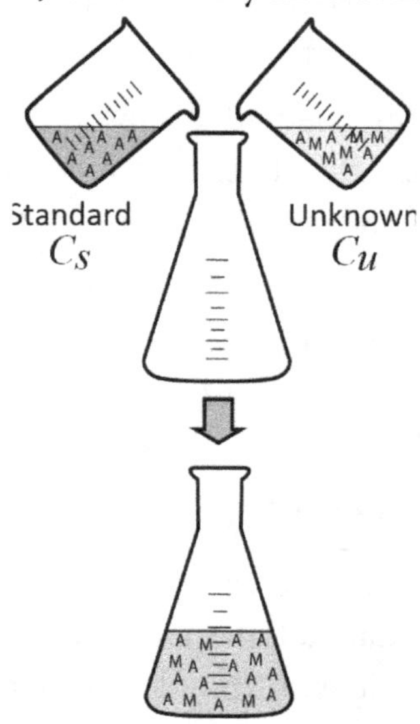

Fig. 2: In standard addition, a standard solution with a known concentration (Cs) of analyte (A) and no matrix components (M) is combined with the sample solution which contains matrix components and an unknown concentration of analyte.

5.4 TWO-POINT STANDARD ADDITION METHOD

The approach used in two-point standard addition methods is simply to use a set of two equations to solve for the concentration of the unknown, Cu. However, it is important to recognize that there is no single equation that can be used for every situation, since this will depend on the actual procedures employed. Therefore, the equations to be used need to be derived in each situation. Fortunately, there is a standard approach that can be applied and this is described below.

- From the procedures described to prepare the two solutions, write expressions for the analyte concentrations in each of the solutions prepared, *C*1 and *C*2, as functions of *C*s and Cu, as well as any volumes specified to account for dilution.
- The measured instrument responses for the two solutions, *R*1 and *R*2, will be given. Write equations for these as functions of *C*1 and *C*2, expanding these further using the relations in step (1) to give functions of *C*s and Cu. Although not strictly required, in virtually all cases, a linear instrument response will be assumed, so the appropriate equations will be

$$R_1 = S, C_1$$

$$R_2 = S, C_2$$

- Recall that the background, *B*, is assumed to be zero. The ellipsis (…) indicates that the full equation in terms of *C*s and Cu follows.
- Solve the equations given in previous step for Cu, substituting in the numerical values for *R*1, *R*2, *C*s and other necessary quantities. Generally, it is not necessary to directly solve for the sensitivity, *S*, unless this is needed for other calculations.

It should be noted that, depending on how quantities are defined, there may be more than one set of equations that give rise to valid solutions.

Although it has been stated that each standard addition problem should be treated individually and there is no universal equation, there are two general procedures that apply to most cases encountered. These are referred to as the *variable volume method* and the *fixed volume method*, and are described in the sections that follow.

5.4.1 The Variable Volume Method

In the variable volume standard addition method, a known amount of standard is added to the unknown. For a two-point calibration, two responses are measured. These are typically the responses for the unknown solution and for a solution to which a known volume of standard, *V*s, has been added to a known volume of the unknown, *V*u. This is illustrated in Figure 3. The principle here is that the addition of the standard to the sample makes the matrix similar between the two samples. It is called the variable volume method because the total volume of the solution changes on the addition of the standard (this terminology will become clearer when we discuss the fixed volume method). Following step (1) in the defined procedure, we can write expressions for the concentration of the analyte in the two solutions.

$$C_1 = C_u \qquad C_2 = \left(\frac{V_u}{V_u + V_s}\right)C_u + \left(\frac{V_s}{V_u + V_s}\right)C_s$$

Note that the concentration in the second solution is affected by both the standard and the unknown, as well as the dilution of each of those solutions to the total volume. Assuming a linear response with no background, the measurements for the two solutions, represented as *R*1 and *R*2, will be given by

$$R_1 = S \cdot C_1 = S \cdot C_u$$

$$R_2 = S \cdot C_2 = S \cdot \left(\frac{V_u}{V_u + V_s}\right)C_u + S \cdot \left(\frac{V_s}{V_u + V_s}\right)C_s$$

Equations3 and 4 represent the two equations needed to solve for the two unknowns, *C*u and *S* (the quantities *R*1, *R*2, *V*u, and *V*s are all known from the measurements and procedure). To solve for *C*u, some mathematical manipulation will be necessary and you need to be proficient at this. Since there are different (equally valid) approaches to the solution, and because the solutions may vary from one situation to another, no general solution is presented.

However, some examples will illustrate how to approach this.

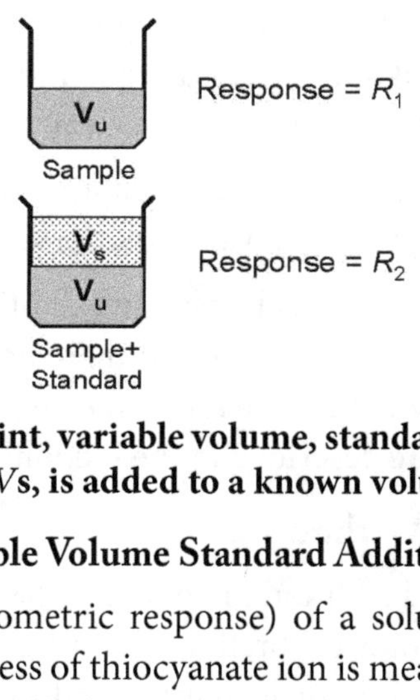

Fig. 3: Principle of the two-point, variable volume, standard addition method. A known volume of standard, *V*s, is added to a known volume of unknown, *V*u.

Example.1: Two-Point Variable Volume Standard Addition

The absorbance (spectrophotometric response) of a solution containing an unknown amount of iron and a large excess of thiocyanate ion is measured as 0.346. When 1.00 mL of a 100. ppm iron standard is added to 10.00 mL of the unknown, the absorbance of this solution is measured as 0.515. Assuming that absorbance is linear in concentration, what is the iron concentration in the sample?

As with all problems of this type, we begin by writing the equations that define the response as a linear function of the concentrations, taking into account any dilution factors. Here, we will use Cu to represent the original concentration of the unknown and Cs to represent the original concentration of the standard. For the two solutions, we have:

$$C_1 = C_u \qquad C_2 = \frac{V_u}{V_s + V_u} C_u + \frac{V_s}{V_s + V_u} C_s = \frac{10.00\text{ mL}}{11.00\text{ mL}} C_u + \frac{1.00\text{ mL}}{11.00\text{ mL}} C_s$$

Substituting these into the linear relationship $R = SC$ gives the two equations:

$$R_1 = 0.346 = S \cdot C_1 = S \cdot C_u$$

$$R_2 = 0.515 = S \cdot C_2 = S\left(\frac{10.00}{11.00} C_u + \frac{1.00}{11.00}(100.\text{ ppm})\right)$$

There are different ways to solve for Cu in these equations. Two solutions are presented below.

Solution 1:

In situations such as this, where one equation only has one term, perhaps the easiest solution is to divide the second equation by the first, thereby eliminating the unknown *S*, as shown below:

$$\frac{R_2}{R_1} = \frac{0.515}{0.346} = \cancel{S}\left(\frac{10.00}{11.00} C_u + \frac{(1.00)(100.\text{ ppm})}{11.00}\right) \Big/ \cancel{S} C_u$$

$$\frac{0.515}{0.346} = \frac{\frac{10.00}{11.00}\cancel{C_u}}{\cancel{C_u}} + \frac{\frac{(1.00)(100.\text{ ppm})}{11.00}}{C_u} = \frac{10.00}{11.00} + \frac{100.\text{ ppm}}{(11.00)C_u}$$

$$\left(\frac{0.515}{0.346} - \frac{10.00}{11.00}\right) = \frac{100.\text{ ppm}}{(11.00)C_u} \quad \text{or} \quad \left(\frac{0.515}{0.346} - \frac{10.00}{11.00}\right) C_u = \frac{100.\text{ ppm}}{11.00}$$

$$C_u = \left(\frac{100.\text{ ppm}}{11.00}\right) \Big/ \left(\frac{0.515}{0.346} - \frac{10.00}{11.00}\right) = \frac{9.0\bar{9}09\text{ ppm}}{1.4\bar{8}844 - 0.909\bar{0}9}$$

$$= \frac{9.0\bar{9}09\text{ ppm}}{0.5\bar{7}93} = 1\bar{5}.7 \text{ or } 16\text{ ppm}$$

Solution 2:

An alternative solution involves rearranging the equation for R1 to give S = R1/Cu and then substituting this into the equation for R2.

$$R_2 = \frac{R_1}{C_u}\left(\frac{10.00}{11.00} C_u + \frac{1.00}{11.00} \cdot (100.\text{ ppm})\right) = \frac{(10.00)R_1}{11.00} + \frac{(100.\text{ ppm})R_1}{(11.00)C_u}$$

$$\left(R_2 - \frac{(10.00)R_1}{11.00}\right) = \frac{(100.\text{ ppm})R_1}{(11.00)C_u} \quad \text{or} \quad \left(R_2 - \frac{(10.00)R_1}{11.00}\right) C_u = \frac{(100.\text{ ppm})R_1}{(11.00)}$$

$$C_u = \frac{(100.\text{ ppm})R_1}{(11.00)} \Big/ \left(R_2 - \frac{(10.00)R_1}{11.00}\right)$$

$$= \frac{(100.\text{ ppm})(0.346)}{(11.00)} \Big/ \left(0.515 - \frac{(10.00)(0.346)}{11.00}\right)$$

$$= \frac{3.1\bar{4}55}{0.515 - 0.31\bar{4}55} = \frac{3.1\bar{4}55}{0.20\bar{0}45} = 15.7\text{ ppm}$$

The two different approaches used to solve for Cu in the previous example produced the same answer and both approaches are equally valid. Attentive readers will note, however, that the second calculation results in three significant digits while the first gives only two.

This is a consequence of the of the fact that rules for significant figures are only approximate and the application of propagation of error methods would produce more consistent results (uncertainty of ±0.3 ppm assuming ±1 in the last digit of all of the variables). In this class, as long as the answer is consistent with the method employed, that is satisfactory.

It is important in standard addition methods that the matrix remain constant between samples, but a drawback of the variable volume method is that the matrix is diluted by the addition of the standard. For this reason, it is important that the volume of the standard solution, often referred to as a "spike" in quantitative analysis, be small relative to the volume of the sample. If the concentration of the standard spiked into the sample is high enough, the volume of the standard can be quite low and possibly even ignored in the calculations, making the solution easier.

Another consideration is the concentration of the standard used to spike the sample, *Cs*. In general, the smaller the concentration of the spike in the diluted sample, the larger will be the errors in the estimated concentration because the difference in the measured responses becomes small. This is illustrated in Figure 4, which shows the effect of *Cs* on the errors in the calculated unknown concentration, *Cu*, under certain conditions (10:1 volume ratio and 1% or 0.1% errors in responses and volumetric quantities). What is important is the ratio of the standard in the *diluted* sample to the concentration of the unknown. Note that there is not much change until this ratio becomes less than unity, but when the concentration of the standard becomes too low, the errors increase rapidly. Choosing a value for *Cs* is complicated by the fact that we don't know *Cu*, so normally we would make an estimate of the upper limit of *Cu*. In general, the concentration of the standard *after dilution* should be around the value of *Cu* (in the above example, the ratio was about 0.6). It is better to make it too high than too low, but the upper limit of *Cs* may be limited by solubility or the range of the instrument response.

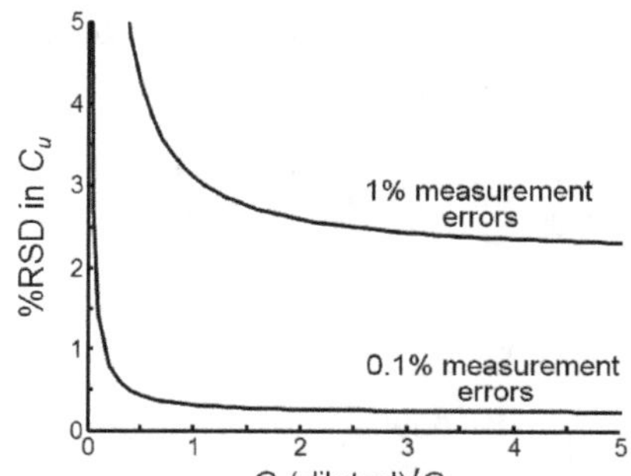

Fig. 4: Effect of standard concentration, *Cs*, on the precision of variable volume standard addition under different measurement error conditions. Note that the errors can increase rapidly when the concentration of the diluted standard becomes less than the concentration of the unknown.

5.4.2 The Fixed Volume Method

As with the variable volume method, the fixed volume standard addition method involves adding a specified amount of standard to a fixed volume of unknown solution. The difference here is that all of the solutions prepared are diluted to the same total final volume, Vt, in contrast to the variable volume method where the total volume depends on the volumes of the standard and unknown (hence the name). With the two-point fixed volume method, one of the solutions prepared typically contains only the unknown and the other contains the unknown and the standard, but this is not a requirement as long as both solutions contain different amounts of standard. This principle is illustrated in Figure 5. A convenient aspect of this approach is that the denominator of the dilution factor does not change, as seen in the derivation below.

Using the same definitions and assumptions (linearity and zero background) as for the variable volume method, equations can be written for the two measured responses.

$$R_1 = S \cdot C_1 = S \cdot C_u^{dil} = S \cdot C_u \cdot \frac{V_u}{V_t} \tag{7.5}$$

$$R_2 = S \cdot C_2 = S \cdot (C_u^{dil} + C_s^{dil}) = S \cdot \left(C_u \cdot \frac{V_u}{V_t} + C_s \cdot \frac{V_s}{V_t} \right) \tag{7.6}$$

Note that the superscript "dil" has been added here to indicate the concentrations of the unknown and standard after dilution. As before, there are two equations and two unknowns (S and Cu) and a variety of methods can be used to solve for the unknown concentration.

These equations can change with the problem under consideration, however, so a general equation will not be given and should be derived for each problem.

A major advantage of the fixed volume method compared to the variable volume method is that the concentration of the matrix is the same for both measurements, so the sensitivity should remain constant. However, the fixed volume method requires more solution manipulation (more pipetting and dilution), and the solutions can be more dilute, resulting in lower sensitivity relative to the original concentration of the unknown.

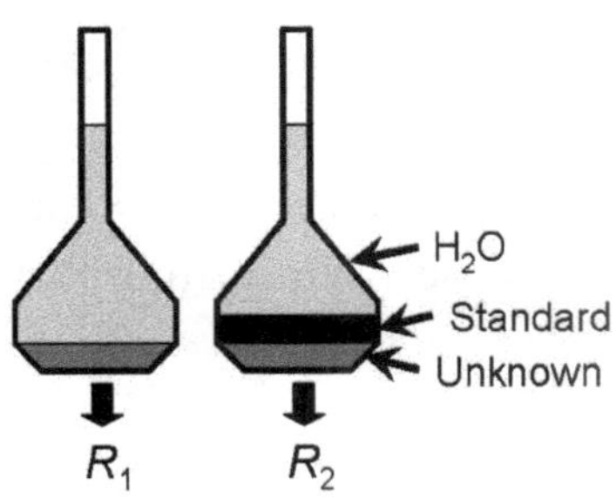

Fig. 5: Principle of the two-point, fixed volume, standard addition method. One solution contains only unknown, while the other contains the unknown plus standard. Both solutions are diluted to the same final volume.

Example 2: Two-Point Fixed Volume Standard Addition

Dextromethorphan, an active ingredient in cough syrups can be determined by fluorescence spectroscopy. A 5.00 mL sample of cough syrup is first diluted to 100.0 mL. A 2.00 mL aliquot of this solution is then diluted to 100.0 mL with distilled water and the fluorescence is recorded as 158. A second 2.00 mL aliquot is combined with 3.00 mL of a standard containing 80.0 ppm of dextromethorpham and the mixture is again diluted to 100.0 mL. This gives a fluorescence reading of 353. Assuming a linear response and zero background, determine the number of milligrams of dextromethorpham in one dose (15 mL) of the cough syrup.

This is a fixed volume standard addition problem since both samples are diluted to 100 mL. To illustrate that there is more than one way to solve these problems, two solutions will be presented.

Solution 1:

In setting up these problems, the definition of the variables is important. In this case, we will define Cu to be the concentration of the analyte in the sample after the first dilution and Cs to be the concentration of the standard. Thus, we can write:

$$R_1 = S \cdot C_u \frac{2.00\ \text{mL}}{100.0\ \text{mL}} \qquad R_2 = S\left(C_u \frac{2.00\ \text{mL}}{100.0\ \text{mL}} + C_s \frac{3.00\ \text{mL}}{100.0\ \text{mL}}\right)$$

In this case, we will take the ratio of the two equations to eliminate S,

$$\frac{R_2}{R_1} = \frac{\cancel{S}\left(C_u \frac{2.00}{100.0} + C_s \frac{3.00}{100.0}\right)}{\cancel{S} \cdot C_u \frac{2.00}{100.0}} = \left(C_u \frac{2.00}{100.0} + C_s \frac{3.00}{100.0}\right) \cdot \frac{100.0}{(2.00)C_u} = 1.00 + \frac{(3.00)C_s}{(2.00)C_u}$$

Rearranging this, we can solve for Cu:

$$C_u = \frac{(3.00)C_s}{(2.00)(R_2/R_1 - 1.00)} = \frac{(3.00)(80.0\ \text{ppm})}{(2.00)(353/158 - 1.00)} = \frac{240.\ \text{ppm}}{(2.00)(2.2\bar{3}42 - 1.00)}$$

$$= \frac{240.\ \text{ppm}}{(2.00)(1.2\bar{3}42)} = 97.\bar{2}3\ \text{ppm}$$

We still need to take into account the initial dilution, so

$$C_u^{\text{original}} = (97.23\ \text{ppm}) \times \frac{100.0\ \text{mL}}{5.00\ \text{mL}} = 19\bar{4}5\ \text{ppm} = 19\bar{4}5\ \text{mg/L} = 1.9\bar{4}5\ \text{mg/mL}$$

Here we have used the equivalence that 1 ppm = 1 mg/L for dilute solutions. Finally, we can calculate the amount in 1 dose. dextromethorphan in one dose = (1.945 mg /mL) (15 mL /dose) = 29 mg

Solution 2:

To illustrate an alternative approach, we will define Cu as the concentration in the cough syrup before the initial dilution and use the first equation to substitute for *S* (these two changes are not related to one another).

$$R_1 = S \cdot C_u \frac{5.00\text{ mL}}{100.0\text{ mL}} \cdot \frac{2.00\text{ mL}}{100.0\text{ mL}} = S \cdot C_u \frac{10.0}{10,0\bar{0}0.} = S \cdot C_u(0.00100)$$

$$R_2 = S\left(C_u \frac{5.00\text{ mL}}{100.0\text{ mL}} \cdot \frac{2.00\text{ mL}}{100.0\text{ mL}} + C_s \frac{3.00\text{ mL}}{100.0\text{ mL}}\right) = S[C_u(0.00100) + C_s(0.0300)]$$

$$= \frac{R_1}{C_u(0.00100)} \cdot [C_u(0.00100) + C_s(0.0300)] = R_1 + \frac{(30.0)R_1C_s}{C_u}$$

$$C_u = \frac{(30.0)R_1C_s}{(R_2 - R_1)} = \frac{(30.0)(158)(80.0\text{ ppm})}{(353-158)} = \frac{37\bar{9},200\text{ ppm}}{195} = 1,9\bar{4}5\text{ ppm}$$

Note that in this case, we do not need to solve for the initial dilution because this has already been included in the definition of Cu.

5.5 MULTIPOINT STANDARD ADDITION METHODS

The use of multiple calibration samples can enhance the quality of traditional calibration methods by improving precision and detecting nonlinearities, and the same is true for standard addition methods. Although multipoint methods exist for both the variable volume and fixed volume approaches, the former is almost never used in practice, so this section will focus on fixed volume multipoint methods.

For fixed volume multipoint calibration, a series of solutions is prepared in which the volume of the unknown added, Vu , and the total volume of the solution, Vt , remain constant, but the concentration of the standard solution varies for each of the solutions prepared. For simplicity, the concentration of the standard is usually varied by adding a different volume of a single standard solution to each of the calibration solutions, as indicated in Figure 7.6. Under these conditions, the equation for the instrument response is:

$$R = \frac{SC_uV_u}{V_t} + \frac{SC_s}{V_t} \cdot V_s$$

From this, it can be seen that a plot of *R vs.* *V*s should be a straight line with a slope of (*SC*s/*V*t) and a *y*-intercept of (*SC*u*V*s/*V*t). Although this is one way to plot the graph, a more common method is based on the equation

$$R = \frac{SC_uV_u}{V_t} + S \cdot C_s^{\text{dil}}$$

In this equation, *C*s represents the concentration of the standard in the *diluted* sample rather than the original. Thus, a plot of *R vs.* Dil *C*s will also give a straight line with the same *y* intercept as before, but with a slope equal to *S*. This is illustrated in Figure 7.6. From this, we can see that the concentration in the original (undiluted) unknown will be

$$C_u = \frac{y\text{-intercept}}{\text{slope}} \cdot \frac{V_t}{V_u}$$

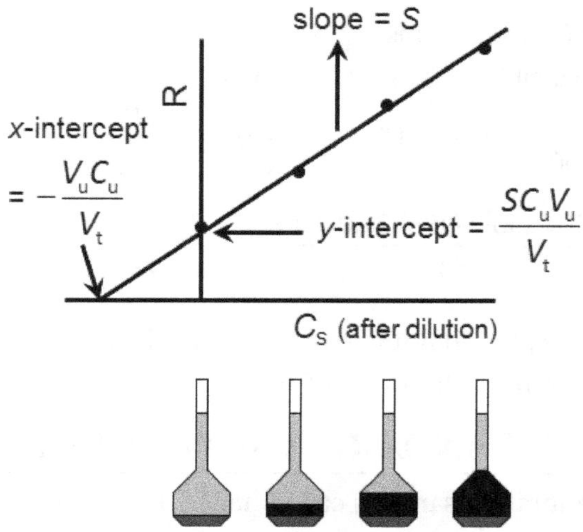

Figure 6: Illustration of the graphical approach to the multipoint, fixed volume standard addition method. Variable volumes of standard solution are added to a fixed volume of unknown, then all solutions are diluted to the same total volume.

Another way to determine the unknown concentration is simply to use the x-intercept, which will be the negative of the concentration of the unknown in the diluted samples. Therefore, correcting for the dilution, we can write

$$C_u = -(x\text{-intercept}) \cdot \frac{V_t}{V_u}$$

- This equation can be easier to use, especially if the x-intercept is read directly from the graph.

Exercise 7

Absorbance spectroscopy is used to determine the amount of phosphate in a urine sample by the standard addition method. Four solutions are prepared, each containing a 1.00 mL of urine, 50.0 mL of colour-forming reagent, varying amounts of a 0.500 mM phosphate standard solution, and enough distilled water to give a total volume of 100.0 mL. The following data were recorded.

Volume of standard (mL)	0	4.00	8.00	12.00
Absorbance	0.205	0.312	0.402	0.534

What is the concentration of phosphate in the urine? (For convenience, the data are plotted in Figure 7.)

Exercise 8

Imagine that 1.00 mL of urine from another subject is treated with 50.0 mL of colourforming reagent and diluted to 100.0 mL. This gives an absorbance reading of 0.242.

(a) Based on the results in the previous exercise estimate the concentration of phosphate in this urine sample.

(b) What important assumption is made in making this estimate?

(c) Is this assumption likely to be valid? Why or why not?

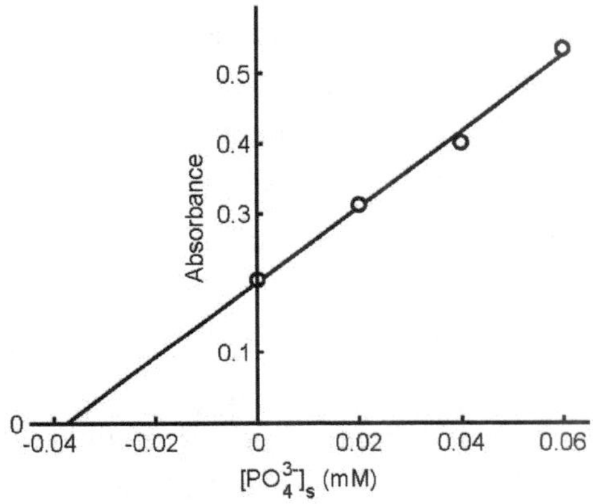

Figure 7: Plot of data for multipoint, fixed volume standard addition problem in Exercise 7.7.

Exercise 9

- In addition to improved precision, an advantage of multipoint methods is that they allow the determination of the standard error of regression, se (see preceding topic), which is an estimate of measurement uncertainty.
- (a) Estimate the value of se in Exercise 7.
- (b) Estimate the limit of detection (LOD) for the measurement (i.e. in the diluted sample).
- (c) Estimate the LOD for the method (i.e. in the original sample).
- (d) Based on this, how many significant figures should be reported in the answers to the two previous exercises?
- (e) Estimate the limit of quantitation (LOQ) for the method.

Summary of Standard Addition Methods

Standard addition is an analytical approach intended to minimize the effect of differences between the matrix of the sample and standards that affect the sensitivity of the measurement.

By adding standards to the sample, the matrix is made similar for all measurements. In all of the methods presented, it was assumed that the instrument response is linear and that the background signal is zero or can be subtracted out. If the matrix affects the background, the standard addition method cannot be used.

It is important to recognize that standard addition is a principle rather than a single method or equation, and the necessary relationships should be derived in each case. Nevertheless, the classifications of "variable volume" and "fixed volume" methods can be useful. The variable volume method has the advantage that fewer solution manipulations are required, since there is no final dilution and the dilution of the sample is minimized, resulting in improved sensitivity. However, a drawback of the variable volume method is that the matrix is not identical between measurements. The extent to which this is a problem depends on the magnitude of the matrix effect and the magnitude of the dilution in each solution, but the volume of standard added should be small relative to the volume of the unknown. This is not a factor in the fixed volume method, since the matrix is diluted to the same extent in every sample, but the sample manipulations are more complicated and the solutions are typically more dilute, resulting in lower method sensitivity.

Multipoint standard addition methods result in greater precision and can establish linearity and limits of detection, but are more time consuming. Traditional multipoint calibration methods assume that matrix effects are negligible and the same calibration curve can be applied to multiple unknowns. For standard addition, this is only possible if the matrix remains essentially constant between samples. If this is not the case, a new calibration curve is necessary for each unknown. This is often impractical, so two-point methods are more commonly used when standard addition is involved. Another approach, referred to as "internal standards" is a way to circumvent this problem.

5.6 INTERNAL STANDARDS

The method of internal standards is another technique to account for changes in the sensitivity of an instrument. It can account not only for matrix effects, but also changes in sensitivity due to drift in instrument components or other factors. To illustrate, imagine that you are going on a long trip and also starting a diet. Weighing yourself along the way is difficult because every scale you use is adjusted slightly differently. Suppose, however, that you had an object, such as a suitcase, that always weighed the same. If you measured the ratio of your weight to that of the object, then you would always be able to tell if you were gaining or losing weight, as long as the scale response was linear. This is the same principle behind the use of internal standards, where the ratio of the measurement for the analyte to the measurement for a fixed amount of a second analyte is used.

An internal standard is a compound which is added to all samples and used to correct for changes in sensitivity. The internal standard must not to be present originally in any of the samples, must not interfere in the measurement of the analyte, must give an independent response, and must mirror the behavior of the analyte as closely as possible. It is usually added in the same amount to all samples and standards. The response of the instrument to both the analyte and the internal standard is recorded. The ratio of these responses is then used in any of the calibration strategies already discussed to calculate the concentration of analyte. This principle is illustrated in Figure 7.8.

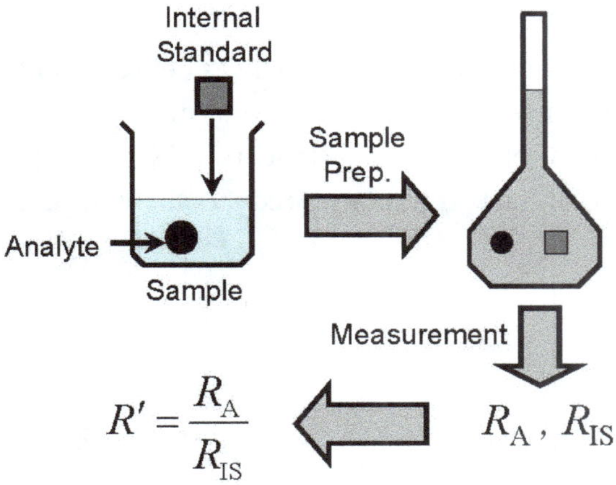

Fig. 8:Principle of the method of internal standards. The internal standard, which is a secondary analyte not present in the sample, is added to all samples, typically in the same amount. Factors affecting the sensitivity for both analytes are removed by using the ratio of responses. Assume that instrument response is linear for both the internal standard and analyte and that the background is zero. Then, using the basic linear response equation, we can write

$R_A = S_A C_A$ for the analyte, and

$R_{IS} = S_{IS} C_{IS}$ for the internal standard

where CIS is the concentration of internal standard in the sample. Note that the sensitivity of the instrument for the analyte and the internal standard is generally different. When the ratio of the responses is calculated

$$\frac{R_A}{R_{IS}} = \frac{S_A C_A}{S_{IS} C_{IS}} = k \frac{C_A}{C_{IS}}$$

where k is a constant (the ratio of two sensitivities). If the concentration of internal standard is always maintained constant, then the above expression becomes even simpler.

$$\frac{R_A}{R_{IS}} = R' = k' C_A$$

Note that, when the ratio of the responses is used in place of the analyte response and the concentration of the internal standard is constant, this becomes an alternative form of the basic response equation, so any of the basic calibration methods can be used (including standard addition).

The use of an internal standard corrects for matrix effects in a more subtle manner than the method of standard addition. A matrix effect that acts upon sensitivity causes SA to change from unknown to standard. If the internal standard is chosen to behave like the analyte, then the matrix effect will change SIS to approximately the same extent as SA. In this case k, the ratio of the two sensitivities, will change very little or not at all. Thus, the equations above involving k or k' will be almost independent of matrix effects. The same can be said for changes in instrument characteristics that affect the sensitivities. This is illustrated in Figure 7.9. If the sensitivity of the instrument changes between calibration standards, this will cause the analyte response, RA, to be erratic and deviate from linearity, as shown in the figure. Since the response of the internal standard, RIS, changes proportionately in the same way, taking the ratio of the two responses reduces these deviations.

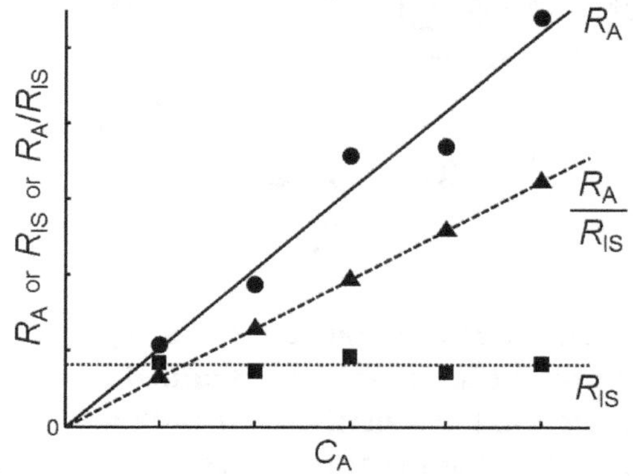

Figure 9: Effect of changes in instrument sensitivity on calibration. The responses of the analyte (•), the internal standard (▪) and their ratio (Δ) are shown, where the concentration of the internal standard is constant. Note that the ratio corrects for deviations in linearity caused by changes in sensitivity.

Internal standards exhibit other advantages as well. Many analytical procedures have steps that involve large systematic errors. For example, some procedures require that the analyte be extracted from the original sample prior to instrumental analysis. This extraction is often incomplete and, worse, the percentage recovery varies considerably from sample to sample. If an internal standard is added to the original sample, the proportion of analyte and internal standard that are lost should be the same, so the ratio of responses should not be affected by this change. Thus, an internal standard can cancel the effects of such problems.

When internal standards are used in a one-point calibration method, a standard solution is prepared containing both the analyte and the internal standard in known concentrations and k (or k') is determined from the measured responses of this standard. The internal standard is then added to the unknown, the responses measured, and the analyte concentration calculated using the value of k (or k'). In this case, it is not necessary that the internal standard be added to all samples in the same concentration.

Example 7.3: Internal Standards in a One-Point Calibration

Cu^{2+} is to be used as an internal standard in the polarographic analysis of Pb^{2+}. A standard solution containing 2.00 ppm of both Cu^{2+} and Pb^{2+} gives currents of 0.0505 mA for Cu^{2+} and 0.0584 mA for Pb^{2+}. A 25.0 mL aliquot of a 2.00 ppm solution of Cu^{2+} is added to 25.0 mL of an unknown containing Pb^{2+} but no copper. The currents measured for the unknown are 0.0226 and 0.0393 mA for Cu^{2+} and Pb^{2+}, respectively. Assuming that measured current is linear in concentration, what is the concentration of Pb^{2+} in the original unknown using: (a) a basic one-point calibration for Pb^{2+}, and (b) using a one-point calibration with copper as the internal standard?

(a) Ignoring the internal standard and considering only the response of the lead, we can write

$$R_{Pb} = S_{Pb} C_{Pb}$$

$$S_{Pb} = \frac{R_{Pb}^{std}}{C_{Pb}^{std}} = \frac{0.0584\ \mu A}{2.00\ ppm} = 0.0292\ \mu A \cdot ppm^{-1}$$

$$C_{Pb}^{unk} = \frac{R_{Pb}^{unk}}{S_{Pb}} = \frac{0.0393\ \mu A}{0.0292\ \mu A \cdot ppm^{-1}} = 1.3\bar{4}59\ ppm$$

Since the second solution was diluted by the internal standard, we need to account for this in determining the concentration of lead in the original unknown.

$$C_{Pb}^{orig} = (1.3\bar{4}59\ ppm) \times \frac{50.0\ mL}{25.0\ mL} = 2.69\ ppm$$

(b) Since the concentration of the internal standard changes between the two solutions, we must use Eqn. (7.11) rather than Eqn. (7.12).

$$R' = \frac{R_{Pb}}{R_{Cu}} = k \frac{C_{Pb}}{C_{Cu}}$$

$$k = \frac{R_{Pb}^{std}}{R_{Cu}^{std}} \cdot \frac{C_{Cu}^{std}}{C_{Pb}^{std}} = \frac{0.0584\ \mu A}{0.0505\ \mu A} \cdot \frac{2.00\ ppm}{2.00\ ppm} = 1.1\bar{5}64$$

$$C_{Pb}^{unk} = \frac{R_{Pb}^{unk}}{R_{Cu}^{unk}} \cdot \frac{C_{Cu}^{unk}}{k} = \frac{0.0393\ \mu A}{0.0226\ \mu A} \cdot \frac{1.00\ ppm}{1.1\bar{5}64} = 1.5\bar{0}37\ ppm$$

Note that both unk Pb *C* and unk Cu *C* represent concentrations in the unknown solution, but the copper concentration is actually known because it is the internal standard. Its concentration in this solution is 1.00 ppm because of the 2:1 dilution. Again, we need to account for dilution of Pb2+ in the original unknown

$$C_{Pb}^{orig} = (1.5\bar{0}37\ ppm) \times \frac{50.0\ mL}{25.0\ mL} = 3.01\ ppm$$

Obviously, since the two results do not agree, some factor changed between the measurements on the standard and those on the unknown. This change will have affected both the response for Cu(II) and Pb(II). Since the responses and the concentrations of the internal standard are known for both solutions, they can be used to correct those of the unknown and provide a more accurate result.

Exercise 7.10

In the preceding example:

- What is the sensitivity for Cu^{2+} in the first solution?
- What is the sensitivity for Cu^{2+} in the second solution?
- If the assumptions for internal standards hold, what would you expect the sensitivity for Pb^{2+} to be in the second solution? Why might it have changed?

Exercise 7.11

In Example 7.3(b), suppose the same measurements were obtained but, unknown to the analyst, there was some Cu^{2+} present in the unknown sample before the addition of the internal standard. Would the true concentration of the Pb^{2+} be higher, lower, or the same as that calculated? Support your answer.

Multipoint Calibration with Internal Standards

As with standard addition, the use of internal standards assumes that the background signal for both the analyte and internal standard responses is zero or can be removed.

Otherwise, the ratio will not be constant. Because of this, a two-point calibration is not really useful, unless it is to check this assumption. However, internal standards can be used in a multipoint calibration to improve reliability and check linearity. A set of standard solutions, each containing the internal standard (at the same concentration) and the analyte (at different known concentrations), are prepared and the responses measured. The ratio of responses, RA/RIS, for these standards is then plotted vs. the concentration of analyte, CA, to produce a graph such as that shown in Figure 10. The internal standard is then added to all unknowns, in the same concentration as in the standards, and the responses measured. The ratio of these responses is then calculated and the concentration of the analyte is determined from this ratio using the calibration graph in the usual manner. The calibration curve method has all the advantages of the use of internal standards plus the usual advantages of multipoint methods.

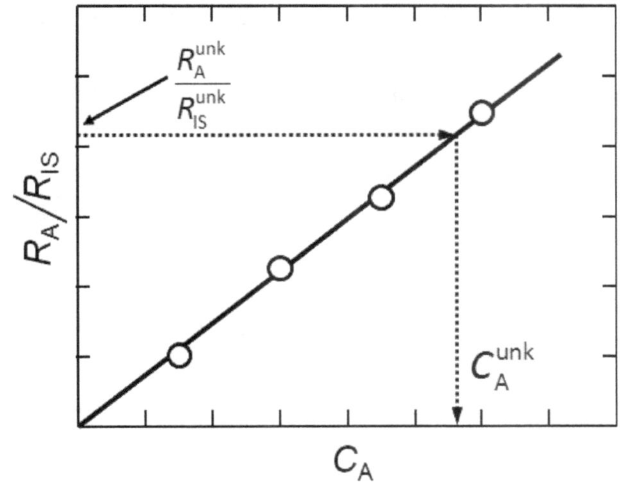

Figure 10: A multipoint calibration using internal standards. The ratio of the analyte response to the response for a fixed concentration of internal standard is plotted against the concentration of the analyte. Likewise, the ratio for the unknown is used to determine the concentration of analyte present.

In terms of ability to reduce matrix effects, the method of standard addition is usually superior to the use of an internal standard, since in the former method the analyte itself is used as the standard. It is almost impossible to select an internal standard that reacts to the matrix in exactly the same fashion as the analyte itself, and in many cases it is impossible to find an internal standard that meets the criteria of being absent in the sample and providing an independent signal. However, as previously noted, the use of internal standards can correct for systematic errors in sample preparation and instrumental drift that cannot be corrected by standard addition. Moreover, standard addition often requires a separate calibration for each sample, whereas the calibration using internal standards can be applied to multiple unknown samples. In certain cases, it is possible to combine the standard addition and internal standard approaches.

Exercise 7.12

The sensitivity of atomic emission spectroscopy (described in a later topic) can be affected by instrumental parameters, such as variations in flame temperature, which can be compensated for by using an internal standard. In the determination of sodium by atomic emission spectroscopy, lithium is used as an internal standard. For each standard solution, a certain volume of a 50.0 ppm Na+ stock solution is mixed with 5.00 mL of a 20.0 ppm Li+ stock solution and diluted to 50.0 mL with distilled water. The following data are obtained.

Volume of Na standard (mL)	1.00	2.00	3.00	5.00
Sodium emission intensity	99	242	336	438
Lithium emission intensity	196	213	200	163

A 2.00 mL sample of an unknown (containing no lithium) is added to 5.00 mL of the lithium stock solution and diluted to 50.0 mL. This gives a sodium intensity of 298 and a lithium intensity of 205. Determine the concentration of sodium in the unknown by (a) a standard multipoint calibration, and (b) a multipoint internal standard calibration. Graphs for the two situations are presented in Figure 7.11.

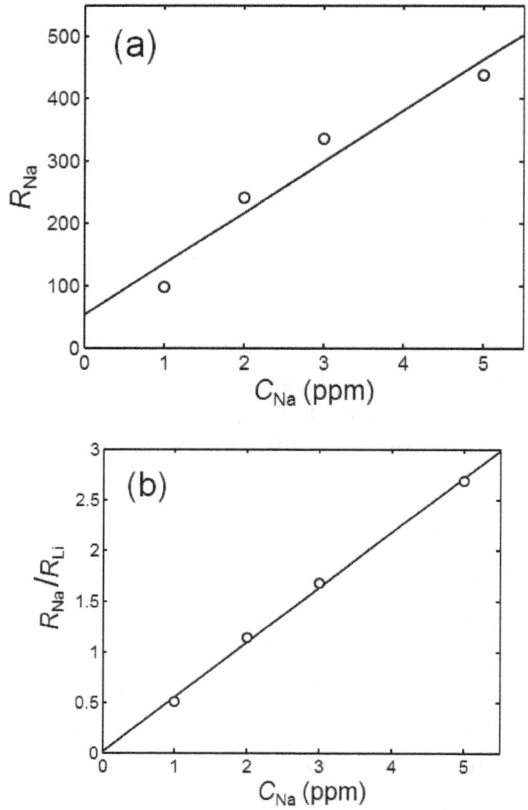

Fig. 11: Calibration curves used for Exercise 7.12: (a) standard multipoint calibration, and (b) multipoint calibration using an internal standard.

Exercise 7.13

Using the data in preceding exercise, calculate the limit of detection for Na for both the conventional calibration and using internal standards. Is there a difference? Which is better? Why is this the case?

Exercise 7.14

Suggest what might make a good internal standard in the following situations.

- (a) the determination of ethanol in beer?
- (b) the determination of sodium in lake water?
- (c) the determination of glucose in blood?
- (d) the determination of a steroid in urine?

5.7 REMOVE ANY DAMAGED OR BENT CULMS AND ANY CULMS WHICH ARE SUMMARY

Matrix effects, which arise from the physical or chemical characteristics of the sample, can be a problem because they cause changes in the sensitivity of the instrument. There are several ways to deal with this problem, two of which were treated in detail here. Standard addition involves adding a known standard to the unknown sample and is commonly applied using fixed or variable volume strategies. The method of internal standards, which employs a secondary analyte, can also reduce matrix effects. It can also correct for certain kinds of systematic errors and instrument drift. For standard addition methods, two-point and multipoint calibration methods are used, while for internal standards, one-point and multipoint calibrations are common. These methods can be used in combination. Both methods assume that the background contribution is zero or can be removed.

7.9 Additional Exercises

Exercise 7.15

Irvina Gratch (Irv's sister) has just purchased a house in a suburb of Halifax and finds that her new white porcelain plumbing is turning brown. She suspects her water has a high iron content and gives a sample of her well water to Irv for analysis. To 10.0 mL of her water sample he adds 10.0 mL of a commercial colour-forming reagent and measures the absorbance in a spectrophotometer as 0.592. To a second 10.0 mL aliquot of her sample he adds 10.0 mL of colour-forming reagent and 5.00 mL of 5.00 ppm Fe(II). This solution reads 0.635 absorbance units. Assuming that absorbance is linear in concentration, what is the concentration of iron in Irvina's well water?

Exercise 7.16

Atomic absorption spectroscopy with graphite furnace atomization is an extremely sensitive technique. The standard addition method is used with this technique to determine the concentration of copper in urine. A 20.00 mL sample of urine is spiked with 1.00 mL of 300. ppb Cu^{2+} and this mixture gives a response of 0.168. Another 1.00 mL aliquot of 300. ppb Cu^{2+} is added to this mixture and gives a reading of 0.235. Determine the concentration of copper in the urine sample.

Exercise 7.17

Suppose you wish to determine calcium in an unknown sample using atomic absorption spectrometry in conjunction with the two-point variable-volume standard addition method. You expect that the maximum concentration of calcium in your sample will be about 1.5 ppm, and you plan to add 1.00 mL of your standard to 100. mL of your sample. For maximum accuracy, the ratio of the diluted standard to the original unknown concentration should not fall below 1 (see Figure 4), but you also know that the maximum linear range of the instrument is 4.0 ppm. What is the allowable range of concentration for your calcium standard solution (in ppm)? If the standard is to be prepared by dissolving calcium carbonate (MM = 100.09 g/mol) in 1.00 L of water, what mass range of CaCO3 could be used?

Exercise 7.18

At concentrations below 2 ppm, the light intensity given off by potassium in a flame is linear in concentration provided that matrix effects are taken into account. A 10.0 mL aliquot of an aqueous sample containing potassium is placed in a 25.0 mL volumetric flask and diluted to the mark with distilled water. This solution reads 86.3 when introduced into the flame photometer. Another 10.0 mL aliquot of the sample and 2.00 mL of a 2.00 ppm solution of K+ are placed in a second 25.0 mL volumetric flask and diluted to the mark with distilled water. This solution reads 98.4. What is the concentration of potassium ion in the sample? Is this concentration in the linear region? Is the answer still valid? Explain.

Exercise 7.19

In a modified two-point standard addition experiment, a student prepares the first solution by adding 10.0 mL of unknown and 10.0 mL of standard and dilutes the mixture to 100.0 mL with distilled water. The second solution is prepared in one of two ways:

(a) 20.0 mL of unknown and 20.0 mL of standard diluted to 100.0 mL, or (b) 20.0 mL of unknown and 10.0 mL of standard diluted to 100.0 mL For each of these cases, can you write an equation to solve for the concentration of the original unknown in terms of the volumes, standard concentration, and measured responses? Are there disadvantages to these approaches, or are they just different?

Exercise 7.20

The flame photometric detector for sulfur measures the light given off by sulfurcontaining molecules in a cool hydrogen/air flame. The response of this detector follows the law (under certain circumstances)

$$R = S \times [RS]2$$

where RS represents a molecule containing a sulfur atom. Thiophene is a sulfur containing impurity often found in benzene. A 2.00 mL benzene sample is diluted to 10.0 mL with hexane, injected into the detector, and reads 42.3. A second 2.00 mL sample of benzene is added to 2.00 mL of a 0.0100 M solution of thiophene in hexane and diluted to 10.0 mL with pure hexane. When injected into the detector, it reads 68.4. What is the concentration of thiophene in the benzene sample, assuming that the response equation above applies?

Exercise 7.21

The response of Ca^{2+} in a flame photometer is known to be linear over a wide concentration range. An aqueous sample with an unknown amount of calcium sample is treated as follows. A 10.0 mL aliquot of the sample and differing amounts of a 20.0 ppm Ca^{2+}

standard are mixed and always diluted to 25.0 mL. The flame photometer readings for these solutions are:

Volume of 20 ppm Ca2+ (mL)	3.00	6.00	9.00	12.00
Response	53.0	62.2	71.5	80.8

What is the concentration of Ca^{2+} in the sample?

Exercise 7.22

Indium is seldom found in lead-containing samples and thus makes a good internal standard for lead analyses. A 10.0 mL aliquot of a 1.00x10-4 M Pb(II) solution is added to 10.0 mL of a 1.00x10-4 M solution of In(III). The resulting currents, measured polarographically, are 0.485 and 0.873 mA for Pb(II) and In(III), respectively. Then, 2.00 mL of a 2.00x10-4 M In(III) solution are added to 10.0 mL of a sample and the currents for Pb(II) and In(III) are 0.186 and 0.504 mA, respectively. Assuming a linear response between current and concentration, calculate the concentration of Pb(II) in the sample by (a) neglecting the information provided by the internal standard, and (b) taking the internal standard into account.

Exercise 7.23

Manual sampling in the technique of gas chromatography is rather imprecise since a very small sample size is produced with a mL syringe. The use of an internal standard might be useful to avoid the imprecision of sampling. The following data are available:

Run #	Response for C_6H_{12}	$[C_6H_{12}]$ (M)	For C_6H_{14}	Response $[C_6H_{14}]$ (M)
1	275	0.100	162	0.0500
2	248	0.100	298	0.100
3	213	0.100	506	0.200

Would cyclohexane (C_6H_{12}) make a good internal standard for hexane (C_6H_{14})? Why? When 1.00 mL of a 0.500 M solution of cyclohexane is added to 4.00 mL of a sample (containing no cyclohexane) and the resulting solution was analyzed chromatographically, the responses for cyclohexane and hexane were 261 and 483, respectively. What is the concentration of hexane in the sample?

Exercise 7.24

Inductively coupled plasma atomic emission spectroscopy (ICP-AES, discussed in a later topic) can detect elemental concentrations at very low levels. The following solutions are prepared in a standard addition experiment to determine the concentration of cadmium, a toxic metal, in urine. Solution 1 contains 5.00 mL of urine and 1.00 mL of a 10.0 mg/mL bismuth standard. Solution 2 contains these components, plus 1.00 mL of a 50.0 mg/mL cadmium standard. Both solutions are diluted to a total volume of 10.00 mL. The ICP-AES measurements for Cd are 11.9 and 42.0 in solutions 1 and 2, respectively.

The corresponding measurements for Bi, used as an internal standard, are 93.7 and 79.6. Calculate the concentration of Cd in the urine sample by (a) considering only the response of Cd, and (2) using Bi as an internal standard.

6

IMPORTANCE OF ANALYTICAL METHODOLOGY

6.1 INTRODUCTION: AN OVERVIEW

The importance of analytical methodology is evident when examining the results of environmental monitoring programs. The purpose of a monitoring program is to determine the present status of an environmental system and to assess long term trends in the quality of the system. These are broad and poorly defined goals. In many cases, such studies are initiated with little thought to the questions the data will be used to answer. This is not surprising since it can be hard to formulate questions in the absence of initial information about the system. Without careful planning, however, a poor experimental design may result in data that has little value.

These concerns are illustrated by the Chesapeake Bay monitoring program. This research program, designed to study nutrients and toxic pollutants in the Chesapeake Bay, was initiated in 1984 as a cooperative venture between the federal government, the state governments of Maryland, Virginia, and Pennsylvania, and the District of Columbia. A 1989 review of some of the problems with this program highlights the difficulties common to many monitoring programs.

At the beginning of the Chesapeake Bay monitoring program, little attention was given to the proper choice of analytical methods, in large part because the intended uses of the monitoring data were not specified. The analytical methods initially chosen were those standard methods already approved by the EPA. In many cases these methods proved to be of little value for this monitoring project. Most of the EPA-approved methods were designed to detect pollutants at their legally mandated maximum allowed concentrations. The concentrations of these contaminants in natural waters, however, are often well below the detection limit of the EPA methods. For example, the EPA-approved standard method for phosphate had a detection limit of 7.5 ppb. Since actual phosphate concentrations in Chesapeake Bay usually were below the EPA detection limit, the EPA method provided no useful information. On the other hand, a non-approved variant of the EPA method commonly used in chemical oceanography had a detection limit of 0.06 ppb. In other cases, such as the elemental analysis for particulate forms of carbon, nitrogen, and phosphorus, EPA-approved procedures provided poorer reproducibility than nonapproved methods.

6.2 EVALUATING ANALYTICAL DATA

A problem dictates the requirements we place on our measurements and results. Regulatory agencies, for example, place stringent requirements on the reliability of measurements and results reported to them. This is the rationale for creating a protocol for regulatory problems. Screening the products of an organic synthesis, on the other hand, places fewer demands on the reliability of measurements, allowing chemists to customize their procedures.

When designing and evaluating an analytical method, we usually make three separate considerations of experimental error. First, before beginning an analysis, errors associated with each measurement are evaluated to ensure that their cumulative effect will not limit the utility of the analysis. Errors known or believed to affect the result can then be minimized. Second, during the analysis the measurement process is monitored, ensuring that it remains under control. Finally, at the end of the analysis the quality of the measurements and the result are evaluated and compared with the original design criteria.

6.2.1 Characterizing Measurements and Results

Let's begin by choosing a simple quantitative problem requiring a single measurement. The question to be answered is—what is the mass of a penny? If you think about how we might answer this question experimentally, you will realize that this problem is too broad. Are we interested in the mass of United State pennies or Canadian pennies, or is the difference in country of importance? Since the composition of a penny probably differs from country to country, let's limit our problem to pennies minted in the United States. There are other considerations. Pennies are minted at several locations in the United States (this is the meaning of the letter, or absence of a letter, below the date stamped on the lower right corner of the face of the coin). Since there is no reason to expect a difference between where the penny was minted, we will choose to ignore this consideration. Is there a reason to expect a difference between a newly minted penny not yet in circulation, and a penny that has been in circulation? The answer to this is not obvious. Let's simplify the problem by narrowing the question to—what is the mass of an average United States penny in circulation? This is a problem that we might expect to be able to answer experimentally.

A good way to begin the analysis is to acquire some preliminary data. Table 1 shows experimentally measured masses for seven pennies from my change jar at home. Looking at these data, it is immediately apparent that our question has no simple answer. That is, we cannot use the mass of a single penny to draw a specific conclusion about the mass of any other penny (although we might conclude that all pennies weigh at least 3 g). We can, however, characterize these data by providing a measure of the spread of the individual measurements around a central value.

Table 1 Masses of Seven United States pennies in Circulation

Penny	Mass (g)
1	3.080
2	3.094
3	3.107
4	3.056
5	3.112
6	3.174
7	3.198

6.2.2 Measures of Central Tendency

One way to characterize the data in Table 1 is to assume that the masses of individual pennies are scattered around a central value that provides the best estimate of a penny's true mass. Two common ways to report this estimate of central tendency are the mean and the median.

6.2.3 Mean

The mean, X–, is the numerical average obtained by dividing the sum of the individual measurements by the number of measurements

$$\overline{X} = \frac{\sum_{i=1}^{n} X_i}{n}$$

where Xi is the ith measurement, and n is the number of independent measurements. The mean is the most common estimator of central tendency. It is not considered a robust estimator, however, because extreme measurements, those much larger or smaller than the remainder of the data, strongly influence the mean's value. For example, mistakenly recording the mass of the fourth penny as 31.07 g instead of 3.107 g, changes the mean from 3.117 g to 7.112 g!

6.2.4 Median

The median, X med, is the middle value when data are ordered from the smallest to the largest value. When the data include an odd number of measurements, the median is the middle value. For an even number of measurements, the median is the average of the n/2 and the (n/2) + 1 measurements, where n is the number of measurements.

As shown by Examples 1 and 2, the mean and median provide similar estimates of central tendency when all data are similar in magnitude. The median, however, provides a more robust estimate of central tendency since it is less sensitive to measurements with extreme values. For example, introducing the transcription error discussed earlier for the mean only changes the median's value from 3.107 g to 3.112 g.

6.2.5 Measures of Spread

If the mean or median provides an estimate of a penny's true mass, then the spread of the individual measurements must provide an estimate of the variability in the masses of individual pennies. Although spread is often defined relative to a specific measure of central tendency, its magnitude is independent of the central value. Changing all measurements in the same direction, by adding or subtracting a constant value, changes the mean or median, but will not change the magnitude of the spread. Three common measures of spread are range, standard deviation, and variance.

Range

The range, w, is the difference between the largest and smallest values in the data set.

$$\text{Range} = w = X_{\text{largest}} - X_{\text{smallest}}$$

The range provides information about the total variability in the data set, but does not provide any information about the distribution of individual measurements. The range for the data in Table 4.1 is the difference between 3.198 g and 3.056 g; thus

$$w = 3.198 \text{ g} - 3.056 \text{ g} = 0.142 \text{ g}$$

6.2.6 Standard Deviation

The absolute standard deviation, s, describes the spread of individual measurements about the mean and is given as

$$s = \sqrt{\frac{\sum_{i=1}^{n}(X_i - \overline{X})^2}{n-1}}$$

where X_i is one of n individual measurements, and X is the mean. Frequently, the relative standard deviation, sr, is reported.

$$s_r = \frac{s}{\overline{X}}$$

The percent relative standard deviation is obtained by multiplying sr by 100%.

6.2.7 Variance

Another common measure of spread is the square of the standard deviation, or the variance. The standard deviation, rather than the variance, is usually reported because the units for standard deviation are the same as that for the mean value.

6.2.8 Characterizing Experimental Errors

Realizing that our data for the mass of a penny can be characterized by a measure of central tendency and a measure of spread suggests two questions. First, does our measure of central tendency agree with the true, or expected value? Second, why are our data scattered around the central value? Errors associated with central tendency reflect the accuracy of the analysis, but the precision of the analysis is determined by those errors associated with the spread.

Accuracy

Accuracy is a measure of how close a measure of central tendency is to the true, or expected value, μ.† Accuracy is usually expressed as either an absolute error

$$E = X - \mu \qquad 2$$

$$E = \overline{X} - \mu$$

or a percent relative error, E_r.

$$E_r = \frac{\overline{X} - \mu}{\mu} \times 100 \qquad 3$$

Although the mean is used as the measure of central tendency in equations 4.2 and 4.3, the median could also be used.

Errors affecting the accuracy of an analysis are called determinate and are characterized by a systematic deviation from the true value; that is, all the individual measurements are either too large or too small. A positive determinate error results in a central value that is larger than the true value, and a negative determinate error leads to a central value that is smaller than the true value. Both positive and negative determinate errors may affect the result of an analysis, with their cumulative effect leading to a net positive or negative determinate error. It is possible, although not likely, that positive and negative determinate errors may be equal, resulting in a central value with no net determinate error.

Determinate errors may be divided into four categories: sampling errors, method errors, measurement errors, and personal errors.

6.2.9 Sampling Errors

We introduce determinate sampling errors when our sampling strategy fails to provide a representative sample. This is especially important when sampling heterogeneous materials. For example, determining the environmental quality of a lake by sampling a single location near a point source of pollution, such as an outlet for industrial effluent, gives misleading results. In determining the mass of a U.S. penny, the strategy for selecting pennies must ensure that pennies from other countries are not inadvertently included in the sample.

6.2.10 Method Errors

Determinate method errors are introduced when assumptions about the relationship between the signal and the analyte are invalid. In terms of the general relationships between the measured signal and the amount of analyte

$$S_{meas} = kn_A + 6S_{reag} \quad \text{(Total analysis method)} \qquad 4$$

$$S_{meas} = kC_A + S_{reag} \quad \text{(concentration method)} \qquad 5$$

Method errors exist when the sensitivity, k, and the signal due to the reagent blank, Sreag, are incorrectly determined. For example, methods in which Smeas is the mass of a precipitate containing the analyte (gravimetric method) assume that the sensitivity is defined by a pure precipitate of known stoichiometry. When this assumption fails, a determinate error will exist. Method errors involving sensitivity are minimized by standardizing the method, whereas method errors due to interferents present in reagents are minimized by using a proper reagent blank. Method errors due to interferents in the sample cannot be minimized by a reagent blank. Instead, such interferents must be separated from the analyte or their concentrations determined independently.

6.2.11 Measurement Errors

Analytical instruments and equipment, such as glassware and balances, are usually supplied by the manufacturer with a statement of the item's maximum measurement error, or tolerance. For example, a 25-mL volumetric flask might have a maximum error of ±0.03 mL, meaning that the actual volume contained by the flask lies within the range of 24.97–25.03 mL. Although expressed as a range, the error is determinate; thus, the flask's true volume is a fixed value within the stated range. A summary of typical measurement errors for a variety of analytical equipment is given in Tables 2–4.

Table 2 Measurement Errors for Selected Glassware[a]

		Measurement Errors for	
Glassware	**Volume (ml)**	**Class A Glassware (±ml)**	**Class B Glassware**
Trasfer Pipets	1	0.006	0.012
	2	0.006	0.012
	5	0.01	0.02
	10	0.02	0.04
	20	0.03	0.06
	25	0.03	0.06
	50	0.05	0.10

Volumetric Flasks	5	0.02	0.04
	10	0.02	0.04
	25	0.03	0.06
	50	0.05	0.10
	100	0.08	0.16
	250	0.12	0.24
	500	0.20	0.40
	1000	0.30	0.60
	2000	0.50	1.0
Burets	10	0.02	0.04
	2 5	0.03	0.06
	50	0.05	0.10

[a]Specifications for class A and class B glassware are taken from American Society for Testing and Materials E288, E542 and E694 standards.

Table 3 Measurement Errors for Selected Balances

Balance	Capacity (g)	Measurement Error
Precisa 160M	160	$\pm$ 1 mg
A & D ER 120M	120	$\pm$0.1 mg
Metler H54	160	$\pm$0.1 mg

Table 4 Measurement Errors for Selected Digital Pipets

Pipet Range	Volume (mL or μL)[a]	Measurement Error ($\pm$ %)
10-100 μL[b]	10	1.0
	50	0.6
	100	0.6
200-1000 μL[c]	200	1.5
	1000	0.8
1-10 mL[d]	1	0.6
	5	0.4
	10	0.3

[b]Units for volume same as for pipet range.

[b]Data fot Eppendorf Digital Pipet 4710.

[C]Data for Oxford Benchmate.

[d]Data for Eppendorf Maxipetter 4720 with Maxitip P.

Volumetric glassware is categorized by class. Class A glassware is manufactured to comply with tolerances specified by agencies such as the National Institute of Standards and Technology. Tolerance levels for class A glassware are small enough that such glassware normally can be used without calibration. The tolerance levels for class B glassware are usually twice those for class A glassware. Other types of volumetric glassware, such as beakers and graduated cylinders, are unsuitable for accurately measuring volumes.

Determinate measurement errors can be minimized by calibration. A pipette can be calibrated, for example, by determining the mass of water that it delivers and using the density of water to calculate the actual volume delivered by the pipette. Al- though glassware and instrumentation can be calibrated, it is never safe to assume that the calibration will remain unchanged during an analysis. Many instruments, in particular, drift out of calibration over time. This complication can be minimized by frequent recalibration.

6.2.12 Personal Errors

Finally, analytical work is always subject to a variety of personal errors, which can include the ability to see a change in the colour of an indicator used to signal the end point of a titration; biases, such as consistently overestimating or underestimating the value on an instrument's readout scale; failing to calibrate glassware and instrumentation; and misinterpreting procedural directions. Personal errors can be minimized with proper care.

6.2.13 Identifying Determinate Errors

Determinate errors can be difficult to detect. Without knowing the true value for an analysis, the usual situation in any analysis with meaning, there is no accepted value with which the experimental result can be compared. Nevertheless, a few strategies can be used to discover the presence of a determinate error.

Some determinate errors can be detected experimentally by analyzing several samples of different size. The magnitude of a constant determinate error is the same for all samples and, therefore, is more significant when analyzing smaller samples. The presence of a constant determinate error can be detected by running several analyses using different amounts of sample, and looking for a systematic change in the property being measured. For example, consider a quantitative analysis in which we separate the analyte from its matrix and determine the analyte's mass. Let's assume that the sample is 50.0% w/w analyte; thus, if we analyze a 0.100-g sample, the analyte's true mass is 0.050 g. The first two columns of Table 5 give the true mass of analyte for several additional samples. If the analysis has a positive constant determinate error of 0.010 g, then the experimentally determined mass for any sample will always be 0.010 g, larger than its true mass (column four of Table 5).

Table 5: Effect of Constant positive Determinate Error on Analysis of Sample containing 50% Analyte (% w/w)

Mass Sample (g)	True Mass of analyte (g)	Constant Error (g)	Mass of Analyte Determined	Percent Analyte Reported (%w/w)
0.100	0.050	0.010	0.060	60.
0.200	0.100	0.010	0.110	55.0
0.400	0.200	0.010	0.210	52.5
0.800	0.400	0.010	0.410	51.2
1.000	0.500	0.010	0.510	51.0

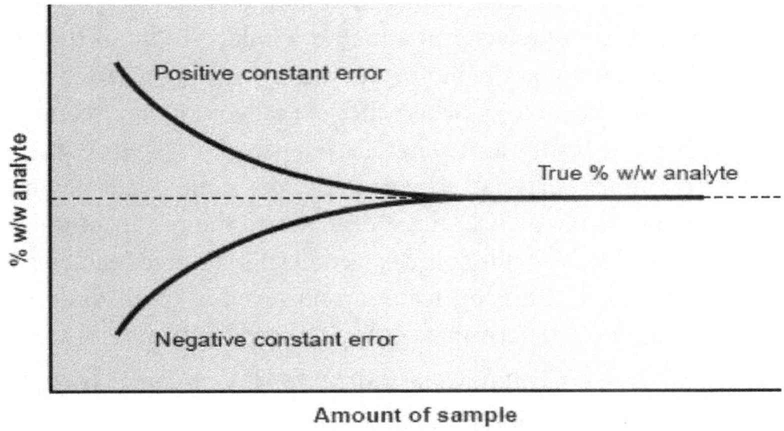

Fig. 1: Effect of a constant determinate error on the reported concentration of analyte.

The analyte's reported weight percent, which is shown in the last column of Table 5, becomes larger when we analyze smaller samples. A graph of % w/w analyte versus amount of sample shows a distinct upward trend for small amounts of sample (Figure 1). A smaller concentration of analyte is obtained when analyzing smaller samples in the presence of a constant negative determinate error.

A proportional determinate error, in which the error's magnitude depends on the amount of sample, is more difficult to detect since the result of an analysis is independent of the amount of sample. Table 6 outlines an example showing the effect of a positive proportional error of 1.0% on the analysis of a sample that is 50.0% w/w in analyte. In terms of equations 4 and 5, the reagent blank, Sreag, is an example of a constant determinate error, and the sensitivity, k, may be affected by proportional errors.

Table 6 Effect of Proportional positive Determinate Error on Analysis of Sample containing 50% Analyte (% w/w)

Mass Sample (g)	True Mass of analyte (g)	Constant Error (g)	Mass of Analyte Determined	Percent Analyte Reported (%w/w)
0.200	0.100	1.00	0.101	50.5
0.400	0.200	1.00	0.202	50.5
0.600	0.300	1.00	0.303	50.5
0.800	0.400	1.00	0.404	50.5
1.000	0.500	1.00	0.505	50.5

Potential determinate errors also can be identified by analyzing a standard sample containing a known amount of analyte in a matrix similar to that of the samples being analyzed. Standard samples are available from a variety of sources, such as the National Institute of Standards and Technology (where they are called standard reference materials) or the American Society for Testing and Materials. For example, Figure 2 shows an analysis sheet for a typical reference material. Alternatively, the sample can be analyzed by an independent method known to give accurate results, and the results of the two methods can be compared. Once identified, the source of a determinate error can be corrected. The best prevention against errors affecting accuracy, however, is a well-designed procedure that identifies likely sources of determinate errors, coupled with careful laboratory work.

Alanysis sheet for simulated Rainwater (SRM 2694a). Adapted from NIST Special Publication 260: Standard Reference

SRM	Type	Unit of issue	
2694a	**Simulated rainwater**	**Set of 4: 2 of 50 mL at each of 2 levels**	
	Constituent element parameter	**2694a.l**	**2694a-II**
	pH. 25°C	4.30	3.60
	Electrolytic Conductivity (S/cm, 25°C)	25.4	129.3
	Acidity. meq/L	0.0544	0.283
	Fluoride. mg&	0.057	0.108
	Chloride. mg/L	(0.23)'	(0.94)'
	Nitrate. mq/L	(0.53)'	7.19
	Sulfate, mg/L	2.69	10.6
	Sodium. mg/L	0.208	0.423
	Potassium. mg/L	0.056	0.108
	Ammonium. mg/L	(0.12)'	(1.06)'
	Calcium. mg/L	0.0126	0.0364
	Magnesium. mg/L	0.0242	0.0484

*Values in parentheses are not certified and are given for information only.

Fig. 2: Analysis sheet for Simulated Rainwater (SRM 2694a).

Table 7 Masses of Seven United States Pennies in Circulation

Penny	Mass (g)Q
1	3.080
2	3.094
3	3.107
4	3.056
5	3.112
6	3.174
7	3.198

The data in Table 7 were obtained using a calibrated balance, certified by the manufacturer to have a tolerance of less than ±0.002 g. Suppose the Treasury Department reports that the mass of a1998 U.S. penny is approximately 2.5 g. Since the mass of every penny in Table 4 exceeds the re-ported mass by an amount significantly greater than the balance's tolerance, we can safely conclude that the error in this analysis is not due to equipment error.

6.2.14 Precision

Precision is a measure of the spread of data about a central value and may be ex- pressed as the range, the standard deviation, or the variance. Precision is commonly divided into two categories: repeatability and reproducibility. Repeatability is the precision obtained when all measurements are made by the same analyst during a single period of laboratory work, using the same solutions and equipment. Reproducibility, on the other hand, is the precision obtained under any other set of conditions, including that between analysts, or between laboratory sessions for a single analyst. Since reproducibility includes additional sources of variability, the reproducibility of an analysis can be no better than its repeatability.

Errors affecting the distribution of measurements around a central value are called indeterminate and are characterized by a random variation in both magnitude and direction. Indeterminate errors need not affect the accuracy of an analysis. Since indeterminate errors are randomly scattered around a central value, positive and negative errors tend to cancel, provided that enough measurements are made. In such situations the mean or median is largely unaffected by the precision of the analysis.

6.2.15 Sources of Indeterminate Error

Indeterminate errors can be traced to several sources, including the collection of samples, the manipulation of samples during the analysis, and the making of measurements.

When collecting a sample, for instance, only a small portion of the available material is taken, increasing the likelihood that small-scale in homogeneities in the sample will affect the repeatability of the analysis. Individual pennies, for example, are expected to show variation from several sources, including the manufacturing process, and the loss of small amounts of metal or the addition of dirt during circulation. These variations are sources of indeterminate error associated with the sampling process.

During the analysis numerous opportunities arise for random variations in the way individual samples are treated. In determining the mass of a penny, for example, each penny should be handled in the same manner. Cleaning some pennies but not cleaning others introduces an indeterminate error.

Finally, any measuring device is subject to an indeterminate error in reading its scale, with the last digit always being an estimate subject to random fluctuations, or background noise. For example, a burette with scale divisions every 0.1 mL has an inherent indeterminate error of ±0.01 – 0.03 mL when estimating the volume to the hundredth of a milliliter (Figure 3). Background noise in an electrical meter (Figure 4) can be evaluated by recording the signal without analyte and observing the fluctuations in the signal over time.

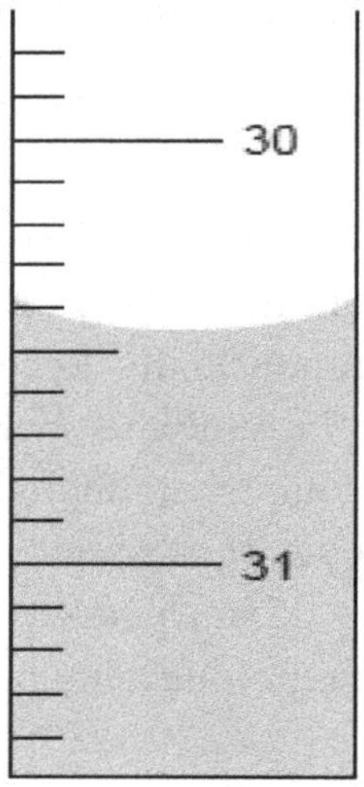

Fig. 3: Close –up buret, showing difficulty in estimating volume .With scale divisions every 0.1 ml it is difficult to read the actual volume to better than ± 0.01-0.03 mL.

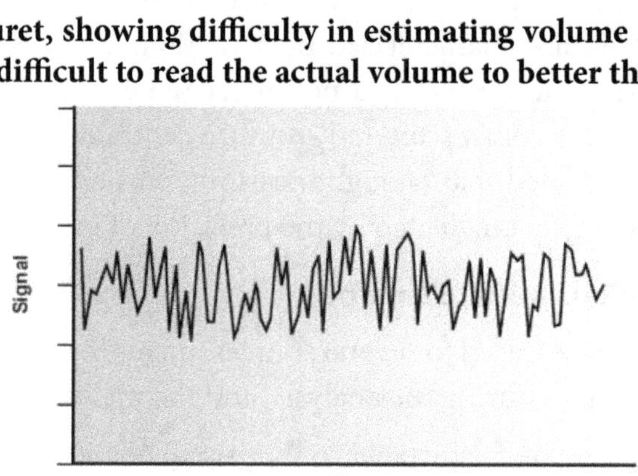

Fig. 4: Background noise in a meter obtained by measuring signal over time in the absence of analyte.

6.2.16 Evaluating Indeterminate Error

Although it is impossible to eliminate indeterminate error, its effect can be minimized if the sources and relative magnitudes of the indeterminate error are known. Indeterminate errors may be estimated by an appropriate measure of spread. Typically, a standard deviation is used, although in some cases estimated values are used. The contribution from analytical instruments and equipment are easily measured or estimated. Indeterminate errors introduced by the analyst, such as in- consistencies in the treatment of individual samples, are more difficult to estimate.

To evaluate the effect of indeterminate error on the data in Table 1, ten replicate determinations of the mass of a single penny were made, with results shown in Table 8. The standard deviation for the data in Table 1 is 0.051, and it is 0.0024 for the data in Table 8. The significantly better precision when determining the mass of a single penny suggests that the precision of this analysis is not limited by the balance used to measure mass, but is due to a significant variability in the masses of individual pennies.

Table 8 Replicate Determination of the mass of a single united States penny in Circulation

Replicate Number	Massq (g)
1	3.025
2	3.024
3	3.028
4	3.027
5	3.028
6	3.023
7	3.022
8	3.021
9	3.076
10	3.024

6.2.17 Error and Uncetainty

Analytical chemists make a distinction between error and uncertainty.3 Error is the difference between a single measurement or result and its true value. In other words, error is a measure of bias. As discussed earlier, error can be divided into determinate and indeterminate sources. Although we can correct for determinate error, the indeterminate portion of the error remains. Statistical significance testing provides a way to determine whether a bias resulting from determinate error might be present.

Uncertainty expresses the range of possible values that a measurement or result might reasonably be expected to have. Note that this definition of uncertainty is not the same as that for precision. The precision of an analysis, whether reported as a range or a standard deviation, is calculated from experimental data and provides an estimation of indeterminate error affecting measurements. Uncertainty accounts for all errors, both determinate and indeterminate, that might affect our result. Al- though we always try to correct determinate errors, the correction itself is subject to random effects or indeterminate errors.

To illustrate the difference between precision and uncertainty, consider the use of a class A 10-mL pipette for delivering solutions.

A pipet's uncertainty is the range of volumes in which its true volume is expected to lie. Sup- pose you purchase a 10-mL class A pipet from a laboratory supply company and use it without calibration. The pipet's tolerance value of ±0.02 mL (see Table 2) represents your uncertainty since your best estimate of its volume is 10.00 mL ±0.02 mL. Precision is determined experimentally by using the pipet several times, measuring the volume of solution delivered each time. Table 8 shows results for ten such trials that have a mean of 9.992 mL and a standard deviation of 0.006. This standard deviation represents the precision with which we expect to be able to deliver a given solution using any class A 10-mL pipet. In this case the uncertainty in using a pipet is worse than its precision. Interestingly, the data in Table 9 allow us to calibrate this specific pipet's delivery volume as 9.992 mL. If we use this volume as a better estimate of this pipet's true volume, then the uncertainty is ±0.006. As expected, calibrating the pipet allows us to lower its uncertainty.

Table 9 Measurement Errors for Selected Glassware[a]

		Measurement Errors for	
Glassware	**Volume (mL)**	**Class A Glassware (±mL)**	**Class B Glassware (±mL)**
Transfer Pipets	1	0.006	0.012
	2	0.006	0.012
	5	0.01	0.02
	10	0.02	0.04
	20	0.03	0.06
	25	0.03	0.06
	50	0.05	0.10
Volumetric Flaks	5	0.02	0.04
	10	0.02	0.04
	25	0.03	0.06
	50	0.05	0.10
	100	0.08	0.16
	250	0.12	0.24
	500	0.20	0.40
	100	0.30	0.60
	2000	0.50	1.0
Burets	10	0.02	0.04
	25	0.03	0.06
	50	0.05	0.10

[a]Specifications for class A and class B glassware are taken from American Society for Testing and materials E288, E542 and E694 standards.

6.2.18 Propagation of Uncertainty

Suppose that you need to add a reagent to a flask by several successive transfers using a class A 10-mL pipette. By calibrating the pipette (see Table 10), you know that it delivers a volume of 9.992 mL with a standard deviation of 0.006 mL. Since the pipette is calibrated, we can use the standard deviation as a measure of uncertainty. This uncertainty tells us that when we use the pipette to repetitively deliver 10 mL of solution, the volumes actually delivered are randomly scattered around the mean of 9.992 mL.

Table 10 Experimentally Determined Volumes Delivered by a 10-ml class a Pipet

Trial	Volume Delivered (mL)	Trial	Volume Delivered (mL)
1	10.002	6	9.983
2	9.993	7	9.991
3	9.984	8	9.990
4	9.996	9	9.988
5	9.989	10	9.999

6.2.19 Propagation of Uncertainty

Suppose that you need to add a reagent to a flask by several successive transfers using a class A 10-mL pipet. By calibrating the pipet (see Table 8), you know that it delivers a volume of 9.992 mL with a standard deviation of 0.006 mL. Since the pipet is calibrated, we can use the standard deviation as a measure of uncertainty. This uncertainty tells us that when we use the pipet to repetitively deliver 10 mL of solution, the volumes actually delivered are randomly scattered around the mean of 9.992 mL.

$$(9.992 \text{ mL} + 9.992 \text{ mL}) \pm (0.006 \text{ mL} - 0.006 \text{ mL}) = 19.984 \pm 0.000 \text{ Ml}$$

underestimating the total uncertainty.

So what is the total uncertainty when using this pipet to deliver two successive volumes of solution? From the previous discussion we know that the total uncerainty is greater than ±0.000 mL and less than ±0.012 mL. To estimate the cumulative effect of multiple uncertainties, we use a mathematical technique known as the propagation of uncertainty. Our treatment of the propagation of uncertainty is based on a few simple rules that we will not derive. A more thorough treatment can be found elsewhere.

6.2.20 A Few Symbols

Propagation of uncertainty allows us to estimate the uncertainty in a calculated result from the uncertainties of the measurements used to calculate the result. In the equations presented in this section the result is represented by the symbol R and the measurements by the symbols A, B, and C. The corresponding uncertainties are sR, sA, sB, and sC. The uncertainties for A, B, and C can be reported in several ways, including calculated standard deviations or estimated ranges, as long as the same form is used for all measurements.

Uncertainty When Adding or Subtracting

When measurements are added or subtracted, the absolute uncertainty in the result is the square root of the sum of the squares of the absolute uncertainties for the individual measurements. Thus, for the equations R = A + B + C or R = A + B – C, or any other combination of adding and subtracting A, B, and C, the absolute uncertainty in R is

6

$$s_R = \sqrt{s_A^2 + s_B^2 + s_C^2}$$

EXAMPLE: 5

The class A 10-mL pipet characterized in Table 4.8 is used to deliver two successive volumes. Calculate the absolute and relative uncertainties for the total delivered volume.

SOLUTION

The total delivered volume is obtained by adding the volumes of each delivery; thus

$$V_{tot} = 9.992 \text{ mL} + 9.992 \text{ mL} = 19.984 \text{ mL}$$

Using the standard deviation as an estimate of uncertainty, the uncertainty in the total delivered volume is

$$\frac{s_R}{R} = \sqrt{\left(\frac{0.01}{0.15}\right)^2 + \left(\frac{1}{120}\right)^2} = \pm 0.0672$$

Thus, we report the volume and its absolute uncertainty as 19.984 ± 0.008 mL. The relative uncertainty in the total delivered volume is

$$s_R = \sqrt{(0.006)^2 + (0.006)^2} = 0.0085$$

Uncertainty When Multiplying or Dividing

When measurements are multiplied or divided, the relative uncertainty in the result is the square root of the sum of the squares of the relative uncertainties for the individual measurements. Thus, for the equations R = A x B x C or R = A x B/C, or any other combination of multiplying and dividing A, B, and C, the relative uncertainty in R is

$$\frac{s_R}{R} = \sqrt{\left(\frac{s_A}{A}\right)^2 + \left(\frac{s_B}{B}\right)^2 + \left(\frac{s_C}{C}\right)^2}$$

EXAMPLE: 6

The quantity of charge, Q in coulombs passing through an electrical circuit is

$$Q = i x t$$

where I is the current in amperes and t is the time in seconds. When a current of 0.15 ± 0.01 A passes through the circuit for 120 ± 1 s, the total charge is

$$Q = (0.15\ A) \times (120\ s) = 18\ C$$

Calculate the absolute and relative uncertainties for the total charge.

SOLUTION

Since charge is the product of current and time, its relative uncertainty is

$$S_{meas} = KC_A + S_{reag}$$

or ±6.7%. The absolute uncertainty in the charge is

$$s_R = R \times 0.0672 = (18) \times (\pm 0.0672) = \pm 1.2$$

Thus, we report the total charge as 18 C ± 1 C.

Uncertain try for Mixed Operations

Many chemical calculations involve a combination of adding and subtracting, and multiply and dividing. As shown in the following example, the propagation of uncertainty is easily calculated by treating each operation separately using equations 6 and 7 as needed.

EXAMPLE: 7

For a concentration technique the relationship between the measured signal and an analyte's concentration is given by equation 4.5

$$\frac{neas - Sreag}{k} = \frac{24.37 - 096}{0.186\,ppm^{-1}} = 12$$

Calculate the absolute and relative uncertainties for the analyte's concentration if S_{meas} is 24.37 ± 0.02, Sun is 0.96 ± 0.02, and k is 0.186 ± 0.003 ppm-[1]

SOLUTION

Rearranging equation 4.5 and solving for C_A

$$s_R = \sqrt{\left(\frac{0.028}{23.41}\right)^2 + \left(\frac{0.003}{0.186}\right)^2} = 0.0162$$

gives the analyte's concentration as 126 ppm. To estimate the uncertainty in CA, we first determine the uncertainty for the numerator, Saws — San, using equation 4.6

$$\frac{s_R}{R} = 2.303 \times s_A = 2.303 \times 0.03 = 0.069$$

The numerator, therefore, is 23.41 ± 0.028 (note that we retain an extra significant figure since we will use this uncertainty in further calculations). To complete the calculation, we estimate the relative uncertainty in CA using equation 4.7, giving

$$\frac{s_R}{R} = \sqrt{\left(\frac{0.028}{23.41}\right)^2 + \left(\frac{0.003}{0.186}\right)^2} = 0.0162$$

or a percent relative uncertainty of 1.6%. The absolute uncertainty in the analyte's concentration is

$$s_R = (125.9 \text{ ppm}) \times (0.0162) = \pm 2.0 \text{ ppm}$$

giving the analyte's concentration as 126 ± 2 ppm.

Uncertainty for Other Mathematical Functions

Many other mathematical operations are commonly used in analytical chemistry, including powers, roots, and logarithms. Equations for the propagation of uncertainty for some of these functions are shown below.

Function	s_R
$R = kA$	$s_R = ks_A$
$R = A + B$	$s_R = \sqrt{s_A^2 + s_B^2}$
$R = A - B$	$s_R = \sqrt{s_A^2 + s_B^2}$
$R = A \times B$	$\frac{s_R}{R} = \sqrt{\left(\frac{s_A}{A}\right)^2 + \left(\frac{s_B}{B}\right)^2}$
$R = \frac{A}{B}$	$\frac{s_R}{R} = \sqrt{\left(\frac{s_A}{A}\right)^2 + \left(\frac{s_B}{B}\right)^2}$
$R = \ln(A)$	$s_R = \frac{s_A}{A}$
$R = \log(A)$	$s_R = 0.4343 \times \frac{s_A}{A}$
$R = e^A$	$\frac{s_R}{R} = s_A$
$R = 10^A$	$\frac{s_R}{R} = 2.303 s_A$
$R = A^k$	$\frac{s_R}{R} = \left[k \frac{s_A}{A}\right]$

These equations assume that the measurements A and B are uncorrelated; that is, s_A is independent of s_g

EXAMPLE: 8

The pH of a solution is defined as

$$pH = -\log[H^+]$$

where [F•I is the molar concentration of H*. If the pH of a solution is 3.72 with an absolute uncertainty of ±0.03, what is the $[H^+]$ and its absolute uncertainty?

SOLUTION

The molar concentration of H^+ for this pH is

$$[H^+] = 10^{-PH} = 10^{-3.72} = 1.91 \times 10^{-4} \text{ M}$$

or 1.9 x 10-4 M to two significant figures. From Table 4.9 the relative uncertainty in $[H^+]$ is

$$\frac{s_R}{R} = 2.303 \times s_A = 2.303 \times 0.03 = 0.069$$

and the absolute uncertainty is

$$(1.91 \text{ x } 10^{-4} \text{ M}) \text{ x } (0.069) = 1.3 \text{ x } 10^{-5} \text{ M}$$

We report the $[H^+]$ and its absolute uncertainty as 1.9 (±0.1) x 10^{-4} M.

Is Calculating Uncertainty Actually Useful?

Given the complexity of determining a result's uncertainty when several measurements are involved, it is worth examining some of the reasons why such calculations are useful. A propagation of uncertainty allows us to estimate an expected uncertainty for an analysis. Comparing the expected uncertainty to that which is actually obtained can provide useful information. For example, in determining the mass of a penny, we estimated the uncertainty in measuring mass as ±0.002 g based on the balance's tolerance. If we measure a single penny's mass several times and obtain a standard deviation of ±0.020 g, we would have reason to believe that our measurement process is out of control. We would then try to identify and correct the problem.

A propagation of uncertainty also helps in deciding how to improve the uncertainty in an analysis. In Example 7, for instance, we calculated the concentration of an analyte, obtaining a value of 126 ppm with an absolute uncertainty of ±2 ppm and a relative uncertainty of 1.6%. How might we improve the analysis so that the absolute uncertainty is only ±1 ppm (a relative uncertainty of 0.8%)? Looking back on the calculation, we find that the relative uncertainty is determined by the relative uncertainty in the measured signal (corrected for the reagent blank)

$$\frac{0.028}{23.41} = \pm 0.0012,\ or \pm 0.12\%$$

and the relative uncertainty in the method's sensitivity, k

$$\frac{0.003}{0.186} = \pm 0.0016,\ or \pm 0.16\%$$

Of these two terms, the sensitivity's uncertainty dominates the total uncertainty. Measuring the signal more carefully will not improve the overall uncertainty of the analysis. On the other hand, the desired improvement in uncertainty can be achieved if the sensitivity's absolute uncertainty can be decreased to ±0.0015 ppm^{-1}.

As a final example, a propagation of uncertainty can be used to decide which of several procedures provides the smallest overall uncertainty. Preparing a solution by diluting a stock solution can be done using several different combinations of volumetric glassware. For instance, we can dilute a solution by a factor of 10 using a 10-mL pipette and a 100-mL volumetric flask, or by using a 25-mL pipette and a 250-mL volumetric flask. The same dilution also can be accomplished in two steps using a 50-mL pipette and a 100-mL volumetric flask for the first dilution, and a 10-mL pipette and a 50-mL volumetric flask for the second dilution. The overall uncertainty, of course, depends on the uncertainty of the glassware used in the dilutions. As shown in the following example, we can use the tolerance values for volumetric glassware to determine the optimum dilution strategy.

EXAMPLE: 9

Which of the following methods for preparing a 0.0010 M solution from a 1.0 M stock solution provides the smallest overall uncertainty?

(a) A one-step dilution using a 1-mL pipet and a 1000-mL volumetric flask.

(b) A two-step dilution using a 20-mL pipet and a 1000-mL volumetric flask for the first dilution and a 25-mL pipet and a 500-mL volumetric flask for the second dilution.

SOLUTION

Letting M_a and M_b represent the molarity of the final solutions from method (a) and method (b), we can write the following equations

$$M_a = 0.0010\,M = \frac{(1.0M)(20.00mL)(25.00mL)}{(1000.0mL)(500.0mL)}$$

$$M_b = 0.0010\,M = \frac{(1.0\,M)(20.00mL)(25.00mL)}{(1000.0\,mL)(500.0\,mL)}$$

Using the tolerance values for pipets and volumetric flasks given in Table 4.2, the overall uncertainties in M_a and M_b are

$$\left(\frac{s_R}{R}\right)M_a = \sqrt{\left(\frac{0.006}{1.000}\right)^2 + \left(\frac{0.3}{1000.0}\right)^2} = 0.006$$

$$\left(\frac{s_R}{R}\right)_{M_b} = \sqrt{\left(\frac{0.03}{20.00}\right)^2 + \left(\frac{0.03}{25.00}\right)^2 + \left(\frac{0.2}{500.0}\right)^2 + \left(\frac{0.3}{1000.0}\right)^2} = 0.002$$

Since the relative uncertainty for M_b ‘s less than that for M_a we find that the two-step dilution provides the smaller overall uncertainty.

The Distribution of Measurements and Results

An analysis, particularly a quantitative analysis, is usually performed on several replicate samples. How do we report the result for such an experiment when results for the replicates are scattered around a central value? To complicate matters further, the analysis of each replicate usually requires multiple measurements that, themselves, are scattered around a central value.

Consider, for example, the data in Table 11 for the mass of a penny. Reporting only the mean is insufficient because it fails to indicate the uncertainty in measuring a penny's mass. Including the standard deviation, or other measure of spread, provides the necessary information about the uncertainty in measuring mass. Nevertheless, the central tendency and spread together do not provide a definitive statement about a penny's true mass. If you are not convinced that this is true, ask yourself how obtaining the mass of an additional penny will change the mean and standard deviation.

Table 11 Results for a Second Determination of the Mass of a United States Penny in Circulation

Penny	Mass (g)
1	3.052
2	3.141
3	3.083
4	3.83
5	3.048
x	3.081
5	0.037

How we report the result of an experiment is further complicated by the need to compare the results of different experiments. For example, Table10 shows results for a second, independent experiment to determine the mass of a U.S. penny in circulation. Although the results shown in Tables 1 and 10 are similar, they are not identical; thus, we are justified in asking whether the results are in agreement. Unfortunately, a definitive comparison between these two sets of data is not possible based solely on their respective means and standard deviations.

Developing a meaningful method for reporting an experiment's result requires the ability to predict the true central value and true spread of the population under investigation from a limited sampling of that population. In this section we will take a quantitative look at how individual measurements and results are distributed around a central value.

Populations and Samples

The terms "population" and "sample" in the context of reporting the result of an experiment. Before continuing, we need to understand the difference between a population and a sample. A population is the set of all objects in the system being investigated. These objects, which also are members of the population, possess qualitative or quantitative characteristics, or values, that can be measured. If we analyze every member of a population, we can deter- mine the population's true central value, μ, and spread, σ.

The probability of occurrence for a particular value, P(V), is given as

$$P(V) = \frac{M}{N}$$

Where V is the value of interest, M is the value's frequency of occurrence in the population, and N is the size of the population. In determining the mass of a circulating United States penny, for instance, the members of the population are all United States pennies currently in circulation, while the values are the possible masses that a penny may have.

In most circumstances, populations are so large that it is not feasible to analyze every member of the population. This is certainly true for the population of circulating U.S. pennies. Instead, we select and analyze a limited subset, or sample, of the population. The data in Tables 1 and 10, for example, give results for two samples drawn at random from the larger population of all U.S. pennies currently in circulation.

Probability Distributions for Populations

To predict the properties of a population on the basis of a sample, it is necessary to know something about the population's expected distribution around its central value. The distribution of a population can be represented by plotting the frequency of occurrence of individual values as a function of the values themselves. Such plots are called probability distributions. Unfortunately, we are rarely able to calculate the exact probability distribution for a chemical system. In fact, the probability distribution can take any shape, depending on the nature of the chemical system being investigated. Fortunately many chemical systems display one of several common probability distributions. Two of these distributions, the binomial distribution and the normal distribution, are discussed next.

Binomial Distribution

The binomial distribution describes a population in which the values are the number of times a particular outcome occurs during a fixed number of trials. Mathematically, the binomial distribution is given as:

$$P(X, N) = \frac{N!}{X!(N - X)!} \times p^X \times (1 - p)^{N-X}$$

Where P(X,N) is the probability that a given outcome will occur X times during N trials, and p is the probability that the outcome will occur in a single trial.* If you flip a coin five times, P(2,5) gives the probability that two of the five trials will turn up "heads."

A binomial distribution has well-defined measures of central tendency and spread. The true mean value, for example, is given as:

$$\mu = Np$$

and the true spread is given by the variance

$$\sigma^2 = Np(1 - p)$$

or the standard deviation

$$\sigma = \sqrt{Np(1 - p)}$$

The binomial distribution describes a population whose members have only certain, discrete values. A good example of a population obeying the binomial distribution is the sampling of homogeneous materials. As shown in Example 10, the binomial distribution can be used to calculate the probability of finding a particular isotope in a molecule.

EXAMPLE 10

Carbon has two common isotopes, ^{12}C and ^{13}C, with relative isotopic abundances of, respectively, 98.89% and 1.11%. (a) What are the mean and standard deviation for the number of 13C atoms in a molecule of cholesterol? (b) What is the probability of finding a molecule of cholesterol ($C_{27}H_{44}O$) containing no atoms of ^{13}C?

SOLUTION

The probability of finding an atom of ^{13}C in cholesterol follows a binomial distribution, where X is the sought for frequency of occurrence of ^{13}C atoms, N is the number of C atoms in a molecule of cholesterol, and p is the probability of finding an atom of ^{13}C.

(a) The mean number of ^{13}C atoms in a molecule of cholesterol is

$$=\sqrt{(27)(0.0111)(1-0.0111)}=0.1$$

with a standard deviation of

$$\sigma=\sqrt{(27)(0.0111)(1-0.0111)}=0.172$$

(b) Since the mean is less than one atom of 13C per molecule, most molecules of cholesterol will not have any "C. To calculate the probability. wr substitute appropriate values into the binomial equation

$$z_{low}=\frac{243-250}{5}=-1.4$$

Thar is therefore a 74.0% probability that a molecuk afdmksterul will not hart an atom Of ^{13}C.

A portion of the binomial distribution for atoms of ^{13}C in cholesterol is shown in Figure 4.5 Note in particular that there is link probability of finding more than two atoms of ^{13}C in any mokrulc of cholesterol.

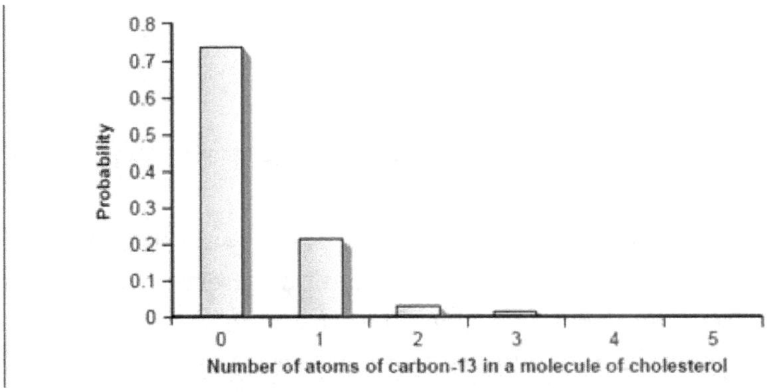

Fig. 5: Portion of the binomial distribution for the number of naturally occurring °C atones in a molecule of cholesterol.

Normal Distribution

The binomial distribution describes a population whose members have only certain, discrete values.

This is the case with the number of 13C toms in a molecule, which must be an integer number no greater then the number of carbon atoms in the molecule. A molecule, for example, cannot have 2.5 atoms of 13C. Other populations are considered continuous, in that members of the population may take on any value.

The most commonly encountered continuous distribution is the Gaussian, or normal distribution, where the frequency of occurrence for a value, X, is given by :

$$f(X) = \frac{1}{\sqrt{2\pi\sigma^2}} \exp\left[\frac{-(X-\mu)^2}{2\sigma^2}\right]$$

The shape of a normal distribution is determined by two parameters, the first of which is the population's central, or true mean value, μ, given as:

$$\mu = \frac{\sum_{i=1}^{N} X_i}{n}$$

where n is the number of members in the population. The second parameter is the population's variance, σ^2, which is calculated using the following equation*

$$\sigma^2 = \frac{\sum_{i=1}^{N} (X_i - \mu)^2}{n}$$

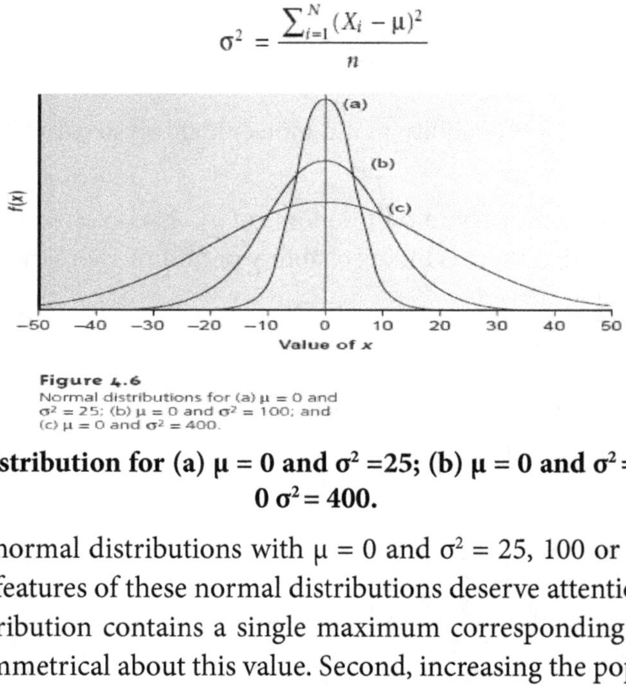

Fig. 6: Normal distribution for (a) $\mu = 0$ and $\sigma^2 = 25$; (b) $\mu = 0$ and $\sigma^2 = 100$; and (c) $\mu = 0$ $\sigma^2 = 400$.

Examples of normal distributions with $\mu = 0$ and $\sigma^2 = 25$, 100 or 400, are shown in Figure 7. Several features of these normal distributions deserve attention. First, note that each normal distribution contains a single maximum corresponding to μ and that the distribution is symmetrical about this value. Second, increasing the population's variance increases the distribution's spread while de- creasing its height. Finally, because the normal distribution depends solely on μ and σ2, the area, or probability of occurrence between any two limits defined in terms of these parameters is the same for all normal distribution curves. For example, 68.26% of the members in a normally distributed population have values within the range μ ±1σ, regardless of the actual values of μ and σ. As shown in Example 11, probability tables can be used to determine the probability of occurrence between any defined limits.

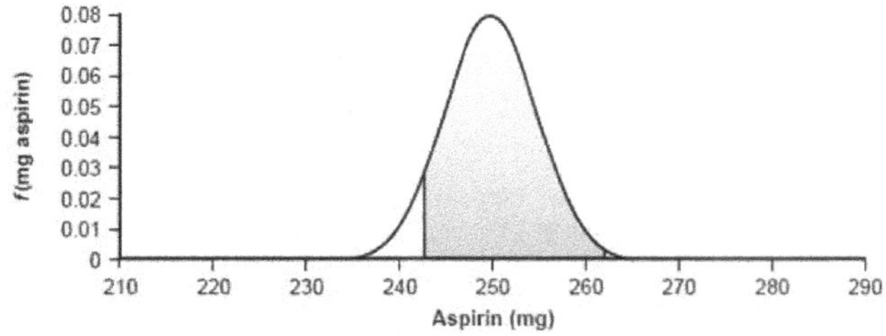

Fig. 7: Normal distribution for population of aspirin tablets with , 250 mg aspirin and (32. 25. The shaded area shows the percentage of tablets containing between 243 and 262 mg of aspirin.

$$z_{low} = \frac{243-250}{5} = -1.4$$

and the deviation for the upper limit is

$$z_{up} = \frac{262-250}{5} = +2.4$$

Using the table in Appendix IA, we find that the percentage of tablets with less than 243 mg of aspirin is 8.08%, and the percentage of tablets with more than 262 mg of aspirin is 0.82%. The percentage of tablets containing between 243 and 262 mg of aspirin is therefore

$$100.00\% - 8.08\% - 0.82\,\% = 91.10\%$$

Confidence Intervals for Populations

If we randomly select a single member from a population, what will be its most likely value? This is an important question, and, in one form or another, it is the fundamental problem for any analysis. One of the most important features of a population's probability distribution is that it provides a way to answer this question.

Earlier we noted that 68.26% of a normally distributed population is found within the range of μ ± 1σ. Stating this another way, there is a 68.26% probability that a member selected at random from a normally distributed population will have a value in the interval of μ ± 1σ. In general, we can write

$$X_i = \mu \pm z\sigma \quad \text{------------- 9}$$

where the factor z accounts for the desired level of confidence. Values reported in this fashion are called confidence intervals. Equation 9, for example, is the confidence interval for a single member of a population. Confidence intervals can be quoted for any desired probability level, several examples of which are shown in Table 12. For reasons that will be discussed later, a 95% confidence interval frequently is reported.

Table 12 Confidence intervals for Normal Distribution Curves Between the Limits $\mu \pm z\sigma$

z	Confidence Interval (%)
0.50	38.30
1.00	68.26
1.50	86.64
1.96	95.00
2.00	95.44
2.50	98.76
3.00	99.73
3.50	99.95

Alternatively, a confidence interval can be expressed in terms of the population's standard deviation and the value of a single member drawn from the population. Thus, equation 9 can be rewritten as a confidence interval for the population mean

$$\mu = X_i \pm z\sigma \qquad 10$$

Confidence intervals also can be reported using the mean for a sample of size n, drawn from a population of known σ. The standard deviation for the mean value, $\sigma_{\bar{X}}$, which also is known as the standard error of the mean, is

$$\sigma_{\bar{X}} = \frac{\sigma}{\sqrt{n}}$$

$$\mu = \bar{X} \pm \frac{z\sigma}{\sqrt{n}}$$

Probability Distributions for Samples

We introduced two probability distributions commonly encountered when studying populations. The construction of confidence intervals for a normally distributed population. We have yet to ad- dress, however, how we can identify the probability distribution for a given population. In Examples 11–14 we assumed that the amount of aspirin in analgesic tablets is normally distributed. We are justified in asking how this can be determined without analyzing every member of the population. When we cannot study the whole population, or when we cannot predict the mathematical form of a population's probability distribution, we must deduce the distribution from a limited sampling of its members.

Sample Distributions and the Central Limit Theorem

Let's return to the problem of determining a penny's mass to explore the relationship between a population's distribution and the distribution of samples drawn from that population. The data shown in Tables 1 and10 are insufficient for our purpose because they are not large enough to give a useful picture of their respective probability distributions. A better picture of the probability distribution requires a larger sample, such as that shown in Table 13, for which X– is 3.095 and s2 is 0.0012.

The data in Table 13 are best displayed as a histogram, in which the frequency of occurrence for equal intervals of data is plotted versus the midpoint of each interval. Table 13 and Figure.9 show a frequency table and histogram for the data in Table 13. Note that the histogram was constructed such that the mean value for the data set is centered within its interval. In addition, a normal distribution curve using X– and s to estimate μ and σ is superimposed on the histogram.

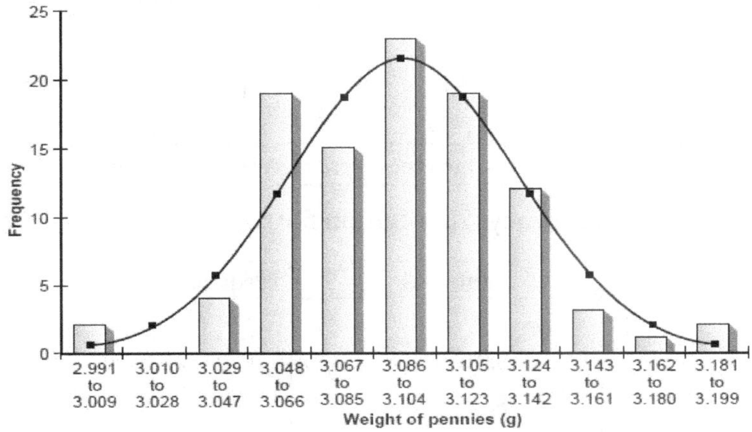

Fig. 9: Histogram for data in table 13. A Normal distribution curve for the data based on and S^2 is superimposed in the histogram.

Table 13 Individual Masses for a Large Sample of U.S. Pennies in Circulation[s]

Penny	Weight(g)	Penny	Weight(g)	Penny	Weight(g)	Penny	Weight(g)
1	3.126	26	3.073	51	3.101	76	3.086
2	3.140	27	3.084	52	3.049	77	3.123
3	3.092	28	3.148	53	3.082	78	3.115
4	3.095	29	3.047	54	3.142	79	3.055
5	3.080	30	3.121	55	3.082	80	3.057
6	3.065	31	3.116	56	3.066	81	3.097
7	3.117	32	3.005	57	3.128	82	3.66
8	3.034	33	3.115	58	3.112	83	3.113
9	3.126	34	3.103	59	3.085	84	3.102
10	3.057	35	3.086	60	3.086	85	3.033
11	3.053	36	3.103	61	3.084	86	3.112
12	3.099	37	3.049	62	3.104	87	3.103
13	3.065	38	2.998	63	3.107	88	3.198

14	3.059	39	3.063	64	3.093	89	3.103
15	3.068	40	3.055	65	3.126	90	3.126
16	3.060	41	3.181	66	3.138	91	3.111
17	3.078	42	3.108	67	3.131	92	3.126
18	3.125	43	3.114	68	3.120	93	3.052
19	3.090	44	3.121	69	3.100	94	3.113
20	3.100	45	3.105	70	3.099	95	3.085
21	3.055	46	3.078	71	3.097	96	3.117
22	3.105	47	3.147	72	3.091	97	3.142
23	3.063	48	3.104	73	3.077	98	3.031
24	3.083	49	3.146	74	3.178	99	3.083
25	3.065	50	3.095	75	3.054	100	3.104

[a]Pennies are identified int he order in which they were sampled and weighed.

Table 14 Frequency Distribution for the Data in Table 13

Interval	Frequency
2.991-3.009	2
3.010-3.028	0
3.029-3.047	4
3.048-3.066	19
3.067-3.085	15
3.086-3.104	23
3.105-3.123	19
3.124-3.142	12
3.143-3.161	13
3.162-3.180	1
3.181-3.199	2

It is noteworthy that the histogram in Figure 9 approximates the normal distribution curve. Although the histogram for the mass of pennies is not perfectly symmetrical, it is roughly symmetrical about the interval containing the greatest number of pennies. In addition, we know from Table 11 that 68.26%, 95.44%, and 99.73% of the members of a normally distributed population are within, respectively, $\pm 1\sigma$, $\pm 2\sigma$ and $\pm 3\sigma$. If we assume that the mean value, 3.095 g, and the sample variance, 0.0012, are good approximations for μ and $\sigma 2$, we find that 73%, 95%, and 100% of the pennies fall within these limits. It is easy to imagine that in- creasing the number of pennies in the sample will result in a histogram that even more closely approximates a normal distribution.

We will not offer a formal proof that the sample of pennies in Table 14 and the population from which they were drawn are normally distributed; how- ever, the evidence we have seen strongly suggests that this is true. Although we cannot claim that the results for all analytical experiments are normally distributed, in most cases the data we collect in the laboratory are, in fact, drawn from a normally distributed population. That this is generally true is a consequence of the central limit theorem. According to this theorem, in systems subject to a variety of indeterminate errors, the distribution of results will be approximately normal. Furthermore, as the number of contributing sources of indeterminate error increases, the results come even closer to approximating a normal distribution. The central limit theorem holds true even if the individual sources of indeterminate error are not normally distributed. The chief limitation to the central limit theorem is that the sources of indeterminate error must be independent and of similar magnitude so that no one source of error dominates the final distribution.

Estimating μ and σ^2

Our comparison of the histogram for the data in Table 12 to a normal distribution assumes that the sample's mean, X–, and variance, s^2 , are appropriate estimators of the population's mean, μ, and variance, σ^2. Why did we select X– and s^2, as opposed to other possible measures of central tendency and spread? The explanation is simple; X– and s^2 are considered unbiased estimators of μ and σ2.If we could analyze every possible sample of equal size for a given population (e.g., every possible sample of five pennies), calculating their respective means and variances, the average mean and the average variance would equal μ and σ^2. Al- though X– and s^2 for any single sample probably will not be the same as μ or σ^2, they provide a reasonable estimate for these values.

Degrees of Freedom

Unlike the population's variance, the variance of a sample includes the term n^{-1} in the denominator, where n is the size of the sample

$$s^2 = \frac{\sum_{i=1}^{n}(X_i - \overline{X})^2}{n-1} \qquad 12$$

Defining the sample's variance with a denominator of n, as in the case of the population's variance leads to a biased estimation of x2. The denominators of the variance equations 9 and 13 are commonly called the degrees of freedom for the population and the sample, respectively. In the case of a population, the degrees of freedom are always equal to the total number of members, n, in the population. For the sample's variance, however, substituting X for μ removes a degree of freedom from the calculation. That is, if there are n members in the sample, the value of the nth member can always be deduced from the remaining n – 1 members and X–. For example, if we have a sample with five members, and we know that four of the members a re 1, 2, 3, and 4, and that the mean is 3, then the fifth member of the sample must be $(X– x n)– X_1 – X_2 – X_3 – X_4 = (3 x 5)–1–2–3–4=5$

Confidence Intervals for Samples

Earlier we introduced the confidence interval as a way to report the most probable value for a population's mean, μ, when the population's standard deviation, σ, is known. Since s2 is an unbiased estimator of σ 2, it should be possible to construct confidence intervals for samples by replacing in equations 4.10 and 4.11 with s. Two complications arise, however. The first is that we cannot define s^2 for a single member of a population. Consequently, equation 10 cannot be extended to situa- tions in which s^2 is used as an estimator of s^2. In other words, when e is unknown, we cannot construct a confidence interval for μ by sampling only a single member of the population.

The second complication is that the values of z shown in Table 11 are derived for a normal distribution curve that is a function of σ^2, not s^2. Although s^2 is an unbiased estimator of σ^2, the value of s^2 for any randomly selected sample may differ significantly from σ^2. To account for the uncertainty in estimating σ^2, the term z in equation 11 is replaced with the variable t, where t is defined such that t >= z at all confidence levels. Thus, equation 11 becomes

$$\mu = \overline{X} \pm \frac{ts}{\sqrt{n}} \quad 13$$

Values for t at the 95% confidence level are shown in Table 15 Note that t be- comes smaller as the number of the samples (or degrees of freedom) increase, approaching z as n approaches infinity. Additional values of t for other confidence levels can be found in Appendix 1B.

Table 15. Values of *t* for the 95 % confidence Interval

Degrees of Freedom	t
1	12.71
2	4.30
3	3.18
4	2.78
5	2.57
6	2.45
7	2.36
8	2.31
9	2.26
10	2.23
12	2.18
14	2.14
16	2.12

18	2.10
20	2.09
30	2.04
50	2.01
∞	1.96

A Cautionary Statement

There is a temptation when analyzing data to plug numbers into an equation, carry out the calculation, and report the result. This is never a good idea, and you should develop the habit of constantly reviewing and evaluating your data. For example, if analyzing five samples gives an analyte's mean concentration as 0.67 ppm with a standard deviation of 0.64 ppm, then the 95% confidence interval is:

$$\mu = 0.67 \pm \frac{(2.78)(0.64)}{\sqrt{5}} = 0.64 \pm 0.80 \text{ ppm}$$

This confidence interval states that the analyte's true concentration lies within the range of −0.16 ppm to 1.44 ppm. Including a negative concentration within the confidence interval should lead you to reevaluate your data or conclusions. On further investigation your data may show that the standard deviation is larger than expected, making the confidence interval too broad, or you may conclude that the analyte's concentration is too small to detect accurately.

A second example is also informative. When samples are obtained from a normally distributed population, their values must be random. If results for several samples show a regular pattern or trend, then the samples cannot be normally distributed. This may reflect the fact that the underlying population is not normally distributed, or it may indicate the presence of a time-dependent determinate error. For example, if we randomly select 20 pennies and find that the mass of each penny exceeds that of the preceding penny, we might suspect that the balance on which the pennies are being weighed is drifting out of calibration.

Statistical Analysis of Data

The result of an analysis is best expressed as a confidence interval. For example, a 95% confidence interval for the mean of five results gives the range in which we expect to find the mean for 95% of all samples of equal size, drawn from the same population. Alternatively, and in the absence of determinate errors, the 95% confidence interval indicates the range of values in which we expect to find the population's true mean.

The probabilistic nature of a confidence interval provides an opportunity to ask and answer questions comparing a sample's mean or variance to either the accepted values for its population or similar values obtained for other samples. For example, confidence intervals can be used to answer questions such as "Does a newly developed method for the analysis of cholesterol in blood give results that are significantly different from those obtained when using a standard method?" or "Is there a significant variation in the chemical composition of rainwater collected at different sites downwind from a coal-burning utility plant?" In this section we introduce a general approach to the statistical analysis of data.

Significance Testing

Let's consider the following problem. Two sets of blood samples have been collected from a patient receiving medication to lower her concentration of blood glucose. One set of samples was drawn immediately before the medication was administered; the second set was taken several hours later. The samples are analyzed and their respective means and variances reported. How do we decide if the medication was successful in lowering the patient's concentration of blood glucose?

One way to answer this question is to construct probability distribution curves for each sample and to compare the curves with each other. Three possible out- comes are shown in Figure 10. In Figure 10a, the probability distribution curves are completely separated, strongly suggesting that the samples are significantly different. In Figure 10b, the probability distributions for the two samples are highly overlapped, suggesting that any difference between the samples is insignificant. Figure 10c, however, presents a dilemma. Although the means for the two samples appear to be different, the probability distributions overlap to an extent that a significant number of possible outcomes could belong to either distribution. In this case we can, at best, only make a statement about the probability that the samples are significantly different.

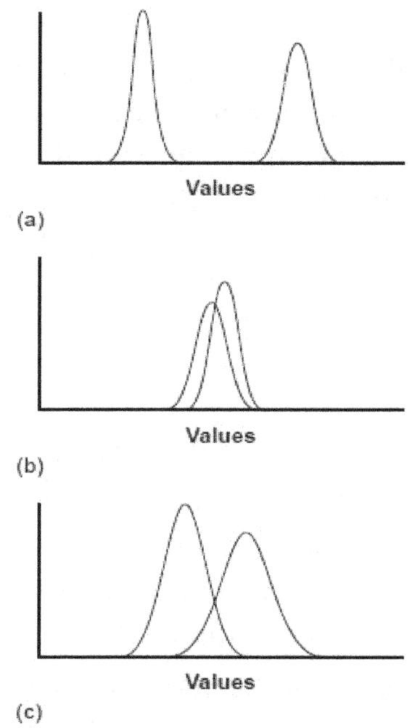

Fig. 10: Three examples of Possible relationships between the probability distributions for two Population (a) completely separate distribution (b) Distribution with a great deal for overlap; (c) Distributions with some overlap

The process by which we determine the probability that there is a significant difference between two samples is called significance testing or hypothesis testing. Before turning to a discussion of specific examples, however, we will first establish a general approach to conducting and interpreting significance tests.

Constructing a Significance Test

A significance test is designed to determine whether the difference between two or more values is too large to be explained by indeterminate error. The first step in constructing a significance test is to state the experimental problem as a yes- or-no question, two examples of which were given at the beginning of this section. A null hypothesis and an alternative hypothesis provide answers to the question. The null hypothesis, H0, is that indeterminate error is sufficient to explain any difference in the values being compared. The alternative hypothesis, HA, is that the difference between the values is too great to be explained by random error and, therefore, must be real. A significance test is conducted on the null hypothesis, which is either retained or rejected. If the null hypothesis is rejected, then the alternative hypothesis must be accepted. When a null hypothesis is not rejected, it is said to be retained rather than accepted. A null hypothesis is retained whenever the evidence is insufficient to prove it is incorrect. Because of the way in which significance tests are conducted, it is impossible to prove that a null hypothesis is true.

The difference between retaining a null hypothesis and proving the null hypothesis is important. To appreciate this point, let us return to our example on determining the mass of a penny. After looking at the data in Table 16, you might pose the following null and alternative hypotheses.

- H0: Any U.S. penny in circulation has a mass that falls in the range of 2.900–3.200 g
- HA: Some U.S. pennies in circulation have masses that are less than 2.900 g or more than 3.200 g.

Table 16 Individual Masses for a Large Sample of U.S. Pennies in Circulation[5]

Penny	Weight(g)	Penny	Weight(g)	Penny	Weight(g)	Penny	Weight(g)
1	3.126	26	3.073	51	3.101	76	3.086
2	3.140	27	3.084	52	3.049	77	3.123
3	3.092	28	3.148	53	3.082	78	3.115
4	3.095	29	3.047	54	3.142	79	3.055
5	3.080	30	3.121	55	3.082	80	3.057
6	3.065	31	3.116	56	3.066	81	3.097
7	3.117	32	3.005	57	3.128	82	3.66
8	3.034	33	3.115	58	3.112	83	3.113
9	3.126	34	3.103	59	3.085	84	3.102
10	3.057	35	3.086	60	3.086	85	3.033
11	3.053	36	3.103	61	3.084	86	3.112

Penny	Weight(g)	Penny	Weight(g)	Penny	Weight(g)	Penny	Weight(g)
12	3.099	37	3.049	62	3.104	87	3.103
13	3.065	38	2.998	63	3.107	88	3.198
14	3.059	39	3.063	64	3.093	89	3.103
15	3.068	40	3.055	65	3.126	90	3.126
16	3.060	41	3.181	66	3.138	91	3.111
17	3.078	42	3.108	67	3.131	92	3.126
18	3.125	43	3.114	68	3.120	93	3.052
19	3.090	44	3.121	69	3.100	94	3.113
20	3.100	45	3.105	70	3.099	95	3.085
21	3.055	46	3.078	71	3.097	96	3.117
22	3.105	47	3.147	72	3.091	97	3.142
23	3.063	48	3.104	73	3.077	98	3.031
24	3.083	49	3.146	74	3.178	99	3.083
25	3.065	50	3.095	75	3.054	100	3.104

[a]Pennies are identified int he order in which they were sampled and weighed.

To test the null hypothesis, you reach into your pocket, retrieve a penny, and determine its mass. If the mass of this penny is 2.512 g, then you have proved that the null hypothesis is incorrect. Finding that the mass of your penny is 3.162 g, how- ever, does not prove that the null hypothesis is correct because the mass of the next penny you sample might fall outside the limits set by the null hypothesis.

After stating the null and alternative hypotheses, a significance level for the analysis is chosen. The significance level is the confidence level for retaining the null hypothesis or, in other words, the probability that the null hypothesis will be incorrectly rejected. In the former case the significance level is given as a percentage (e.g., 95%), whereas in the latter case, it is given as α, where α is defined as

$$\alpha = 1 - \frac{\text{confidence level}}{100}$$

Thus, for a 95% confidence level, α is 0.05.

Next, an equation for a test statistic is written, and the test statistic's critical value is found from an appropriate table. This critical value defines the breakpoint between values of the test statistic for which the null hypothesis will be retained or rejected. The test statistic is calculated from the data, compared with the critical value, and the null hypothesis is either rejected or retained. Finally, the result of the significance test is used to answer the original question.

One-Tailed and Two-Tailed Significance Tests

Consider the situation when the accuracy of a new analytical method is evaluated by analyzing a standard reference material with a known μ. A sample of the standard is analyzed, and the sample's mean is determined. The null hypothesis is that the sample's mean is equal to μIf the significance test is conducted at the 95% confidence

$$H_0:\ \overline{X} = \mu$$

level (α = 0.05), then the null hypothesis will be retained if a 95% confidence interval around X– contains μ. If the alternative hypothesis is

$$H_A:\ \overline{X} \neq \mu$$

then the null hypothesis will be rejected, and the alternative hypothesis accepted if μ lies in either of the shaded areas at the tails of the sample's probability distribution (Figure 11a). Each of the shaded areas accounts for 2.5% of the area under the probability distribution curve. This is called a two-tailed significance test because the null hypothesis is rejected for values of μ at either extreme of the sample's probability distribution.

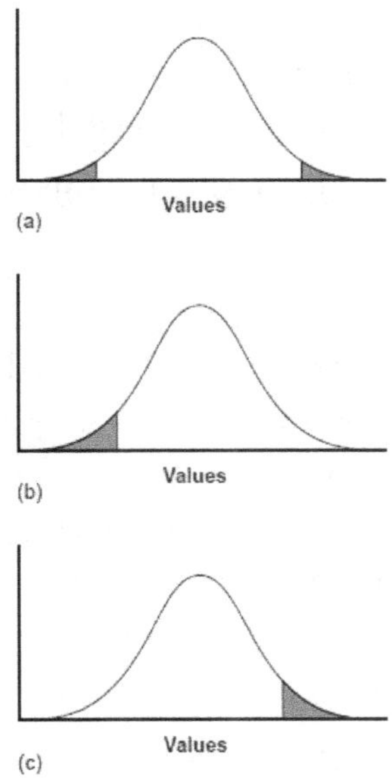

Figure 11 Examples of (a) two-tailed (b) and (c) one tailed significance tests. The shaded areas in each curve represent the value for which the null hypothesis is rejected.

The alternative hypothesis also can be stated in one of two additional ways

$$H_A:\ \overline{X} > \mu$$

$$H_A:\ \overline{X} < \mu$$

For which the null hypothesis is rejected if μ falls within the shaded areas shown in Figure 4.10(b) and Figure 4.10(c), respectively. In each case the shaded area represents 5% of the area under the probability distribution curve. These are examples of one-tailed significance tests.

For a fixed confidence level, a two-tailed test is always the more conservative test because it requires a larger difference between X– and μ to reject the null hypothesis. Most significance tests are applied when there is no a priori expectation about the relative magnitudes of the parameters being compared. A two-tailed significance test, therefore, is usually the appropriate choice. One-tailed significance tests are reserved for situations when we have reason to expect one parameter to be larger or smaller than the other. For example, a one-tailed significance test would be appropriate for our earlier example regarding a medication's effect on blood glucose levels since we believe that the medication will lower the concentration of glucose.

Errors in Significance Testing

Since significance tests are based on probabilities, their interpretation is naturally subject to error. As we have already seen, significance tests are carried out at a significance level, α, that defines the probability of rejecting a null hypothesis that is true. For example, when a significance test is conducted at α = 0.05, there is a 5% probability that the null hypothesis will be incorrectly rejected. This is known as a type 1 error, and its risk is always equivalent to α. Type 1 errors in two-tailed and one-tailed significance tests are represented by the shaded areas under the probability distribution curves in Figure 10.

The second type of error occurs when the null hypothesis is retained even though it is false and should be rejected. This is known as a type 2 error, and its probability of occurrence is β. Unfortunately, in most cases β cannot be easily calculated or estimated.

The probability of a type 1 error is inversely related to the probability of a type 2 error. Minimizing a type 1 error by decreasing α, for example, increases the likelihood of a type 2 error. The value of α chosen for a particular significance test, therefore, represents a compromise between these two types of error. Most of the examples in this text use a 95% confidence level, or α = 0.05, since this is the most frequently used confidence level for the majority of analytical work. It is not unusual, however, for more stringent (e.g. α = 0.01) or for more lenient (e.g. α = 0.10) confidence levels to be used.

Statistical Methods for Normal Distributions

The most commonly encountered probability distribution is the normal, or Gaussian, distribution. A normal distribution is characterized by a true mean, μ, and variance c2, which are estimated using X– and s. Since the area between any two limits of a normal distribution is well defined, the construction and evaluation of significance tests are straightforward.

Comparing X– to μ

One approach for validating a new analytical method is to analyze a standard sample containing a known amount of analyte, μ. The method's accuracy is judged by determining the average amount of analyte in several samples, X, and using a significance test to compare it with μ The null hypothesis is that X and μ are the same and that any difference between the two values can be explained by in- determinate errors affecting the determination of X–. The alternative hypothesis is that the difference between X– and μ is too large to be explained by indeterminate error. The equation for the test (experimental) statistic, texp, is derived from the confidence interval for μ.

$$\mu = \overline{X} \pm \frac{t_{\exp} s}{\sqrt{n}} \qquad 14$$

Rearranging equation 4.14

$$t_{\exp} = \frac{\left|\mu - \overline{X}\right| \times \sqrt{n}}{s} \qquad 15$$

Gives the value of texp when μ is at either the right or left edge of the sample's apparent confidence interval (Figure 12a). The value of texp is compared with a critical value, t(α ,v), which is determined by the chosen significance level, α , the degrees of freedom for the sample, v, and whether the significance test is one- tailed or two-tailed. Values for t(α ,v) are found in Appendix 1B. The critical value t(α ,v) defines the confidence interval that can be explained by indeterminate errors. If texp is greater than t(α ,v), then the confidence interval for the data is wider than that expected from indeterminate errors (Figure 12b). In this case, the null hypothesis is rejected and the alternative hypothesis is accepted. If texp is less than or equal to t(α ,v), then the confidence interval for the data could be attributed to indeterminate error, and the null hypothesis is retained at the stated significance level (Figure 12c).

If evidence for a determinate error is found, as in Example 16, its source should be identified and corrected before analyzing additional samples. Failing to reject the null hypothesis, however, does not imply that the method is accurate, but only indicates that there is insufficient evidence to prove the method inaccurate at the stated confidence level.

The utility of the t-test for X– and μ is improved by optimizing the conditions used in determining X–. Examining equation 15 shows that increasing the number of replicate determinations, n, or improving the precision of the analysis enhances the utility of this significance test. A t-test can only give useful results, however, if the standard deviation for the analysis is reasonable. If the standard deviation is substantially larger than the expected standard deviation, σ, the confidence interval around X– will be so large that a significant difference between X– and μ may be difficult to prove. On the other hand, if the standard deviation is significantly smaller than expected, the confidence interval around X will be too small, and a significant difference between X and μ may be found when none exists. A significance test that can be used to evaluate the standard deviation is the subject of the next section.

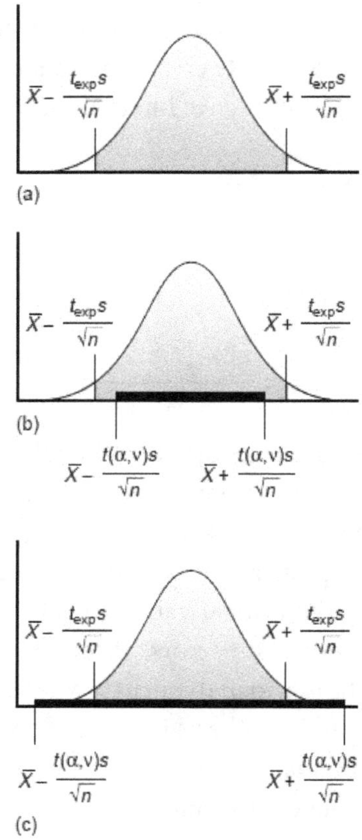

Fig. 12: Relationship between confidence intervals and result of a significance test (a) the shaded area under the normal distribution curves shows the apparent confidence intervals for the sample based in t exp. the solid bars in (b) and (c) show the actual confidence intervals that can be explained by indeterminate error using the critical value of (a, v) In part (b) the null hypothesis is rejected and the alternative hypothesis is accepted in part (c) the null hypothesis is retained. A typical application of this significance test, which is known as a t-test μ, is outlined in the following example.

Comparing s^2 to σ^2

When a particular type of sample is analyzed on a regular basis, it may be possible to determine the expected, or true variance, σ^2, for the analysis. This often is the case in clinical labs where hundreds of blood samples are analyzed each day. Replicate analyses of any single sample, however, results in a sample variance, s^2. A statistical comparison of s^2 to σ^2 provides useful information about whether the analysis is in a state of "statistical control." The null hypothesis is that σ^2 and s^2 are identical, and the alternative hypothesis is that they are not identical.

The test statistic for evaluating the null hypothesis is called an F-test, and is given as either

$$F_{exp} = \frac{s^2}{\sigma^2} \quad \text{or} \quad F_{exp} = \frac{\sigma^2}{s^2} \qquad 16$$

$$(s^2 > \sigma^2) \qquad\qquad (\sigma^2 > s^2)$$

Depending on whether s^2 is larger or smaller than σ^2. Note that F_{exp} is defined such that its value is always greater than or equal to 1.

If the null hypothesis is true, then F_{exp} should equal 1. Due to indeterminate errors, however, the value for F_{exp} usually is greater than 1. A critical value, F(α, vnum, vden), gives the largest value of F that can be explained by indeterminate error. It is chosen for a specified significance level, α, and the degrees of freedom for the variances in the numerator, vnum, and denominator, vden. The degrees of freedom for s^2 is n^{-1}, where n is the number of replicates used in determining the sample's variance. Critical values of F for α = 0.05 are listed in Appendix 1C for both one-tailed and two-tailed significance tests.

Comparing Two Sample Variances

The F-test can be extended to the comparison of variances for two samples, A and B, by rewriting equation 16 as;

$$F_{exp} = \frac{s_A^2}{s_B^2}$$

Where A and B are defined such that s^2_A is greater than or equal to s^2_B. An example of this application of the *F*-test is shown in the following example.

Comparing Two Sample Means

The result of an analysis is influenced by three factors: the method, the sample, and the analyst. The influence of these factors can be studied by conducting a pair of experiments in which only one factor is changed. For example, two methods can be compared by having the same analyst apply both methods to the same sample and examining the resulting means. In a similar fashion, it is possible to compare two analysts or two samples.

Significance testing for comparing two mean values is divided into two categories depending on the source of the data. Data are said to be unpaired when each mean is derived from the analysis of several samples drawn from the same source. Paired data are encountered when analyzing a series of samples drawn from different sources.

Unpaired Data

Consider two samples, A and B, for which mean values, X–A, X–B,, and standard deviations, sA and sB, have been measured. Confidence intervals for μ_A and μ_B can be written for both samples

$$\mu_A = \overline{X}_A \pm \frac{ts_A}{\sqrt{n_A}} \tag{17}$$

$$\mu_B = \overline{X}_B \pm \frac{ts_B}{\sqrt{n_B}} \tag{18}$$

Where n_A and n_B are the number of replicate trials conducted on samples A and B. A comparison of the mean values is based on the null hypothesis that X–A and X–B are identical, and an alternative hypothesis that the means are significantly different. A test statistic is derived by letting μ_A equal μ_B, and combining equations 17.

$$\overline{X}_A \pm \frac{ts_A}{\sqrt{n_A}} = \overline{X}_B \pm \frac{ts_B}{\sqrt{n_B}}$$

Solving for $|\overline{X}_A - \overline{X}_B|$ and using a propagation of uncertainty, gives

$$|\overline{X}_A - \overline{X}_B| = t \times \sqrt{\frac{s_A^2}{n_A} + \frac{s_B^2}{n_B}}$$

Finally, solving for *t*, which we replace with t_{exp}, leaves us with

$$t_{exp} = \frac{|\overline{X}_A - \overline{X}_B|}{\sqrt{(s_A^2/n_A) + (s_B^2/n_B)}} \tag{19}$$

The value of texp is compared with a critical value, t(α, v), as determined by the chosen significance level, α, the degrees of freedom for the sample, v, and whether the significance test is one-tailed or two-tailed.

It is unclear, however, how many degrees of freedom are associated with t(α, v) since there are two sets of independent measurements. If the variances sA2 and sB2 estimate the same σ2, then the two standard deviations can be factored out of equation 19 and replaced by a pooled standard deviation, spool, which provides a better estimate for the precision of the analysis. Thus, equation 19 becomes

$$t_{exp} = \frac{|\overline{X}_A - \overline{X}_B|}{s_{pool}\sqrt{(1/n_A) + (1/n_B)}} \tag{20}$$

with the pooled standard deviation given as

$$s_{pool} = \sqrt{\frac{(n_A - 1)s_A^2 + (n_B - 1)s_B^2}{n_A + n_B - 2}} \tag{21}$$

As indicated by the denominator of equation 21, the degrees of freedom for the pooled standard deviation is $n_A + n_B^{-2}$.

If s_A and s_B are significantly different, however, then t_{exp} must be calculated using equation 19. In this case, the degrees of freedom is calculated using the following imposing equation.

$$v = \frac{[(s_A^2/n_A) + (s_B^2/n_B)]^2}{[(s_A^2/n_A)^2/(n_A + 1)] + [(s_B^2/n_B)^2/(n_B + 1)}$$ 22

Since the degrees of freedom must be an integer, the value of v obtained using equation 22 is rounded to the nearest integer.

Regardless of whether equation 19 or 20 is used to calculate texp, the null hypothesis is rejected if texp is greater than t(α, v), and retained if texp is less than or equal to t(α, v).

Paired Data

In some situations the variation within the data sets being compared is more significant than the difference between the means of the two data sets. This is commonly encountered in clinical and environmental studies, where the data being compared usually consist of a set of samples drawn from several populations. For example, a study designed to investigate two procedures for monitoring the concentration of glucose in blood might involve blood samples drawn from ten patients. If the variation in the blood glucose levels among the patients is significantly larger than the anticipated variation between the methods, then an analysis in which the data are treated as unpaired will fail to find a significant difference between the methods. In general, paired data sets are used whenever the variation being investigated is smaller than other potential sources of variation.

In a study involving paired data the difference, d , between the paired values for each sample is calculated. The average difference, d', and standard deviation of the differences, sd, are then calculated. The null hypothesis is that d is 0, and that there is no difference in the results for the two data sets. The alternative hypothesis is that the results for the two sets of data are significantly different, and, therefore, d is not equal to 0.

The test statistic, t_{exp}, is derived from a confidence interval around $\bar{d}$

$$0 = \bar{d} \pm \frac{ts_d}{\sqrt{n}}$$

Where n is the number of paired samples. Replacing t with t_{exp} and rearranging gives

$$t_{exp} = \frac{|\bar{d}|\sqrt{n}}{s_d}$$

The value of texp is then compared with a critical value, t(α, v), which is determined by the chosen significance level, α, the degrees of freedom for the sample, v, and whether the significance test is one-tailed or two-tailed. For paired data, the degrees of freedom is n– 1. If texp is greater than t(α, v), then the null hypothesis is rejected and the alternative hypothesis is accepted. If texp is less than or equal to t(α, v), then the null hypothesis is retained, and a significant difference has not been demonstrated at the stated significance level. This is known as the paired t-test.

A paired t-test can only be applied when the individual differences, di, belong to the same population. This will only be true if the determinate and indeterminate errors affecting the results are independent of the concentration of analyte in the samples. If this is not the case, a single sample with a larger error could result in a value of d that is substantially larger than that for the remaining samples. Including this sample in the calculation of d– and sd leads to a biased estimate of the true mean and standard deviation. For samples that span a limited range of analyte concentra- tions, such as that in Example 21, this is rarely a problem. When paired data span a wide range of concentrations, however, the magnitude of the determinate and in- determinate sources of error may not be independent of the analyte's concentration. In such cases the paired t-test may give misleading results since the paired data with the largest absolute determinate and indeterminate errors will dominate d–. In this situation a comparison is best made using a linear regression.

Outliers

On occasion, a data set appears to be skewed by the presence of one or more data points that are not consistent with the remaining data points. Such values are called outliers. The most commonly used significance test for identifying outliers is Dixon's Q-test. The null hypothesis is that the apparent outlier is taken from the same population as the remaining data. The alternative hypothesis is that the outlier comes from a different population, and, therefore, should be excluded from consideration.

The Q-test compares the difference between the suspected outlier and its nearest numerical neighbor to the range of the entire data set. Data are ranked from smallest to largest so that the suspected outlier is either the first or the last data point. The test statistic, Q_{exp}, is calculated using equation 23 if the suspected outlier is the smallest value (X_1).

$$Q_{exp} = \frac{X_2 - X_1}{X_n - X_1} \qquad 23$$

or using equation 4.24 if the suspected outlier is the largest value (X_n)

$$Q_{exp} = \frac{X_n - X_{n-1}}{X_n - X_1} \qquad 24$$

Where n is the number of members in the data set, including the suspected outlier. It is important to note that equations 23 and 24 are valid only for the detection of a single outlier. Other forms of Dixon's Q-test allow its extension to the detection of multiple outliers. The value of Qexp is compared with a critical value, Q(α, n), at a significance level of α. The Q-test is usually applied as the more conservative two- tailed test, even though the outlier is the smallest or largest value in the data set. Values for Q(α, n) can be found If Qexp is greater than Q(α, n), then the null hypothesis is rejected and the outlier may be rejected. When Qexp is less than or equal to Q(α, n) the suspected outlier must be retained.

The Q-test should be applied with caution since there is a probability, equal to α, that an outlier identified by the Q-test actually is not an outlier. In addition, the Q-test should be avoided when rejecting an outlier leads to a precision that is un- reasonably better than the expected precision determined by a propagation of un- certainty. Given these two concerns it is not surprising that some statisticians caution against the removal of outliers. On the other hand, testing for outliers can provide useful information if we try to understand the source of the suspected outlier. For example, the outlier identified in Example 22 represents a significant change in the mass of a penny (an approximately 17% decrease in mass), due to a change in the composition of the U.S. penny. In 1982, the composition of a U.S. penny was changed from a brass alloy consisting of 95% w/w Cu and 5% w/w Zn, to a zinc core covered with copper. The pennies in Example 4.22 were therefore drawn from different populations.

Detection Limits

A method's detection limit is the smallest amount or concentration of analyte that can be detected with statistical confidence. The International Union of Pure and Applied Chemistry (IUPAC) defines the detection limit as the smallest concentration or absolute amount of analyte that has a signal significantly larger than the signal arising from a reagent blank. Mathematically, the analyte's signal at the detection limit, $(S_A)_{DL}$, is

$$(S_A)_{DL} = S_{reag} + z\sigma_{reag} \qquad 25$$

Where Sreag is the signal for a reagent blank, σreag is the known standard devia- tion for the reagent blank's signal, and z is a factor accounting for the desired confidence level. The concentration, (CA)DL, or absolute amount of analyte, (nA)DL, at the detection limit can be determined from the signal at the detection limit.

$$(C_A)_{DL} = \frac{(S_A)_{DL}}{k}$$

$$(n_A)_{DL} = \frac{(S_A)_{DL}}{k}$$

The value for z depends on the desired significance level for reporting the detection limit. Typically, z is set to 3, which, from Appendix 1A, corresponds to a significance level of α = 0.00135. Consequently, only 0.135% of measurements made on the blank will yield signals that fall outside this range (Figure 13a). When σreag is unknown, the term zσreag may be replaced with tsreag, where t is the appropriate value from a t-table for a one-tailed analysis.

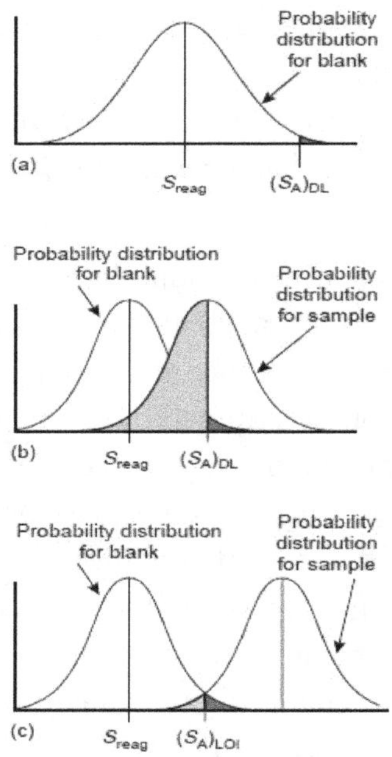

Fig. 13: Normal distribution curves showing the definition of detection limit and limit of identification (LOI). The Probability of a type 1 error is indicated by the dark shading and the probability of a type 2 error is indicated by light shading

In analyzing a sample to determine whether an analyte is present, the signal for the sample is compared with the signal for the blank. The null hypothesis is that the sample does not contain any analyte, in which case (SA)DL and Sreag are identical. The alternative hypothesis is that the analyte is present, and (SA)DL is greater than Sreag. If (SA)DL exceeds Sreag by $z\sigma$ (or ts), then the null hypothesis is rejected and there is evidence for the analyte's presence in the sample. The probability that the null hypothesis will be falsely rejected, a type 1 error, is the same as the significance level. Selecting z to be 3 minimizes the probability of a type 1 error to 0.135%.

Significance tests, however, also are subject to type 2 errors in which the null hypothesis is falsely retained. Consider, for example, the situation shown in Figure 4.12b, where SA is exactly equal to (SA)DL. In this case the probability of a type 2 error is 50% since half of the signals arising from the sample's population fall below the detection limit. Thus, there is only a 50:50 probability that an analyte at the IUPAC detection limit will be detected. As defined, the IUPAC definition for the detection limit only indicates the smallest signal for which we can say, at a significance level of α, that an analyte is present in the sample. Failing to detect the analyte, however, does not imply that it is not present.

An alternative expression for the detection limit, which minimizes both type 1 and type 2 errors, is the limit of identification, (SA)LOI, which is defined as:

$$(S_A)_{LOI} = S_{reag} + z\sigma_{reag} + z\sigma_{samp}$$

As shown in Figure 12c, the limit of identification is selected such that there is an equal probability of type 1 and type 2 errors. The American Chemical Society's Committee on Environmental Analytical Chemistry recommends the limit of quantisation, (SA)LOQ, which is defined as

$$(S_A)_{LOQ} = S_{reag} + 10\sigma_{reag}$$

Other approaches for defining the detection limit have also been developed. The detection limit is often represented, particularly when used in debates over public policy issues, as a distinct line separating analytes that can be detected from that cannot be detected. This use of a detection limit is incorrect. Defining the detection limit in terms of statistical confidence levels implies that there may be a gray area where the analyte is sometimes detected and sometimes not detected. This is shown in Figure 14 where the upper and lower confidence limits are de- fined by the acceptable probabilities for type 1 and type 2 errors. Analytes producing signals greater than that defined by the upper confidence limit are always de- tected, and analytes giving signals smaller than the lower confidence limit are never detected. Signals falling between the upper and lower confidence limits, however, are ambiguous because they could belong to populations representing either the reagent blank or the analyte. Figure 13c represents the smallest value of SA for which no such ambiguity exists.

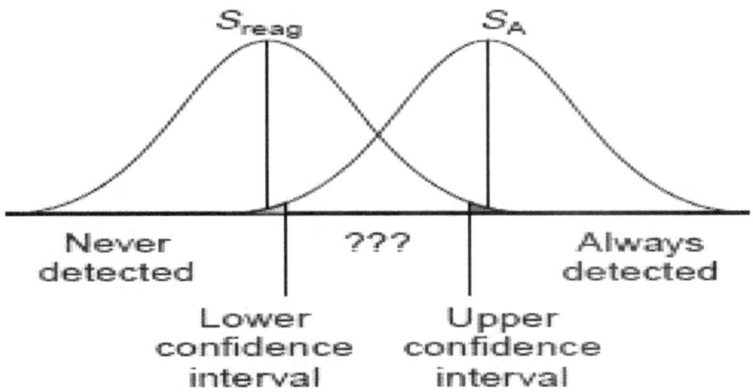

Fig. 14: Establishment of areas where the signal is never detected, always detected and where results are ambiguous . the upper and lower confidence limits are defined by the probability of a type 1 error (dark shading) and the probability of a type 2 error (light shading).

7

STANDARDIZING ANALYTICAL METHODS

7.1 STANDARDIZATION:

The American Chemical Society's Committee on Environmental Improvement defines standardization as the process of determining the relationship between the signal and the amount of analyte in a sample. We defined this relationship as:

$$St_{otal=kAnA + Sreag}or_{Stotal = kACA + Sreag}$$

Where S_{total} the signal, n_A is the moles of analyte, C_A is the analyte's concentration, k_A is the method's sensitivity for the analyte, and S_{reag} is the contribution to S_{total} from sources other than the sample. To standardize a method we must determine values for k_A and S_{reag}. Strategies for accomplishing this are the subject of this chapter.

7.2 ANALYTICAL STANDARDS

To standardize an analytical method we use standards that contain known amounts of analyte. The accuracy of a standardization, therefore, depends on the quality of the reagents and the glassware we use to prepare these standards. For example, in an acid–base titration the stoichiometry of the acid–base reaction defines the relationship between the moles of analyte and the moles of titrant. In turn, the moles of titrant is the product of the titrant's concentration and the volume of titrant used to reach the equivalence point. The accuracy of a titrimetric analysis, therefore, is never better than the accuracy with which we know the titrant's concentration.

7.2.1 Primary and Secondary Standards

There are two categories of analytical standards: primary standards and secondary standards. A primary standard is a reagent that we can use to dispense an accurately known amount of analyte. For example, a 0.1250-g sample of $K_2Cr_2O_7$ contains 4.249 × 10–4 moles of $K_2Cr_2O_7$. If we place this sample in a 250-mL volumetric flask and dilute to volume, the concentration of $K_2Cr_2O_7$ in the resulting solution is 1.700×10^{-3}

M. A primary standard must have a known stoichiometry, a known purity (or assay), and it must be stable during long-term storage. Because it is difficult to establishing accurately the degree of hydration, even after drying, a hydrated reagent usually is not a primary standard. Reagents that do not meet these criteria are secondary standards. The concentration of a secondary standard is determined relative to a primary standard. Lists of acceptable primary standards are available. Appendix 8 provides examples of some common primary standards.

7.2.2 Other Reagents

Preparing a standard often requires additional reagents that are not primary standards or secondary standards, such as a suitable solvent or reagents needed to adjust the standard's matrix. These solvents and reagents are potential sources of additional analyte, which, if not accounted for, produce a determinate error in the standardization. If available, reagent grade chemicals that conform to standards set by the American Chemical Society are used. The label on the bottle of a reagent grade chemical lists either the limits for specific impurities or provides an assay for the impurities. We can improve the quality of a reagent grade chemical by purifying it, or by conducting a more accurate assay. We can correct for contributions to S *total* from reagents used in an analysis by including an appropriate blank determination in the analytical procedure.

7.2.3 Preparing a Standard Solution

It often is necessary to prepare a series of standards, each with a different concentration of analyte. We can prepare these standards in two ways. If the range of concentrations is limited to one or two orders of magnitude, then each solution is best prepared by transferring a known mass or volume of the pure standard to a volumetric flask and diluting to volume.

When working with a larger range of concentrations, particularly a range that extends over more than three orders of magnitude, standards are best prepared by a serial dilution from a single stock solution. In a serial dilution we prepare the most concentrated standard and then dilute a portion of that solution to prepare the next most concentrated standard. Next, we dilute a portion of the second standard to prepare a third standard, continuing this process until we have prepared all of our standards. Serial dilutions must be prepared with extra care because an error in preparing one standard is passed on to all succeeding standards.

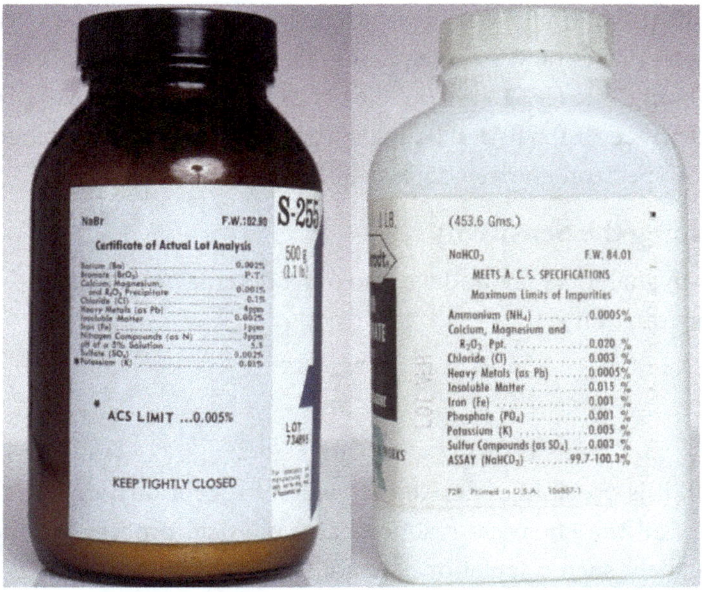

Fig. 1: Two examples of packaging labels for reagent grade chemicals. The label in (a) provides the manufacturer's assay for the reagent, NaBr. Note that potassium is flagged with an asterisk (*) because its assay exceeds the limit established by the American Chemical Society (ACS). The label in (b) does not provide an assay for impurities; however it indicates that the reagent meets ACS specifications by providing the maximum limits for impurities. An assay for the reagent, $NaHCO_3$, is provided.

7.2.4 Calibrating the Signal (*S total*)

The accuracy with which we determine k_A and S_{reag} depends on how accurately we can measure the signal, S_{total}. We measure signals using equipment, such as glassware and balances, and instrumentation, such as spectrophotometers and pH meters. To minimize determinate errors that might affect the signal, we first calibrate our equipment and instrumentation by measuring S_{total} for a standard with a known response of S_{std}, adjusting S_{total} until

$$S_{total} = S_{std}$$

Here are two examples of how we calibrate signals; other examples are provided in later chapters that focus on specific analytical methods.

When the signal is a measurement of mass, we determine S_{total} using an analytical balance. To calibrate the balance's signal we use a reference weight that meets standards established by a governing agency, such as the National Institute for Standards and Technology or the American Society for Testing and Materials. An electronic balance often includes an internal calibration weight for routine calibrations, as well as programs for calibrating with external weights. In either case, the balance automatically adjusts S_{total} to match S_{std}.

We also must calibrate our instruments. For example, we can evaluate a spectrophotometer's accuracy by measuring the absorbance of a carefully prepared solution of 60.06 mg/L $K_2Cr_2O_7$ in 0.0050 M H_2SO_4, using 0.0050 M H_2SO_4 as a reagent blank. An absorbance of 0.640 ± 0.010 absorbance units at a wavelength of 350.0 nm indicates that the spectrometer's signal is calibrated properly.

7.2.5 Determining the Sensitivity (*kA*)

To standardize an analytical method we also must determine the analyte's sensitivity, *kA*, in equation 1 or equation 2.

$$S_{total} = k_A n_A + S_{reag} \quad 1$$

$$S_{total} = k_A C_A + S_{reag} \quad 2$$

In principle, it is possible to derive the value of kA for any analytical method if we understand fully all the chemical reactions and physical processes responsible for the signal. Unfortunately, such calculations are not feasible if we lack a sufficiently developed theoretical model of the physical processes or if the chemical reaction's evince nonideal behavior. In such situations we must determine the value of kA by analyzing one or more standard solutions, each of which contains a known amount of analyte. In this section we consider several approaches for determining the value of kA. For simplicity we assume that Sreag is accounted for by a proper reagent blank, allowing us to replace Stotal in equation 1 and equation 2 with the analyte's signal, SA.

$$S_A = k_A n_A \quad 3$$

$$S_A = k_A C_A \quad 4$$

7.2.6 Single-Point versus Multiple-Point Standardizations

The simplest way to determine the value of k_A in equation 4 is to use a single-point standardization in which we measure the signal for a standard, S_{std}, that contains a known concentration of analyte, C_{std}. Substituting these values into equation 4

$$k_A = \frac{S_{std}}{C_{std}} \quad 5$$

gives the value for *k*A. Having determined the value for *k*A, we can calculate the concentration of analyte in any sample by measuring its signal, Ssamp, and calculating CA using equation 6

$$C_A = \frac{S_{samp}}{k_A} \quad 6$$

A single-point standardization is the least desirable method for standardizing a method. There are at least two reasons for this. First, any error in our determination of k_A carries over into our calculation of C_A. Second, our experimental value for k_A is for a single concentration of analyte. Extending this value of k_A to other concentrations of analyte requires us to assume a linear relationship between the signal and the analyte's concentration, an assumption that often is not true.5 Figure 5.2 shows how assuming a constant value of k_A may lead to a determinate error in the analyte's concentration. Despite these limitations, single-point standardizations find routine use when the expected range for the analyte's concentrations is small. Under these conditions it is often safe to assume that k_A is constant (although you should verify this assumption experimentally). This is the case, for example, in clinical labs where many automated analyzers use only a single standard.

The preferred approach to standardizing a method is to prepare a series of standards, each containing the analyte at a different concentration.

Standards are chosen such that they bracket the expected range for the analyte's concentration. A multiple-point standardization should include at least three standards, although more are referable. A plot of S_{std} versus C_{std} is known as a calibration curve. The exact standardization, or calibration relationship is determined by an appropriate curve-fitting algorithm.

There are at least two advantages to a multiple-point standardization. First, although a determinate error in one standard introduces a determinate error into the analysis, its effect is minimized by the remaining standards. Second, by measuring the signal for several concentrations of analyte we no longer must assume that the value of k_A is independent of the analyte's concentration. Constructing a calibration curve similar to the "actual relationship" in Figure 2, is possible.

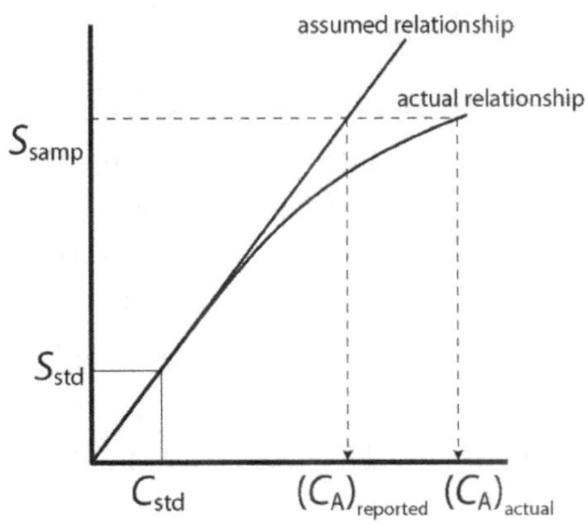

Fig. 2: Example showing how a single-point standardization leads to a determinate error in an analyte's reported concentration if we incorrectly assume that the value of k_A is constant.

7.2.7 External Standards

The most common method of standardization uses one or more external standards, each containing a known concentration of analyte. We call them "external" because we prepare and analyze the standards separate from the samples.

7.2.7.1 Single External Standard

A quantitative determination using a single external standard was described at the beginning of this section, with k_A given by equation 5. After determining the value of k_A, the concentration of analyte, C_A, is calculated using equation 6

Example 1

A spectrophotometric method for the quantitative analysis of Pb^{2+} in blood yields an S_{std} of 0.474 for a single standard whose concentration of lead is 1.75 ppb What is the concentration of Pb^{2+} in a sample of blood for which S_{samp} is 0.361?

SOLUTION

Equation 5.5 allows us to calculate the value of k_A for this method using the data for the standard.

$$k_A = \frac{S_{std}}{C_{std}} = \frac{0.474}{1.75 \text{ ppb}} = 0.2709 \text{ ppb}^{-1}$$

Having determined the value of k_A, the concentration of Pb^{2+} in the sample of blood is calculated using equation 6.

$$C_A = \frac{S_{samp}}{k_A} = \frac{0.361}{0.2709 \text{ ppb}^{-1}} = 1.33 \text{ ppb}$$

7.2.7.2 Multiple External Standards

Figure 3 shows a typical multiple-point external standardization. The volumetric flask on the left is a reagent blank and the remaining volumetric flasks contain increasing concentrations of Cu^{2+}. Shown below the volumetric flasks is the resulting calibration curve. Because this is the most common method of standardization the resulting relationship is called a normal calibration curve. When a calibration curve is a straight line, as it is in Figure 3, the slope of the line gives the value of kA. This is the most desirable situation since the method's sensitivity remains constant throughout the analyte's concentration range. When the calibration curve is not a straight-line, the method's sensitivity is a function of the analyte's concentration. In Figure 2, for example, the value of kA is greatest when the analyte's concentration is small and decreases continuously for higher concentrations of analyte.

The value of k_A at any point along the calibration curve in Figure 2 is given by the slope at that point. In either case, the calibration curve provides a means for relating S_{samp} to the analyte's concentration.

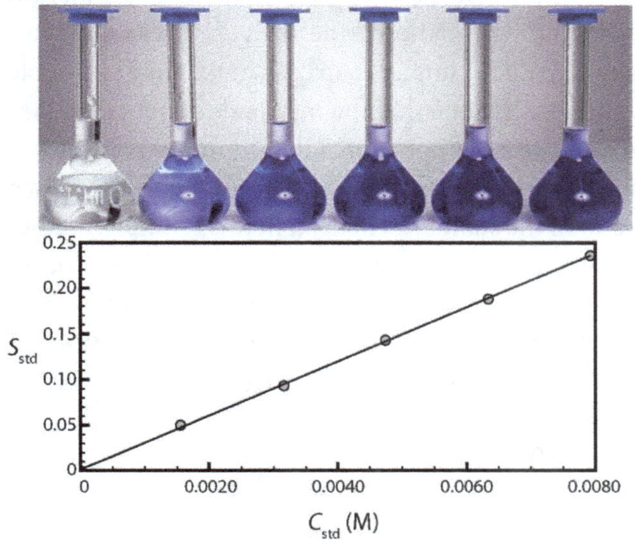

Fig. 3: Shown at the top is a reagent blank (far left) and a set of five external standards for Cu^{2+} with concentrations increasing from left to right. Shown below the external standards is the resulting normal calibration curve. The absorbance of each standard, S_{std}, is shown by the filled circles.

A second spectrophotometric method for the quantitative analysis of Pb^{2+} in blood has a normal calibration curve for which

$$S_{std} = \left(0.296\ \text{ppb}^{-1}\right) \times C_{std} + 0.003$$

What is the concentration of Pb^{2+} in a sample of blood if S_{samp} is 0.397?

SOLUTION

To determine the concentration of Pb^{2+} in the sample of blood we replace S_{std} in the calibration equation with Ssamp and solve for C_A.

$$C_A = \frac{S_{samp} - 0.003}{0.296\ \text{ppb}^{-1}} = \frac{0.397 - 0.003}{0.296\ \text{ppb}^{-1}} = 1.33\ \text{ppb}$$

It is worth noting that the calibration equation in this problem includes an extra term that does not appear in equation 6. Ideally we expect the calibration curve to have a signal of zero when CA is zero. This is the purpose of using a reagent blank to correct the measured signal. The extra term of +0.003 in our calibration equation results from the uncertainty in measuring the signal for the reagent blank and the standards.

An external standardization allows us to analyze a series of samples using a single calibration curve. This is an important advantage when we have many samples to analyze.

Not surprisingly, many of the most common quantitative analytical methods use an external standardization.

There is a serious limitation, however, to an external standardization. When we determine the value of k_A using equation 5, the analyte is present in the external standard's matrix, which usually is a much simpler matrix than that of our samples. When using an external standardization we assume that the matrix does not affect the value of k_A. If this is not true, then we introduce a proportional determinate error into our analysis. This is not the case in Figure 4, for instance, where we show calibration curves for the analyte in the sample's matrix and in the standard's matrix. In this example, a calibration curve using external standards results in a negative determinate error. If we expect that matrix effects are important, then we try to match the standard's matrix to that of the sample. This is known as matrix matching. If we are unsure of the sample's matrix, then we must show that matrix effects are negligible, or use an alternative method of standardization. Both approaches are discussed in the following section.

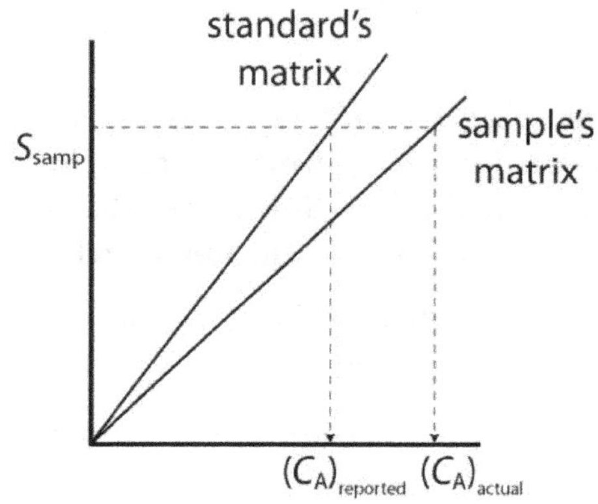

Fig. 4: Calibration curves for an analyte in the standard's matrix and in the sample's matrix. If the matrix affects the value of k_A, as is the case here, then we introduce a determinate error into our analysis if we use a normal calibration curve.

7.2.8 Standard Additions

We can avoid the complication of matching the matrix of the standards to the matrix of the sample by conducting the standardization in the sample. This is known as the method of standard additions.

7.2.8.1 Single Standard Addition

The simplest version of a standard addition is shown in Figure 5. First we add a portion of the sample, Vo, to a volumetric flask, dilute it to volume, Vf, and measure its signal, Ssamp. Next, we add a second identical portion of sample to an equivalent volumetric flask along with a spike, Vstd, of an external standard whose concentration is Cstd. After diluting the spiked sample to the same final volume, we measure its signal, Sspike. The following two equations relate Ssamp and Sspike to the concentration of analyte, CA, in the original sample.

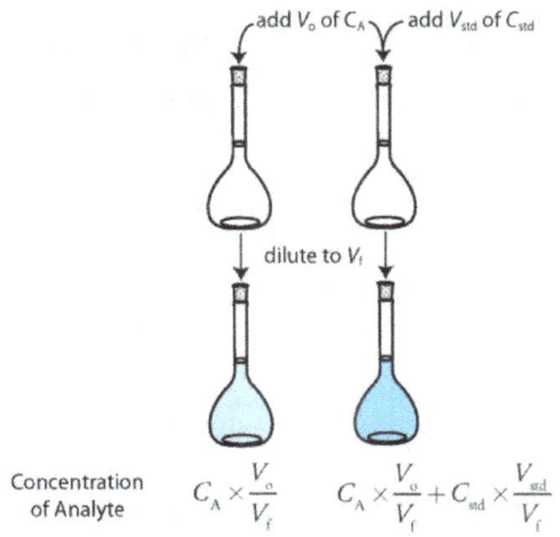

Fig. 5: Illustration showing the method of standard additions. The volumetric flask on the left contains a portion of the sample, Vo, and the volumetric flask on the right contains an identical portion of the sample and a spike, V_{std}, of a standard solution of the analyte. Both flasks are diluted to the same final volume, V_f. The concentration of analyte in each flask is shown at the bottom of the figure where C_A is the analyte's concentration in the original sample and C_{std} is the concentration of analyte in the external standard

$$S_{samp} = k_A C_A \frac{V_o}{V_f} \qquad 5.7 \qquad 7$$

$$S_{spike} = k_A \left(C_A \frac{V_o}{V_f} + C_{std} \frac{V_{std}}{V_f} \right) \qquad 5.8 \qquad 8$$

As long as V_{std} is small relative to V_o, the effect of the standard's matrix on the sample's matrix is insignificant. Under these conditions the value of k_A is the same in equation 7 and equation 8. Solving both equations for k_A and equating gives

$$\frac{S_{samp}}{C_A \frac{V_o}{V_f}} = \frac{S_{spike}}{C_A \frac{V_o}{V_f} + C_{std} \frac{V_{std}}{V_f}} \qquad 5.9 \qquad 9$$

which we can solve for the concentration of analyte, C_A, in the original sample.

Example 3

A third spectrophotometric method for the quantitative analysis of Pb^{2+} in blood yields an S_{samp} of 0.193 when a 1.00 mL sample of blood is diluted to 5.00 mL. A second 1.00 mL sample of blood is spiked with 1.00 μL of a 1560-ppb Pb^{2+} external standard and diluted to 5.00 mL, yielding an Sspike of 0.419. What is the concentration of Pb^{2+} in the original sample of blood?

SOLUTION

We begin by making appropriate substitutions into equation 9 and solving for C_A. Note that all volumes must be in the same units; thus, we first covert *V*std from 1.00 μL to 1.00 × 10⁻³ mL.

$$\frac{0.193}{C_A \frac{1.00 \text{ mL}}{5.00 \text{ mL}}} = \frac{0.419}{C_A \frac{1.00 \text{ mL}}{5.00 \text{ mL}} + 1560 \text{ ppb} \frac{1.00 \times 10^{-3} \text{ mL}}{5.00 \text{ mL}}}$$

$$\frac{0.193}{0.200 C_A} = \frac{0.419}{0.200 C_A + 0.3120 \text{ ppb}}$$

$$0.0386 C_A + 0.0602 \text{ ppb} = 0.0838 C_A$$

$$0.0452 C_A = 0.0602 \text{ ppb}$$

$$C_A = 1.33 \text{ ppb}$$

The concentration of Pb^{2+} in the original sample of blood is 1.33 ppb. It also is possible to make a standard addition directly to the sample, measuring the signal both before and after the spike (Figure 6). In this case the final volume after the standard addition is $V_o + V_{std}$ and equation 7, equation 8, and equation 9 become

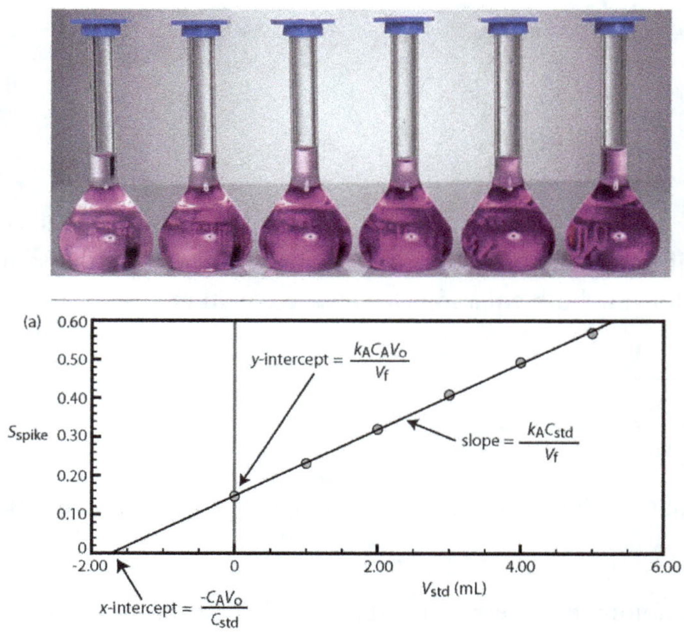

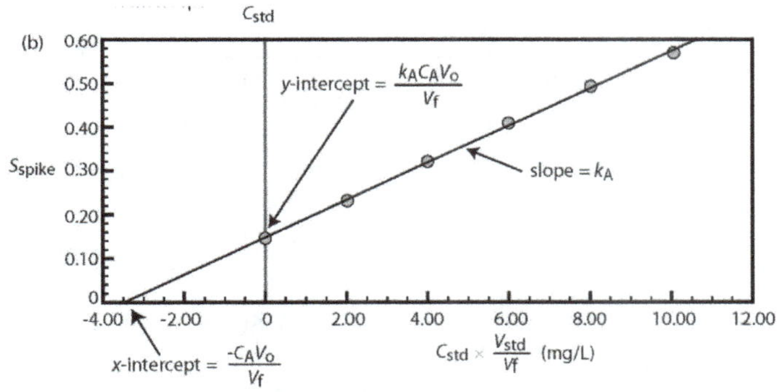

Fig. 6: Illustration showing an alternative form of the method of standard additions. In this case we add a spike of the external standard directly to the sample without any further adjust in the volume.

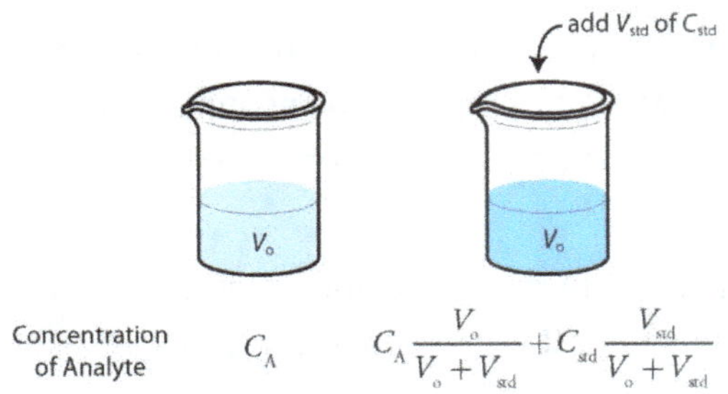

$$S_{samp} = k_A C_A$$

$$S_{spike} = k_A \left(C_A \frac{V_o}{V_o + V_{std}} + C_{std} \frac{V_{std}}{V_o + V_{std}} \right) \quad 10$$

$$\frac{S_{samp}}{C_A} = \frac{S_{spike}}{C_A \frac{V_o}{V_o + V_{std}} + C_{std} \frac{V_{std}}{V_o + V_{std}}} \quad 11$$

Example 5

A fourth spectrophotometric method for the quantitative analysis of Pb2+ in blood yields an Ssamp of 0.712 for a 5.00 mL sample of blood. After spiking the blood sample with 5.00 μL of a 1560-ppb Pb^{2+} external standard, an S_{spike} of 1.546 is measured. What is the concentration of Pb^{2+} in the original sample of blood.

SOLUTION

To determine the concentration of Pb^{2+} in the original sample of blood, we make appropriate substitutions into equation 11 and solve for CA.

$$\frac{0.712}{C_A} = \frac{1.546}{C_A \dfrac{5.00 \text{ mL}}{5.005 \text{ mL}} + 1560 \text{ ppb} \dfrac{5.00 \times 10^{-3} \text{ mL}}{5.005 \text{ mL}}}$$

$$\frac{0.712}{C_A} = \frac{1.546}{0.9990 C_A + 1.558 \text{ ppb}}$$

$$0.7113 C_A + 1.109 \text{ ppb} = 1.546 C_A$$

$$C_A = 1.33 \text{ ppb}$$

The concentration of Pb^{2+} in the original sample of blood is 1.33 ppb.

7.2.8.2 Multiple Standard Additions

We can adapt the single-point standard addition into a multiple-point standard addition by preparing a series of samples containing increasing amounts of the external standard. Figure 7 shows two ways to plot a standard addition calibration curve based on equation 8. In Figure 7a

We plot *S*spike against the volume of the spikes, *V*std. If *k*A is constant, then the calibration curve is a straight-line. It is easy to show that the *x*-intercept is equivalent to $-C_A V_o / C_{std}$.

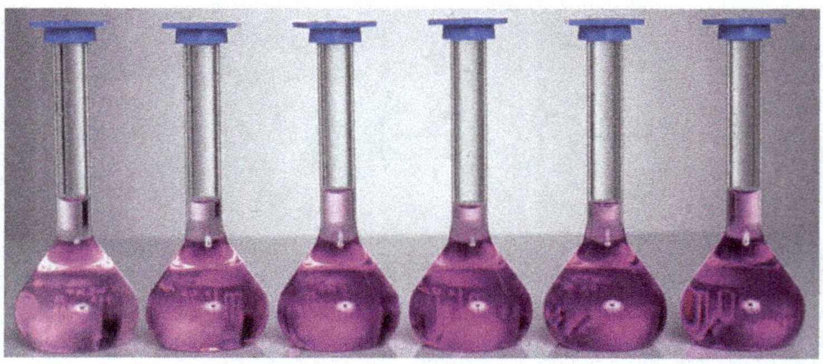

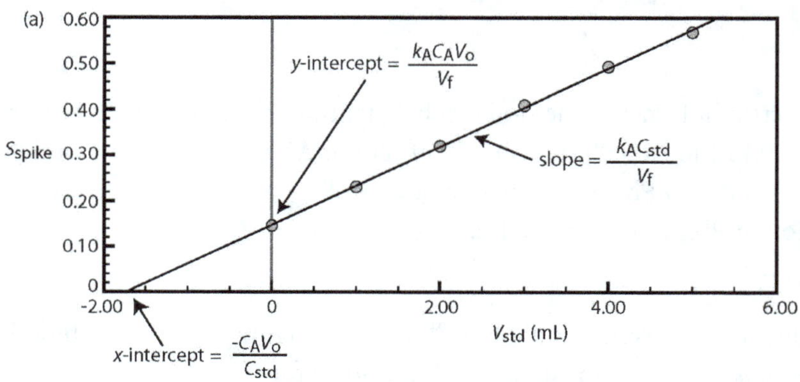

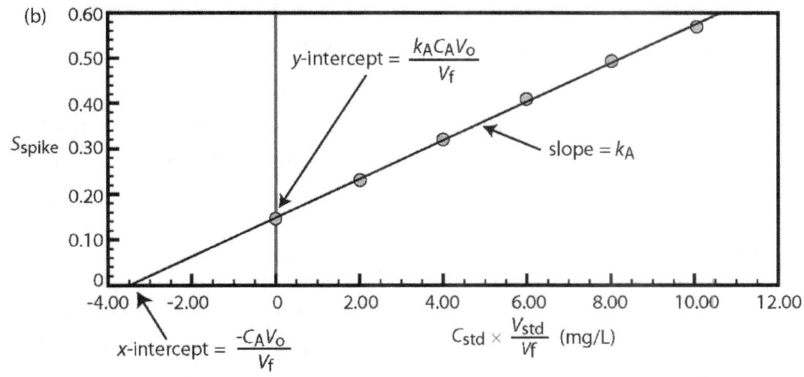

Fig. 7: Shown at the top is a set of six standard additions for the determination of Mn^{2+}. The flask on the left is a 25.00 mL sample diluted to 50.00 mL. The remaining flasks contain 25.00 mL of sample and, from left to right, 1.00, 2.00, 3.00, 4.00, and 5.00 mL of an external standard of 100.6 mg/L Mn^{2+}. Shown below are two ways to plot the standard additions calibration curve. The absorbance for each standard addition, S_{spike}, is shown by the filled circles.

Because we know the volume of the original sample, Vo, and the concentration of the external standard, Cstd, we can calculate the analyte's concentrations from the x-intercept of a multiple-point standard additions.

Example 6

A fifth spectrophotometric method for the quantitative analysis of Pb^{2+} in blood uses a multiple-point standard addition based on equation 8. The original blood sample has a volume of 1.00 mL and the standard used for spiking the sample has a concentration of 1560 ppb Pb^{2+}. All samples were diluted to 5.00 mL before measuring the signal. A calibration curve of *S*spike versus *V*std has the following equation

$$S_{spike} = 0.266 + 312 \text{ mL}-1 \times V_{std}$$

What is the concentration of Pb2+ in the original sample of blood.

SOLUTION

To find the *x*-intercept we set *S*spike equal to zero.

$$0 = 0.266 + 312 \text{ mL}^{-1} \times V_{std}$$

Solving for V_{std}, we obtain a value of -8.526×10^{-4} mL for the *x* - intercept. Substituting the *x*-interecpt's value into the equation from Figure 7a and solving for *CA* gives the concentration of Pb2+ in the blood sample as 1.33 ppb.

$$-8.526 \times 10^{-4} \text{ mL} = -\frac{C_A V_o}{C_{std}} = -\frac{C_A \times 1.00 \text{ mL}}{1560 \text{ ppb}}$$

Since we construct a standard additions calibration curve in the sample, we can not use the calibration equation for other samples. Each sample, therefore, requires its own standard additions calibration curve. This is a serious drawback if you have many samples. For example, suppose you need to analyze 10 samples using a three-point calibration curve. For a normal calibration curve you need to analyze only 13 solutions (three standards and ten samples). If you use the method of standard additions, however, you must analyze 30 solutions (each of the ten samples must be analyzed three times, once before spiking and after each of two spikes).

7.2.8.3 Using A Standard Addition to Identify Matrix Effects

We can use the method of standard additions to validate an external standardization when matrix matching is not feasible. First, we prepare a normal calibration curve of S_{std} versus C_{std} and determine the value of *k*A from its slope. Next, we prepare a standard additions calibration curve using equation 8, plotting the data as shown in Figure 7b. The slope of this standard additions calibration curve provides an independent determination of k_A. If there is no significant difference between the two values of *k*A, then we can ignore the difference between the sample's matrix and that of the external standards. When the values of *k*A are significantly different, then using a normal calibration curve introduces a proportional determinate error.

7.2.9 Internal Standards

To successfully use an external standardization or the method of standard additions, we must be able to treat identically all samples and standards. When this is not possible, the accuracy and precision of our standardization may suffer. For example, if our analyte is in a volatile solvent, then its concentration increases when we lose solvent to evaporation. Suppose we have a sample and a standard with identical concentrations of analyte and identical signals. If both experience the same proportional loss of solvent then their respective concentrations of analyte and signals continue to be identical. In effect, we can ignore evaporation if the samples and standards experience an equivalent loss of solvent. If an identical standard and sample lose different amounts of solvent, however, then their respective concentrations and signals will no longer be equal. In this case a simple external standardization or standard addition is not possible.

We can still complete a standardization if we reference the analyte's signal to a signal from another species that we add to all samples and standards. The species, which we call an internal standard, must be different than the analyte.

Because the analyte and the internal standard in any sample or standard receive the same treatment, the ratio of their signals is unaffected by any lack of reproducibility in the procedure. If a solution contains an analyte of concentration *C*A, and an internal standard of concentration, *C*IS, then the signals due to the analyte, *S*A, and the internal standard, *S*IS, are

$$S_a = k_A C_A$$

$$S_{IS} = k_{IS}C_{IS}$$

where k_A and k_{IS} are the sensitivities for the analyte and internal standard. Taking the ratio of the two signals gives the fundamental equation for an internal standardization.

$$\frac{S_A}{S_{IS}} = \frac{k_A C_A}{k_{IS} C_{IS}} = K \times \frac{C_A}{C_{IS}} \qquad 12$$

Because *K* is a ratio of the analyte's sensitivity and the internal standard's sensitivity, it is not necessary to independently determine values for either *k*A or *k*IS.

7.2.9.1 Single Internal Standard

In a single-point internal standardization, we prepare a single standard containing the analyte and the internal standard, and use it to determine the value of *K* in equation 12.

$$K = \left(\frac{C_{IS}}{C_A}\right)_{std} \times \left(\frac{S_A}{S_{IS}}\right)_{std} \qquad 13$$

Having standardized the method, the analyte's concentration is given by

$$C_A = \frac{C_{IS}}{K} \times \left(\frac{S_A}{S_{IS}}\right)_{samp}$$

Example 7

A sixth spectrophotometric method for the quantitative analysis of Pb^{2+} in blood use Cu^{2+} as an internal standard. A standard containing 1.75 ppb Pb^{2+} and 2.25 ppb Cu^{2+} yields a ratio of $(S_A/S_{IS})_{std}$ of 2.37. A sample of blood is spiked with the same concentration of Cu^{2+} giving a signal ratio, $(S_A/S_{IS})_{samp,}$ of 1.80. Determine the concentration of Pb2+ in the sample of blood.

SOLUTION

Equation allows us to calculate the value of K using the data for the standard

$$K = \left(\frac{C_{IS}}{C_A}\right)_{std} \times \left(\frac{S_A}{S_{IS}}\right)_{std} = \frac{2.25 \text{ ppb Cu}}{1.75 \text{ ppb Pb}^{2+}} \times 2.37 = 3.05 \frac{\text{ppb Cu}^{2+}}{\text{ppb Pb}^{2+}}$$

The concentration of Pb^{2+}, therefore, is

$$C_A = \frac{C_{IS}}{K} \times \left(\frac{S_A}{S_{IS}}\right)_{samp} = \frac{2.25 \text{ ppb Cu}^{2+}}{3.05 \frac{\text{ppb Cu}^{2+}}{\text{ppb Pb}^{2+}}} \times 1.80 = 1.33 \text{ ppb Cu}^{2+}$$

7.2.9.2 Multiple Internal Standards

A single-point internal standardization has the same limitations as a singlepoint normal calibration. To construct an internal standard calibration curve we prepare a series of standards, each containing the same concentration of internal standard and a different concentrations of analyte. Under these conditions a calibration curve of (*SA*/*SIS*)std versus *CA* is linear with a slope of *K*/*CIS*.

Example 7

A seventh spectrophotometric method for the quantitative analysis of Pb^{2+} in blood gives a liner internal standards calibration curve for which in blood gives a linear internal standards calibration curve for which

$$\left(\frac{S_A}{S_{IS}}\right)_{std} = (2.11\ \text{ppb}^{-1}) \times C_A - 0.006$$

What is the ppb Pb^{2+} in a sample of blood if $(S_A/S_{IS})_{samp}$ is 2.80?

SOLUTION

To determine the concentration of Pb^{2+} in the sample of blood we replace $(S_A/S_{IS})_{std}$ in the calibration equation with $(S_A/S_{IS})_{samp}$ and solve for C_A.

$$C_A = \frac{\left(\frac{S_A}{S_{IS}}\right)_{samp} + 0.006}{2.11\ \text{ppb}^{-1}} = \frac{2.80 + 0.006}{2.11\ \text{ppb}^{-1}} = 1.33\ \text{ppb}$$

The concentration of Pb^{2+} in the sample of blood is 1.33ppb.

In some circumstances it is not possible to prepare the standards so that each contains the same concentration of internal standard. This is the case, for example, when preparing samples by mass instead of volume. We can still prepare a calibration curve, however, by plotting (SA/SIS)std versus CA/ CIS, giving a linear calibration curve with a slope of K.

7.3 LINEAR REGRESSION AND CALIBRATION CURVES

In a single-point external standardization we determine the value of kA by measuring the signal for a single standard containing a known concentration of analyte. Using this value of kA and the signal for our sample, we then calculate the concentration of analyte in our sample (see Example 5.1). With only a single determination of kA, a quantitative analysis using a single-point external standardization is straightforward.

A multiple-point standardization presents a more difficult problem. Consider the data in Table 5.1 for a multiple-point external standardization. What is our best estimate of the relationship between Sstd and Cstd? It is tempting to treat this data as five separate single-point standardizations, determining kA for each standard, and reporting the mean value. Despite it simplicity, this is not an appropriate way to treat a multiple-point standardization.

So why is it inappropriate to calculate an average value for kA as done in Table .1? In a single-point standardization we assume that our reagent blank (the first row in Table .1) corrects for all constant sources of deter-minate error. If this is not the case, then the value of kA from a single-point standardization has a determinate error. Table .2 demonstrates how an mean value for kA = 122.5

Table 1 Data for a Hypothetical Multiple-Point ExternalStandardization

C_{std} (arbitrary units)	S_{std} (arbitrary units)	$k_A = S_{std}/C_{std}$
0.000	0.00	—
0.100	12.36	123.6
0.200	24.83	124.2
0.300	35.91	119.7
0.400	48.79	122.0
0.500	60.42	122.8

Table 2 Effect of a Constant Determinate Error on the Value of k_A From a Single-Point Standardization 5.2 Effect of a Constant Determinate Error on the Value of *k*A From a Single-Point Standardization

C_{std}	(Without constant Error)	(Actual)	(With constant error)	(Apparent)
1.00	1.00	1.00	1.50	1.50
2.00	2.00	1.00	2.50	1.25
3.00	3.00	1.00	3.50	1.17
4.00	4.00	1.00	4.50	1.13
5.00	5.00	1.00	5.50	1.10
	mean kA (true) =	1.00	mean kA (apparent) =	1.23

Uncorrected constant error affects our determination of kA. The first three columns show the concentration of analyte in the standards, Cstd, the signal without any source of constant error, Sstd, and the actual value of kA for five standards. As we expect, the value of kA is the same for each standard. In the fourth column we add a constant determinate error of +0.50 to the signals, (Sstd)e. The last column contains the corresponding apparent values of kA. Note that we obtain a different value of kA for each standard and that all of the apparent kA values are greater than the true value.

How do we find the best estimate for the relationship between the signal and the concentration of analyte in a multiple-point standardization? Figure 8 shows the data in Table 5.1 plotted as a normal calibration curve. Although the data certainly appear to fall along a straight line, the actual calibration curve is not intuitively obvious. The process of mathematically determining the best equation for the calibration curve is called linear regression.

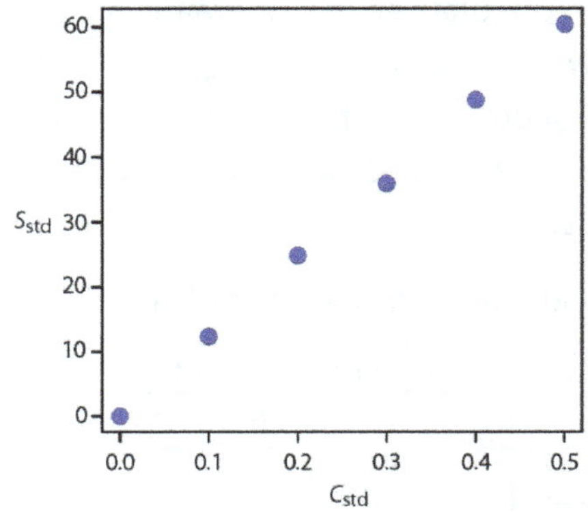

Fig. 8: Normal calibration curve for the hypothetical multiple-point external standardization in Table 1.

7.3.1 Linear Regression of Straight Line Calibration Curves

When a calibration curve is a straight-line, we represent it using the follow- ing mathematical equation

$$y = \beta_0 + \beta_1 x \qquad 14$$

where y is the signal, S_{std}, and x is the analyte's concentration, C_{std}. The constants β_0 and β_1 are, respectively, the calibration curve's expected y-in- tercept and its expected slope. Because of uncertainty in our measurements, the best we can do is to estimate values for β_0 and β_1, which we represent as β_0 and β_1. The goal of a lINEAR REGRESSION analysis is to determine the best estimates for b0 and b1. How we do this depends on the uncertainty in our measurements.

7.3.2 Unweighted Linear Regression with Errors in y

The most common approach to completing a linear regression for equation 14 makes three assumptions:

1. That any difference between our experimental data and the calculated regression line is the result of indeterminate errors affecting y,
2. That indeterminate errors affecting y are normally distributed, and
3. That the indeterminate errors in y are independent of the value of x.

Because we assume that the indeterminate errors are the same for all standards, each standard contributes equally in estimating the slope and the y-intercept. For this reason the result is considered an UNWEIGHTED LINEAR REGRESSION.

The validity of the two remaining assumptions is less obvious and you should evaluate them before accepting the results of a linear regression. In particular the first assumption is always suspect since there will certainly be some indeterminate errors affecting the values of x. When preparing a calibration curve, however, it is not unusual for the uncertainty in the signal, S_{std}, to be significantly larger than that for the concentration of analyte in the standards C_{std}. In such circumstances the first assumption is usually reasonable.

HOW A LINEAR REGRESSION WORKS

To understand the logic of an linear regression consider the example shown in Figure 9, which shows three data points and two possible straight-lines that might reasonably explain the data. How do we decide how well these straight-lines fits the data, and how do we determine the best straight- line?

Let's focus on the solid line in Figure 9. The equation for this line is

$$\hat{y} = b_0 + b_1x \qquad 15$$

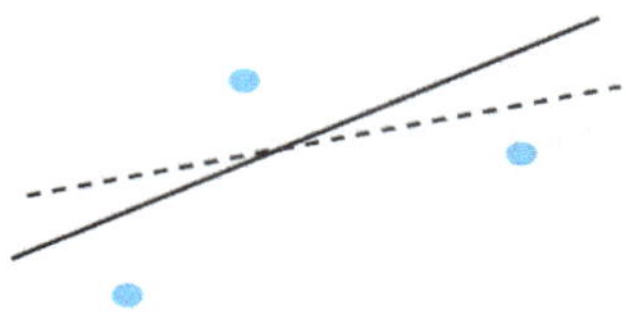

Fig. 9: Illustration showing three data points and two possible straight-lines that might explain the data. The goal of a linear regression is to find the mathematical model, in this case a straight-line, that best explains the data

Where b_0 and b_1 are our estimates for the *y*-intercept and the slope, and $\hat{y}$ is our prediction for the experimental value of *y* for any value of *x*. Because we assume that all uncertainty is the result of indeterminate errors affecting *y*, the difference between *y* and $\hat{y}$ for each data point is the residual error, *r*, in the our mathematical model for a particular value of *x*.

$$r_i = (y_i - \hat{y}_i)$$

Fig. 10: shows the residual errors for the three data points. The smaller the total residual error, *R*, which we define as

$$R = \sum_i (y_i - \hat{y}_i)^2 \qquad 16$$

The better the fit between the straight-line and the data. In a linear regression analysis, we seek values of *b*0 and *b*1 that give the smallest total residual error.

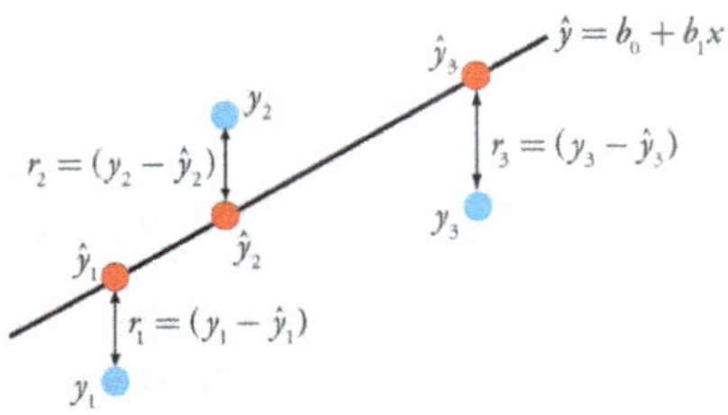

Fig. 11: Illustration showing the evaluation of a linear regression in which we assume that all uncertainty is the result of indeterminate errors affecting *y*. The points in blue, *yi*, are the original data and the points in red, ˆ*yi* , are the predicted values from the regression equation, *y*ˆ =*b* +*b x* 0 1.

The smaller thetotal residual error (equation 5.16), the better the fit of the straight-line to the data.

7.4 FINDING THE SLOPE AND Y-INTERCEPT

Although we will not formally develop the mathematical equations for a linear regression analysis, you can find the derivations in many standard statistical texts.6 The resulting equation for the slope, b1, is

$$b_1 = \frac{n\sum_i x_i y_i - \sum_i x_i \sum_i y_i}{n\sum_i x_i^2 - \left(\sum_i x_i\right)^2} \quad 17$$

and the equation for the y-intercept, b0, is

$$b_0 = \frac{\sum_i y_i - b_1 \sum_i x_i}{n} \quad 18$$

Although equation 5.17 and equation 5.18 appear formidable, it is only necessary to evaluate the following four summations

$$\sum_i x_i \quad \sum_i y_i \quad \sum_i x_i y_i \quad \sum_i x_i^2$$

Many calculators, spreadsheets, and other statistical software packages are capable of performing a linear regression analysis based on this model. To save time and to avoid tedious calculations, learn how to use one of these tools.

7.5 UNCERTAINTY IN THE REGRESSION ANALYSIS

As shown in Figure12, because of indeterminate error affecting our signal, the regression line may not pass through the exact center of each data point. The cumulative deviation of our data from the regression line—that is, the total residual error—is proportional to the uncertainty in the regression.

We call this uncertainty the standard deviation about the regression, sr, which is equal to

$$s_r = \sqrt{\frac{\sum_i \left(y_i - \hat{y}_i\right)^2}{n-2}} \quad 19$$

Where yi is the ith experimental value, and ˆyi is the corresponding value predicted by the regression line in equation 15. Note that the denominator of equation 19 indicates that our regression analysis has n^{-2} degrees of freedom—we lose two degree of freedom because we use two parameters, the slope and the y-intercept, to calculate ˆyi .

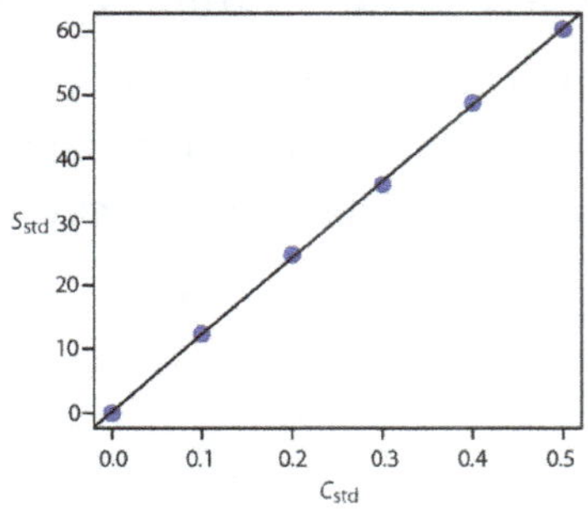

Fig. 12: Calibration curve for the data in Table 1 and Example 9.

A more useful representation of the uncertainty in our regression is to consider the effect of indeterminate errors on the slope, b_1, and the yintercept, b_0, which we express as standard deviations.

$$s_{b_1} = \sqrt{\frac{ns_r^2}{n\sum_i x_i^2 - \left(\sum_i x_i\right)^2}} = \sqrt{\frac{s_r^2}{\sum_i (x_i - \bar{x})^2}} \qquad 20$$

$$s_{b_0} = \sqrt{\frac{s_r^2\sum_i x_i^2}{n\sum_i x_i^2 - \left(\sum_i x_i\right)^2}} = \sqrt{\frac{s_r^2\sum_i x_i^2}{n\sum_i (x_i - \bar{x})^2}} \qquad 21$$

We use these standard deviations to establish confidence intervals for the expected slope, β_1, and the expected y-intercept, β_0

$$\beta_1 = b_1 \pm ts_b \qquad 22$$

$$\beta_0 = b_0 \pm ts_b \qquad 23$$

Where we select t for a significance level of α and for n^{-2} degrees of freedom. Note that equation 22 and equation 23 do not contain a factor of $(n)^{-1}$ because the confidence interval is based on a single regression line. Again, many calculators, spreadsheets, and computer software packages provide the standard deviations and confidence intervals for the slope and y-intercept. Example 10 illustrates the calculations.

7.5.1 Minimizing Uncertainty In Calibration Curves

To minimize the uncertainty in a calibration curve's slope and *y*-intercept, you should evenly space your standards over a wide range of analyte concentrations.

A close examination of equation 20 and equation 21 will help you appreciate why this is true. The denominators of both equations include the term $\Sigma (x_l - x)^2$. The larger the value of this term which you accomplish by increasing the range of *x* around its mean value—the smaller the standard deviations in the slope and the *y*-intercept. Furthermore, to minimize the uncertainty in the *y*-intercept, it also helps to decrease the value of the term Σx_i in equation 21, which you accomplish by including standards for lower concentrations of the analyte.

7.5.2 Obtaining the Analyte's Concentration from A Regression Equation

Once we have our regression equation, it is easy to determine the concentration of analyte in a sample. When using a normal calibration curve, for example, we measure the signal for our sample, S_{samp}, and calculate the analyte's concentration, C_A, using the regression equation.

$$C_A = \frac{S_{samp} - b_o}{b} \quad 24$$

What is less obvious is how to report a confidence interval for CA that expresses the uncertainty in our analysis. To calculate a confidence interval we need to know the standard deviation in the analyte's concentration, sCA

$$s_{C_A} = \frac{s_r}{b_1}\sqrt{\frac{1}{m} + \frac{1}{n} + \frac{\left(\bar{S}_{samp} - \bar{S}_{std}\right)^2}{(b_1)^2 \sum_i \left(C_{std_i} - \bar{C}_{std}\right)^2}} \quad 25$$

where m is the number of replicate used to establish the sample's average signal (S_{samp}), n is the number of calibration standards, S_{std} is the average signal for the calibration standards, and $C_{std i}$ and C_{std} are the individual and mean concentrations for the calibration standards.7 Knowing the value of s_{CA} , the confidence interval for the analyte's concentration is:

$$\mu_{CA} = CA = C \pm ts_{CA}$$

where μ_{CA} is the expected value of CA in the absence of determinate errors, and with the value of t based on the desired level of confidence and n–2 degrees of freedom.

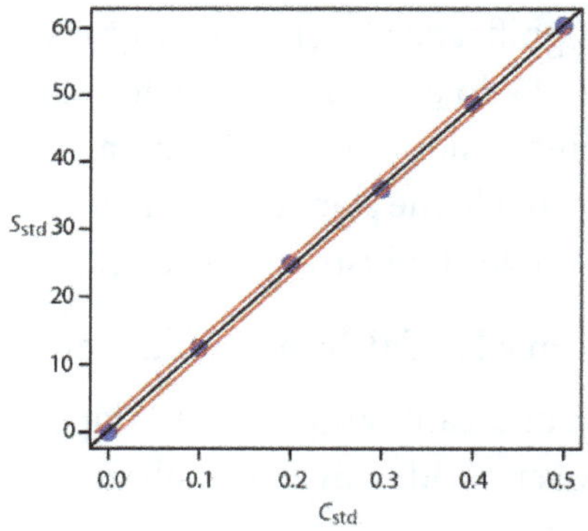

Fig. 13: Example of a normal calibration curve with a superimposed confidence interval for the analyte's concentration. The points in blue are the original data from Table 1. The black line is the normal calibration curve as determined in Example 9. The red lines show the 95% confidence interval for CA assuming a single determination of S_{samp}. In a standard addition we determine the analyte's concentration by extrapolating the calibration curve to the *x*-intercept. In this case the value of CA is:

$$C_A = x\text{-intercept} = \frac{-b_0}{b_1}$$

and the standard deviation in CA is

$$s_{C_A} = \frac{s_r}{b_1}\sqrt{\frac{1}{n} + \frac{(\bar{S}_{std})^2}{(b_1)^2 \sum_i (C_{std_i} - \bar{C}_{std})^2}}$$

where n is the number of standard additions (including the sample with no added standard), and Sstd is the average signal for the n standards. Because we determine the analyte's concentration by extrapolation, rather than by interpolation, sCA for the method of standard additions generally is larger than for a normal calibration curve.

7.5.3 Evaluating a Linear Regression Model

You should never accept the result of a linear regression analysis without evaluating the validity of the your model. Perhaps the simplest way to evaluate a regression analysis is to examine the residual errors. As we saw earlier, the residual error for a single calibration standard, r_i, is

$$r_i = (y_1 - \hat{y}_i)$$

If your regression model is valid, then the residual errors should be randomly distributed about an average residual error of zero, with no apparent trend toward either smaller or larger residual errors (Figure 5.13a). Trends such as those shown in Figure 13b and Figure 13c provide evidence that at least one of the model's assumptions is incorrect. For example, a trend toward larger residual errors at higher concentrations, as shown in Figure 13b, suggests that the indeterminate errors affecting the signal are not independent of the analyte's concentration. In Figure 13c, the residual errors are not random, suggesting that the data can not be modeled with a straight-line relationship. Regression methods for these two cases are discussed in the following sections.

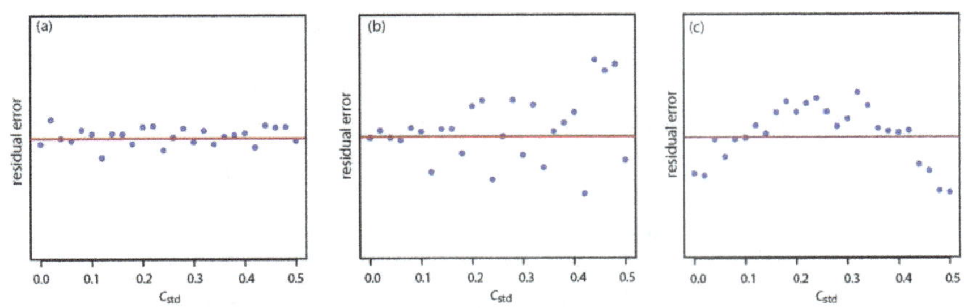

Fig. 14: Plot of the residual error in the signal, Sstd, as a function of the concentration of analyte, Cstd for an unweighted straight-line regression model. The red line shows a residual error of zero. The distribution of the residual error in (a) indicates that the unweighted linear regression model is appropriate. The increase in the residual errors in (b) for higher concentrations of analyte, suggest that a weighted straight-line regression

is more appropriate. For (c), the curved pattern to the residuals suggests that a straight-line model is inappropriate; linear regression using a quadratic model might produce a better fit.

7.5.4 Weighted Linear Regression with Errors in y

Our treatment of linear regression to this point assumes that indeterminate errors affecting *y* are independent of the value of *x*. If this assumption is false, as is the case for the data in Figure 13b, then we must include the variance for each value of *y* into our determination of the *y*-intercept, *b*o, and the slope, b_1; thus

$$b_0 = \frac{\sum_i w_i y_i - b_1 \sum_i w_i x_i}{n} \quad 26$$

$$b_1 = \frac{n\sum_i w_i x_i y_i - \sum_i w_i x_i \sum_i w_i y_i}{n\sum_i w_i x_i^2 - \left(\sum_i w_i x_i\right)^2} \quad 27$$

where *wi* is a weighting factor that accounts for the variance in *yi*

$$w_i = \frac{n\left(s_{y_i}\right)^{-2}}{\sum_i \left(s_{y_i}\right)^{-2}} \quad 28$$

and sy_i is the standard deviation for y_i. In a weighted linear regression,each x_y-pair's contribution to the regression line is inversely proportional to the precision of y_i—that is, the more precise the value of *y*, the greater its contribution to the regression.

Equations for calculating confidence intervals for the slope, the *y*-intercept, and the concentration of analyte when using a weighted linear regression are not as easy to define as for an unweighted linear regression. The confidence interval for the analyte's concentration, however, is at its

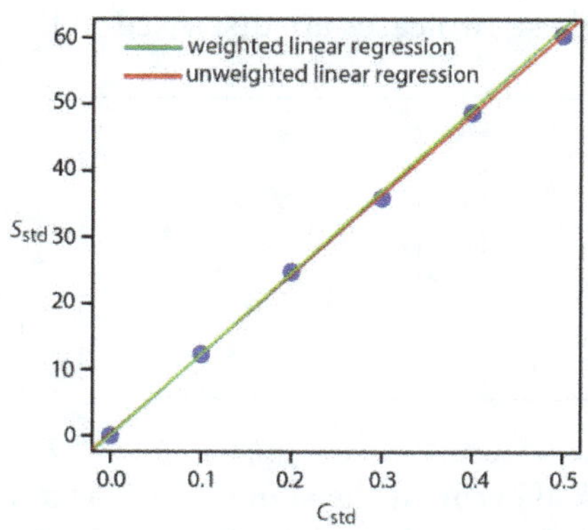

Fig. 15: A comparison of unweighted and weighted normal calibration curves. See Example 5.9 for details of the unweighted linear regression and Example 5.12 for details of the weighted linear regression.

Optimum value when the analyte's signal is near the weighted centroid, *yc* , of the calibration curve.

$$y_c = \frac{1}{n}\sum_i w_i x_i$$

7.5.5 Weighted Linear Regression with Errors in Both x and y

If we remove our assumption that the indeterminate errors affecting a calibration curve exist only in the signal (*y*), then we also must factor into the regression model the indeterminate errors affecting the analyte's concentration in the calibration standards (*x*). The solution for the resulting regression line is computationally more involved than that for either the unweighted or weighted regression lines. Although we will not consider the details in this textbook, you should be aware that neglecting the presence of indeterminate errors in *x* can bias the results of a linear regression.

7.5.6 Curvilinear and Multivariate Regression

A straight-line regression model, despite its apparent complexity, is the simplest functional relationship between two variables. What do we do if our calibration curve is curvilinear—that is, if it is a curved-line instead of a straight-line? One approach is to try transforming the data into a straight-line.

Logarithms, exponentials, reciprocals, square roots, and trigonometric functions have been used in this way. A plot of log(*y*) versus *x* is a typical example. Such transformations are not without complications. Perhaps the most obvious complication is that data with a uniform variance in *y* will not maintain that uniform variance after the transformation.

Another approach to developing a linear regression model is to fit a polynomial equation to the data, such as $y = a + b_x + c_x^2$. You can use linear regression to calculate the parameters *a*, *b*, and *c*, although the equations are different than those for the linear regression of a straight line. If you cannot fit your data using a single polynomial equation, it may be possible to fit separate polynomial equations to short segments of the calibration curve. The result is a single continuous calibration curve known as a spline function.

The regression models in this chapter apply only to functions containing a single independent variable, such as a signal that depends upon the analyte's concentration. In the presence of an interferent, however, the signal may depend on the concentrations of both the analyte and the interferent

$$S = k_A C_A + k_I C_I + S_{reag}$$

where *k*I is the interferent's sensitivity and CI is the interferent's concentration. Multivariate calibration curves can be prepared using standards that contain known amounts of both the analyte and the interferent, and modelled using multivariate regression.

7.5.7 Blank Corrections

Thus far in our discussion of strategies for standardizing analytical methods, we have assumed the use of a suitable reagent blank to correct for signals arising from sources other than the analyte. We did not, however ask an important question—"What constitutes an appropriate reagent blank?"

Surprisingly, the answer is not immediately obvious.

In one study, approximately 200 analytical chemists were asked to evaluate a data set consisting of a normal calibration curve, a separate analyte-free blank, and three samples of different size but drawn from the same source.

The first two columns in Table 3 shows a series of external standards and their corresponding signals. The normal calibration curve for the data is

$$S_{std} = 0.0750 \times w_{std} + 0.1250$$

where the y-intercept of 0.1250 is the calibration blank. A separate reagent blank gives the signal for an analyte-free sample.

In working up this data, the analytical chemists used at least four different approaches for correcting signals: (a) ignoring both the calibration blank, CB, and the reagent blank, RB, which clearly is incorrect; (b) using the calibration blank only; (c) using the reagent blank only; and (d) using both the calibration blank and the reagent blank. Table 4 shows the equations for calculating the analyte's concentration using each approach, along with the resulting concentration for the analyte in each sample.

Table 3 Data Used to Study the Blank in an Analytical Method

1.6667	0.2500	1	62.4746	0.8000
5.0000	0.5000	2	82.7915	1.0000
8.3333	0.7500	3	103.1085	1.2000
11.6667	0.8413			
18.1600	1.4870		reagent blank	0.1000
19.9333	1.6200			

That all four methods give a different result for the analyte's concentration underscores the importance of choosing a proper blank, but does not tell us which blank is correct. Because all four methods fail to predict the same concentration of analyte for each sample, none of these blank corrections properly accounts for an underlying constant source of determinate error. To correct for a constant method error, a blank must account for signals from any reagents and solvents used in the analysis, as well as any bias resulting from interactions between the analyte and the sample's matrix. Both the calibration blank and the reagent blank compensate for signals from reagents and solvents. Any difference in their values is due to indeterminate errors in preparing and analyzing the standards

Table 4 Equations and Resulting Concentrations of Analyte for Different Approaches to Correcting for the Blank

Approach for Correcting Signal	Equation	Concentration of Analyte in... Sample 1	Sample 2	Sample 3
ignore calibration and reagent blank	$C_A = \frac{W_A}{W_{samp}} = \frac{S_{samp}}{k_A W_{samp}}$	0.1707	0.1610	0.1552
use calibration blank only	$C_A = \frac{W_A}{W_{samp}} = \frac{S_{samp} - CB}{k_A W_{samp}}$	0.1441	0.1409	0.1390
use reagent blank only	$C_A = \frac{W_A}{W_{samp}} = \frac{S_{samp} - RB}{k_A W_{samp}}$	0.1494	0.1449	0.1422
use both calibration and reagent blank	$C_A = \frac{W_A}{W_{samp}} = \frac{S_{samp} - CB - RB}{k_A W_{samp}}$	0.1227	0.1248	0.1261
use total Youden blank	$C_A = \frac{W_A}{W_{samp}} = \frac{S_{samp} - TYB}{k_A W_{samp}}$	0.1313	0.1313	0.1313

Unfortunately, neither a calibration blank nor a reagent blank can correct for a bias resulting from an interaction between the analyte and the sample's matrix. To be effective, the blank must include both the sample's matrix and the analyte and, consequently, must be determined using the sample itself. One approach is to measure the signal for samples of different size, and to determine the regression line for a plot of S_{samp} versus the amount of sample. The resulting *y*-intercept gives the signal in the absence of sample, and is known as the total youden blank. This is the true blank correction. The regression line for the three samples in Table 3 is

$$S_{samp} = 0.009844 \times W_{samp} + 0.185$$

giving a true blank correction of 0.185. As shown by the last row of Table 4, using this value to correct Ssamp gives identical values for the concentration of analyte in all three samples.

The use of the total Youden blank is not common in analytical work, with most chemists relying on a calibration blank when using a calibration curve, and a reagent blank when using a single-point standardization.

As long we can ignore any constant bias due to interactions between the analyte and the sample's matrix, which is often the case, the accuracy of an analytical method will not suffer. It is a good idea, however, to check for constant sources of error before relying on either a calibration blank or a reagent blank.

7.6 USING EXCEL AND R FOR A REGRESSION ANALYSIS

Although the calculations in this chapter are relatively straightforward— consisting, as they do, mostly of summations—it can be quite tedious to work through problems using nothing more than a calculator. Both Excel and R include functions for completing a linear regression analysis and for visually evaluating the resulting model.

7.6.1 Excel

Let's use Excel to fit the following straight-line model to the data in Example 9.

$$y = \beta_0 + \beta_1 x$$

Enter the data into a spreadsheet, as shown in Figure 5.15. Depending upon your needs, there are many ways that you can use Excel to complete a linear regression analysis. We will consider three approaches here.

7.6.2 Use Excel's Built-In Functions

If all you need are values for the slope, β1, and the *y*-intercept, β0, you can use the following functions:

=intercept(*known_y's*, *known_x's*)

=slope(*known_y's*, *known_x's*)

where *known_y's* is the range of cells containing the signals (*y*), and *known_x's* is the range of cells containing the concentrations (*x*). For example, clicking on an empty cell and entering returns Excel's exact calculation for the slope (120.705 714 3).

=slope(B2:B7, A2:A7)

7.6.3 Use Excel's Data Analysis Tools

To obtain the slope and the y-intercept, along with additional statistical details, you can use the data analysis tools in the Analysis ToolPak. The ToolPak is not a standard part of Excel's instillation. To see if you have access to the Analysis ToolPak on your computer, select **Tools** from the menu bar and look for the **Data Analysis...** option. If you do not see **Data Analysis...**, select **Add-ins...** from the **Tools** menu. Check the box for the **Analysis ToolPak** and click on **OK** to install them.

Select **Data Analysis...** from the **Tools** menu, which opens the *Data Analysis* window. Scroll through the window, select **Regression** from the available options, and press **OK**. Place the cursor in the box for *Input Y range* and then click and drag over cells B1:B7. Place the cursor in the box for *Input X range* and click and drag over cells A1:A7. Because cells A1 and B1 contain labels, check the box for *Labels*. Select the radio button for *Output range* and click on any empty cell; this is where Excel will place the results.

There are three parts to Excel's summary of a regression analysis. At the top of Figure 16 is a table of *Regression Statistics*. The *standard error* is the standard deviation about the regression, *sr*. Also of interest is the value for *Multiple R*, which is the model's correlation coefficient, *r*, a term with which

you may already by familiar. The correlation coefficient is a measure of the extent to which the regression model explains the variation in *y*. Values of *r* range from –1 to +1. The closer the correlation coefficient is to ±1, the better the model is at explaining the data. A correlation coefficient of 0 means that there is no relationship between *x* and *y*.

In developing the calculations for linear regression, we did not consider the correlation coefficient. There is a reason for this. For most straight-line calibration curves the correlation coefficient will be very close to +1, typically 0.99 or better. There is a tendency, however, to put too much faith in the correlation coefficient's significance, and to assume that an r greater than 0.99 means the linear regression model is appropriate. Figure 16 provides a counterexample.

SUMMARY OUTPUT

Regression Statistics	
Multiple R	0.99987244
R Square	0.9997449
Adjusted R Square	0.99968113
Standard Error	0.40329713
Observations	6

ANOVA

	df	*SS*	*MS*	*F*	*Significance F*
Regression	1	2549.727156	2549.72716	15676.296	2.4405E-08
Residual	4	0.650594286	0.16264857		
Total	5	2550.37775			

	Coefficients	*Standard Error*	*t Stat*	*P-value*	*Lower 95%*	*Upper 95%*	*Lower 95.0%*	*Upper 95.0%*
Intercept	0.20857143	0.29188503	0.71456706	0.51436267	-0.60183133	1.01897419	-0.60183133	1.01897419
Cstd	120.705714	0.964064525	125.205016	2.4405E-08	118.029042	123.382387	118.029042	123.382387

Fig. 16: Output from Excel's Regression command in the Analysis Tool Pak. See the text for a discussion of how to interpret the information in these tables.

Although the regression line has a correlation coefficient of 0.993, the data clearly shows evidence of being curvilinear. The take-home lesson here is: don't fall in love with the correlation coefficient!

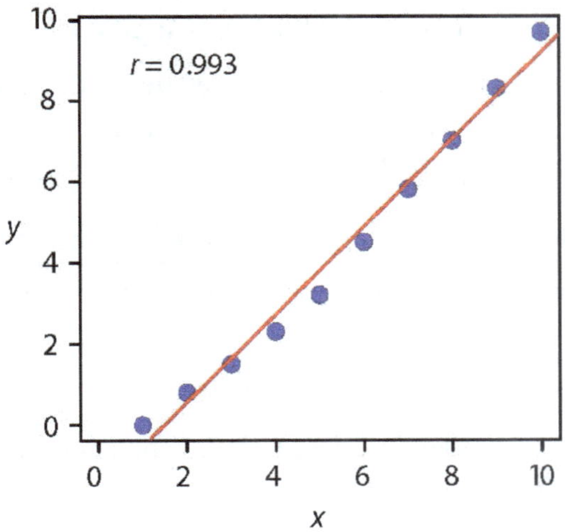

Fig. 17:Example of fitting a straight-line to curvilinear data.

Although the regression line has a correlation coefficient of 0.993, the data clearly shows evidence of being curvilinear. The take-home lesson here is: don't fall in love with the correlation coefficient!

The second table in Figure 16 is entitled *ANOVA*, which stands for analysis of variance. We will take a closer look at ANOVA in Chapter 14. For now, it is sufficient to understand that this part of Excel's summary provides information on whether the linear regression model explains a significant portion of the variation in the values of y. The value for *F* is the result of an *F*-test of the following null and alternative hypotheses.

*H*0: regression model does not explain the variation in *y*

HA: regression model does explain the variation in *y*

The value in the column for *Significance F* is the probability for retaining the null hypothesis. In this example, the probability is 2.5×10–6%, suggesting that there is strong evidence for accepting the regression model. As is the case with the correlation coefficient, a small value for the probability is a likely outcome for any calibration curve, even when the model is inappropriate.

The probability for retaining the null hypothesis for the data in Figure 5.17, for example, is

9.0 × 10 – 7%.

The third table in Figure 16 provides a summary of the model itself. The values for the model's coefficients—the slope, β1, and the *y*-intercept, β0—are identified as *intercept* and with your label for the *x*-axis data, which in this example is *Cstd*. The standard deviations for the coefficients, *s*b0 and *s*b1, are in the column labeled *Standard error*. The column *t Stat* and the column *P-value* are for the following *t*-tests.

slope *H*0: β1 = 0, *HA*: β1 ≠ 0

y-intercept *H*0: β0 = 0, *HA*: β0 ≠ 0

The results of these t-tests provide convincing evidence that the slope is not zero, but no evidence that the y-intercept significantly differs from zero. Also shown are the 95% confidence intervals for the slope and the y-intercept (lower 95% and upper 95%).

7.7 PROGRAM THE FORMULAS YOURSELF

A third approach to completing a regression analysis is to program a spreadsheet using Excel's built-in formula for a summation

=sum(*first cell:last cell*)

and its ability to parse mathematical equations. The resulting spreadsheet is shown in Figure 18.

C7, and cells D3 to D7. Next, enter the formulas for cells A9 to D9. Finally, enter the formulas in cells F2 and F3. When you enter a formula, Excel replaces it with the resulting calculation. The values in these cells should agree with the results in Example 9. You can simplify the entering of formulas by copying and pasting. For example, enter the formula in cell C2. Select **Edit: Copy**, click and drag your cursor over cells C3 to C7, and select **Edit: Paste**. Excel automatically updates the cell referencing.

	A	B	C	D	E	F
1	x	y	xy	x^2	n =	6
2	0.000	0.00	=A2*B2	=A2^2	slope =	=(F1*C8 - A8*B8)/(F1*D8-A8^2)
3	0.100	12.36	=A3*B3	=A3^2	y-int =	=(B8-F2*A8)/F1
4	0.200	24.83	=A4*B4	=A4^2		
5	0.300	35.91	=A5*B5	=A5^2		
6	0.400	48.79	=A6*B6	=A6^2		
7	0.500	60.42	=A7*B7	=A7^2		
8						
9	=sum(A2:A7)	=sum(B2:B7)	=sum(C2:C7)	=sum(D2:D7)	<--sums	

Fig. 18: Spreadsheet showing the formulas for calculating the slope and the y-intercept for the data in Example 9. The cells with the shading contain formulas that you must enter. Enter the formulas in cells C3 to

7.8 USING EXCEL TO VISUALIZE THE REGRESSION MODEL

You can use Excel to examine your data and the regression line. Begin by plotting the data. Organize your data in two columns, placing the *x* values in the left-most column. Click and drag over the data and select **Insert:**

Chart... from the main menu. This launches Excel's *Chart Wizard*. Select **xy-chart**, choosing the option without lines connecting the points. Click on **Next** and work your way through the screens, tailoring the plot to meet your needs. To add a regression line to the chart, click on the chart and

select **Chart: Add Trendline...** from the main men. Pick the straight-line model and click **OK** to add the line to your chart. By default, Excel displays the regression line from your first point to your last point. Figure 19 shows the result for the data in Figure 15.

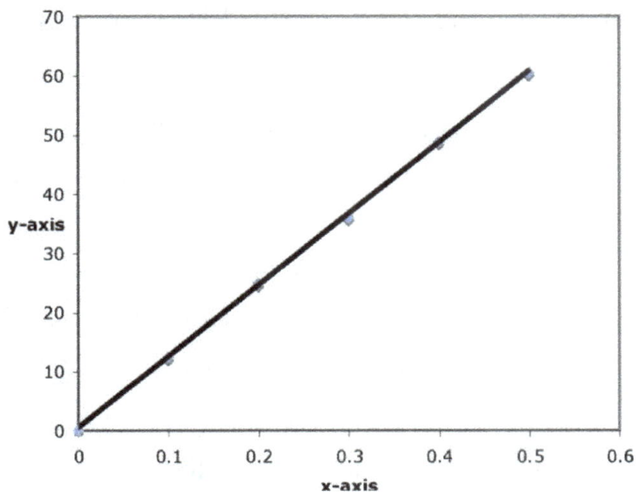

Fig. 19: Example of an Excel scatterplot showing the data and a regression line.

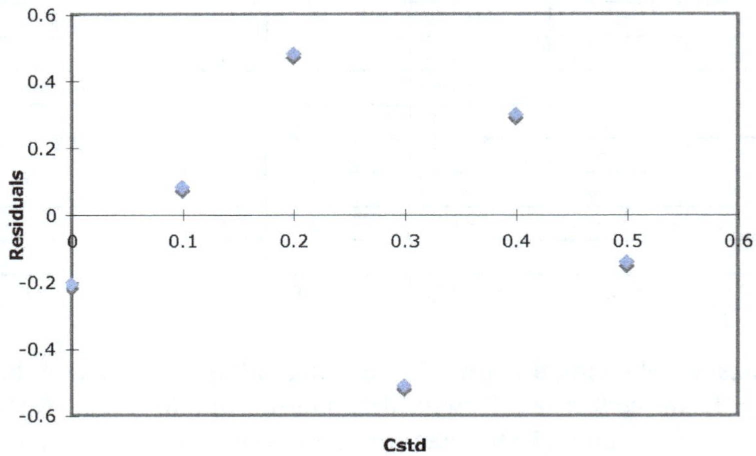

Fig. 20: Example of Excel's plot of a regression model's residual errors.

Excel also will create a plot of the regression model's residual errors. To create the plot, build the regression model using the Analysis ToolPak, as described earlier. Clicking on the option for *Residual plots* creates the plot shown in Figure20.

7.8.1 Limitations to Using Excel for A Regression Analysis

Excel's biggest limitation for a regression analysis is that it does not provide a function for calculating the uncertainty when predicting values of *x*. In terms of this chapter, Excel cannot calculate the uncertainty for the analyte's concentration, CA, given the signal for a sample, Ssamp. Another limitation is that Excel does not include a built-in function for a weighted linear regression. You can, however, program a spreadsheet to handle these calculations.

5F.2 R

Let's use Excel to fit the following straight-line model to the data in Example 9.

$$y = \beta_0 + \beta_1 x$$

7.8.2 Entering Data and Creating the Regression Model

To begin, create objects containing the concentration of the standards and their corresponding signals.

```
> conc = c(0, 0.1, 0.2, 0.3, 0.4, 0.5)
> signal = c(0, 12.36, 24.83, 35.91, 48.79, 60.42)
```

The command for creating a straight-line linear regression model is

```
lm(y ~ x)
```

where *y* and *x* are the objects containing our data. To access the results of the regression analysis, we assign them to an object using the following command

> model = lm(signal ~ conc) where *model* is the name we assign to the object.

7.8.3 Evaluating the Linear Regression Model

To evaluate the results of a linear regression we need to examine the data and the regression line, and to review a statistical summary of the model. To examine our data and the regression line, we use the **plot** command, which takes the following general form plot(*x*, *y*, optional arguments to control style) where *x* and *y* are objects containing our data, and the **abline** comm. and abline(*object*, optional arguments to control style) where *object* is the object containing the results of the linear regression.

Entering the commands creates the plot shown in Figure 21.

> plot(conc, signal, pch = 19, col = "blue", cex = 2)

> abline(model, col = "red")

To review a statistical summary of the regression model, we use the summary command.> summary(model)

The resulting output, shown in Figure 21, contains three sections. The first section of R's summary of the regression model lists the residual errors. To examine a plot of the residual errors, use the command > plot(model, which=1)

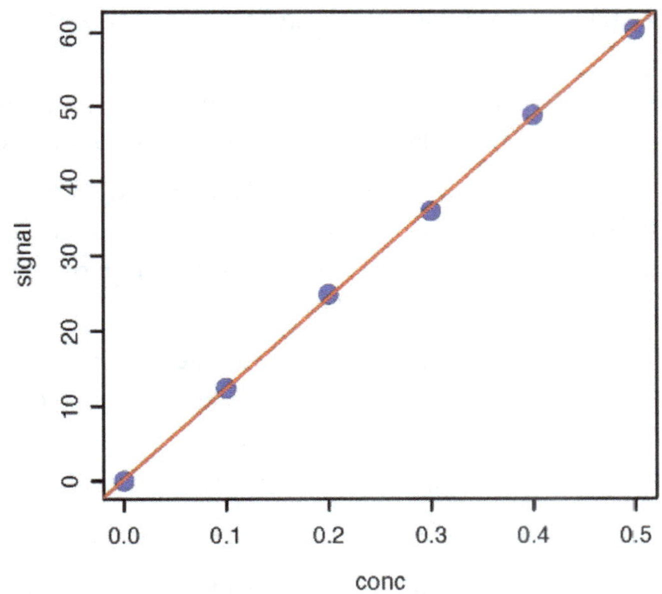

Fig. 21: Example of a regression plot in R showing the data and the regression line. You can customize your plot by adjusting the plot command's optional arguments. The argument *pch* controls the symbol used for plotting points, the argument *col* allows you to select a color for the points or the line, and the argument *cex* sets the size for the points. You can use the command help(plot) to learn more about the options for plotting data in R.

```
> model=lm(signal~conc)
> summary(model)

Call:
lm(formula = signal ~ conc)

Residuals:
       1        2        3        4        5        6
-0.20857  0.08086  0.48029 -0.51029  0.29914 -0.14143

Coefficients:
            Estimate   Std. Error   t value   Pr(>|t|)
(Intercept)   0.2086       0.2919     0.715      0.514
conc        120.7057       0.9641   125.205   2.44e-08 ***
---
Signif. codes: 0 '***' 0.001 '**' 0.01 '*' 0.05 '.' 0.1 ' ' 1

Residual standard error: 0.4033 on 4 degrees of freedom
Multiple R-Squared: 0.9997,  Adjusted R-squared: 0.9997
F-statistic: 1.568e+04 on 1 and 4 DF, p-value: 2.441e-08
```

Fig. 22: The summary of R's regression analysis. See the text for a discussion of how to interpret the information in the output's three sections.

which produces the result shown in Figure 23. Note that R plots the residuals against the predicted (fitted) values of y instead of against the known values of x. The choice of how to plot the residuals is not critical, as you can see by comparing Figure 23 to Figure 20. The line in Figure 23 is a smoothed fit of the residuals.

The second section of Figure 22 provides the model's coefficients— the slope, $\beta 1$, and the y-intercept, $\beta 0$—along with their respective standard deviations (*Std. Error*). The column *t value* and the column *Pr(>|t|)* are for the following t-tests.

$$\text{slope } H_0\text{: } \beta_1 = 0,\ H_A\text{: } \beta_1 \neq 0$$

$$y\text{-intercept } H_0\text{: } \beta_0 = 0,\ H_A\text{: } \beta_0 \neq 0$$

The results of these t-tests provide convincing evidence that the slope is not zero, but no evidence that the y-intercept significantly differs from zero.

The last section of the regression summary provides the standard deviation about the regression (*residual standard error*), the square of the correlation coefficient (*multiple R-squared*), and the result of an F-test on the model's ability to explain the variation in the y values. For a discussion of the correlation coefficient and the F-test of a regression model, as well as their limitations, refer to the section on using Excel's data analysis tools.

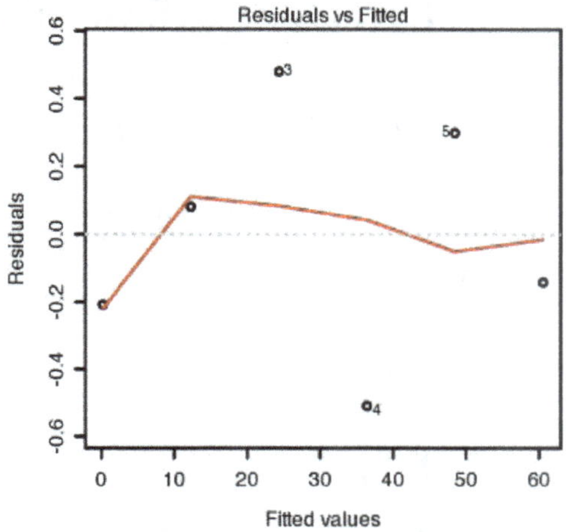

Fig. 23: Example showing R's plot of a regression model's residual error.

7.9 PREDICTING THE UNCERTAINTY IN *CA* GIVEN *SSAMP*

Unlike Excel, R includes a command for predicting the uncertainty in an analyte's concentration, *CA*, given the signal for a sample, S_{samp}. This command is not part of R's standard installation. To use the command you need to install the "chemical" package by entering the following command (*note:*

you will need an internet connection to download the package).

> install.packages("chemCal")

After installing the package, you will need to load the functions into R using the following command. (*note: you will need to do this step each time you begin a new R session as the package does not automatically load when you start R*).

> library("chemCal")

The command for predicting the uncertainty in *CA* is **inverse.predict**, which takes the following form for an unweighted linear regression inverse.predict(*object*, *newdata*, alpha = *value*) where *object* is the object containing the regression model's results, *newdata* is an object containing values for Ssamp, and *value* is the numerical value for the significance level. Let's use this command to complete Example 11.

First, we create an object containing the values of S_{samp}

> sample = c(29.32, 29.16, 29.51)

and then we complete the computation using the following command > inverse. predict(model, sample, alpha = 0.05)

producing the result shown in Figure 24. The analyte's concentration, CA, is given by the value $Prediction, and its standard deviation, sCA, is shown as $`Standard Error`. The value for $Confidence is the confidence interval, ±tsCA, for the analyte's concentration, and $`Confidence Limits` provides the lower limit and upper limit for the confidence interval for CA.

7.10 USING R FOR A WEIGHTED LINEAR REGRESSION

R's command for an unweighted linear regression also allows for a weighted linear regression by including an additional argument, *weights*, whose value is an object containing the weights.

$$\text{lm}(y \sim x, \textit{weights} = \textit{object})$$

```
> inverse.predict(model, sample, alpha = 0.05)
$Prediction
[1] 0.2412597

$`Standard Error`
[1] 0.002363588

$Confidence
[1] 0.006562373

$`Confidence Limits`
[1] 0.2346974 0.2478221
```

Fig. 24: Output from R's command for predicting the analyte's concentration, *CA*, from the sample's signal, *Ssamp*.

First, we need to create an object containing the weights, which in R are the reciprocals of the standard deviations in *y*, (*syi*)−2. Using the data from Example 5.12, we enter > syi=c(0.02, 0.02, 0.07, 0.13, 0.22, 0.33)

> w=1/syi^2

to create the object containing the weights. The commands > modelw = lm(signal ~ conc, weights = w)

> summary(modelw) generate the output shown in Figure 25. Any difference between the results shown here and the results shown in Example 12 are the result of round-off errors in our earlier calculations.

```
> modelw=lm(signal~conc, weights = w)
> summary(modelw)

Call:
lm(formula = signal ~ conc, weights = w)

Residuals:
     1      2      3      4      5      6
-2.223  2.571  3.676 -7.129 -1.413 -2.864

Coefficients:
             Estimate   Std. Error  t value  Pr(>|t|)
(Intercept)  0.04446    0.08542     0.52     0.63
conc         122.64111  0.93590     131.04   2.03e-08 ***
---
Signif. codes:  0 '***' 0.001 '**' 0.01 '*' 0.05 '.' 0.1 ' ' 1

Residual standard error: 4.639 on 4 degrees of freedom
Multiple R-Squared: 0.9998,  Adjusted R-squared: 0.9997
F-statistic: 1.717e+04 on 1 and 4 DF,  p-value: 2.034e-08
```

Fig. 25: The summary of R's regression analysis for a weighted linear regression. The types of information shown here is identical to that for the unweighted linear regression in Figure 22.

In a quantitative analysis we measure a signal, S_{total}, and calculate the amount of analyte, n_A or C_A, using one of the following equations.

$$S_{total} = k_A n_A + S_{reag}$$

$$S_{total} = k_A C_A + S_{reag}$$

To obtain an accurate result we must eliminate determinate errors affecting the signal, S_{total}, the method's sensitivity, k_A, and the signal due to the reagenSts, S_{reag}.

To ensure that we accurately measure S_{total}, we calibrate our equipment and instruments. To calibrate a balance, for example, we a standard weight of known mass. The manufacturer of an instrument usually suggests appropriate calibration standards and calibration methods.

To standardize an analytical method we determine its sensitivity. There are several standardization strategies, including external standards, the method of standard addition and internal standards. The most common strategy is a multiple-point external standardization, resulting in a normal calibration curve. We use the method of standard additions, in which known amounts of analyte are added to the sample, when the sample's matrix complicates the analysis. When it is difficult to reproducibly handle samples and standards, we may choose to add an internal standard.

Single-point standardizations are common, but are subject to greater uncertainty. Whenever possible, a multiple-point standardization is preferred, with results displayed as a calibration curve. A linear regression analysis can provide an equation for the standardization. A reagent blank corrects for any contribution to the signal from the reagents used in the analysis. The most common reagent blank is one in which an analyte - free sample is taken through the analysis. When a simple reagent blank does not compensate for all constant sources of determinate error, other types of blanks, such as the total Youden blank, can be used.

8

THE ASSESSMENT OF ANALYTICAL DATA

8.1 INTRODUCTION: AN OVERVIEW

A critical attitude towards the results obtained in analysis is necessary in order to appreciate their meaning and limitations. Precision is dependent on the practical method and beyond a certain degree cannot be improved. Inevitably there must be a compromise between the reliability of the results obtained and the use of the analyst's time. To reach this compromise requires an assessment of the nature and origins of errors in measurements; relevant statistical tests may be applied in the appraisal of the results. With the development of microcomputers and their ready availability, access to complex statistical methods has been provided. These complex methods of data handling and analysis have become known collectively as *chemo metrics*.

8.2 DEFINITIONS AND BASIC CONCEPTS

True Result

The 'correct' value for a measurement which remains unknown except when a standard sample is being analysed. It can be estimated from the results with varying degrees of precision depending on the experimental method.

Accuracy

The nearness of a measurement or result to the true value. Expressed in terms of error.

Error

The difference between the true result and the measured value. It is conveniently expressed as an *absolute error*, defined as the actual difference between the true result and the experimental value in the same units. Alternatively, the *relative error* may be computed, i.e. the error expressed as a percentage of the measured value or in 'parts per thousand'.

Mean

The arithmetic average of a replicate set of results.

Median

The middle value of a replicate set of results.

Degree of Freedom

An independent variable. The number of degrees of freedom possessed by a replicate set of results equals the total number of results in the set. When another quantity such as the mean is derived from the set, the degrees of freedom are reduced by one, and by one again for each subsequent derivation made.

Precisiosn

The variability of a measurement. As in the case of error, above, it may be expressed as an absolute or relative quantity. Standard deviations are the most valuable precision indicators (*vide infra*).

Spread

The numerical difference between the highest and lowest results in a set. It is a measure of precision.

Deviation (e.g. From the Mean or Median)

The numerical difference, with respect to sign, between an individual result and the mean or median of the set. It is expressed as a relative or absolute value.

Standard Deviation, s

A valuable parameter derived from the normal error curve (p. 16) and expressed by:

$$\sigma = \left[\frac{\sum_{i=1}^{i=N}(x_i - \mu)^2}{N}\right]^{\frac{1}{2}} \qquad 1$$

where xi is a measured result, μ is the true mean and N is the number of results in the set. Unfortunately, μ is never known and the mean derived from the set of results has to be used. In these circumstances the degrees of freedom are reduced by one and an estimate of the true standard deviation is calculated from:

$$s = \left[\frac{\sum_{i=1}^{i=N}(x_i - \bar{x})^2}{N-1}\right]^{\frac{1}{2}} \qquad 2$$

A better estimate of the standard deviation may often be obtained by the pooling of results from more than one set. Thus, s may be calculated from K sets of data.

$$s = \left[\frac{\sum_{i=1}^{N_1}(x_i - \bar{x}_1)^2 + \sum_{i=1}^{N_2}(x_i - \bar{x}_2)^2 + \ldots \sum_{i=1}^{N_K}(x_i - \bar{x}_K)^2}{M - K} \right]^{\frac{1}{2}} \quad 3$$

where $M = N_1 + N_2 + \ldots NK$. One degree of freedom is lost with each set pooled. A common requirement is the computation of the pooled value for two sets of data only. In this case the simplified equation (4) may conveniently be used:

$$S^2 = [(n_1-1) + n_2-1] / (n_1+n_2-2) \quad 4$$

Standard deviations for results obtained by the arithmetic combination of data will be related to the individual standard deviations of the data being combined. The exact relation will be determined by the nature of the arithmetic operation (Table 1).

Table 1 Standard deviations from arithmetically combined data

Arithmetic quantity	Standard deviation
x	σ_x
y	σ_y
$1/x$	σ_x/x^2
$x \pm y$	$(\sigma_x^2 + \sigma_y^2)^{\frac{1}{2}}$
x^2	$2\sigma_x x$
$x^{\frac{1}{2}}$	$a_x / 2x_2^1$
xy	$(\sigma_x^2 y^2 + \sigma_y^2 x^2)^{\frac{1}{2}}$
x/y	$x/y[(\sigma_x^2/y^2 + \sigma_y^2/x^2)^{\frac{1}{2}}$

The Relative Standard Deviation

Also known as the *coefficient of variation* , this is often used in comparing precisions.

Variance

The square of the standard deviation (s^2 or s^2). This is often of practical use as the values are additive, e.g.

= ++

8.3 THE NATURE AND ORIGIN OF ERRORS

On the basis of their origin, errors may usually be classified as *determinate* or *indeterminate.* The first are those having a value which is (in principle at least) measurable and for which a correction may be made. The second fluctuate in a random manner and do not have a definite measurable value.

Determinate errors may be *constant* or *proportional.* The former have a fixed value and the latter increase with the magnitude of the measurement. Thus their overall effects on the results will differ.

These effects are summarized in Figure .1. The errors usually originate from one of three major sources: operator error; instrument error; method error. They may be detected by blank determinations, the analysis of standard samples, and independent analyses by alternative and dissimilar methods.

Proportional variation in error will be revealed by the analysis of samples of varying sizes. Proper training should ensure that operator errors are eliminated. However, it may not always be possible to eliminate instrument and method errors entirely and in these circumstances the error must be assessed and a correction applied.

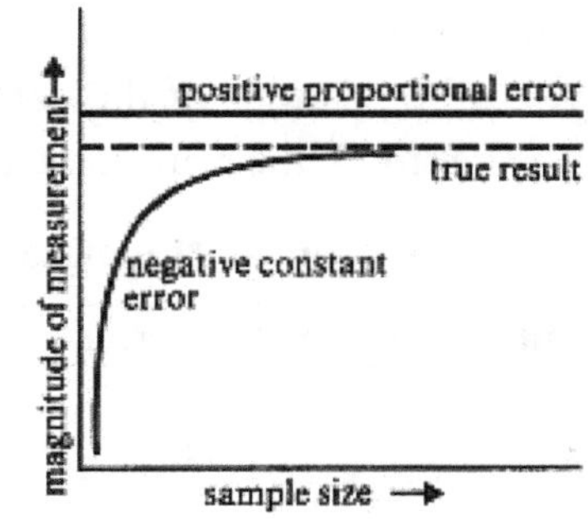

Fig. 1: The effects of constant and proportional errors on a measurement of concentration.

Indeterminate errors arise from the unpredictable minor inaccuracies of the individual manipulations in a procedure. A degree of uncertainty is introduced into the result which can be assessed only by statistical tests. The deviations of a number of measurements from the mean of the measurements should show a symmetrical or *Gaussian distribution* about that mean. Figure 2 represents this graphically and is known as *a normal error curve.* The general equation for such a curve is

$$y = \frac{\exp\left[-(x-\mu)^2/2\sigma^2\right]}{\sigma(2\pi)^{\frac{1}{2}}} \qquad 5$$

where μ is the mean and s is the standard deviation. The width of the curve is determined by s, which is a useful measure of the spread or precision of a set of results, and is unique for that set of data. An interval of μ ± s will contain 68.3% of the statistical sample, whilst the intervals μ ± 2s and μ ± 3s will contain 95.5% and 99.7% respectively.

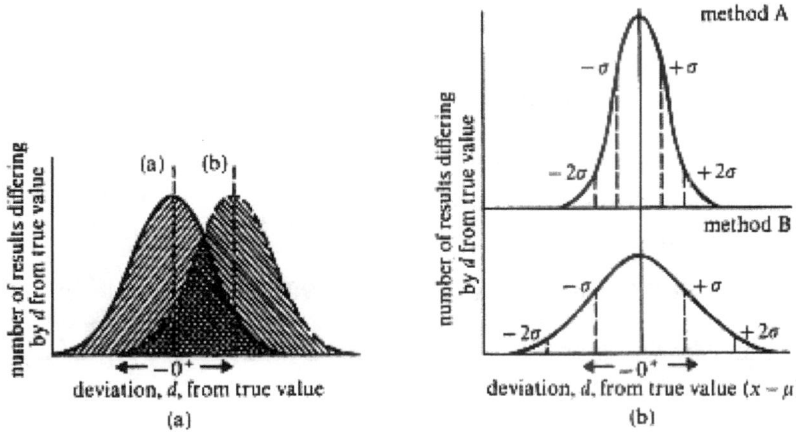

Fig. 2: Normal error curves. (a) Curve (a) shows a normal distribution about the true value. Curve (b) shows the effect of a determinate error on the normal distribution. (b) Curves showing the results of the analysis of a sample by two methods of differing precision. Method A is the more precise, or reliable.

8.4 THE EVALUATION OF RESULTS AND METHODS

A set of replicate results should number at least twenty-five if it is to be a truly representative 'statistical sample'. The analyst will rarely consider it economic to make this number of determinations and therefore will need statistical methods to enable him to base his assessment on fewer data, or data that have been accumulated from the analysis of similar samples. Any analytical problem should be examined at the outset with respect to the precision, accuracy and reliability required of the results. Analysis of the results obtained will then be conveniently resolved into two stages – an examination of the reliability of the results themselves and an assessment of the meaning of the results.

8.4.1 The Reliability of Measurements

When considering the reliability of the results, any determination which deviates rather widely from the mean should be first investigated for gross experimental or arithmetic error. Except in cases where such errors are revealed, questionable data should only be rejected when a proper statistical test has been applied. This process of *data rejection* presents the analyst with an apparent paradox. If the limits for acceptance are set too narrowly, results which are rightly part of a statistical sample may be rejected and narrow limits may therefore only be applied with a low *confidence* of containing all statistically relevant determinations. Conversely wide limits may be used with a high confidence of including all relevant data, but at a risk of including some that have been subject to gross error. A practical compromise is to set limits at a *confidence level* of 90% or 95%.

There are two criteria which are commonly used to gauge the rejection of results. Of these, the most convenient to use is based on the $x \pm 2\sigma$ interval which contains 95.5% of the relevant measurements. Some workers believe this limit is too wide, and regard the

Q-test at a 90% confidence level to be a more acceptable alternative. A *rejection quotient* *Q* is defined as:

$$Q = (X_n - x_{n-1}) / (X_n - x_1) \qquad 6$$

where x_n is the questionable result in a set $x_1, x_2, x_3, \ldots, x_n$. *Q* is calculated for the questionable data and compared with a table of critical values (Table 2). The result is rejected if *Q* (experimental) exceeds *Q* (critical).

Table 2 Critical values of *Q* at the 90% confidence level

Number of results	Qcrit (90% confidence)
2	—
3	0.94
4	0.76
5	0.64
6	0.56
7	0.51
8	0.47
9	0.44
10	0.41

8.4.2 The Analysis of Data

Once the reliability of a replicate set of measurements has been established the mean of the set may be computed as a measure of the true mean. Unless an infinite number of measurements is made this true mean will always remain unknown. However, the *t-factor* may be used to calculate a *confidence interval* about the experimental mean, within which there is a known (90%) confidence of finding the true mean. The limits of this confidence interval are given by:

$$\bar{x} \pm ts/N^{\frac{1}{2}} \qquad (2.7) \qquad 7$$

where is *x* the experimental mean, *t* is a statistical factor derived from the normal error curve (values in Table 3), *s* is the estimated standard deviation and *N* is the number of results.

Example 1

If the analysis of a sample for iron content yields a mean result of 35.40% with a standard deviation of 0.30%, the size of the confidence interval will vary inversely with the number of measurements made.

For two measurements, the confidence interval (90%) is

$$35.40 \pm 6.31 \times 0.30 \times 2^{-\frac{1}{2}} = 35.40\% \pm 1.34\%$$

and for five measurements

$$35.40 \pm 2.13 \times 0.30 \times 5^{-\frac{1}{2}} = 35.40\% \pm 0.29\%$$

(*N.B. s* has been derived from the set of data and $N - 1$ degrees of freedom

Table 3 Values of *t* for various levels of probability

	Confidence level (%)				
Degrees of freedom	***80***	***90***	***95***	***99***	***99.9***
1	3.08	6.31	12.7	63.7	637
2	1.89	2.92	4.30	9.92	31.6
3	1.64	2.35	3.18	5.84	12.9
4	1.53	2.13	2.78	4.60	8.60
5	1.48	2.02	2.57	4.03	6.86
6	1.44	1.94	2.45	3.71	5.96
7	1.42	1.90	2.36	3.50	5.40
8	1.40	1.86	2.31	3.36	5.04
9	1.38	1.83	2.26	3.25	4.78
10	1.37	1.81	2.23	3.17	4.59
11	1.36	1.80	2.20	3.11	4.44
12	1.36	1.78	2.18	3.06	4.32
13	1.35	1.77	2.16	3.01	4.22
14	1.34	1.76	2.14	2.98	4.14
∞	1.29	1.64	1.96	2.58	3.29

Are used in evaluating *t*.) The essential conclusion, here, is that five analyses at most are required to get a reasonable estimate of the true mean.

When a comparison of two separate replicate sets of data is required, the first stage is normally to compare their respective precisions by means of the *F-test*. This test uses the ratio of the variances of the two sets to establish any statistically significant difference in precision. *F* is calculated from

$$F = s_x^2 / s_y^2 \qquad (2.8)$$ 8

(By convention the larger variance is always taken as numerator.) The value of *F* thus obtained is compared with critical values computed on the assumption that they will be exceeded purely on a probability basis in only 5% of cases (Table .4). When the experimental value of *F* exceeds the critical value then the difference in variance or precision is deemed to be statistically significant.

Table .4 Critical values for *F* at the 5% level

Degrees of freedom (Denominator)	Degrees of freedom (numerator)						
	3	4	5	6	12	20	∞
3	9.28	9.12	9.01	8.94	8.74	8.64	8.53
4	6.59	6.39	6.26	6.16	5.91	5.80	5.63
5	5.41	5.19	5.05	4.95	4.68	4.56	4.36
6	4.76	4.53	4.39	4.28	4.00	3.87	3.67
12	3.49	3.26	3.11	3.00	2.69	2.54	2.30
20	3.10	2.87	2.71	2.60	2.28	2.12	1.84
∞	2.60	2.37	2.21	2.10	1.75	1.57	1.00

Having established that the standard deviations of two sets of data agree at a reasonable confidence level it is possible to proceed to a comparison of the mean results derived from the two sets, using the *t* test in one of its forms. As in the previous case, the factor is calculated from the experimental set of results and compared with the table of critical values (Table 2.3). If *t*exp exceeds the critical value for the appropriate number of degrees of freedom, the difference between the means is said to be significant.

When there is an accepted value for the result based on extensive previous analysis *t* is computed from equation (2.9)

$$t = [(\bar{x} - \mu)/s]N^{\frac{1}{2}} \qquad (2.9) \qquad 9$$

where is *x* the mean of the experimental set, μ the accepted value, *s* the experimental standard deviation and *N* the number of results.

If there is no accepted value and two experimental means are to be compared, *t* can be obtained from equation (10) with $(M + N - 2)$ degrees of freedom.

$$t = [(\bar{x} - \bar{y})/s][MN/M + N)]^{\frac{1}{2}} \qquad (2.10) \qquad 10$$

where is the mean of *M* determinations, the mean of *N* determinations and *s* the pooled standard deviation (equation (3)).

8.4.3 The Application of Statistical Tests

Table 5, together with the subsequent worked examples, illustrates the application of the statistical tests to real laboratory situations. Equation (10) is a simplified expression derived on the assumption that the precisions of the two sets of data are not significantly different. Thus the application of the *F*test (equation (8)) is a prerequisite for its use. The evaluation of *t* in more general circumstances is of course possible, but from a much more complex expression requiring tedious calculations. Recent and rapid developments in desk top computers are removing the tedium and making use of the general expression more acceptable. The references at the end of the chapter will serve to amplify this point.

Example 2

In a series of replicate analyses of a sample the following data (%) were obtained:

4.20, 7.01, 7.31, 7.54, 7.55, 759

In the assessment of these data for reliability the first step of arranging them in rank order has been carried out.

On inspection 20 is rejected as having a gross error, and 7.01 is seen as questionable and requiring to be tested by the Q-test.

$$Q_{exp} = \frac{7.31 - 7.01}{7.59 - 7.01} = 0.52$$

Table 5 Some practical problems with relevant statistical tests

Practical problems Relevant tests	Practical problems Relevant tests
One result in a replicate set differs rather widely from the rest. Is it a significant result?	Examine for gross error. Apply Q-test (equation (2.6))
Two operators analysing the same sample by the same method obtain results with different spreads. Is there a significant difference in precision between the results	Examine data for unreliable results. Apply F-test (equation (2.8))
A new method of analysis is being tested by the analysis of a standard sample with an accurately known composition. Is the difference between the experimental value and the accepted value significant?	Examine data for unreliable results. Apply t-test (equation (2.9))
Two independent methods of analysis have been used to analyse a sample of unknown composition. Is the difference between the two results significant and thus indicative of an error in one method?	Examine data for unreliable results. Establish that both sets have similar precisions by F-test. Apply T-test (equation (10))
With what confidence can the mean of a set of experimental results be quoted as a measure of the true mean?	Calculate the confidence interval (equation (.7))
If the standard deviation for a method is known, how many results must be obtained to provide a reasonable estimate of the true mean?	Use the confidence interval method (equation (7))
Is a determinate error fixed or proportional?	Graphical plot of results against sample weight (Figure 1)

Comparison with critical values in Table 2 shows Qcrit for 6 results to be 0.56. Thus Qexp < Qcrit and 7.01 is retained.

Example 3

The accepted value for the chloride content of a standard sample obtained from extensive previous analysis is 54.20%. Five analyses of the same sample are carried out by a new instrumental procedure, 54.01, 54.24, 54.05, 54.27, 54.11% being the results obtained. Is the new method giving results consistent with the accepted value?

(a) A preliminary examination shows no unreliable results.

(b) The mean and the standard deviations are then calculated (equation (2)).

x	(x-ẋ)	$(x-\dot{x})^2$	
54.01	–0.13	0.0169	$s = (0.0528/4)1/2 = 0.115$
54.24	+0.10	0.0100	
54.05	–0.09	0.0089	
54.27	+0.13	0.0169	
54.11	–0.03	0.0009	
Sx = 270.68		$\Sigma\ (x-\dot{x})^2$	
ẋ=54.14		=0. 0528	

(c) Equation (9) is then applied to compare the means

$$t = [(54.20 - 54.14)/0.115]5^{\frac{1}{2}}$$
$$= 0.06 \times 2.236/0.115 = 1.17$$

From *t*-criteria the confidence level for this difference to be significant is less than 80%.

Conclusion: The new instrumental method is giving the same results as the accepted method.

Example .4

The technique of a trainee operator is being assessed by comparing his results with those obtained by an experienced operator. Do the results obtained indicate a significant difference between the skill of the two operators?

The trainee operator carried out six determinations yielding a mean of 35.25% with a standard deviation of 0.34%. The experienced operator obtained a mean of 35.35% and a standard deviation of 0.25% from five determinations.

(a) The *F*-test is used to compare the standard deviations (equation (8))]

$$F_{exp} = 0.34^2/\ 0.25^2 = 1.85$$

*F*crit from Table 4 is 6.26 and there is no significant difference in the standard deviations (at the 95% level).

(b) Equation (10) now enables the two means to be compared. If *s* is first computed from the pooled data (equation (3)) and found to be 0.29% then

$$t=[(35.35-35.25)/0.29]\ (30/11)^{½}$$
$$= 0.57$$

***Conclusion*:** The probability of the difference in means being significant is very low, and there is no difference between the skill of the two operators.

Example 5

In an investigation of a determinate error a series of replicate measurements were made using a range of sample weights. The results obtained are tabulated below.

Sample weight/g	Analyte/% 0.113 9.67
0.351	9.96
0.483	10.04
0.501	10.03
0.711	10.09
0.867	10.12
0.904	10.13

In order to decide whether the error is constant or proportional plot a graph of results against sample weight. Take care in selecting the scale to ensure that the trends are not obscured. The graph (Figure 3) shows clearly that the error is negative and constant (see Figure 1).

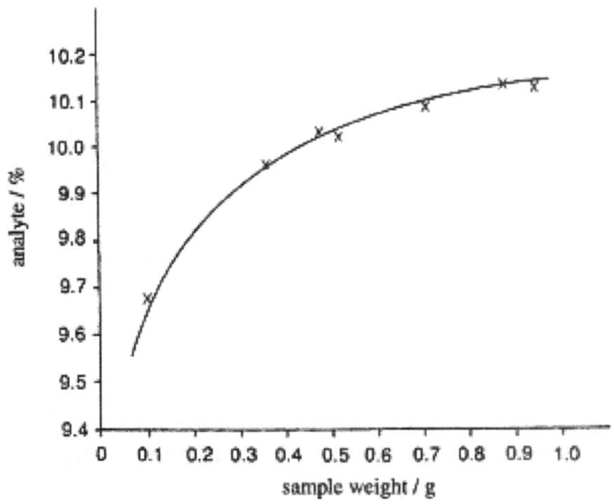

Fig. 3: Plot of analyte content against sample weight to investigate a determinate error.

The estimation of the overall precision of a method from its unit operations A frequent problem in analysis is the estimation of the overall precision of a method before it has been used or when insufficient data are available to carry out a statistical analysis. In these circumstances the known precision limits for the unit operations of the method (volume measurement, weighing, etc.) may be used to indicate its precision. Table 2.6 gives the normal precision limits for 'Grade A' volumetric equipment.

If the absolute standard deviations for a set of unit operations are *a, b, c*, . . . , then *s*, the overall standard deviation for the method is given by:

$$S = (a^2+b^2+c^2+.....)^{1/2} \qquad 11$$

when the individual measurements are combined as a sum or difference (Table 1). Conveniently, the relative standard deviations may be used to express the overall standard deviation when the individual measurements are combined as a product or quotient.

$$s_r = (a_r^2 + b_r^2 + c_r^2 + \ldots)^{\frac{1}{2}} \qquad (2.12) \qquad 12$$

Example 6

Consider as an example the standardization of a solution of hydrochloric acid by titration against a weighed amount of sodium carbonate. The strength of the hydrochloric acid will be computed from

Table 6 Precision limits for laboratory equipment used in volumetric analysis

Burettes	**± 0.02 cm3 for each reading**
Pipettes (to deliver)	(5 ± 0.01) cm3
	(10 ± 0.02) cm3
	(25 ± 0.03) cm3
	(50 ± 0.05) cm3
	(100 ± 0.08) cm3
Volumetric flasks (to contain)	(25 ± 0.03) cm3
	(50 ± 0.05) cm3
	(100 ± 0.08) cm^3
	(250 ± 0.12) cm^3
	(500 ± 0.15) cm^3
	(1 000 ± 0.30) cm^3
	(2 000 ± 0.50) cm^3
Analytical balances (to weigh to)	± 0.0001 g

$$\text{concentration of HCl} = \frac{\text{weight of } Na_2CO_3}{\text{vol of HCl used}} \cdot \frac{2M_r(HCl)}{M_r(Na_2CO_3)}$$

Where *Mr* represents the relative molecular mass of the compound.

The results of an analysis are as follows:

Weight of bottle + Na_2CO_3 (1)	16.254 1 ± 0.0001 g
Weight of bottle + Na_2CO_3 (2)	16.041 9 ± 0.0001 g
Weight of Na_2CO_3 used	0.212 2
Final burette reading	45.21 ± 0.02 cm3
Initial burette reading	0.52 ± 0.02 cm3
Volume of HCl used	44.69 cm3

The overall precision of the weighing is now computed using equation (11).

$$s = [(0.0001)^2 + (0.0001)^2]^{\frac{1}{2}} = 0.000\ 14\ g$$

The relative standard deviation for the weighing is then Similarly, the overall precision of the volume measurement is obtained. An allowance for the uncertainty in the colour change observation at the end point must be included (e.g. ± 0.03 cm3).

Thus

$$s = [(0.02(^2+(0.02)^2 + (0.03)^2]^{\frac{1}{2}} = 0.041\ cm^3$$

Whence the relative standard deviation is

$$0.041/44.69 = 0.00092 \quad \text{i.e. about } 0.09\%$$

Finally, the estimated precision for the determination of the concentration of hydrochloric acid is obtained using equation (12).

$$s_r = [(0.07)^2 + (0.09)^2]^{\frac{1}{2}} = 0.11\%$$

One important point to remember is that the absolute standard deviations for the unit processes are constant, but the relative standard deviations will decrease with the magnitude of the sample and the titre. In other words, within limits the larger the sample taken the better the precision of the results.

Significant Figures

Results are normally given to a certain number of *significant figures.* All the digits in a number that are known with certainty plus the first that is uncertain, constitute the significant figures of the number. In the case of a zero it is taken as significant when it is part of the number but not where it merely indicates the magnitude. Thus a weight of 1.0421 g which is known within the limits of ± 0.0001 g has five significant figures whilst one of 0.0421 g which is known within the same absolute limits has only three. When a derived result is obtained from addition or subtraction of two numbers, its significant figures are determined from the *absolute uncertainties.* Consider the numbers 155.5 ± 0.1 and 0.085 ± 0.001 which are added together to give 155.585. Uncertainty appears at the fourth digit, whence the result should be rounded off to 155.6. If the derived result is a product or quotient of the two quantities, the *relative uncertainty* of the least certain quantity dictates the significant figures. 0.085 has the greatest relative uncertainty at 12 parts per thousand. The product 155.5 × 0.085 = 13.3275 has an absolute deviation of 13.3275 × 0.012 = 0.16. Uncertainty thus appears in the third digit and the result is rounded off to 13.3.

Limits of Detection

It is important in analysis at trace levels to establish the smallest concentration or absolute amount of an analyte that can be detected. The problem is one of discerning a difference between the response given by a 'blank' and that given by the sample, i.e. detecting a weak signal in the presence of background noise. All measurements are subject to random errors, the distribution of which should produce a normal error curve. The spread of replicate measurements from the blank and from the sample will therefore overlap as the two signals approach each other in magnitude. It follows that the chances of mistakenly identifying the analyte as present when it is not, or vice versa, eventually reach an unacceptable level. The detection limit must therefore be defined in statistical terms and be related to the probability of making a wrong decision.

Figure 4 (a) shows normal error curves (B and S) with true means μB and μS for blank and sample measurements respectively. It is assumed that for measurements made close to the limit of detection, the standard deviations of the blank and sample are the same, i.e. sB = sS = s. In most cases, a 95% confidence level is a realistic basis for deciding if a given response arises from the presence of the analyte or not, i.e. there is a 5% risk in reporting the analyte 'detected' when it is not present and vice versa. Thus, point *L* on curve B represents an upper limit above which only 5% of blank measurements with true mean Mb will lie, whilst point *L* on curve S represents a lower limit below which only 5% of sample measurements with true mean μS will lie. If μS now approaches μB until points *L* on each curve coincide (Figure 2.4(b)), the point of coincidence represents the *practical detection limit* for a single measurement, i.e. if a measurement falls at or below *L*, it has a 95% probability of arising from background sources or random noise only, whilst if it falls above *L* it has a 95% probability of arising from the presence of the analyte. Furthermore, it follows that μS must now represent the *theoretical detection limit* because a true mean lying below μS would have a normal distribution with more than 5% of values below *L*. Because the chances of making an incorrect decision were chosen to be equal in this case (5% probability), then μS is given by:

$$\mu_S = 2L \qquad (2.13)$$ **13**

Individual practical results falling between *L* and μS must be regarded as 'detected' but should be reported as 'less than μS'.

The value of *L* and hence of μS is related to s and is given by

$$L = \mu_B + 1.64\sigma \qquad (2.14)$$ **14**

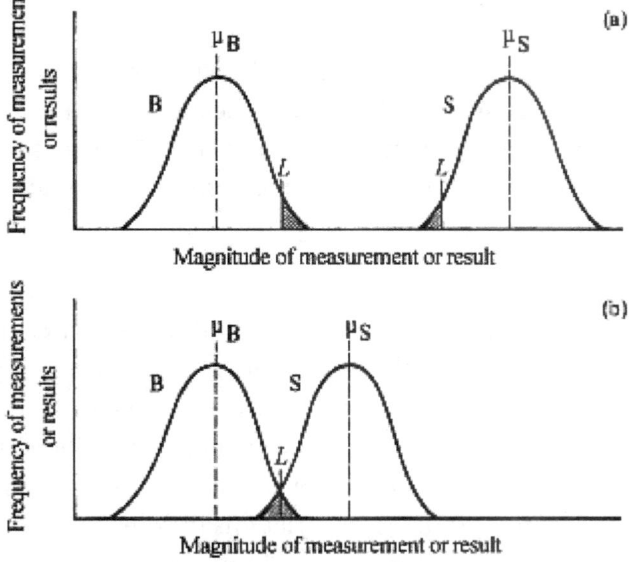

Fig. 4: Normal error curves for blank B and sample S measurements.

Where a large number of blanks (preferably more than 20) have been measured or

$$L = \mu_B + 2.33\sigma \qquad (2.15)$$

15

Where a smaller number of paired measurements of sample and blank have been made.

Example 7

The determination of iron at trace levels can be accomplished by the spectrophotometric measurement of its complex with the reagent *o*-phenanthroline. The sensitivity of a particular method is 53 ppm per unit absorbance and the standard deviation of the blank estimated from 25 measurements is 0.002 of an absorbance unit. The practical detection limit is therefore 1.64sB or 0.0033 of an absorbance unit which corresponds to 0.17 ppm, and the theoretical detection limit is 3.28sB or 0.0066 of an absorbance unit which corresponds to 0.35 ppm. (The value of μB in this case is assumed to be zero.) Hence if a result is less than 0.17 ppm, the conclusion is that iron is 'not detected'. If the value lies between 0.17 ppm and 0.35 ppm, the iron content should be reported as 'less than 0.35 ppm'.

Quality Control Charts

Chemical analysis finds important applications in the quality control of industrial processes. In an ideal situation a continuous analysis of the process stream is made. However, such continuous analysis is by no means always possible, and it is common to find a process being monitored by the analysis of separate samples taken at regular intervals. The analytical data thus obtained need to be capable of quick and simple interpretation, so that rapid warning is available if a process is going out of control and effective corrective action can be taken.

One method of data presentation which is in widespread use is the *control chart*. A number of types of chart are used but where chemical data are concerned the most common types used are *Shewhart charts* and *cusum charts*. Only these types are discussed here. The charts can also be used to monitor the performance of analytical methods in analytical laboratories.

Shewhart Charts

In an explanation of the construction and operation of control charts it is helpful to consider a simple example such as the mixing of two materials in a process stream. It is important to recognize that there are two aspects of the composition of the process stream which need to be controlled. Firstly, the overall composition of the mixture may be assessed by averaging the results on a run of about five samples and plotting the results on an *averages chart*. Secondly, short term variations in composition are reflected by the range of results in a run plotted on a *ranges chart* . Shewhart charts are thus used in pairs (Figure5(a) and 5(b)).

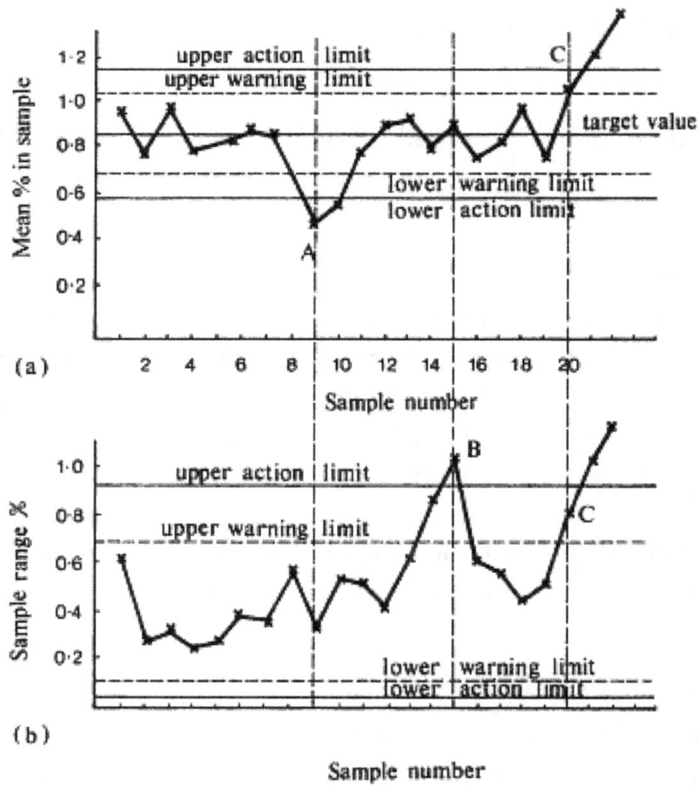

Fig. 5: A typical pair of Shewhart charts. (a) Averages chart and (b) ranges chart. Point A shows a lack of control of averages only, point B of ranges only and point C of both together.

It is usual to mark the charts with *warning limits* (*inner control limits*) and *action limits* (*outer control limits*). These represent the values at which there is respectively 95% (1.96s) or 99.9% (3.09s) confidence that the process is going out of control. The use of the averages chart is thus based on an estimate of the variability and an assumption that this remains constant. It is hence invalidated if a significant change in variability occurs. Therefore the ranges chart fulfils a second important function in providing a check on such changes in variability.

Figure 5 shows a pair of Shewhart charts. Point A illustrates the pattern which occurs when the averages are going out of control but the ranges remain in control. This would suggest perhaps that the supply of one of the feedstocks for the process is being interrupted. Point B shows the ranges going out of control with the averages remaining under control. One explanation of this would be that the mixing had become inefficient. Finally, point C shows both averages and ranges going out of control which implies a serious fault in the process.

Cusum Charts

An alternative method of monitoring the behaviour of a process is the *cumulative sum technique* (*cusum*). The cusum plot provides a simple method of diagnosing when a process begins to move out of control. A sequence of analyses is made on the process stream at regular intervals. The values thus obtained are compared with the target value for that particular component and the deviations aggregated to provide the *cumulative sum*. For a process in control, the deviations will have small positive or negative values and the cusum will remain close to zero, with the plot having a roughly horizontal pattern. If the process goes out of control, the cusum will become increasingly positive or negative with the corresponding change in slope signifying the point at which loss of control ensues. One major advantage of such charts is that the computations required are minimal and the plotting very simple.

Thus charts may be constructed and used by operators of limited technical knowledge. Figure 6 shows a cusum plot for a process going out of control.

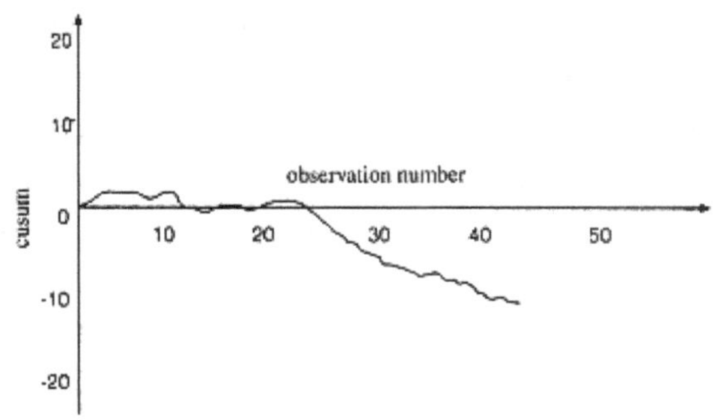

Fig. 6: A typical cusum plot showing a lack of control with a negative deviation developing after 20 observations.

Standardization of Analytical Methods

Quantitative analysis demands that an analytical measurement can be accurately and reliably related to the composition of the sample in a strict proportionality (p. 2). The complexity of relationships, especially for instrumental techniques, means that the proportionalities need to be practically established in *calibration procedures*. For a typical simple calibration, a range of standards is prepared containing varying amounts of the analyte. These are then analysed by the standard method and a *calibration curve* of signal *vs* amount of analyte is plotted. Results for unknowns are then interpolated from this graph (Figure 7). An important convention is that, to avoid confusion, the analytical signal is plotted on the *y*-axis and the amount of analyte on the *x*axis.

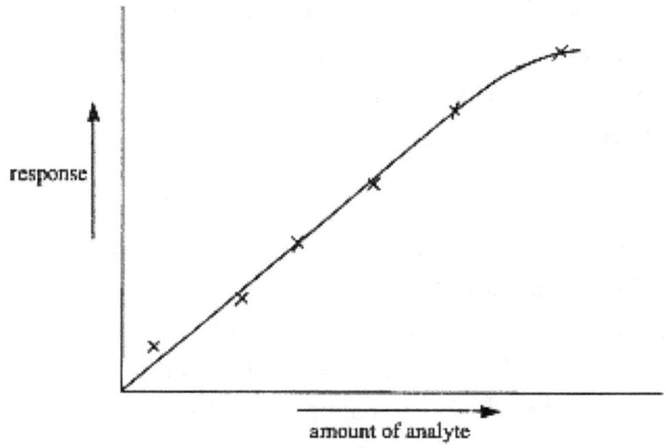

Fig. 7: A typical calibration curve.

Various examples of calibration curves will be found in later pages of this text. A good calibration curve will be linear over a good range of analyte quantities. Ultimately curvature must be anticipated at the higher ranges and uncertainty and deteriorating precision at the low ones. Ideally the calibration line should pass through the origin but does not always do so. It is not invalidated if it does not but nevertheless, it is prudent to consider carefully the possible reasons for this.

A particular issue that must be considered for all calibration procedures is the possibility of *matrix effects* on the analyte signal. If such effects are present they may be allowed for in many cases by *matrix matching* of the standard to the sample. This of course requires an accurate knowledge of the sample matrix. Where this is not available, the method of *standard addition* is often effective. This involves 'spiking' at least three equal aliquots of the sample with different amounts of the analyte, and then measuring the response for both spiked and unspiked aliquots. A plot of response *vs* analyte, extrapolated back, will give abscissae intercepts from which the amount of analyte in the sample may be deduced (Figure 8).

Even this method, however, does not guarantee to overcome all matrix effects, and is often limited where the matrix is extremely complex and remains largely unknown, as in naturally occurring materials such as rocks and biological specimens. The use of *recognized standards* has proved invaluable for the standardization and assessment of methods in these circumstances. Pioneered in geochemistry, this involves the collection and thorough mixing of suitable standard materials. These are then distributed for analysis by recognized laboratories using as many different techniques

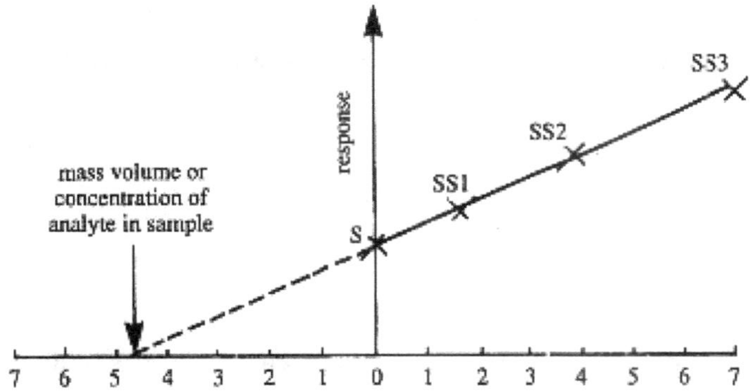

Fig. 8: Calibration curve for standard addition; S = unspiked sample; SS1, SS2, SS3 = spiked samples.

as possible. These ensuing results ultimately converge to give an accepted composition for the standard, which may subsequently be used for the assessment and calibration of various analytical methods.

Standards are exemplified by a *range of rock types* for geochemistry and *bovine liver* and *kale leaves* for biological analysis.

Plotting Calibration (Reference) Graphs Using the 'Least Squares' Regression Method

An ideal calibration curve (Figure 7) is a straight line with a slope of about 45 degrees. It is prepared by making a sequence of measurements on reference materials which have been prepared with known analyte contents. The curve is fundamental to the accuracy of the method. It is thus vitally important that it represents the *best fit* for the calibration data. Many computer software packages, supplied routinely with various analytical instruments, provide this facility. It is, however, useful to review briefly the principles on which they are based.

If the relationship between the signal and the amount of analyte is linear, the *method of least squares* may be used to obtain the best straight line through the points. Two important assumptions are made in doing this. Firstly, that the amount of analyte in the references is known accurately, and that any variations observed in the measurements originate in the intrinsic variability of the analytical signal.

Secondly, that the convention of plotting the signal on the y-axis, and the amount of analyte on the $x$axis, is followed. It is then possible to calculate the best line on the basis of minimizing the sum of the squares of the deviations in y from the calculated line. (In mathematically orientated texts these variations are known as *residuals*.)

If the straight line is represented by

$$y = bx + a \qquad (2.16) \qquad 16$$

the values of a, *the intercept* on the y-axis, and b, *the slope*, can be computed from equations (17) and (18) respectively. Commonly a and b are referred to as *regression coefficients*.

a=_y^--b_x^- 17

$$a = \bar{y} - b\bar{x} \qquad (2.17)$$

$$b = \frac{\sum_i [(x_i - \bar{x})(y_i - \bar{y})]}{\sum_i (x_i - \bar{x})^2} \qquad (2.18) \qquad 18$$

In the evaluation of equation (16) it is appropriate first to calculate b (equation (18)) and then a (equation (17)).

Example 8

A series of measurements obtained in the preparation of a calibration curve for an analytical method is shown in Table 7. From equations (18), (17) and (16)

b=0.270 ppm per signal unit, a= 0.175 signal units, and y= 0.270x + 0.715

Table 7 Calibration data obtained in the assessment of a new method of analysis

Analyte concentration (ppm)	Signal (arbitary units)
3.0	1.4
4.0	1.5
5.0	2.2
6.0	2.4
8.0	3.1
9.0	3.2
10.0	3.2
11.0	3.9
12.0	4.1
14.0	4.7
15.0	4.5
16.0	5.2
17.0	5.0
n = 13	

The final 'best fit' calibration curve can be seen at Figure 9. Note the limitations on the use of the lower, extrapolated, portion of the curve. The line obtained will always pass through the *centroid* of all the points, i.e. the point on the graph corresponding to $\bar{x}$, $\bar{y}$. A further aspect is the uncertainty associated with the values of *a* and *b*. These may be estimated in terms of the variability of *x* and *y*.

Details can be found in the references at the end of this chapter. Where relationships are not straight lines, polynomial curve fitting routines can be used. This is more complex than dealing with straight lines, and a number of pitfalls may be encountered. A specialist text needs to be consulted in these circumstances.

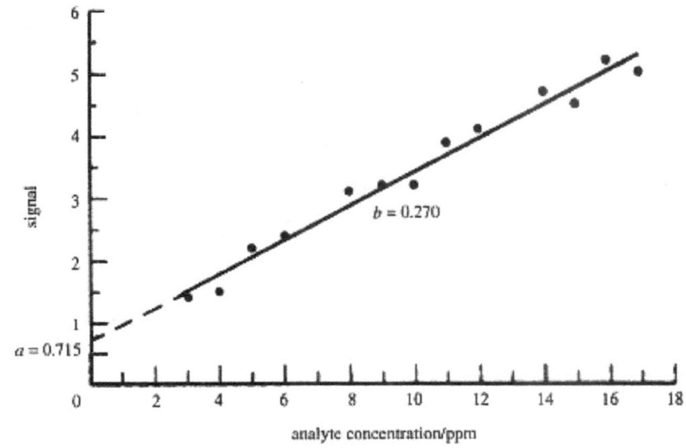

Fig. 9: 'Best fit' plot of data from Table 7 obtained by 'least squares' regression analysis. (*Important note*: This graph implies a straight line relationship down to zero concentration. It is, however, unsafe to use the extrapolated portion as there are no experimental data for this part of the curve).

Chemometrics:

Advanced mathematical and statistical techniques used in analytical chemistry are often referred to under the umbrella term of *chemometrics*. This is a loose definition, and chemometrics are not readily distinguished from the more rudimentary techniques discussed in the earlier parts of this chapter, except in terms of sophistication. The techniques are applied to the development and assessment of analytical methods as well as to the assessment and interpretation of results. Once the province of the mathematician, the computational powers of the personal computer now make such techniques routinely accessible to analysts. Hence, although it would be inappropriate to consider the detail of the methods in a book at this level, it is nevertheless important to introduce some of the salient features to give an indication of their value. Two important applications in analytical chemistry are in method optimization and pattern recognition of results.

Analytical methods often contain many different variables which need to be optimized to attain best performance. The different variables are not always independent. For example, pH and polarity in a solution may be interdependent. Optimization by changing one variable at a time, while keeping the others nominally constant is neither efficient nor necessarily successful. This point is well illustrated in Chapter 4 where the optimization of chromatographic separations is discussed.

A sample may be characterized by the determination of a number of different analytes. For example, a hydrocarbon mixture can be analysed by use of a series of UV absorption peaks. Alternatively, in a sediment sample a range of trace metals may be determined. Collectively, these data represent patterns characteristic of the samples, and similar samples will have similar patterns. Results may be compared by vectorial presentation of the variables, when the variables for similar samples will form clusters.

Hence the term *cluster analysis*. Where only two variables are studied, clusters are readily recognized in a two-dimensional graphical presentation. For more complex systems with more variables, i.e. n, the clusters will be in n-dimensional space. *Principal component analysis* (PCA) explores the interdependence of pairs of variables in order to reduce the number to certain principal components. A practical example could be drawn from the sediment analysis mentioned above. Trace metals are often attached to sediment particles by sorption on to the hydrous oxides of Al, Fe and Mn that are present.

The Al content could be a principal component to which the other metal contents are related. *Factor analysis* is a more sophisticated form of principal component analysis.

Problems

1. Define the terms *precision* and *accuracy* as they are used in analytical chemistry. Indicate how they may be estimated quantitatively. To what extent is the estimate of accuracy dependent upon precision?
2. In an experiment the following replicate set of volume measurements (cm3) was recorded:

   25.35, 25.25, 25.28, 25.50, 25.45, 25.43

a. Calculate the mean of the raw data.
b. Using the rejection quotient (*Q*-test) reject any questionable results.
c. Recalculate the mean and compare it with the value obtained in 2(a).

3. Calculate the absolute standard deviation and the relative standard deviation (%) of the following replicate set of data. Comment upon the precision of the measurement. Weight of component/g:

   5.346, 5.362, 5.362, 5.379, 5.335, 5.341

4. The three sets of replicate results below were accumulated for the analysis of the same sample. Pool these data to obtain the most efficient estimate of the mean analyte content and the standard deviation. Lead content/ppm:

Set 1	Set 2	Set 3
9.76	9.87	9.85
9.42	9.64	9.91
9.53	9.71	9.42
9.81	9.49	

1. Calculate the confidence interval (90%) about the mean of the data in question 3.
2. An exercise was carried out to compare the precisions of analytical measurements being made in two different laboratories. A completely homogeneous sample was supplied to both laboratories and the following analytical results (%) obtained by the same method.

Laboratory 1	Laboratory 2
34.97	35.02
34.85	34.96
34.94	34.99
34.88	35.07
34.85	

Comment upon the precisions attained in the two laboratories.

1. On the basis of extensive past analysis the iron content of an homogeneous ore sample is accepted to be 19.85%. An analysis of seven samples gave an average of 20.06% Fe. Within what probability levels is a determinate error indicated in the analysis if the absolute standard deviation of the results is 0.14%?
2. The efficiency of mixing of powdered solids in a process stream is being investigated. Two samples of the mixture, each of about 1 kg, are taken from different parts of the same hopper. After grinding to a fine powder, analytical samples were obtained using the coning and quartering technique. Subsequent analysis gave the following results for one component:

Sample A – six analyses, mean 12.32%
Sample B – four analyses, mean 12.63%.

The precisions were found to be similar and the pooled standard deviation (absolute) computed to be 0.071%. Comment upon the efficiency of the mixing.

1. Replicate analysis by a new method for a constituent of a standard sample gave the following results (%):

0.45,0.48,0.46,0.60,0.42,0.50,0.41.

The content of the standard is accepted to be 0.47%.

Assess these data and draw conclusions concerning the suitability of the method of analysis for routine use.

1. A standard reference material with an accepted analyte content of 5.85 ppm was used to compare the performance of two alternative methods of analysis. Solutions were prepared using a range of sample weights. Each solution was analysed by the two methods. Use the results obtained to assess any determinate errors exhibited by the methods.

Sample weight/g *Result, Method 1*/ppm *Result, Method 2*/ppm

0.125	6.37	5.72
0.612	5.97	5.69
0.334	6.10	5.70
0.781	5.93	5.71
0.271	6.20	5.68
0.457	6.03	5.74

(11) Find the equation for the regression line of y on x for the following pairs of values. Plot the graphand comment upon its characteristics.

X	2	4	6	8	10
y	3	1	7	5	9

(12) A calibration exercise for an atomic absorption method to be used in the determination of a metal yielded the following results:

Reference/mg kg^{-1}	*Absorbance*
0	0.021
2	0.050
4	0.090
6	0.126
8	0.173
10	0.210
12	0.247

Use the least squares method to plot the 'best fit' calibration curve. Comment upon its suitability for use.

9

GRAVIMETRIC METHODS

9.1 USING MASS AS AN ANALYTICAL SIGNAL

Suppose we are to determine the total suspended solids in the water released by a sewage-treatment facility. Suspended solids are just that: solid matter that has yet to settle out of its solution matrix. The analysis is easy. After collecting a sample, we pass it through a preweighed filter that retains the suspended solids, and then dry the filter and solids to remove any residual moisture. The mass of suspended solids is the difference between the filter's final mass and its original mass. We call this a direct analysis because the analyte—the suspended solids in this example—is the species that is weighed.

Method 2540D in *Standard Methods for the Examination of Waters and Wastewaters*, 20th Edition (American Public Health Association, 1998) provides an approved method for determining total suspended solids. The method uses a glass-fiber filter to retain the suspended solids. After filtering the sample, the filter is dried to a constant weight at 103–105°C.

What if our analyte is an aqueous ion, such as Pb^{2+}? Because the analyte is not a solid, we cannot isolate it by filtration. We can still measure the analyte's mass directly if we first convert it into a solid form. If we suspend a pair of Pt electrodes in the sample and apply a sufficiently positive potential between them for a long enough time, we can convert the Pb^2+ to Pb^{O2}, which deposits on the Pt anode.

$$Pb^2+(aq)+4H_2O(l) \rightleftharpoons PbO_2(s)+H_2(g)+2H_3O+(aq)$$

If we weigh the anode before and after we apply the potential, its change in mass gives the mass of PbO_2 and, from the reaction's stoichiometry, the amount of Pb^{2+} in the sample. This is a direct analysis because PbO_2 contains the analyte.

Sometimes it is easier to remove the analyte and let a change in mass serve as the analytical signal. Suppose we need to determine a food's moisture content. One approach is to heat a sample of the food to a temperature that will vaporize water and capture the water vapor using a preweighed absorbent trap. The change in the absorbent's mass provides a direct determination of the amount of water in the sample. An easier approach is to weigh the sample of food before and after we heat it and use the change in its mass to determine the amount of water originally present. We call this an indirect analysis because we determine the analyte, H_2O in this case, using a signal that is proportional its disappearance.

Method 925.10 in *Official Methods of Analysis*, 18th Edition (AOAC International, 2007) provides an approved method for determining the moisture content of flour. A preweighed sample is heated for one hour in a 130°C oven and transferred to a desiccator while it cools to room temperature. The loss in mass gives the amount of water in the sample.

The indirect determination of a sample's moisture content is made by measuring a change in mass. The sample's initial mass includes the water, but its final mass does not. We can also determine an analyte indirectly without its being weighed. For example, phosphite, $PO_3^{-3}PO_3^{3-}$, reduces Hg^{2+} to $Hg^{2+}2Hg_2^{2+}$, which in the presence of Cl– precipitates as Hg_2Cl_2.

$$2HgCl_2(aq)+PO_3-3(aq)+3H_2O(l)\rightleftharpoons Hg_2Cl_2(s)+2H_3O+(aq)+2Cl-(aq)+PO^{3-}_4(aq)$$

If we add $HgCl_2$ in excess to a sample that contains phosphite, each mole of $PO_3^{-3}PO_3^{3-}$ will produce one mole of Hg_2Cl_2. The precipitate's mass, therefore, provides an indirect measurement of the amount of $PO_3^{-3}PO_3^{3-}$ in the original sample.

9.2 TYPES OF GRAVIMETRIC METHODS

The examples in the previous section illustrate four different ways in which a measurement of mass may serve as an analytical signal. When the signal is the mass of a precipitate, we call the method precipitation gravimetry. The indirect determination of PO3–3PO33– by precipitating Hg_2Cl_2 is an example, as is the direct determination of Cl– by precipitating AgCl.

In electrogravimetry, we deposit the analyte as a solid film on an electrode in an electrochemical cell. The deposition as PbO_2 at a Pt anode is one example of electrogravimetry. The reduction of Cu^{2+} to Cu at a Pt cathode is another example of electrogravimetry.

We will not consider electrogravimetry in this chapter. See Chapter 11 on electrochemical methods of analysis for a further discussion of electrogravimetry.

When we use thermal or chemical energy to remove a volatile species, we call the method volatilization gravimetry. In determining the moisture content of bread, for example, we use thermal energy to vaporize the water in the sample. To determine the amount of carbon in an organic compound, we use the chemical energy of combustion to convert it to CO_2.

Finally, in particulate gravimetry we determine the analyte by separating it from the sample's matrix using a filtration or an extraction. The determination of total suspended solids is one example of particulate gravimetry.

9.3 CONSERVATION OF MASS

An accurate gravimetric analysis requires that the analytical signal—whether it is a mass or a change in mass—is proportional to the amount of analyte in our sample. For all gravimetric methods this proportionality involves a conservation of mass. If the method relies on one or more chemical reactions, then we must know the stoichiometry of the reactions. In the analysis of $PO_3^{-3}PO_3^{3-}$ described earlier, for example, we know that each mole of Hg_2Cl_2 corresponds to a mole of $PO_3^{-3}PO_3^{3-}$. If we remove the analyte from its matrix, then the separation must be selective for the analyte. When determining the moisture content in bread, for example, we know that the mass of H_2O in the bread is the difference between the sample's final mass and its initial mass.

We will return to this concept of applying a conservation of mass later in the chapter when we consider specific examples of gravimetric methods.

9.4 WHY GRAVIMETRY IS IMPORTANT

Except for particulate gravimetry, which is the most trivial form of gravimetry, you probably will not use gravimetry after you complete this course. Why, then, is familiarity with gravimetry still important? The answer is that gravimetry is one of only a small number of definitive techniques whose measurements require only base SI units, such as mass or the mole, and defined constants, such as Avogadro's number and the mass of 12C. Ultimately, we must be able to trace the result of any analysis to a definitive technique, such as gravimetry, that we can relate to fundamental physical properties [Valacárcel, M.; Ríos, A. Analyst 1995, 120, 2291–2297]. Although most analysts never use gravimetry to validate their results, they often verifying an analytical method by analyzing a standard reference material whose composition is traceable to a definitive technique [(a) Moody, J. R.; Epstein, M. S. Spectrochim. Acta 1991, 46B, 1571–1575; (b) Epstein, M. S. Spectrochim. Acta 1991, 46B, 1583–1591].

9.4.1 Precipitation Gravimetry

In precipitation gravimetry an insoluble compound forms when we add a precipitating reagent, or precipitant, to a solution that contains our analyte. In most cases the precipitate is the product of a simple metathesis reaction between the analyte and the precipitant; however, any reaction that generates a precipitate potentially can serve as a gravimetric method.

Most precipitation gravimetric methods were developed in the nineteenth century, or earlier, often for the analysis of ores.

9.4.2 Theory and Practice

All precipitation gravimetric analyses share two important attributes. First, the precipitate must be of low solubility, of high purity, and of known composition if its mass is to reflect accurately the analyte's mass. Second, it must be easy to separate the precipitate from the reaction mixture.

9.4.3 Solubility Considerations

To provide an accurate result, a precipitate's solubility must be minimal. The accuracy of a total analysis technique typically is better than ±0.1%, which means the precipitate must account for at least 99.9% of the analyte. Extending this requirement to 99.99% ensures the precipitate's solubility will not limit the accuracy of a gravimetric analysis.

A total analysis technique is one in which the analytical signal—mass in this case—is proportional to the absolute amount of analyte in the sample.

We can minimize solubility losses by controlling the conditions under which the precipitate forms. This, in turn, requires that we account for every equilibrium reaction that might affect the precipitate's solubility. For example, we can determine Ag+ gravimetrically by adding NaCl as a precipitant, forming a precipitate of AgCl.

$Ag^{+}(aq)+Cl^{-}(aq)\rightleftharpoons AgCl(s)$ 1

If this is the only reaction we consider, then we predict that the precipitate's solubility, SAgCl, is given by the following equation.

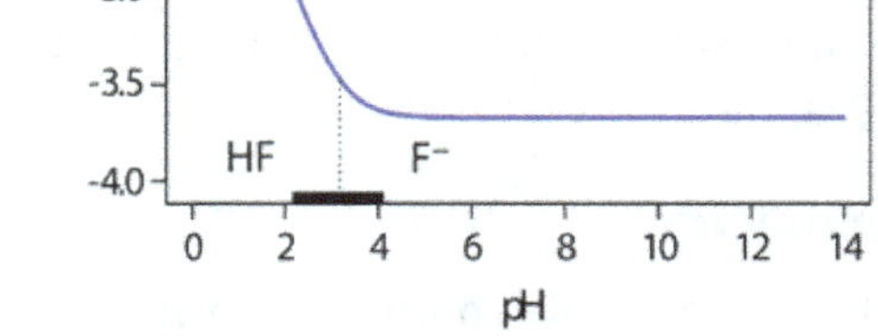

Equation 2 suggests that we can minimize solubility losses by adding a large excess of Cl–. In fact, as shown in Figure 1 , adding a large excess of Cl– increases the precipitate's solubility.

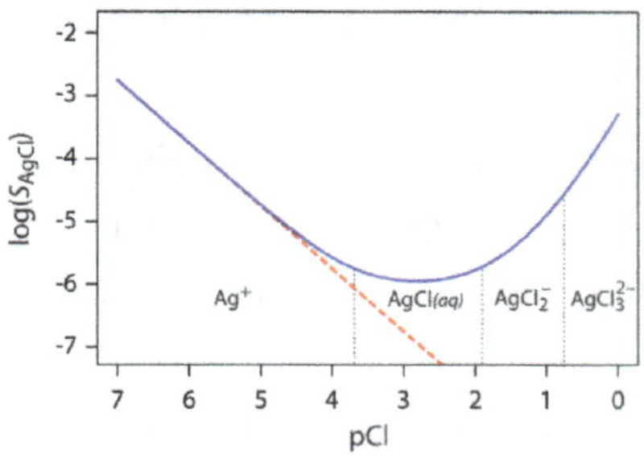

Fig. 1: Solubility of AgCl as a function of pCl. The dashed red line shows our prediction for SAgCl if we assume incorrectly that only reaction 1 and Equation2 affect silver chloride's solubility. The solid blue curve is calculated using Equation7, which accounts for reaction 1 and reactions 5. Because the solubility of AgCl spans several orders of magnitude, SAgCl is displayed on the y-axis in logarithmic form. The vertical dotted lines are a ladder diagram for the silver-chloro complexes.

To understand why the solubility of AgCl is more complicated than the relationship suggested by Equation 2, we must recall that Ag^+ also forms a series of soluble silver-chloro metal–ligand complexes.

$$Ag^+(aq)+Cl^-(aq) \rightleftharpoons AgCl\ (aq)\ \log K_1 = 3.70 \quad (3)$$

$$AgCl(aq)+ Cl^-(aq) \rightleftharpoons AgCl^2(aq)\log K_2=1.92 \quad (4)$$

$$AgCl^-_2(aq)+Cl^-(aq) \rightleftharpoons Ag\ Cl^{2-}_3(aq)\log K_3 =0.78 \quad (5)$$

Note the difference between reaction 3, in which we form AgCl(*aq*) as a product, and reaction 1, in which we form AgCl(*s*) as a product. The formation of AgCl(*aq*) from AgCl(*s*)

$$AgCl(s) \rightleftharpoons AgCl(aq)$$

is called AgCl's intrinsic solubility.

The actual solubility of AgCl is the sum of the equilibrium concentrations for all soluble forms of Ag+.

$$SgCl=[Ag^+]+[AgCl(aq)]+[AgCl^-_2]+[AgCl^{2-}_3] \quad (6)$$

By substituting into Equation 8.2.68.2.6 the equilibrium constant expressions for reaction 8.2.18.2.1 and reactions 8.2.38.2.3–8.2.58.2.5, we can define the solubility of AgCl as

$$S_{AgCl} = \frac{K_{sp}}{[Cl^-]} + K_1K_{sp} + K_1K_2K_{sp}[Cl^-] + K_1K_2K_3K_{sp}[Cl^-]^2 \quad (7)$$

Equation7 explains the solubility curve for AgCl shown in Figure 1. As we add NaCl to a solution of Ag+, the solubility of AgCl initially decreases because of reaction1. Under these conditions, the final three terms in Equation 7 are small and Equation2 is sufficient to describe AgCl's solubility. For higher concentrations of Cl–, reaction 4 and reaction 5 increase the solubility of AgCl. Clearly the equilibrium concentration of chloride is important if we wish to determine the concentration of silver by precipitating AgCl. In particular, we must avoid a large excess of chloride.

The predominate silver-chloro complexes for different values of pCl are shown by the ladder diagram along the x-axis in Figure 1. Note that the increase in solubility begins when the higher-order soluble complexes of AgCl−2AgCl2− and AgCl2−3AgCl32− are the predominate species.

Another important parameter that may affect a precipitate's solubility is pH. For example, a hydroxide precipitate, such as $Fe(OH)_3$, is more soluble at lower pH levels where the concentration of OH−is small. Because fluoride is a weak base, the solubility of calcium fluoride, SCaF2SCaF2, also is pH-dependent. We can derive an equation for $SCaF_2SCaF_2$ by considering the following equilibrium reactions

$CaF^2(s) \rightleftharpoons Ca^{2+}(aq)+2F-(aq)Ksp=3.9 \times 10^{-11}$ (8)

$HF(aq) + H_2O(l) \rightleftharpoons H_3O+(aq) + F-(aq)\ Ka=6.8\times10^{-4}$ (9)

and the following equation for the solubility of CaF_2.

$$S_{CaF_2} = [Ca^{2+}] = \frac{1}{2}\{[F^-] + [HF]\} \quad 10$$

Be sure that Equation 10 makes sense to you. Reaction 8 tells us that the dissolution of CaF_2 produces one mole of Ca^{2+} for every two moles of F−, which explains the term of 1/2 in Equation 10. Because F− is a weak base, we must account for both chemical forms in solution, which explains why we include HF.

Substituting the equilibrium constant expressions for reaction 8 and reaction 9 into Equation 10 allows us to define the solubility of CaF_2 in terms of the equilibrium concentration of H_3O^+.

$$S_{CaF_2} = [Ca^{2+}] = \left\{\frac{K_p}{4}\left(1+\frac{[H_3O^+]}{K_a}\right)^2\right\}^{1/3}$$

Fig. 2: shows how pH affects the solubility of CaF_2. Depending on the solution's pH, the predominate form of fluoride is either HF or F−. When the pH is greater than 4.17, the predominate species is F− and the solubility of CaF_2 is independent of pH because only reaction 8.2.88.2.8 occurs to an appreciable extent. At more acidic pH levels, the solubility of CaF_2 increases because of the contribution of reaction 9.

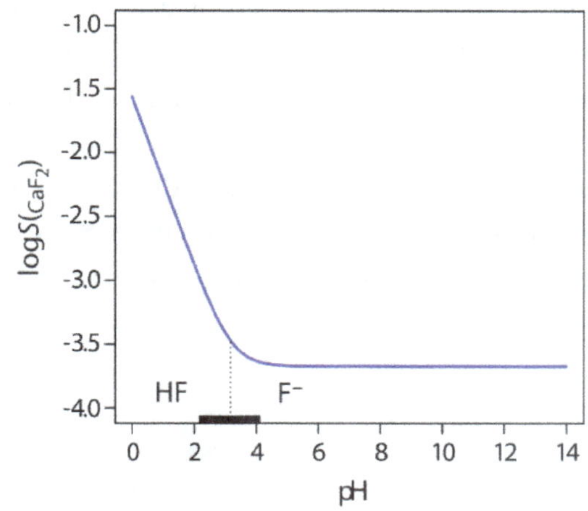

Fig. 3:. Solubility of CaF_2 as a function of pH. The solid blue curve is a plot of Equation 11. The predominate form of fluoride in solution is shown by the ladder diagram along the x-axis, with the black rectangle indicating the region where both HF and F- are important species. Note that the solubility of CaF_2 is independent of pH for pH levels greater than 4.17, and that its solubility increases dramatically at lower pH levels where HF is the predominate species. Because the solubility of CaF_2 spans several orders of magnitude, its solubility is shown in logarithmic form.

Exercise 1

You can use a ladder diagram to predict the conditions that will minimize a precipitate's solubility. Draw a ladder diagram for oxalic acid, H_2C2O_4, and use it to predict the range of pH values that will minimize the solubility of CaC_2O_4. Relevant equilibrium constants are in the appendices.

Answer

When solubility is a concern, it may be possible to decrease solubility by using a non-aqueous solvent. A precipitate's solubility generally is greater in an aqueous solution because of water's ability to stabilize ions through solvation. The poorer solvating ability of a non-aqueous solvent, even those that are polar, leads to a smaller solubility product. For example, the K_{sp} of $PbSO_4$ is 2×10−82×10 8 in H_2O and 2.6×10−122.6×10−12 in a 50:50 mixture of H_2O and ethanol.

9.4.4 Avoiding Impurities

In addition to having a low solubility, a precipitate must be free from impurities. Because precipitation usually occurs in a solution that is rich in dissolved solids, the initial precipitate often is impure. To avoid a determinate error, we must remove these impurities before we determine the precipitate's mass.

The greatest source of impurities are chemical and physical interactions that take place at the precipitate's surface. A precipitate generally is crystalline—even if only on a microscopic scale—with a well-defined lattice of cations and anions. Those cations and anions at the precipitate's surface carry, respectively, a positive or a negative charge because they have incomplete coordination spheres. In a precipitate of AgCl, for example, each silver ion in the precipitate's interior is bound to six chloride ions. A silver ion at the surface, however, is bound to no more than five chloride ions and carries a partial positive charge (Figure 8.2.3). The presence of these partial charges makes the precipitate's surface an active site for the chemical and physical interactions that produce impurities.

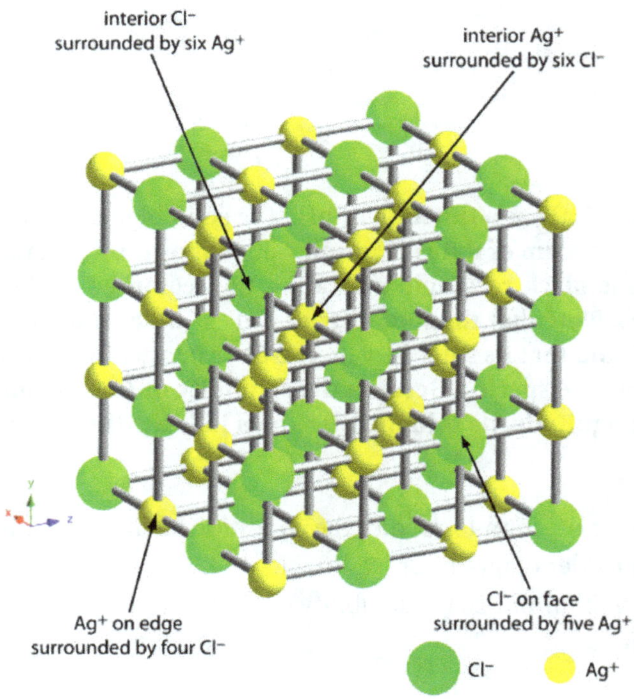

Fig. 4: Ball-and-stick model showing the lattice structure of AgCl. Each silver ion in the lattice's interior binds with six chloride ions and each chloride ion in the interior binds with six silver ions. Those ions on the lattice's surface or edges bind to fewer than six ions and carry a partial charge. A silver ion on the surface, for example, carries a partial positive charge. These charges make a precipitate's surface an active site for chemical and physical interactions.

One common impurity is an inclusion, in which a potential interferent, whose size and charge is similar to a lattice ion, can substitute into the lattice structure if the interferent precipitates with the same crystal structure (Figure 8.2.4 a). The probability of forming an inclusion is greatest when the interfering ion's concentration is substantially greater than the lattice ion's concentration. An inclusion does not decrease the amount of analyte that precipitates, provided that the precipitant is present in sufficient excess. Thus, the precipitate's mass always is larger than expected.

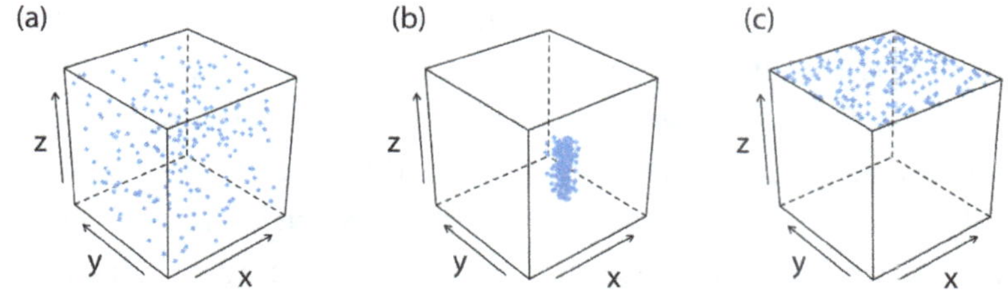

Fig. 5: Three examples of impurities that may form during precipitation. The cubic frame represents the precipitate and the blue marks are impurities present as (a) inclusions, (b) occlusions, and (c) surface adsorbates. Inclusions are randomly distributed throughout the precipitate. Occlusions are localized within the interior of the precipitate and surface adsorbates are localized on the precipitate's exterior. For ease of viewing, in (c) adsorption is shown on only one surface.

An inclusion is difficult to remove since it is chemically part of the precipitate's lattice. The only way to remove an inclusion is through reprecipitation in which we isolate the precipitate from its supernatant solution, dissolve the precipitate by heating in a small portion of a suitable solvent, and then reform the precipitate by allowing the solution to cool. Because the interferent's concentration after dissolving the precipitate is less than that in the original solution, the amount of included material decreases upon reprecipitation. We can repeat the process of reprecipitation until the inclusion's mass is insignificant. The loss of analyte during reprecipitation, however, is a potential source of determinate error.

Suppose that 10% of an interferent forms an inclusion during each precipitation. When we initially form the precipitate, 10% of the original interferent is present as an inclusion. After the first reprecipitation, 10% of the included interferent remains, which is 1% of the original interferent. A second reprecipitation decreases the interferent to 0.1% of the original amount.

An occlusion forms when an interfering ions is trapped within the growing precipitate. Unlike an inclusion, which is randomly dispersed within the precipitate, an occlusion is localized, either along flaws within the precipitate's lattice structure or within aggregates of individual precipitate particles. An occlusion usually increases a precipitate's mass; however, the precipitate's mass is smaller if the occlusion includes the analyte in a lower molecular weight form than that of the precipitate.

We can minimize an occlusion by maintaining the precipitate in equilibrium with its supernatant solution for an extended time, a process called digestion. During a digestion, the dynamic nature of the solubility–precipitation equilibria, in which the precipitate dissolves and reforms, ensures that the occlusion eventually is reexposed to the supernatant solution. Because the rates of dissolution and reprecipitation are slow, there is less opportunity for forming new occlusions.

After precipitation is complete the surface continues to attract ions from solution (Figure 8.2.4 c). These surface adsorbates comprise a third type of impurity. We can minimize surface adsorption by decreasing the precipitate's available surface area. One benefit of digestion is that it increases a precipitate's average particle size. Because the probability that a particle will dissolve completely is inversely proportional to its size, during digestion larger particles increase in size at the expense of smaller particles. One consequence of forming a smaller number of larger particles is an overall decrease in the precipitate's surface area. We also can remove surface adsorbates by washing the precipitate, although we cannot ignore the potential loss of analyte.

Inclusions, occlusions, and surface adsorbates are examples of coprecipitates—otherwise soluble species that form along with the precipitate that contains the analyte. Another type of impurity is an interferent that forms an independent precipitate under the conditions of the analysis. For example, the precipitation of nickel dimethylglyoxime requires a slightly basic pH. Under these conditions any Fe^{3+} in the sample will precipitate as $Fe(OH)_3$. In addition, because most precipitants rarely are selective toward a single analyte, there is a risk that the precipitant will react with both the analyte and an interferent.

In addition to forming a precipitate with Ni^{2+}, dimethylglyoxime also forms precipitates with Pd^{2+} and Pt^{2+}. These cations are potential interferents in an analysis for nickel.

We can minimize the formation of additional precipitates by controlling solution conditions. If an interferent forms a precipitate that is less soluble than the analyte's precipitate, we can precipitate the interferent and remove it by filtration, leaving the analyte behind in solution. Alternatively, we can mask the analyte or the interferent to prevent its precipitation. Both of the approaches outline above are illustrated in Fresenius' analytical method for the determination of Ni in ores that contain Pb^{2+}, Cu^{2+}, and Fe^{3+}. Dissolving the ore in the presence of H_2SO_4 selectively precipitates Pb^{2+} as $PbSO_4$. Treating the resulting supernatant with H_2S precipitates Cu^{2+} as CuS. After removing the CuS by filtration, ammonia is added to precipitate Fe^{3+} as $Fe(OH)_3$. Nickel, which forms a soluble amine complex, remains in solution.

9.4.5 Controlling Particle Size

Size matters when it comes to forming a precipitate. Larger particles are easier to filter and, as noted earlier, a smaller surface area means there is less opportunity for surface adsorbates to form. By controlling the reaction conditions we can significantly increase a precipitate's average particle size.

The formation of a precipitate consists of two distinct events: nucleation, the initial formation of smaller, stable particles of the precipitate, and particle growth. Larger particles form when the rate of particle growth exceeds the rate of nucleation. Understanding the conditions that favor particle growth is important when we design a gravimetric method of analysis.

We define a solute's ***relative supersaturation***, *RSS*, as

$$RSS = Q - SS \qquad 12$$

where Q is the solute's actual concentration and S is the solute's concentration at equilibrium. The numerator of Equation 12, Q – S, is a measure of the solute's supersaturation. A solution with a large, positive value of RSS has a high rate of nucleation and produces a precipitate with many small particles. When the RSS is small, precipitation is more likely to occur by particle growth than by nucleation.

A supersaturated solution is one that contains more dissolved solute than that predicted by equilibrium chemistry. A supersaturated solution is inherently unstable and precipitates solute to reach its equilibrium position. How quickly precipitation occurs depends, in part, on the value of RSS.

Equation-12 suggests that we can minimize RSS if we decrease the solute's concentration, Q, or if we increase the precipitate's solubility, S. A precipitate's solubility usually increases at higher temperatures and adjusting pH may affect a precipitate's solubility if it contains an acidic or a basic ion. Temperature and pH, therefore, are useful ways to increase the value of S. Forming the precipitate in a dilute solution of analyte or adding the precipitant slowly and with vigorous stirring are ways to decrease the value of Q.

There are practical limits to minimizing RSS. Some precipitates, such as Fe(OH)3 and PbS, are so insoluble that S is very small and a large RSS is unavoidable. Such solutes inevitably form small particles. In addition, conditions that favor a small RSS may lead to a relatively stable supersaturated solution that requires a long time to precipitate fully.

A visible precipitate takes longer to form when RSS is small both because there is a slow rate of nucleation and because there is a steady decrease in RSS as the precipitate forms. One solution to the latter problem is to generate the precipitant in situ as the product of a slow chemical reaction, which effectively maintains a constant RSS. Because the precipitate forms under conditions of low RSS, initial nucleation produces a small number of particles. As additional precipitant forms, particle growth supersedes nucleation, which results in larger particles of precipitate. This process is called a ***homogeneous precipitation.***

Two general methods are used for homogeneous precipitation. If the precipitate's solubility is pH-dependent, then we can mix the analyte and the precipitant under conditions where precipitation does not occur, and then increase or decrease the pH by chemically generating OH– or H_3O^+. For example, the hydrolysis of urea, $CO(NH_2)_2$, is a source of OH– because of the following two reactions.

$$CO(NH_2)^2(aq)+H_2O(l) \rightleftharpoons 2NH_3(aq)+CO_2(g)$$

$$NH_3(aq)+H_2O(\mathrm{l}) \rightleftharpoons OH-(\underline{aq})+NH+4(aq)$$

Because the hydrolysis of urea is temperature-dependent—the rate is negligible at room temperature—we can use temperature to control the rate of hydrolysis and the rate of precipitate formation. Precipitates of CaC_2O_4, for example, have been produced by this method. After dissolving a sample that contains Ca^{2+}, the solution is made acidic with HCl before adding a solution of 5% w/v $(NH_4)2C_2O_4$. Because the solution is acidic, a precipitate of CaC_2O_4 does not form. The solution is heated to approximately 50oC and urea is added. After several minutes, a precipitate of CaC_2O_4 begins to form, with precipitation reaching completion in about 30 min.

In the second method of homogeneous precipitation, the precipitant is generated by a chemical reaction. For example, Pb_{2+} is precipitated homogeneously as PbCrO4 by using bromate, BrO−3BrO3−, to oxidize Cr^{3+} to $CrO_2^{-4}CrO_4^{2-}$.

$$6BrO^-_3(aq)+10Cr^{3+}(aq)+22H_2O(l)\rightleftharpoons 3Br_2(aq)+10CrO^{2-}_4(aq)+44H+(aq)$$

Figure 5 shows the result of preparing PbCrO4 by direct addition of K2CrO4 (Beaker A) and by homogenous precipitation (Beaker B). Both beakers contain the same amount of PbCrO4. Because the direct addition of K2CrO4 leads to rapid precipitation and the formation of smaller particles, the precipitate remains less settled than the precipitate prepared homogeneously. Note, as well, the difference in the color of the two precipitates.

Fig. 6: Two precipitates of PbCrO4. In Beaker A, mixing together 0.1 M Pb(NO3)2 and 0.1 M K_2CrO_4 forms the precipitate under conditions of high RSS. The precipitate forms rapidly and consists of very small particles. In Beaker B, heating a solution of 0.1 M Pb(NO3)2, 0.1 M Cr(NO3)3, and 0.1 M $KBrO_3$ slowly oxidizes Cr^{3+} to $CrO_2^{-4}CrO_4^{2-}$, precipitating $PbCrO_4$ under conditions of low RSS. The precipitate forms slowly and consists of much larger particles.

The effect of particle size on color is well-known to geologists, who use a streak test to help identify minerals. The color of a bulk mineral and its color when powdered often are different. Rubbing a mineral across an unglazed porcelain plate leaves behind a small streak of the powdered mineral. Bulk samples of hematite, Fe_2O_3, are black in color, but its streak is a familiar rust-red. Crocite, the mineral $PbCrO_4$, is red-orange in color; its streak is orange-yellow.

A homogeneous precipitation produces large particles of precipitate that are relatively free from impurities. These advantages, however, are offset by the increased time needed to produce the precipitate and by a tendency for the precipitate to deposit as a thin film on the container's walls. The latter problem is particularly severe for hydroxide precipitates generated using urea.

An additional method for increasing particle size deserves mention. When a precipitate's particles are electrically neutral they tend to coagulate into larger particles that are easier to filter. Surface adsorption of excess lattice ions, however, provides the precipitate's particles with a net positive or a net negative surface charge. Electrostatic repulsion between particles of similar charge prevents them from coagulating into larger particles.

Let's use the precipitation of AgCl from a solution of $AgNO_3$ using NaCl as a precipitant to illustrate this effect. Early in the precipitation, when NaCl is the limiting reagent, excess Ag+ ions chemically adsorb to the AgCl particles, forming a positively charged primary adsorption layer (Figure 6 a). The solution in contact with this layer contains more inert anions, NO−3 in this case, than inert cations, Na+, giving a secondary adsorption layer with a negative charge that balances the primary adsorption layer's positive charge. The solution outside the secondary adsorption layer remains electrically neutral. Coagulation cannot occur if the secondary adsorption layer is too thick because the individual particles of AgCl are unable to approach each other closely enough.

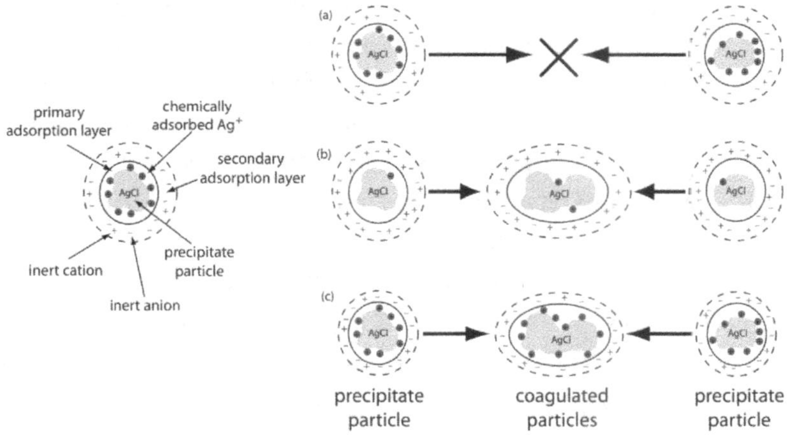

Fig. 7: Two methods for coagulating a precipitate of AgCl. (a) Coagulation does not occur due to the electrostatic repulsion between the positively charged particles. (b) Decreasing the charge within the primary adsorption layer, by adding additional NaCl, decreases the electrostatic repulsion and allows the particles to coagulate. (c) Adding additional inert ions decreases the thickness of the secondary adsorption layer. Because the particles can approach each other more closely, they are able to coagulate.

We can induce coagulation in three ways: by decreasing the number of chemically adsorbed Ag+ ions, by increasing the concentration of inert ions, or by heating the solution. As we add additional NaCl, precipitating more of the excess Ag+, the number of chemically adsorbed silver ions decreases and coagulation occurs. Adding too much NaCl, however, creates a primary adsorption layer of excess Cl− with a loss of coagulation.

The coagulation and decoagulation of AgCl as we add NaCl to a solution of AgNO can serve as an endpoint for a titration.

A second way to induce coagulation is to add an inert electrolyte, which increases the concentration of ions in the secondary adsorption layer. With more ions available, the thickness of the secondary absorption layer decreases. Particles of precipitate may now approach each other more closely, which allows the precipitate to coagulate. The amount of electrolyte needed to cause spontaneous coagulation is called the critical coagulation concentration.

Heating the solution and the precipitate provides a third way to induce coagulation. As the temperature increases, the number of ions in the primary adsorption layer decreases, which lowers the precipitate's surface charge. In addition, heating increases the particles' kinetic energy, allowing them to overcome the electrostatic repulsion that prevents coagulation at lower temperatures.

9.4.6 Filtering the Precipitate

After precipitating and digesting a precipitate, we separate it from solution by filtering. The most common filtration method uses filter paper, which is classified according to its speed, its size, and its ash content on ignition. Speed, or how quickly the supernatant passes through the filter paper, is a function of the paper's pore size. A larger pore size allows the supernatant to pass more quickly through the filter paper, but does not retain small particles of precipitate. Filter paper is rated as fast (retains particles larger than 20–25 μm), medium–fast (retains particles larger than 16 μm), medium (retains particles larger than 8 μm), and slow (retains particles larger than 2–3 μm). The proper choice of filtering speed is important. If the filtering speed is too fast, we may fail to retain some of the precipitate, which causes a negative determinate error. On the other hand, the precipitate may clog the pores if we use a filter paper that is too slow.

A filter paper's size is just its diameter. Filter paper comes in many sizes, including 4.25 cm, 7.0 cm, 11.0 cm, 12.5 cm, 15.0 cm, and 27.0 cm. Choose a size that fits comfortably into your funnel. For a typical 65-mm long-stem funnel, 11.0 cm and 12.5 cm filter paper are good choices.

Because filter paper is hygroscopic, it is not easy to dry it to a constant weight. When accuracy is important, the filter paper is removed before we determine the precipitate's mass. After transferring the precipitate and filter paper to a covered crucible, we heat the crucible to a temperature that coverts the paper to $CO_2(g)$ and $H_2O(g)$, a process called ***ignition.***

Igniting a poor quality filter paper leaves behind a residue of inorganic ash. For quantitative work, use a low-ash filter paper. This grade of filter paper is pretreated with a mixture of HCl and HF to remove inorganic materials. Quantitative filter paper typically has an ash content of less than 0.010% w/w.

Gravity filtration is accomplished by folding the filter paper into a cone and placing it in a long-stem funnel (Figure 7). To form a tight seal between the filter cone and the funnel, we dampen the paper with water or supernatant and press the paper to the wall of the funnel. When prepared properly, the funnel's stem fills with the supernatant, increasing the rate of filtration.

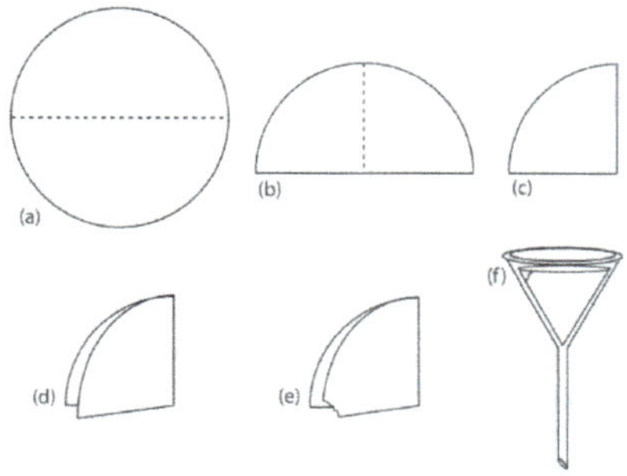

Fig. 8: To prepare a filter paper cone the filter paper circle is (a) folded in half (b), folded in half a second time (c), parted (d) and a small corner is torn off (e) before the paper is opened up into a cone and placed in the funnel (f).

The precipitate is transferred to the filter in several steps. The first step is to decant the majority of the supernatant through the filter paper without transferring the precipitate (Figure 8). This prevents the filter paper from clogging at the beginning of the filtration process. The precipitate is rinsed while it remains in its beaker, with the rinsings decanted through the filter paper. Finally, the precipitate is transferred onto the filter paper using a stream of rinse solution. Any precipitate that clings to the walls of the beaker is transferred using a rubber policeman (a flexible rubber spatula attached to the end of a glass stirring rod).

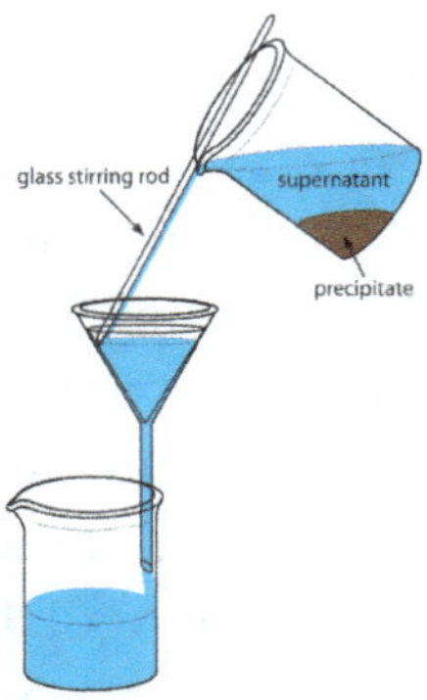

Fig. 9: Proper procedure for transferring the supernatant to the filter paper cone. The glass stirring rod allows the supernatant to trickle into the funnel without splashing.

An alternative method for filtering a precipitate is to use a filtering crucible. The most common option is a fritted-glass crucible that contains a porous glass disk filter. Fritted-glass crucibles are classified by their porosity: coarse (retaining particles larger than 40–60 μm), medium (retaining particles greater than 10–15 μm), and fine (retaining particles greater than 4–5.5 μm). Another type of filtering crucible is the Gooch crucible, which is a porcelain crucible with a perforated bottom. A glass fiber mat is placed in the crucible to retain the precipitate. For both types of crucibles, the pre- cipitate is transferred in the same manner described earlier for filter paper. Instead of using gravity, the supernatant is drawn through the crucible with the assistance of suction from a vacuum aspirator or pump (Figure 9).

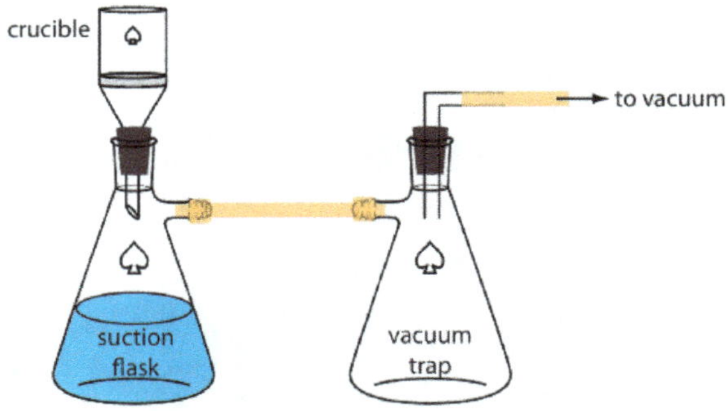

Fig. 10: Procedure for filtering a precipitate through a filtering crucible. The trap prevents water from the aspirator from back-washing into the suction flask.

9.4.7 Rinsing the Precipitate

Because the supernatant is rich with dissolved inert ions, we must remove residual traces of supernatant without incurring loss of analyte due to solubility. In many cases this simply involves the use of cold solvents or rinse solutions that contain organic solvents such as ethanol. The pH of the rinse solution is critical if the precipitate contains an acidic or a basic ion. When coagulation plays an important role in determining particle size, adding a volatile inert electrolyte to the rinse solution prevents the precipitate from reverting into smaller particles that might pass through the filter. This process of reverting to smaller particles is called peptization. The volatile electrolyte is removed when drying the precipitate.

In general, we can minimize the loss of analyte if we use several small portions of rinse solution instead of a single large volume. Testing the used rinse solution for the presence of an impurity is another way to guard against over-rinsing the precipitate. For example, if Cl– is a residual ion in the supernatant, we can test for its presence using $AgNO_3$. After we collect a small portion of the rinse solution, we add a few drops of $AgNO_3$ and look for the presence or absence of a precipitate of AgCl. If a precipitate forms, then we know Cl– is present and continue to rinse the precipitate. Additional rinsing is not needed if the $AgNO_3$ does not produce a precipitate.

9.4.8 Drying the Precipitate

After separating the precipitate from its supernatant solution, we dry the precipitate to remove residual traces of rinse solution and to remove any volatile impurities. The temperature and method of drying depend on the method of filtration and the precipitate's desired chemical form. Placing the precipitate in a laboratory oven and heating to a temperature of 110°C is sufficient to remove water and other easily volatilized impurities. Higher temperatures require a muffle furnace, a Bunsen burner, or a Meker burner, and are necessary if we need to decompose the precipitate before its weight is determined.

Because filter paper absorbs moisture, we must remove it before we weigh the precipitate. This is accomplished by folding the filter paper over the precipitate and transferring both the filter paper and the precipitate to a porcelain or platinum crucible. Gentle heating first dries and then chars the filter paper. Once the paper begins to char, we slowly increase the temperature until there is no trace of the filter paper and any remaining carbon is oxidized to CO_2.

Fritted-glass crucibles cannot withstand high temperatures and are dried in an oven at a temperature below 200°C. The glass fiber mats used in Gooch crucibles can be heated to a maximum temperature of approximately 500°C.

9.4.9 Composition of the Final Precipitate

For a quantitative application, the final precipitate must have a well-defined composition. A precipitate that contains volatile ions or substantial amounts of hydrated water, usually is dried at a temperature that completely removes these volatile species. For example, one standard gravimetric method for the determination of magnesium involves its precipitation as $MgNH_4PO_4{\cdot}6H_2O$. Unfortunately, this precipitate is difficult to dry at lower temperatures without losing an inconsistent amount of hydrated water and ammonia. Instead, the precipitate is dried at a temperature greater than 1000°C where it decomposes to magnesium pyrophosphate, $Mg_2P_2O_7$.

An additional problem is encountered if the isolated solid is nonstoichiometric. For example, precipitating Mn^{2+} as $Mn(OH)_2$ and heating frequently produces a nonstoichiometric manganese oxide, MnOx, where x varies between one and two. In this case the nonstoichiometric product is the result of forming a mixture of oxides with different oxidation state of manganese. Other nonstoichiometric compounds form as a result of lattice defects in the crystal.

9.5 DETERMINATION OF MG IN WATER AND WASTEWATER

The best way to appreciate the theoretical and practical details discussed in this section is to carefully examine a typical precipitation gravimetric method. Although each method is unique, the determination of Mg^{2+} in water and wastewater by precipitating $MgNH_4PO_4H_2O$ and isolating Mg_2P2O_7 provides an instructive example of a typical procedure. The description here is based on Method 3500-Mg D in Standard Methods for the Examination of Water and Wastewater, 19th Ed., American Public Health Association: Washington, D. C., 1995. With the publication of the 20th Edition in 1998, this method is no longer listed as an approved method.

9.5.1 Description of Method

Magnesium is precipitated as $MgNH_4PO_46H_2O$ using $(NH4)_2HPO_4$ as the precipitant. The precipitate's solubility in a neutral solution is relatively high (0.0065 g/100 mL in pure water at 10°C), but it is much less soluble in the presence of dilute ammonia (0.0003 g/100 mL in 0.6 M NH_3). Because the precipitant is not selective, a preliminary separation of Mg^{2+} from potential interferents is necessary. Calcium, which is the most significant interferent, is removed by precipitating it as CaC_2O_4. The presence of excess ammonium salts from the precipitant, or from the addition of too much ammonia, leads to the formation of Mg $(NH_4)_4(PO_4)_2$, which forms $Mg(PO_3)_2$ after drying. The precipitate is isolated by gravity filtration, using a rinse solution of dilute ammonia. After filtering, the precipitate is converted to $Mg_2P_2O_7$ and weighed.

9.5.2 Procedure

Transfer a sample that contains no more than 60 mg of Mg^{2+} into a 600-mL beaker. Add 2–3 drops of methyl red indicator, and, if necessary, adjust the volume to 150 mL. Acidify the solution with 6 M HCl and add 10 mL of 30% w/v $(NH_4)_2HPO_4$. After cooling and with constant stirring, add concentrated NH3 dropwise until the methyl red indicator turns yellow (pH > 6.3). After stirring for 5 min, add 5 mL of concentrated NH_3 and continue to stir for an additional 10 min. Allow the resulting solution and precipitate to stand overnight. Isolate the precipitate by filtering through filter paper, rinsing with 5% v/v NH_3. Dissolve the precipitate in 50 mL of 10% v/v HCl and precipitate a second time following the same procedure. After filtering, carefully remove the filter paper by charring. Heat the precipitate at 500°C until the residue is white, and then bring the precipitate to constant weight at 1100°C.

9.6 QUANTITATIVE APPLICATIONS

Although no longer a common analytical technique, precipitation gravimetry still provides a reliable approach for assessing the accuracy of other methods of analysis, or for verifying the composition of standard reference materials. In this section we review the general application of precipitation gravimetry to the analysis of inorganic and organic compounds.

9.6.1 Inorganic Analysis

Table 1 provides a summary of precipitation gravimetric methods for inorganic cations and anions. Several methods for the homogeneous generation of precipitants are shown in Table 2 . The majority of inorganic precipitants show poor selectivity for the analyte. Many organic precipitants, however, are selective for one or two inorganic ions. Table 8.2.3 lists examples of several common organic precipitants.

Table 1 . Selected Precipitation Gravimetric Methods for Inorganic Cations and Anions (Arranged by Precipitant)

Analyte	Precipitant	Precipitate formed	Precipitate weighed
Ba^{2+}	$(NH_4)_2CrO_4$	$BaCrO_4$	$BaCrO_4$
Pb^{2+}	K_2CrO_4	$PbCrO_4$	$PbCrO_4$
Ag^+	HCl	AgCl	AgCl
Hg2+2Hg22+	HCl	Hg_2Cl_2	Hg_2Cl_2
Al^{3+}	NH_3	$Al(OH)_3$	Al_2O_3
Be^{2+}	NH_3	$Be(OH)_2$	BeO
Fe^{3+}	NH_3	$Fe(OH)_3$	Fe_2O_3
Ca^{2+}	$(NH_4)_2CrO_4$	CaC_2O_4	$CaCO_3$ or CaO
Sb^{3+}	H_2S	Sb_2S_3	Sb_2S_3
As^{3+}	H_2S	As_2S_3	As_2S_3
Hg^{2+}	H_2S	HgS	HgS
Ba^{2+}	H_2SO_4	$BaSO_4$	$BaSO_4$
Pb^{2+}	H_2SO_4	$PbSO_4$	PbSO4
Sr^{2+}	H_2SO_4	$SrSO_4$	$SrSO_4$
Be^{3+}	$(NH_4)_2HPO_4$	NH_4BePO_4	$Be_2P_2O_7$
Mg^{2+}	$(NH_4)_2HPO_4$	NH_4MgPO_4	$Mg_2P_2O_7$
Zn^{2+}	$(NH_4)_2HPO_4$	NH_4ZnPO_4	$Zn_2P_2O_7$
Sr^{2+}	KH_2PO_4	$SrHPO_4$	$Sr_2P_2O_7$
CN^-	$AgNO_3$	AgCN	AgCN
I–	$AgNO_3$	AgI	AgI
Br^-	$AgNO_3$	AgBr	AgBr
Cl^-	$AgNO_3$	AgCl	AgCl
ClO–3ClO3–	$FeSO_4/AgNO_3$	AgCl	AgCl
SCN^-	$SO_2/CuSO_4$	CuSCN	CuSCN
SO24–SO42–	$BaCl_2$	$BaSO_4$	$BaSO_4$

Table 2 . Reactions for the Homogeneous Preparation of Selected Inorganic Precipitants

Precipitant	Reaction
OH–	$(NH_2)2CO(aq)+3H_2O(l)\rightleftharpoons 2NH^{+4}(aq)+CO^2(g)+2OH–(aq)$
SO2–4SO42–	$NH_2HSO^3(aq)+2H_2O(l)\rightleftharpoons NH+4(aq)+H_3O+(aq)+SO^{2-}{}_4(aq)$
S^{2-}	$CH_3CSNH_2(aq)+H_2O(l)\rightleftharpoons CH_3CONH_2(aq)+H_2S(aq)$
IO–3IO3–	$HOCH_2CH_2OH(aq)+IO^{-4}(aq)\rightleftharpoons 2HCHO(aq)+H_2O(l)+IO_{-3}$
PO3–4PO43–	$(CH_3O)3PO(aq)+3H_2O(l)\rightleftharpoons 3CH_3OH(aq)+H_3PO4(aq)$
C2O2–4C2O42–	$(C_2H5)2C_2O4(aq)+2H_2O(l)\rightleftharpoons 2C_2H_5OH(aq)+H_2C_2O_4(aq)$
CO2–3CO32–	$Cl_3CCOOH(aq)+2OH–(aq)\rightleftharpoons CHCl_3(aq)+CO_2{}^{-3}(aq)+H_2O(l)$

Analyte	Precipitant	Structure	Precipirate formed	Precipitate weighed
Ni^{2+}	Dimethylglyoxime	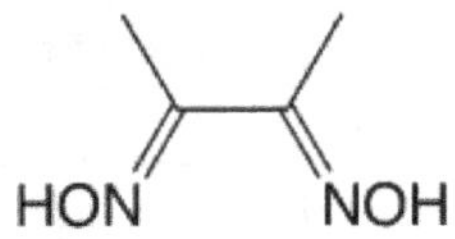	$Ni(C_4H_7O_2N_2)_2$	$Ni(C_4H_7O_2N_2)_2$
Fe^{3+}	Cupferron	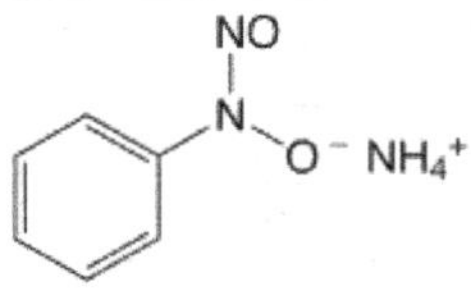	$Fe(C_6H_5N_2O_2)_3$	Fe_2O_3
Cu^{2+}	Cupron	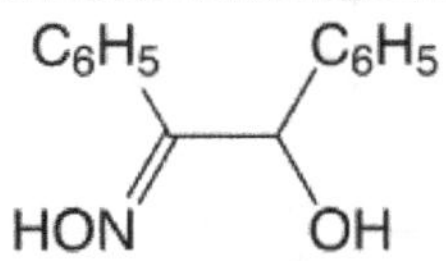	$CuC_{14}H_{11}O_2N$	$CuC_{14}H_{11}O_2N$
Co^{2+}	1-nitrso-2-napthol	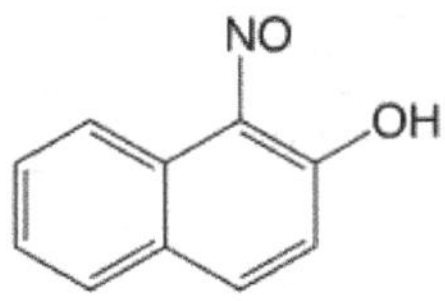	$Co(C_{10}H_6O_2N)_3$	Co or $CoSO_4$
K^+	Sodium tetraphenylborate	$Na[B(C_6H_5)_4]$	$K[B(C_6H_5)_4]$	$K[B(C_6H_5)_4]$
NO− 3NO3−	Nitron	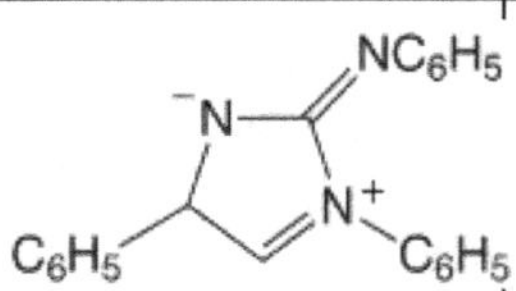	$C_{20}H_{16}N_4HNO_3$	$C_{20}H_{16}N_4HNO_3$

Precipitation gravimetry continues to be listed as a standard method for the determination of SO2−4SO42− in water and wastewater analysis [Method 4500-SO42− C and Method 4500-SO42− D as published in *Standard Methods for the Examination of Waters and Wastewaters*, 20th Ed., American Public Health Association: Wash- ington, D. C., 1998]. Precipitation is carried out using $BaCl_2$ in an acidic solution (adjusted with HCl to a pH of 4.5–5.0) to prevent the precipitation of $BaCO_3$ or $Ba_3(PO_4)_2$, and at a temperature near the solution's boiling point. The precipitate is digested at 80–90°C for at least two hours. Ashless filter paper pulp is added to the precipitate to aid in its filtration. After filtering, the precipitate is ignited to constant weight at 800°C. Alternatively, the precipitate is filtered through a fine porosity fritted glass crucible (without adding filter paper pulp), and dried to constant weight at 105°C. This procedure is subject to a variety of errors, including occlusions of $Ba(NO_3)_2$, $BaCl_2$, and alkali sulfates.

Other standard methods for the determination of sulfate in water and wastewater include ion chromatography, capillary ion electrophoresis, turbidimetry and flow injection analysis.

9.6.2 Organic Analysis

Several organic functional groups or heteroatoms can be determined using precipitation gravimetric methods. Table 4 provides a summary of several representative examples. Note that the determination of alkoxy functional groups is an indirect analysis in which the functional group reacts with and excess of HI and the unreacted I^- determined by precipitating as AgCl.

Table .4 . Selected Precipitation Gravimetric Methods for the Analysis of Organic Functional Groups and Heteroatoms

Analyte	Treatment	Precipitant	Precipitate
Organic halides (r-*x*) where *x* is cl, br, or i	Oxidation with hno_3 in the presence of ag^+	$AgNO_3$	Ag*X*
Organic halides (r-x) where *x* is cl, br, or i	Combustion in o_2 (with a pt catalyst) in the presence of ag+	$AgNO_3$	AgX
Organic sulfur	Oxidation with hno_3 in the presence of ba^{2+}	$BaCl_2$	$BaSO_4$
Organic sulfur	Combustion in o_2 (with pt catalyst) with so_2 and so_3 collected in dilute h_2o_2	$BaCl_2$	$BaSO_4$
alkoxy groups (–O-R or –COO-*R*) where R is $-CH_3$ or $-C_2H_5$	Reaction with HI to produce RI	$AgNO_3$	AgI

9.6.3 Quantitative Calculations

The stoichiometry of a precipitation reaction provides a mathematical relationship between the analyte and the precipitate. Because a precipitation gravimetric method may involve additional chemical reactions to bring the analyte into a different chemical form, knowing the stoichiometry of the precipitation reaction is not always sufficient. Even if you do not have a complete set of balanced chemical reactions, you can use a conservation of mass to deduce the mathematical relationship between the analyte and the precipitate. The following example demonstrates this approach for the direct analysis of a single analyte.

9.7 QUALITATIVE APPLICATIONS

A precipitation reaction is a useful method for identifying inorganic and organic analytes. Because a qualitative analysis does not require quantitative measurements, the analytical signal is simply the observation that a precipitate forms. Although qualitative applications of precipitation gravimetry have been replaced by spectroscopic methods of analysis, they continue to find application in spot testing for the presence of specific analytes.

Any of the precipitants listed in Table 1 , Table .3 , and Table.4 can be used for a qualitative analysis.

9.7.1Evaluating Precipitation Gravimetry

9.7.1.1 Scale of Operation

The scale of operation for precipitation gravimetry is limited by the sensitivity of the balance and the availability of sample. To achieve an accuracy of ±0.1% using an analytical balance with a sensitivity of ±0.1 mg, we must isolate at least 100 mg of precipitate. As a consequence, precipitation gravimetry usually is limited to major or minor analytes, in macro or meso samples. The analysis of a trace level analyte or a micro sample requires a microanalytical balance.

9.7.1.2 Accuracy

For a macro sample that contains a major analyte, a relative error of 0.1– 0.2% is achieved routinely. The principal limitations are solubility losses, impurities in the precipitate, and the loss of precipitate during handling. When it is difficult to obtain a precipitate that is free from impurities, it often is possible to determine an empirical relationship between the precipitate's mass and the mass of the analyte by an appropriate calibration.

9.7.1.2 Precision

The relative precision of precipitation gravimetry depends on the sample's size and the precipitate's mass. For a smaller amount of sample or precipitate, a relative precision of 1–2 ppt is obtained routinely. When working with larger amounts of sample or precipitate, the relative precision extends to several ppm. Few quantitative techniques can achieve this level of precision.

9.7.1.3 Sensitivity

For any precipitation gravimetric method we can write the following general equation to relate the signal (grams of precipitate) to the absolute amount of analyte in the sample

g precipitate =k×g analyte (8.2.13)(8.2.13) g precipitate =k×g analyte

where *k*, the method's sensitivity, is determined by the stoichiometry between the precipitate and the analyte.

Equation 13 assumes we used a suitable blank to correct the signal for any contributions of the reagent to the precipitate's mass.

Consider, for example, the determination of Fe as Fe_2O_3. Using a conservation of mass for iron, the precipitate's mass is g Fe2O3=g Fe×1 mol FeAW Fe×FW Fe2O32 mol Feg Fe2O3=g Fe×1 mol FeAW Fe×FW Fe2O32 mol Fe and the value of *k* is

k=12×FW Fe2O3AW Fe(8.2.14)(8.2.14)k=12×FW Fe2O3AW Fe

As we can see from Equation 8.2.148.2.14, there are two ways to improve a method's sensitivity. The most obvious way to improve sensitivity is to increase the ratio of the precipitate's molar mass to that of the analyte. In other words, it helps to form a precipitate with the largest possible formula weight. A less obvious way to improve a method's sensitivity is indicated by the term of 1/2 in Equation 8.2.148.2.14, which accounts for the stoichiometry between the analyte and precipitate. We can also improve sensitivity by forming a precipitate that contains fewer units of the analyte.

9.7.1.4 Selectivity

Due to the chemical nature of the precipitation process, precipitants usually are not selective for a single analyte. For example, silver is not a selective precipitant for chloride because it also forms precipitates with bromide and with iodide. Interferents often are a serious problem and must be considered if accurate results are to be obtained.

9.7.1.5 Time, Cost, and Equipment

Precipitation gravimetry is time intensive and rarely practical if you have a large number of samples to analyze; however, because much of the time invested in precipitation gravimetry does not require an analyst's immediate supervision, it is a practical alternative when working with only a few samples. Equipment needs are few—beakers, filtering devices, ovens or burners, and balances—inexpensive, routinely available in most laboratories, and easy to maintain.

9.8 VOLATILIZATION GRAVIMETRY

A second approach to gravimetry is to thermally or chemically decompose the sample and measure the resulting change in its mass. Alternatively, we can trap and weigh a volatile decomposition product. Because the release of a volatile species is an essential part of these methods, we classify them collectively as volatilization gravimetric methods of analysis.

9.8.1 Theory and Practice

Whether an analysis is direct or indirect, volatilization gravimetry usually requires that we know the products of the decomposition reaction. This rarely is a problem for organic compounds, which typically decompose to form simple gases such as CO_2, H_2O, and N_2. For an inorganic compound, however, the products often depend on the decomposition temperature.

9.8.2 Thermogravimetry

One method for determining the products of a thermal decomposition is to monitor the sample's mass as a function of temperature, a process called ***thermogravimetry***. Figure 8.3.1 shows a typical ***thermogram*** in which each change in mass—each "step" in the thermogram—represents the loss of a volatile product. As the following example illustrates, we can use a thermogram to identify a compound's decomposition reactions.

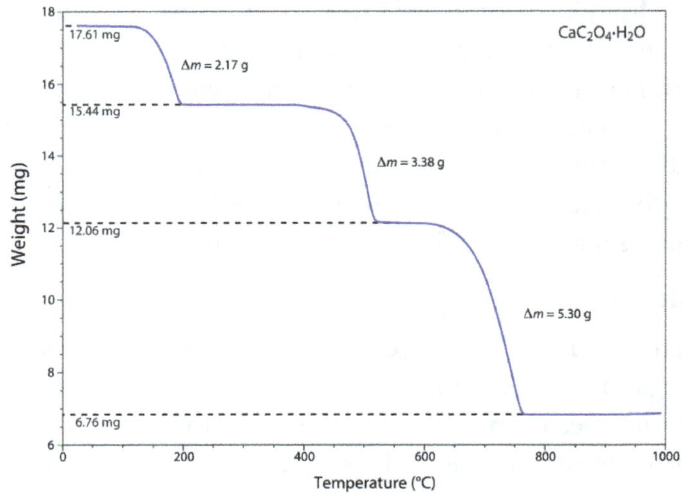

Fig. 11: Thermogram for $CaC_2O_4H_2O$ obtained by heating a sample from room temperature to 1000°C at a rate of 20°C/min. Each change in mass results from the loss of a volatile product. The sample's initial mass and its mass after each loss are shown by the dotted lines. See Example.1 for information on interpreting this thermogram.

Example.1

The thermogram in Figure 8.3.1 shows the mass of a sample of calcium oxalate monohydrate, $CaC_2O_4H_2O$, as a function of temperature. The original sample of 17.61 mg was heated from room temperature to 1000°C at a rate of 20°C per minute. For each step in the thermogram, identify the volatilization product and the solid residue that remains.

Solution

From 100–250°C the sample loses 17.61 mg – 15.44 mg, or 2.17 mg, which is 2.17 mg17.61 mg×100=12.3%2.17 mg17.61 mg×100=12.3% of the sample's original mass. In terms of $CaC_2O_4H_2O$, this corresponds to a decrease in the molar mass of 0.123×146.11 g/mol=18.0 g/mol0.123×146.11 g/mol=18.0 g/mol.

The product's molar mass and the temperature range for the decomposition, suggest that this is a loss of H_2O*(g)*, leaving a residue of CaC_2O_4.

The loss of 3.38 mg from 350–550oC is a 19.2% decrease in the sample's original mass, or a decrease in the molar mass of 0.192×146.11 g/mol=28.1 g/mol0.192×146.11 g/mol=28.1 g/mol which is consistent with the loss of CO(g) and a residue of $CaCO_3$.

Finally, the loss of 5.30 mg from 600-800oC is a 30.1% decrease in the sample's original mass, or a decrease in molar mass of 0.301×146.11 g/mol=44.0 g/mol0.301×146.11 g/mol=44.0 g/mol.

This loss in molar mass is consistent with the release of CO2(g), leaving a final residue of CaO. The three decomposition reactions are

$CaC_2O_4.H_2O(s)$→ CaC2O4(s)+2H2O(l)$CaC_2O_4(s)$→ $CaCO_3(s)$+CO(g)$CaCO_3(s)$→ CaO(s)+$CO_2(g)$$CaC_2O_4.H_2O(s)$→ $CaC_2O_4(s)^{+2}H_2O(l)$$CaC_2O_4(s)$→ $CaCO_3(s)$+CO(g) $CaCO_3(s)$→ CaO(s)+$CO_2(g)$

Identifying the products of a thermal decomposition provides information that we can use to develop an analytical procedure. For example, the thermogram in Figure 8.3.1 shows that we must heat a precipitate of $CaC_2O_4H_2O$ to a temperature between 250 and 400oC if we wish to isolate and weigh CaC_2O_4. Alternatively, heating the sample to 1000oC allows us to isolate and weigh CaO.

Exercise.1

Under the same conditions as Figure 8.3.1 , the thermogram for a 22.16 mg sample of $MgC_2O_4H_2O$ shows two steps: a loss of 3.06 mg from 100–250°C and a loss of 12.24 mg from 350–550oC. For each step, identify the volatilization product and the solid residue that remains. Using your results from this exercise and the results from Example 8.3.1 , explain how you can use thermogravimetry to analyze a mixture that contains CaC_2O4H_2O and $MgC_2O_4H_2O$. You may assume that other components in the sample are inert and thermally stable below 1000oC.

Equipment

Depending on the method of analysis, the equipment for volatilization gravimetry may be simple or complex. In the simplest experimental design, we place the sample in a crucible and decompose it at a fixed temperature using a Bunsen burner, a Meker burner, a laboratory oven, or a muffle furnace. The sample's mass and the mass of the residue are measured using an analytical balance.

Trapping and weighing the volatile products of a thermal decomposition requires specialized equipment. The sample is placed in a closed container and heated. As decomposition occurs, a stream of an inert purge-gas sweeps the volatile products through one or more selective absorbent traps. In a thermogravimetric analysis, the sample is placed on a small balance pan attached to one arm of an electromagnetic balance (Figure 8.3.2). The sample is lowered into an electric furnace and the furnace's temperature is increased at a fixed rate of few degrees per minute while monitoring continuously the sample's weight. The instrument usually includes a gas line for purging the volatile decomposition products out of the furnace, and a heat exchanger to dissipate the heat emitted by the furnace.

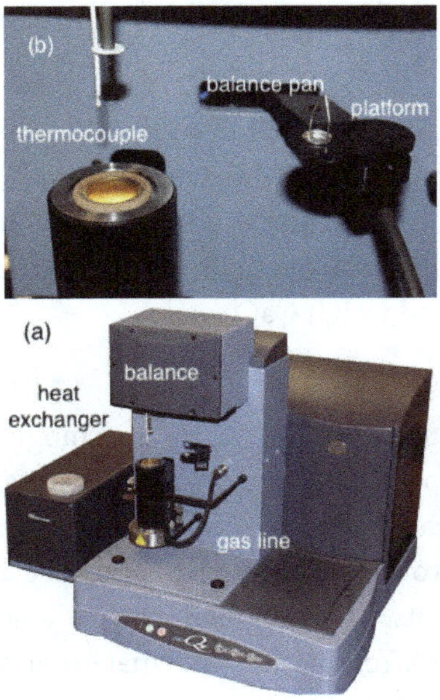

Fig. 12:(a) Instrumentation for conducting a thermogravimetric analysis.

The balance sits on the top of the instrument with the sample suspended below. A gas line supplies an inert gas that sweeps the volatile decomposition products out of the furnace. The heat exchanger dissipates the heat from the furnace to a reservoir of water. (b) Close-up showing the balance pan, which sits on a moving platform, the thermocouple for monitoring temperature, a hook for lowering the sample pan into the furnace, and the opening to the furnace. After placing a small portion of the sample on the balance pan, the platform rotates over the furnace and transfers the balance pan to a hook that is suspended from the balance. Once the balance pan is in place, the platform rotates back to its initial position. The balance pan and the thermocouple are then lowered into the furnace.

9.9 DETERMINATION OF Si IN ORES AND ALLOYS

The best way to appreciate the theoretical and practical details discussed in this section is to carefully examine a typical volatilization gravimetric method. Although each method is unique, the determination of Si in ores and alloys by forming volatile SiF_4 provides an instructive example of a typical procedure.

9.9.1 Description of Method

Silicon is determined by dissolving the sample in acid and dehydrating to precipitate SiO_2. Because a variety of other insoluble oxides also form, the precipitate's mass is not a direct measure of the amount of silicon in the sample. Treating the solid residue with HF forms volatile SiF_4. The decrease in mass following the loss of SiF_4 provides an indirect measure of the amount of silicon in the original sample.

9.9.2 Procedure

Transfer a sample of between 0.5 g and 5.0 g to a platinum crucible along with an excess of Na_2CO_3, and heat until a melt forms. After cooling, dissolve the residue in dilute HCl. Evaporate the solution to dryness on a steam bath and heat the residue, which contains SiO_2 and other solids, for one hour at 110°C. Moisten the residue with HCl and repeat the dehydration. Remove any acid soluble materials from the residue by adding 50 mL of water and 5 mL of concentrated HCl. Bring the solution to a boil and filter through #40 filter paper (note: #40 filter paper is a medium speed, ashless filter paper for filtering crystalline solids). Wash the residue with hot 2% v/v HCl followed by hot water. Evaporate the filtrate to dryness twice and, following the same procedure, treat to remove any acid-soluble materials. Combine the two precipitates and dry and ignite to a constant weight at 1200°C. After cooling, add 2 drops of 50% v/v H_2SO_4 and 10 mL of HF. Remove the volatile SiF4 by evaporating to dryness on a hot plate. Finally, bring the residue to constant weight by igniting at 1200°C.

9.9.3 Quantitative Applications

Unlike precipitation gravimetry, which rarely is used as a standard method of analysis, volatilization gravimetric methods continue to play an important role in chemical analysis. Several important examples are discussed below.

9.9.3. 1 Inorganic Analysis

Determining the inorganic ash content of an organic material, such as a polymer, is an example of a direct volatilization gravimetric analysis. After weighing the sample, it is placed in an appropriate crucible and the organic material carefully removed by combustion, leaving behind the inorganic ash. The crucible that contains the residue is heated to a constant weight using either a burner or an oven before the mass of the inorganic ash is determined.

Another example of volatilization gravimetry is the determination of dissolved solids in natural waters and wastewaters. In this method, a sample of water is transferred to a weighing dish and dried to a constant weight at either 103–105°C or at 180°C. Samples dried at the lower temperature retain some occluded water and lose some carbonate as CO_2; the loss of organic material, however, is minimal at this temperature. At the higher temperature, the residue is free from occluded water, but the loss of carbonate is greater. In addition, some chloride, nitrate, and organic material is lost through thermal decomposition. In either case, the residue that remains after drying to a constant weight at 500°C is the amount of fixed solids in the sample, and the loss in mass provides an indirect measure of the sample's volatile solids.

Indirect analyses based on the weight of a residue that remains after volatilization are used to determine moisture in a variety of products and to determine silica in waters, waste waters, and rocks. Moisture is determined by drying a preweighed sample with an infrared lamp or a low temperature oven. The difference between the original weight and the weight after drying equals the mass of water lost.

9.9.3.2 Organic Analysis

The most important application of volatilization gravimetry is for the elemental analysis of organic materials. During combustion with pure O_2, many elements, such as carbon and hydrogen, are released as gaseous combustion products, such as $CO_2(g)$ and $H_2O(g)$. Passing the combustion products through preweighed tubes that contain selective absorbents and measuring the increase in each tube's mass provides a direct analysis for the mass of carbon and hydrogen in the sample.

Instead of measuring mass, modern instruments for completing an elemental analysis use gas chromatography (Chapter 12) or infrared spectroscopy (Chapter 10) to monitor the gaseous decomposition products.

Alkaline metals and earths in organic materials are determined by adding H_2SO_4 to the sample before combustion. After combustion is complete, the metal remains behind as a solid residue of metal sulfate. Silver, gold, and platinum are determined by burning the organic sample, leaving a metallic residue of Ag, Au, or Pt. Other metals are determined by adding HNO_3 before combustion, which leaves a residue of the metal oxide.

Volatilization gravimetry also is used to determine biomass in waters and wastewaters. Biomass is a water quality index that provides an indication of the total mass of organisms contained within a sample of water. A known volume of the sample is passed through a preweighed 0.45-μm membrane filter or a glass-fiber filter and dried at 105°C for 24 h. The residue's mass provides a direct measure of biomass. If samples are known to contain a substantial amount of dissolved inorganic solids, the residue is ignited at 500°C for one hour, which volatilizes the biomass. The resulting inorganic residue is wetted with distilled water to rehydrate any clay minerals and dried to a constant weight at 105°C. The difference in mass before and after ignition provides an indirect measure of biomass.

9.9.4 Quantitative Calculations

For some applications, such as determining the amount of inorganic ash in a polymer, a quantitative calculation is straightforward and does not require a balanced chemical reaction. For other applications, however, the relationship between the analyte and the analytical signal depends upon the stoichiometry of any relevant reactions. Once again, a conservation of mass is useful when solving problems.

Example .2

A 101.3-mg sample of an organic compound that contains chlorine is combusted in pure O_2. The volatile gases are collected in absorbent traps with the trap for CO_2 increasing in mass by 167.6 mg and the trap for H_2O increasing in mass by 13.7-mg. A second sample of 121.8 mg is treated with concentrated HNO_3, producing Cl_2 that reacts with Ag+ to form 262.7 mg of AgCl. Determine the compound's composition, as well as its empirical formula.

Solution

A conservation of mass requires that all the carbon in the organic compound is in the CO_2 produced during combustion; thus 0.1676 g CO_2×1 mol C44.010 g CO_2×12.011 g Cmol C=0.04574 g C0.1676 g CO_2×1 mol C44.010 g CO_2×12.011 g Cmol C=0.04574 g C 0.04574 g C0.1013 g sample ×100=45.15%w/w C0.04574 g C0.1013 g sample ×100=45.15%w/w C Using the same approach for hydrogen and chlorine, we find that 0.0137 g H_2O×2 mol H18.015 gH_2O×1.008 g Hmol H=1.533×10−3g H0.0137 g H_2O×2 mol H18.015 g H_2O×1.008 g Hmol H=1.533×10−3g H 1.533 ×10−3g H0.1003 g sample ×100=1.53%w/w H1.533 ×10−3g H0.1003 g sample ×100=1.53%w/w H 0.2627 g AgCl×1 mol Cl143.32 g AgCl×35.455 g Clmol Cl=0.06498 g Cl0.2627 g AgCl×1 mol Cl143.32 g AgCl×35.455 g Clmol Cl=0.06498 g Cl 0.06498 g Cl0.1218 g sample ×100=53.35%w/w Cl0.06498 g Cl0.1218 g sample ×100=53.35%w/w Cl.

Adding together the weight percents for C, H, and Cl gives a total of 100.03%; thus, the compound contains only these three elements. To determine the compound's empirical formula we note that a gram of sample contains 0.4515 g of C, 0.0153 g of H and 0.5335 g of Cl. Expressing each element in moles gives 0.0376 moles C, 0.0152 moles H and 0.0150 moles Cl. Hydrogen and chlorine are present in a 1:1 molar ratio. The molar ratio of C to moles of H or Cl is mol Cmol H=mol Cmol Cl=0.03760.0150=2.51≈2.5mol Cmol H=mol Cmol Cl=0.03760.0150=2.51≈2.5

Thus, the simplest, or empirical formula for the compound is $C_5H_2Cl_2$.

In an indirect volatilization gravimetric analysis, the change in the sample's weight is proportional to the amount of analyte in the sample. Note that in the following example it is not necessary to apply a conservation of mass to relate the analytical signal to the analyte.

Example 3

A sample of slag from a blast furnace is analyzed for SiO_2 by decomposing a 0.5003-g sample with HCl, leaving a residue with a mass of 0.1414 g. After treating with HF and H_2SO_4, and evaporating the volatile SiF4, a residue with a mass of 0.0183 g remains. Determine the %w/w SiO_2 in the sample.

Solution

The difference in the residue's mass before and after volatilizing SiF_4 gives the mass of SiO_2 in the sample; thus the sample contains

0.1414 g−0.0183 g=0.1231 g SiO20.1414 g−0.0183 g=0.1231 g SiO2

and the %w/w SiO_2 is

0.1231 g SiO20.5003 g sample ×100=24.61%w/w SiO20.1231 g SiO20.5003 g sample ×100=24.61%w/w SiO2

Example 4

A 26.23-mg sample of $MgC_2O_4H_2O$ and inert materials is heated to constant weight at 1200°C, leaving a residue that weighs 20.98 mg. A sample of pure $MgC_2O_4H_2O$, when treated in the same fashion, undergoes a 69.08% change in its mass. Determine the %w/w $MgC_2O_4H_2O$ in the sample.

Solution

The change in the sample's mass is 5.25 mg, which corresponds to 5.25 mglost×100.0 mg MgC2O4·H2O69.08 mg lost =7.60 mg MgC2O4·H2O5.25 mglost×100.0 mg MgC2O4·H2O69.08 mg lost =7.60 mg MgC2O4·H2O

The %w/w $MgC_2O_4{\bullet}H_2O$ in the sample is 7.60 mg MgC2O4·H2O26.23 mg sample ×100=29.0%w/w MgC2O4·H2O7.60 mg MgC2O4·H2O26.23 mg sample ×100=29.0%w/w MgC2O4·H2O.

9.10 EVALUATING VOLATILIZATION GRAVIMETRY

The scale of operation, accuracy, and precision of a gravimetric volatilization method is similar to that described in the last section for precipitation gravimetry. The sensitivity of a direct analysis is fixed by the analyte's chemical form following combustion or volatilization. We can improve the sensitivity of an indirect analysis by choosing conditions that give the largest possible change in mass. For example, the thermogram in Figure 8.3.1 shows us that an indirect analysis for $CaC_2O_4H_2O$ is more sensitive if we measure the change in mass following ignition at 1000°C than if we ignite the sample at 300°C.

Selectivity is not a problem for a direct analysis if we trap the analyte using a selective absorbent trap. A direct analysis based on the residue's weight following combustion or volatilization is possible if the residue contains only the analyte of interest. As noted earlier, an indirect analysis only is feasible when the change in mass results from the loss of a single volatile product that contains the analyte.

Volatilization gravimetric methods are time and labor intensive. Equipment needs are few, except when combustion gases must be trapped, or for a thermogravimetric analysis, when specialized instrumentation is needed.

9.10.1 Particulate Gravimetry

Precipitation and volatilization gravimetric methods require that the analyte, or some other species in the sample, participates in a chemical reaction. In a direct precipitation gravimetric analysis, for example, we convert a soluble analyte into an insoluble form that precipitates from solution. In some situations, however, the analyte already is present in a particulate form that is easy to separate from its liquid, gas, or solid matrix. When such a separation is possible, we can determine the analyte's mass without relying on a chemical reaction.

A particulate is any tiny portion of matter, whether it is a speck of dust, a globule of fat, or a molecule of ammonia. For particulate gravimetry we simply need a method to collect the particles and a balance to measure their mass.

9.10.1.1 Theory and Practice

There are two methods for separating a particulate analyte from its matrix. The most common method is filtration, in which we separate solid particulates from their gas, liquid, or solid matrix. A second method, which is useful for gas particles, solutes, and solids, is an extraction.

9.10.1.2 Filtration

To separate solid particulates from their matrix we use gravity or apply suction from a vacuum pump or an aspirator to pull the sample through a filter. The type of filter we use depends upon the size of the solid particles and the sample's matrix. Filters for liquid samples are constructed from a variety of materials, including cellulose fibers, glass fibers, cellulose nitrate, and polytetrafluoroethylene (PTFE). Particle retention depends on the size of the filter's pores. Cellulose fiber filter papers range in pore size from 30 μm to 2–3 μm. Glass fiber filters, manufactured using chemically inert borosilicate glass, are available with pore sizes between 2.5 μm and 0.3 μm. Membrane filters, which are made from a variety of materials, including cellulose nitrate and PTFE, are available with pore sizes from 5.0 μm to 0.1 μm.

Solid aerosol particulates are collected using either a single-stage or a multiple-stage filter. In a single-stage system, we pull the gas through a single filter, which retains particles larger than the filter's pore size. To collect samples from a gas line, we place the filter directly in the line. Atmospheric gases are sampled with a high volume sampler that uses a vacuum pump to pull air through the filter at a rate of approximately 75 m3/h. In either case, we can use the same filtering media for liquid samples to collect aerosol particulates. In a multiple-stage system, a series of filtering units separates the particles into two or more size ranges.

The particulates in a solid matrix are separated by size using one or more sieves (Figure 8.4.1). Sieves are available in a variety of mesh sizes, ranging from approximately 25 mm to 40 μm. By stacking together sieves of different mesh size, we can isolate particulates into several narrow size ranges. Using the sieves in Figure 8.4.1 , for example, we can separate a solid into particles with diameters >1700 μm, with diameters between 1700 μm and 500 μm, with diameters between 500 μm and 250 μm, and those with a diameter <250 μm.

Fig. 13: Three sieves with, from left to right, mesh sizes of 1700 μm, 500 μm, and 250 μm. Source: BMK (commons.wikimedia.com).

9.10.1.3 Extraction

Filtering limits particulate gravimetry to solid analytes that are easy to separate from their matrix. We can extend particulate gravimetry to the analysis of gas phase analytes, solutes, and solids that are difficult to filter if we extract them with a suitable solvent. After the extraction, we evaporate the solvent and determine the analyte's mass. Alternatively, we can determine the analyte indirectly by measuring the change in the sample's mass after we extract the analyte.

For a more detailed review of extractions, particularly solid-phase extractions, see Chapter 7.

Another method for extracting an analyte from its matrix is by adsorption onto a solid substrate, by absorption into a thin polymer film or chemical film coated on a solid substrate, or by chemically binding to a suitable receptor that is covalently bound to a solid substrate (Figure 8.4.2). Adsorption, absorption, and binding occur at the interface between the solution that contains the analyte and the substrate's surface, the thin film, or the receptor. Although the amount of extracted analyte is too small to measure using a conventional balance, it can be measured using a quartz crystal microbalance.

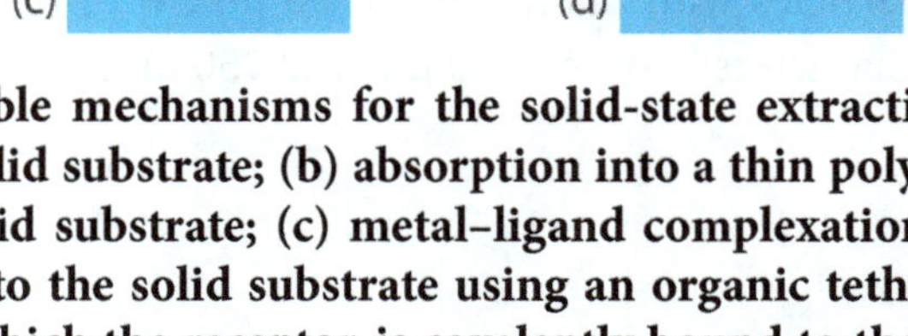

Fig. 14: Four possible mechanisms for the solid-state extraction of an analyte: (a) adsorption onto a solid substrate; (b) absorption into a thin polymer film or chemical film coated on a solid substrate; (c) metal–ligand complexation in which the ligand is covalently bound to the solid substrate using an organic tether; and (d) antibody–antigen binding in which the receptor is covalently bound to the solid substrate using an organic tether.

9.10.2 Quantitative Applications

Particulate gravimetry is important in the environmental analysis of water, air, and soil samples. The analysis for suspended solids in water samples, for example, is accomplished by filtering an appropriate volume of a well-mixed sample through a glass fiber filter and drying the filter to constant weight at 103–105°C.

The microbiological testing of water also uses particulate gravimetry. One example is the analysis for coliform bacteria in which an appropriate volume of sample is passed through a sterilized 0.45-μm membrane filter. The filter is placed on a sterilized absorbent pad that is saturated with a culturing medium and incubated for 22–24 hours at 35 ± 0.5°C. Coliform bacteria are identified by the presence of individual bacterial colonies that form during the incubation period (Figure 8.4.3). As with qualitative applications of precipitation gravimetry, the signal in this case is a visual observation of the number of colonies rather than a measurement of mass.

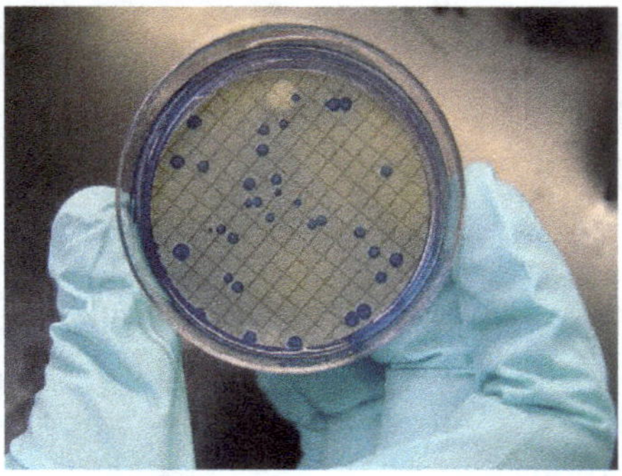

Fig. 15: Colonies of fecal coliform bacteria from a water supply. Source: Susan Boyer. Photo courtesy of ARS–USDA (www.ars.usda.gov).

Total airborne particulates are determined using a high-volume air sampler equipped with either a cellulose fiber or a glass fiber filter. Samples from urban environments require approximately 1 h of sampling time, but samples from rural environments require substantially longer times.

Grain size distributions for sediments and soils are used to determine the amount of sand, silt, and clay in a sample. For example, a grain size of 2 mm serves as the boundary between gravel and sand. The grain size for the sand–silt and the silt–clay boundaries are 1/16 mm and 1/256 mm, respectively.

Several standard quantitative analytical methods for agricultural products are based on measuring the sample's mass following a selective solvent extraction. For example, the crude fat content in chocolate is determined by extracting with ether for 16 hours in a Soxhlet extractor. After the extraction is complete, the ether is allowed to evaporate and the residue is weighed after drying at 100°C. This analysis also can be accomplished indirectly by weighing a sample before and after extracting with supercritical CO_2.

Quartz crystal microbalances equipped with thin film polymer films or chemical coatings have found numerous quantitative applications in environmental analysis. Methods are reported for the analysis of a variety of gaseous pollutants, including ammonia, hydrogen sulfide, ozone, sulfur dioxide, and mercury. Biochemical particulate gravimetric sensors also have been developed. For example, a piezoelectric immunosensor has been developed that shows a high selectivity for human serum albumin, and is capable of detecting microgram quantities [Muratsugu, M.; Ohta, F.; Miya, Y.; Hosokawa, T.; Kurosawa, S.; Kamo, N.; Ikeda, H. Anal. Chem. 1993, 65, 2933–2937].

9.10.3 Quantitative Calculations

The result of a quantitative analysis by particulate gravimetry is just the ratio, using appropriate units, of the amount of analyte relative to the amount of sample.

Example.1

A 200.0-mL sample of water is filtered through a pre-weighed glass fiber filter. After drying to constant weight at 105°C, the filter is found to have increased in mass by 48 2 mg. Determine the sample's total suspended solids.

Solution

One ppm is equivalent to one mg of analyte per liter of solution; thus, the total suspended solids for the sample is 48.2 mg solids 0.2000 L sample =241 ppm solids 48.2 mg solids 0.2000 L sample =241 ppm solids filtration generally require less time, labor and capital than other gravimetric methods.

10

INTRODUCTION TO ELECTROANALYTICAL TECHNIQUES

10.1 INTRODUCTION: AN OVERVIEW

Electroanalytical techniques are concerned with the interplay between electricity and chemistry, namely the measurement of electrical quantities such as current, potential or charge and their relationship to chemical parameters such as concentration. Although there are only three principal sources for the analytical signals, that is, potential, current, and charge, a wide variety of experimental designs are possible. The simplest division is between bulk methods, which measure properties of the whole solution, and interfacial methods, in which the signal is a function of phenomena occurring at the interface between an electrode and the solution in contact with the electrode.

The measurement of a solution's conductivity, which is proportional to the total concentration of dissolved ions, is one example of a bulk electrochemical method. A determination of pH using a pH electrode is one example of an interfacial electrochemical method. The use of electrical measurements for analytical purposes has found large range of applications including environmental monitoring, industrial quality control and biomedical analysis.

In static methods, no current passes between the electrodes, and the concentrations of species in the electrochemical cell remain unchanged, or static. The largest division of interfacial electrochemical methods is the group of dynamic methods, in which current flows and concentrations change as the result of a redox reaction.

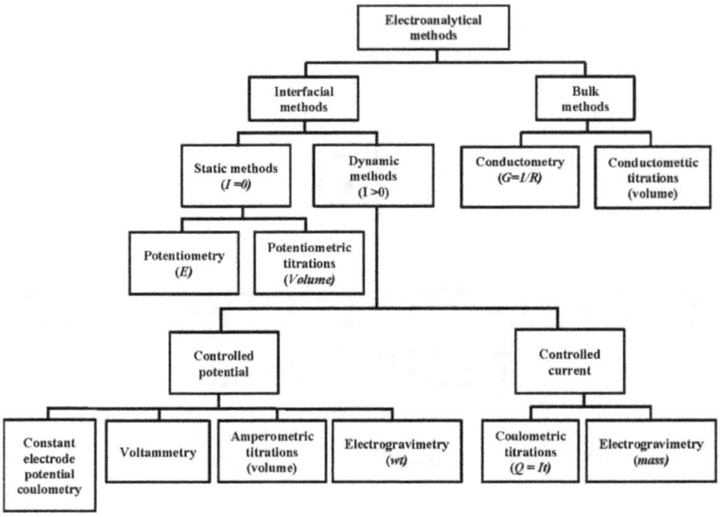

Classification of Electroanalytical Techniques

10.2 IMPORTANCE OF ELECTROANALYTICAL TECHNIQUES

Electroanalytical methods have certain advantages over other analytical methods:

- Electrochemical analysis allows for the determination of different oxidation states of an element in a solution, not just the total concentration of the element.
- Electroanalytical techniques are capable of producing exceptionally low detection limits and an abundance of characterization information including chemical kinetics information.
- Selective for particular redox state of a species e.g. Ce^{3+} vs. Ce^{4+}.
- Its low cost.
- Fastness

10.3 CONTROLLING AND MEASURING CURRENT AND POTENTIAL

An electroactive species is one that can be oxidized or reduced at an electrode. We regulate the potential of the working electrode to control which electroactive species react and which do not. Metal electrodes are said to be polarizable, which means that their potentials are easily changed when small currents flow. A reference electrode such as calomel or is said to be nonpolarizable, because its potential does not vary much unless a significant current is flowing.

Ideally, we want to measure the potential of a polarizable working electrode with respect to a nonpolarizable reference electrode. How can we do this, if there is to be significant current at the working electrode and negligible current at the reference electrode?

The answer is to introduce a third electrode. The working electrode is the one at which the reaction of interest occurs. A calomel or other reference electrode is used to measure the potential of the working electrode. The auxiliary electrode (the counter electrode) is the current-supporting partner of the working electrode. Current flows between the working and the auxiliary electrodes. Negligible current flows through the reference electrode, so its potential is unaffected by ohmic potential, concentration polarization, and overpotential. It truly maintains a constant reference potential. In controlled-potential electrolysis, the voltage difference between working and reference electrodes in a three-electrode cell is regulated by an electronic device called a potentiostat.

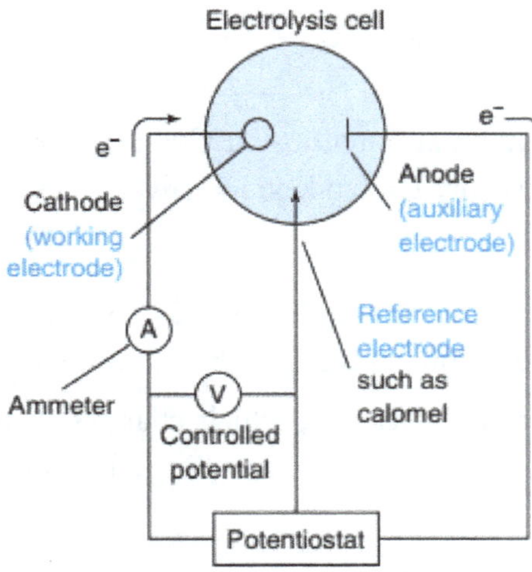

Controlled potential electrolysis with a three-electrode cell

Although many different electrochemical methods of analysis are possible, there are only three basic experimental designs:

- Measuring the potential under static conditions of no current flow;
- Measuring the potential while controlling the current; and
- Measuring the current while controlling the potential.

Each of these experimental designs, however, is based on Ohm's law that a current, *i*, passing through an electric circuit of resistance, *R*, generates a potential, *E*; thus

$$E = IR$$

10.3.1 Potentiometer

A potentiometer is a device for measuring the potential of electrochemical cell without drawing a current or altering the cell's composition.

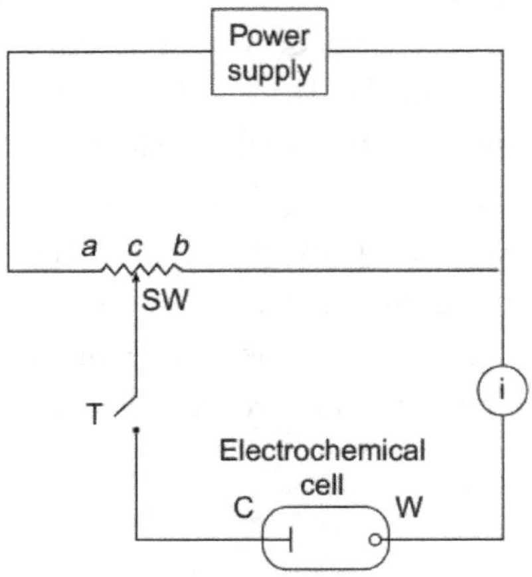

Schematic diagram of a manual potentiostat: C = counter electrode; W = working electrode; SW = slide-wire resistor; T = tap key; i = galvanometer

10.3.2 Galvanostats

A galvanostat is used for dynamic methods, such as constant-current, in which it is necessary to control the current flowing through an electrochemical cell.

The potential of the working electrode, which changes as the composition of the electrochemical cell changes, is monitored by including a reference electrode and a high-impedance potentiometer

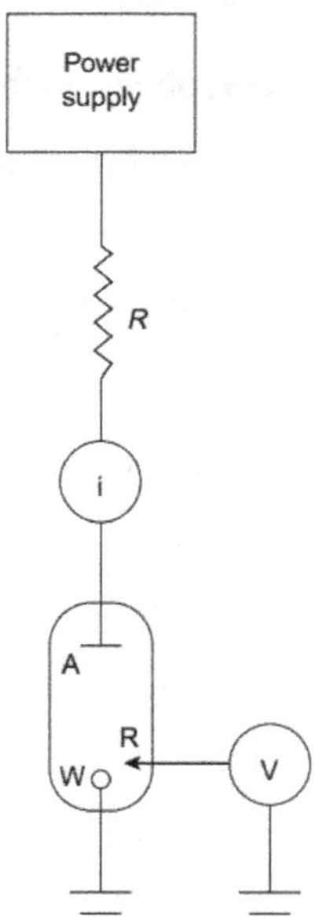

Schematic diagram of a galvanostat: R = resistor; i = galvanometer; A = auxiliary electrode; W = working electrode; R = reference electrode; V = voltmeter or potentiometer (optional)

10.3.3 Potentiostats

A potentiostat is used for dynamic methods when it is necessary to control the potential of the working electrode. The current flowing between the auxiliary and working electrodes is measured with a galvanostat.

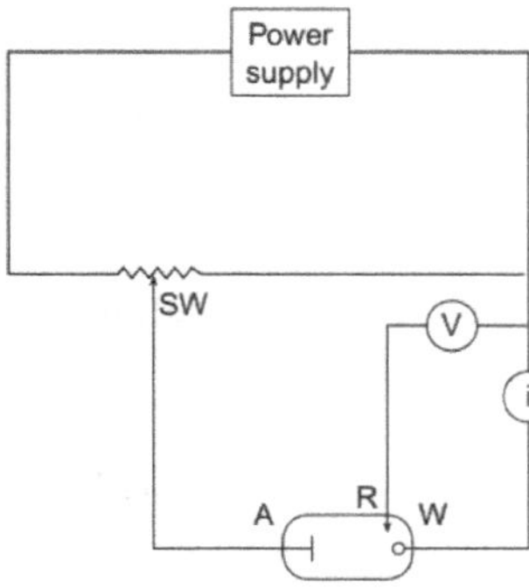

Schematic diagram of a manual potentiostat: SW = slide-wire resistor; A = auxiliary electrode; R = reference electrode; W = working electrode; V = voltmeter or potentiometer; i = galvanometer

10.4 POTENTIOMETRIC METHOD OF ANALYSIS

In potentiometry, the potential of an electrochemical cell is measured under static conditions. Because no current, or only a negligible current, flows while measuring a solution's potential, its composition remains unchanged. For this reason, potentiometry is a useful quantitative method

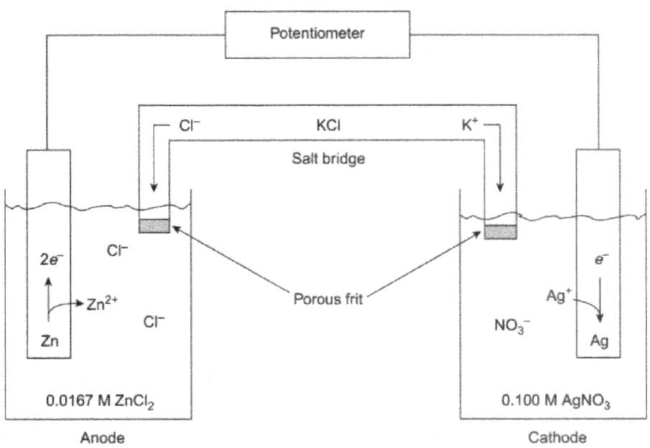

Electrochemical cell for potentiometry

Since no significant current flows in potentiometer, the role of the counter electrode is reduced to that of supplying a reference potential; thus, the counter electrode is usually called the reference electrode.

By convention, the electrode on the left is considered to be the anode, where oxidation occurs

Zn(s) Zn2+(aq) + 2e– and the electrode on the right is the cathode, where reduction occurs

Ag+(aq) + e– Ag(s)

The electrochemical cell's potential, therefore, is for the reaction 6

Zn(s) + 2Ag+(aq) 2Ag(s) + Zn2+(aq)

The electrochemical cell is described in shorthand notation as

Zn(s) | ZnCl2 (aq, 0.0167 M) || AgNO3 (aq, 0.100 M) | Ag(s)

Thus, the potential of a potentiometric electrochemical cell is given as

*E*cell = *E*c – *E*a

However, at the interface between the salt bridge and the solution in which it is immersed, there is a liquid junction potential.

Liquid junction potential (*E*lj) is potential that develops at the interface between two ionic solutions that differ in composition, because of a difference in the mobilities of the ions.

For instance, solutions of 0.1 M HCl and 0.01 M separated by a porous membrane will have a net diffusion of H+ and Cl- in the direction of the arrows. The mobility of H+ is greater than that for Cl-, as shown by the difference in the lengths of their respective arrows. Thus, a potential is produced at the membrane.

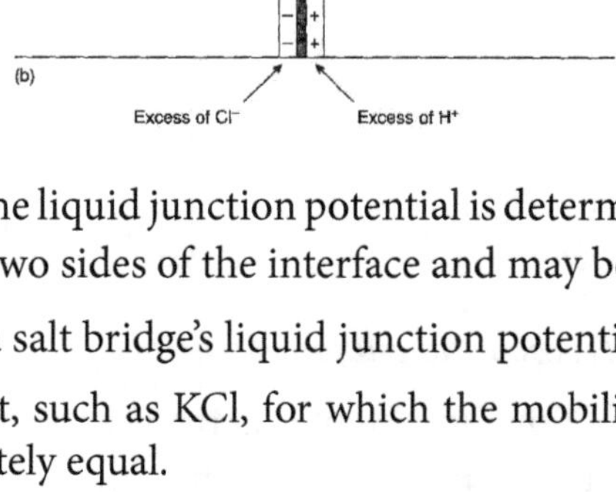

The magnitude of the liquid junction potential is determined by the ionic composition of the solutionson the two sides of the interface and may be as large as 30–40 mV.

The magnitude of a salt bridge's liquid junction potential is minimized by:

1. Using a salt, such as KCl, for which the mobility of the cation and anion are approximately equal.
2. Incorporating a high concentration of the salt in the salt bridge. For this reason salt bridges are frequently constructed using solutions that are saturated with KCl nevertheless, a small liquid junction potential, generally of unknown magnitude, is always present.

When the potential of an electrochemical cell is measured, the contribution of the liquid junction potential must be included. Thus, potential of a potentiometric electrochemical cell is rewritten as

$$E\text{cell} = E\text{c} - E\text{a} + E\text{lj}$$

Reference Electrodes for Potentiometric Method

Potentiometric electrochemical cells are constructed such that one of the half-cells provides a known reference potential, and the potential of the other half-cell indicates the analyte's concentration.

By convention, the reference electrode is taken to be the anode; thus, the shorthand notation for a potentiometric electrochemical cell is Reference || Indicator and the cell potential is

$$E\text{cell} = E\text{ind} - E\text{ref} + E\text{lj}$$

The ideal reference electrode

1. Must provide a stable potential so that any change in *E*cell is attributed to the indicator electrode, and, therefore, to a change in the analyte's concentration.
2. Should be easy to make and to use.

The reference electrodes commonly used in potentiometric method of analysis are:

1. Standard hydrogen electrode (SHE)

$2H+(aq) + e– H2(g)$

2. Saturated calomel electrode (SCE)

$Hg_2Cl2(s) + 2e– 2Hg(l) + 2Cl–(aq)$

3. 3. Silver–silver chloride electrode (SSCE)

$AgCl(s) + e– Ag(s) + Cl–(aq)$

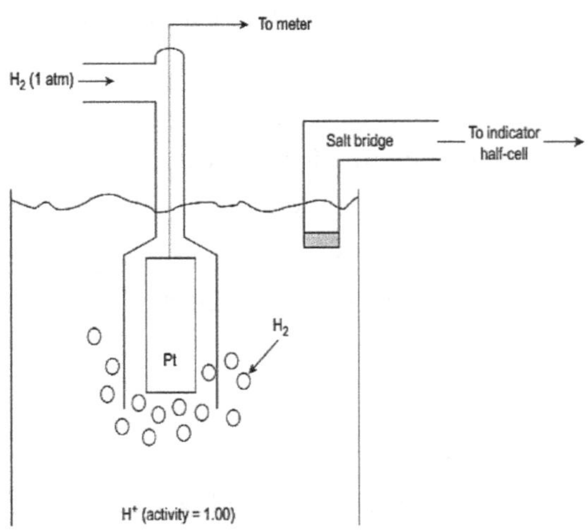

Schematic diagram of SHE

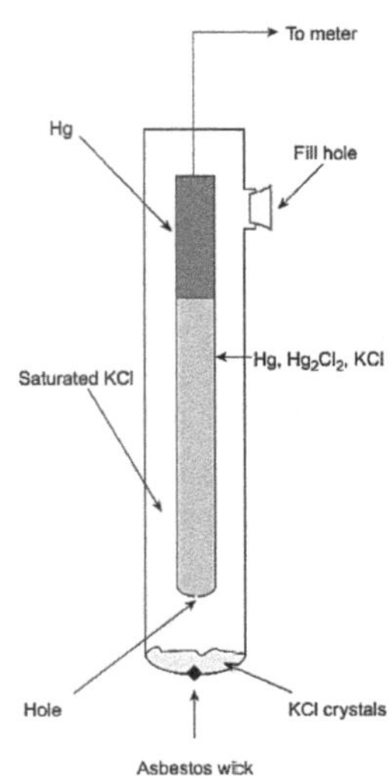

Schematic diagram of SCE

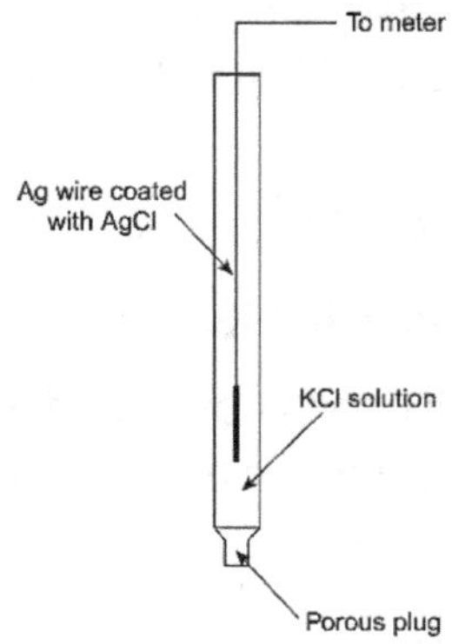

Schematic diagram of SSCE

Indicator Electrode for Potentiometric Method of Analysis

The potential of the indicator electrode in a potentiometric electrochemical cell is proportional to the concentration of analyte.

Two classes of indicator electrodes are used in potentiometry:

1. Metallic electrodes and
2. Ion-selective electrodes

$$E_{cell} = K - \frac{0.05916}{n}\log\frac{1}{[M^{n+}]} = K + \frac{0.05916}{n}\log[M^{n+}]$$

Metallic Electrodes

The potential of a metallic electrode is determined by the position of a redox reaction at the electrode–solution interface.

Electrodes of the first kind

Metallic indicator electrodes in which a metal is in contact with a solution containing its ion are called electrodes of the first kind.

In general, for a metal M, in a solution of Mn+, the cell potential is given as where, K is a constant that includes the standard-state potential for the Mn+/M redox couple, the potential of the reference electrode, and the junction potential.

For a variety of reasons, including slow kinetics for electron transfer, the existence of surface oxides and interfering reactions, electrodes of the first kind are limited to Ag, Bi, Cd, Cu, Hg, Pb, Sn, Tl, and Zn.

Electrodes of the Second Kind

When the potential of an electrode of the first kind responds to the potential of another ion that is in equilibrium with M*n*+, it is called an electrode of the second kind.

Two common electrodes of the second kind are the calomel and silver/silver chloride reference electrodes. Electrodes of the second kind also can be based on complexation reactions.

For example, an electrode for EDTA is constructed by coupling a Hg2+/Hg electrode of the first kind to EDTA by taking advantage of its formation of a stable complex with Hg2+.

Redox Electrodes

Metallic electrodes that serve simply as a source of, or a sink for, electrons in other redox reactions are called redox electrodes. They are inert electrodes. An example is when platinum is used as a cathode.

Ion-Selective Electrodes (ISEs)

ISE is an electrode in which the membrane potential is a function of the concentration of a particular ion in solution. Membrane potential is developed across a conductive membrane whose opposite sides are in contact with solutions of different composition. ISEs, such as the glass pH electrode, function by using a membrane that reacts selectively with a single ion. Interaction of the analyte with the membrane results in a membrane potential, if there is a difference in the analyte's concentration on opposite sides of the membrane.

One side of the membrane is in contact with an internal solution containing a fixed concentration of analyte, while the other side of the membrane is in contact with the sample.

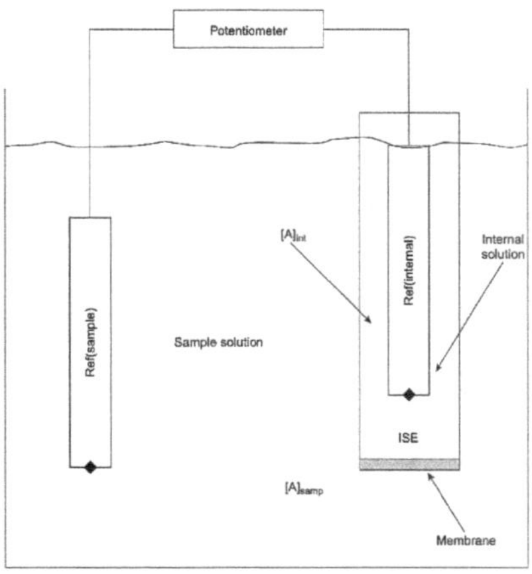

Two reference electrodes are used; one positioned within the internal solution, and one in the sample solution. The cell potential, therefore, is

$$E\text{cell} = E\text{Ref(int)} - E\text{Ref(samp)} + E\text{mem} + E\text{lj}$$

where, *Eref(int)* = potential of reference electrode within the internal solution, *Eref(samp)*= potential of reference electrode within the sample, *E*mem is the potential across the membrane and *Elj* = liquid junction potential.

The potential of the membrane is given as

$$E_{mem} = E_{asy} - \frac{0.05916}{Z} log\left(\frac{[A]_{int}}{[A]_{samp}}\right)$$

where, [A]samp and [A]int are the concentrations of analyte in the sample and the internal solution, respectively, and *z* is the analyte's charge.

$$E_{cell} = K + \frac{0.05916}{Z} log[A]_{samp}$$

Ideally, *E*mem should be zero when the concentrations of analyte on both sides of the membrane are equal. The term *E*asym, which is called an asymmetry potential, accounts for the fact that the membrane potential is usually not zero under these conditions. The potential of potentiometer with an ISE is written as

Types of ISEs

1. Glass Ion-Selective Electrodes

An ion-selective electrode based on a glass membrane in which the potential develops from an ion-exchange reaction on the membrane's surface. The first commercial glass electrodes were manufactured using Corning 015, a glass with a composition of approximately 22% Na2O, 6% CaO, and 72% SiO2.

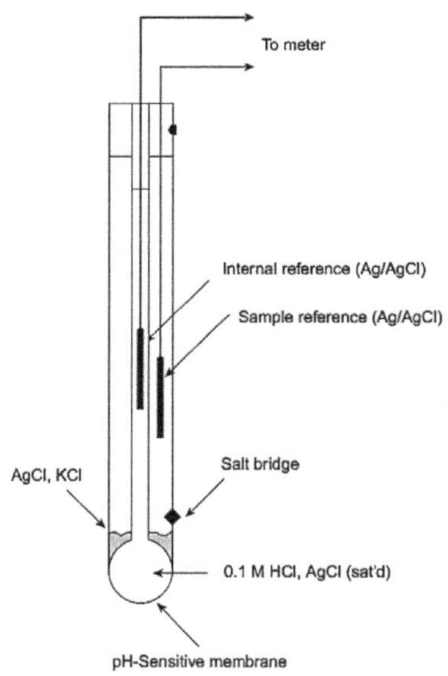

Schematic diagram of a combination glass electrode for measuring pH

2. Solid state Ion-Selective Electrode

An ion-selective electrode based on a sparingly soluble inorganic crystalline material. Polycrystalline ion-selective electrodes are made by forming a thin pellet of Ag2S, or a mixture of Ag2S and either a second silver salt or another metal sulphide.

3. Liquid-Based Ion-Selective Electrodes

An ion-selective electrode in which a chelating agent is incorporated into a hydrophobic membrane. Three types of organic liquids have been used: cation exchangers, anion exchangers, and neutral ionophores. A neutral ligand whose exterior is hydrophobic and whose interior is hydrophilic is called ionophores.

One example of a liquid-based ion-selective electrode is that for Ca2+, which uses a porous plastic membrane saturated with di-(n-decyl) phosphate.

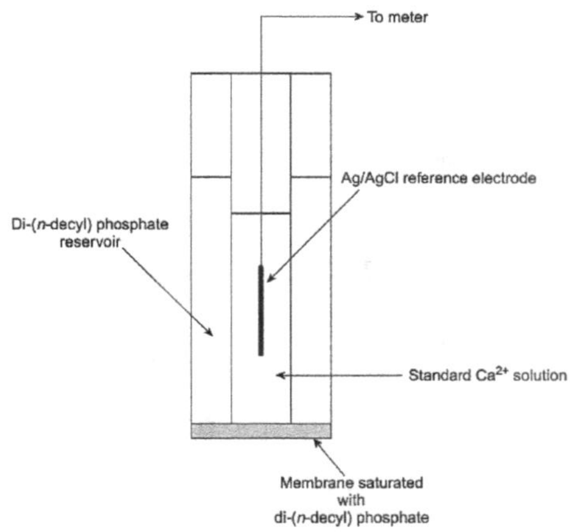

Schematic diagram of a Ca^{2+} liquid-based ion-selective electrode

4. Gas-Sensing Electrodes

A number of membrane electrodes have been developed that respond to the concentration of dissolved gases. The membrane is permeable to the gaseous analyte, but is not permeable to non-volatile components in the sample matrix.

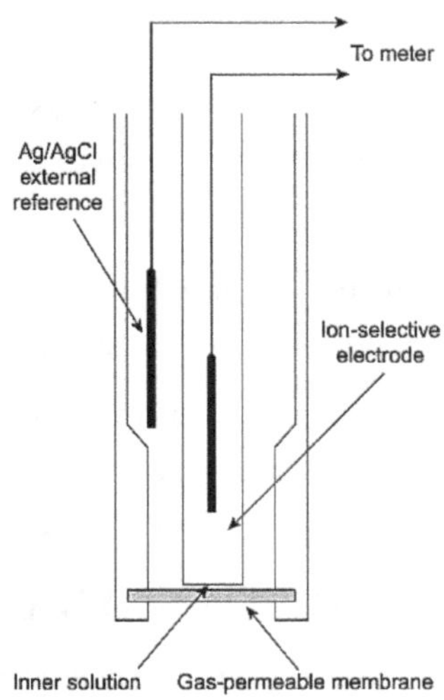

Schematic diagram of a gas-sensing membrane electrode

5. Potentiometric Biosensors

Potentiometric electrodes for the analysis of molecules of biochemical importance can be constructed in a fashion similar to that used for gas-sensing electrodes. The most common class of potentiometric biosensors are the so-called enzyme electrodes, in which an enzyme is trapped or immobilized at the surface of an ion-selective electrode.

Potentiometric biosensors have also been designed around other biologically active species, including antibodies, bacterial particles, tissue, and hormone receptors.

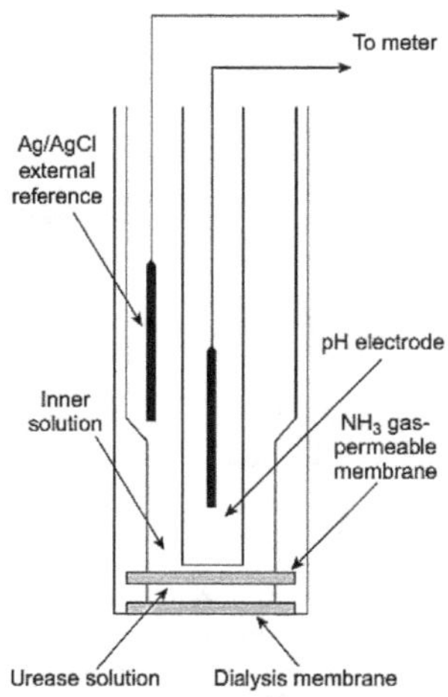

Schematic diagram of an enzyme-based potentiometric biosensor for urea in which urease is trapped between two membranes

10.5 APPLICATIONS OF POTENTIOMETRY

- **Measurement of pH** :With the availability of inexpensive glass pH electrodes and pH meters, the determination of pH has become one of the most frequent quantitative analytical measurements.
- **Clinical Applications:** For extracellular fluids, such as blood and urine, the analysis can be made in vitro with conventional electrodes, provided that sufficient sample is available.
- **Environmental Applications:** Standard methods have been developed for the analysis of CN–, F–, NH3, and NO3– in water and wastewater.
- **Potentiometric Titrations:** The potentiometric determination of equivalence points is feasible for acid–base, complexation, redox, and precipitation titrations, as well as for titrations in aqueous and non-aqueous solvents.

10.6 COULOMETRIC METHOD OF ANALYSIS

Coulometric methods of analysis are based on an exhaustive electrolysis of the analyte. By exhaustive we mean that the analyte is quantitatively oxidized or reduced at the working electrode or reacts quantitatively with a reagent generated at the working electrode.

There are two forms of coulometry namely:

3. Controlled-potential coulometry, in which a constant potential is applied to the electrochemical cell, and
1. Controlled-current coulometry, in which a constant current is passed through the electrochemical cell.

The total charge, Q, in coulombs, passed during electrolysis is related to the absolute amount of analyte by **Faraday's law**

$$Q = nFN$$

where, n = number of electrons transferred per mole of analyte, F = Faraday's constant (96487 C mol−1), and N = moles of analyte.

$$Q = ite = nFN$$

where, te = electrolysis time and i = current.

If current varies with time, as it does in controlled-potential coulometry, then the total charge is given by

$$Q = \int_{t=0}^{t=t_e} i(t)dt$$

To obtain an accurate value for N, therefore, all the current must result in the analyte's oxidation or reduction. In other words, coulometry requires 100% **current efficiency**, which is percentage of current that actually leads to the analyte's oxidation or reduction.

Controlled-Potential Coulometry

The easiest method for ensuring 100% current efficiency is to maintain the working electrode at a constant potential that allows for the analyte's quantitative oxidation or reduction, without simultaneously oxidizing or reducing an interfering species. The current flowing through an electrochemical cell under a constant potential is proportional to the analyte's concentration. As electrolysis progresses the analyte's concentration decreases, as does the current.

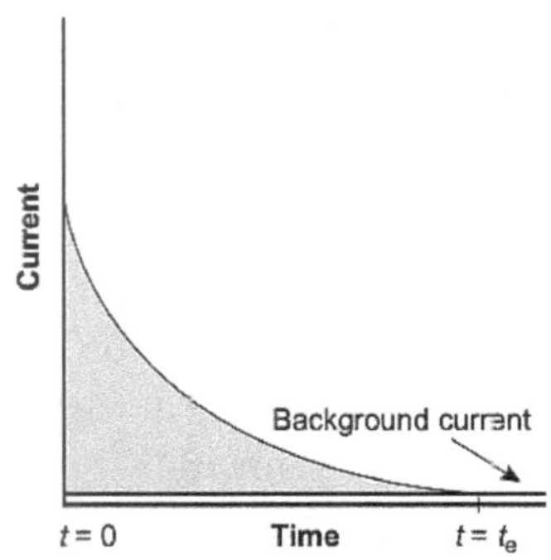

Current–time curve for controlled-potential coulometry

The potential in controlled-potential coulometry is set using a three-electrode potentiostat. Two types of working electrodes are commonly used: a Pt electrode manufactured from platinum-gauze and fashioned into a cylindrical tube, and an Hg pool electrode.

The auxiliary electrode, which is often a Pt wire, is separated by a salt bridge from the solution containing the analyte. This is necessary to prevent electrolysis products generated at the auxiliary electrode from reacting with the analyte and interfering in the analysis. A saturated calomel or Ag/AgCl electrode serves as the reference electrode.

Controlled-Current Coulometry

A second approach to coulometry is to use a constant current in place of a constant potential.

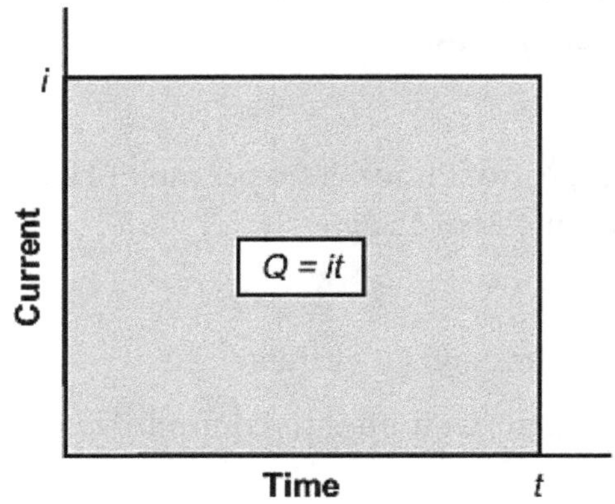

Current–time curve for controlled-current coulometry

Controlled-current coulometry, also known as amperostatic coulometry or coulometric titrimetry, has two advantages over controlled-potential coulometry:

1. more rapid analysis since the current does not decrease over time. Thus, a typical analysis time for controlled-current coulometry is less than 10 min, as opposed to approximately 30–60 min for controlled-potential coulometry.
2. the total charge is simply the product of current and time. A method for integrating the current–time curve, therefore, is not necessary.

There are two important experimental problems that must be solved if accurate results are to be obtained:

1. First, as electrolysis occurs the analyte's concentration and, therefore, the current due to its oxidation or reduction steadily decreases. To maintain a constant current the cell potential must change until another oxidation or reduction reaction can occur at the working electrode. Unless the system is carefully designed, these secondary reactions will produce a current efficiency of less than 100%.
2. There is the need for a method of determining when the analyte has been exhaustively electrolyzed. In controlled-potential coulometry this is signaled by a decrease in the current to a constant background or residual current.

Controlled-current coulometry normally is carried out using a galvanostat and an electrochemical cell consisting of a working electrode and a counterelectrode. The working electrode, which often is constructed from Pt, is also called the generator electrode since it is where the mediator reacts to generate the species reacting with the analyte. A species, such as Ce3+, which is used to maintain 100% current efficiency, is called a **mediator**.

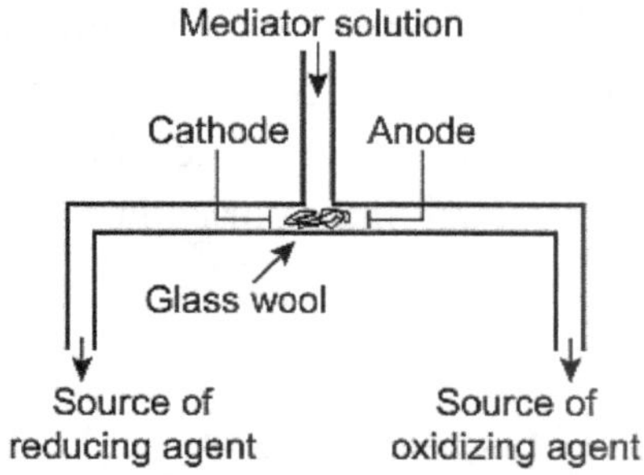

Method for the external generation of oxidizing and reducing agents in coulometric titrations

10.7 APPLICATIONS OF COULOMETRY

- ***Quantitative analysis:*** Coulometry may be used for the quantitative analysis of both inorganic and organic compounds.
- ***Coulumetric titrations***: This is advantageous over conventional titrimetry, because electrochemical generation of a "titrant" that reacts immediately with the analyte allows the use of reagents whose instability prevents their preparation and storage as a standard solution.
- ***Characterization Applications:*** Studies aimed at characterizing the mechanisms of electrode reactions often make use of coulometry for determining the number of electrons involved in the reaction.

10.8 KARL FISCHER TITRATION OF WATER

The Karl Fischer titration, which measures traces of water in transformer oil, solvents, foods, polymers, and other substances, is performed half a million times each day.

The titration is usually performed by delivering titrant from an automated burette or by coulometric generation of titrant. The volumetric procedure tends to be appropriate for larger amounts of water (but can go as low as) and the coulometric procedure tends to be appropriate for smaller amounts of water.

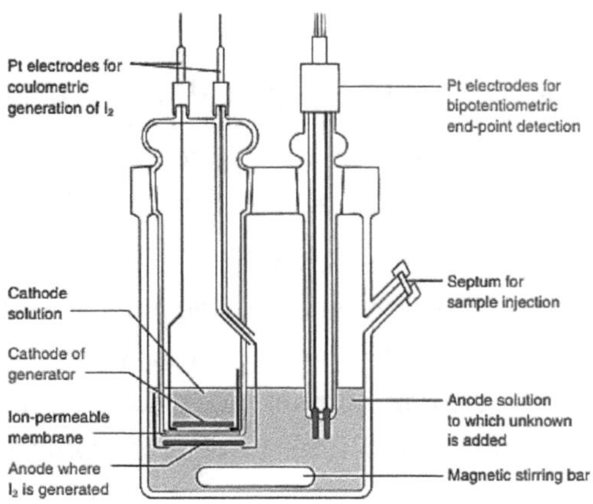

Diluted to volume with distilled water. A 10.00-mL portion is transferred to an electrochemical cell along with 25 mL of 1 M KI, 75 mL of a pH 7.0 phosphate buffer, and several drops of a starch indicator solution. Electrolysis at a constant current of 36.45 mA required 221.8 s to reach the starch indicator end point. Determine the purity of the sample. [Na = 23, S = 32, O = 16, 127, F = 96,487 C]

Solution

$$2S_2O_3^{2-}(aq) + I^{3-}(aq)\ S_4O_6^{2-}(aq) + 3I-(aq)$$

Using

$$Q = ite = nFN$$

where, i = 36.45 mA = 3.645 x10^{-2}, te = 221.8 s, n = 1 and N = number of moles of the sample =

$$\frac{Mass\ of\ substance}{Molar\ mass\ of\ the\ substance}$$

$$\frac{Mass\ of\ Na_2S_2O_3}{Molar\ mass\ of\ Na_2S_2O_3} = \frac{it_e}{nF}$$

$$Mass\ of\ Na_2S_2O_3 = \frac{Molar\ mass\ of\ Na_2S_2O_3 \times it_e}{nF}$$

$$Mass\ of\ Na_2S_2O_3 = \frac{Molar\ mass\ of\ Na_2S_2O_3 \times it_e}{nF}$$

$$Mass\ of\ Na_2S_2O_3 = \frac{158 \times 0.03645 \times 221.8}{1 \times 96487} = 0.01324\ g\ Na_2S_2O_3$$

This represents the amount of in a 10.00 ml portion of a 100.00 ml sample. Thus, 0.1325 g of is present in the original sample.

Then,

$$\%\ \text{Purity} = \frac{0.1325\ \text{g of}\ Na_2S_2O_3}{0.1342\ \text{g of sample}} \times 100$$

$$\%\ \text{Purity} = 98.73\%\ Na_2S_2O_3$$

10.9 VOLTAMMETRIC METHODS OF ANALYSIS

In **voltammetry** a time-dependent potential is applied to an electrochemical cell, and the current flowing through the cell is measured as a function of that potential.

Although early voltammetric methods relied on the use of only two electrodes, modern voltammetry makes use of a three-electrode potentiostat. The working electrode is commonly mercury electrode, the auxiliary electrode, which is a platinum wire, and the SCE and Ag/AgCl electrode are common reference electrodes. Since mercury is a liquid, the working electrode often consists of a drop suspended from the end of a capillary tube, which exists as:

1. 1. ***Hanging mercury drop electrode (HMDE):*** In HMDE, a drop of the desired size is formed by the action of a micrometer screw that pushes the mercury through a narrow capillary tube.
2. 2. ***Dropping mercury electrode (DME):*** In DME, mercury drops form at the end of the capillary tube as a result of gravity. Unlike the HMDE, the mercury drop of a DME grows continuously and has a finite lifetime of several seconds. At the end of its lifetime the mercury drop is dislodged, either manually or by gravity, and replaced by a new drop.
3. **3. Static mercury drop electrode (SMDE):** SMDE uses a solenoid-driven plunger to control the flow of mercury. The SMDE can be used as either a hanging mercury drop electrode or as a dropping mercury electrode.

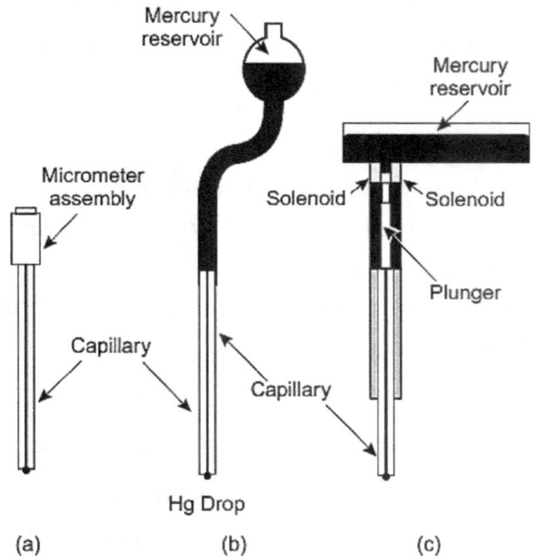

Mercury electrodes: (a) hanging mercury drop electrode; (b) dropping mercury electrode; (c) static mercury drop electrode

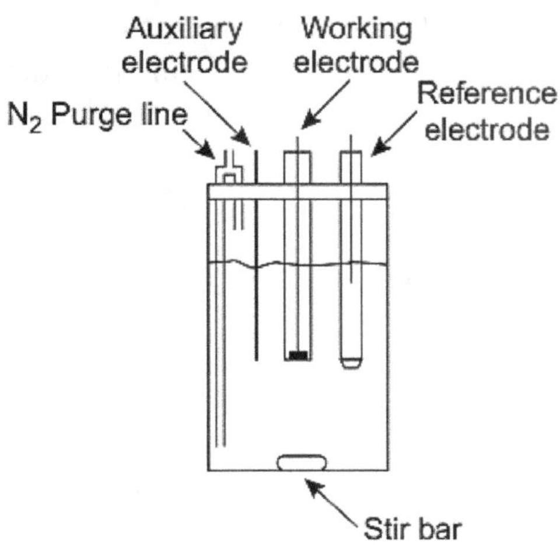

Typical electrochemical cell for use in voltammetry

When an analyte is oxidized at the working electrode, a current passes electrons through the external electric circuitry to the auxiliary electrode, where reduction of the solvent or other components of the solution matrix occurs. Reducing an analyte at the working electrode requires a source of electrons, generating a current that flows from the auxiliary electrode to the cathode. In either case, a current resulting from redox reactions at the working and auxiliary electrodes is called a **faradaic current**.

There are three modes of **mass transport** that influence the rate at which reactants and products are transported to and from the electrode surface:

1. Diffusion

Diffusion from a region of high concentration to a region of low concentration occurs whenever the concentration of an ion or molecule at the surface of the electrode is different from that in bulk solution

2. Migration

Migration occurs when charged particles in solution are attracted or repelled from an electrode that has a positive or negative surface charge. Thus, when the electrode is positively charged, negatively charged particles move toward the electrode, while positively charged particles move toward the bulk solution. Unlike diffusion and convection, migration only affects the mass transport of charged particles.

3. Convection

Convection occurs when a mechanical means is used to carry reactants toward the electrode and to remove products from the electrode. The most common means of convection is to stir the solution using a stir bar. Voltammetry that involves convection by stirring is called hydrodynamic voltammetry.

A plot of current as a function of applied potential is called a **voltammogram** and is the electrochemical equivalent of a spectrum in spectroscopy, providing quantitative and qualitative information about the species involved in the oxidation or reduction reaction.

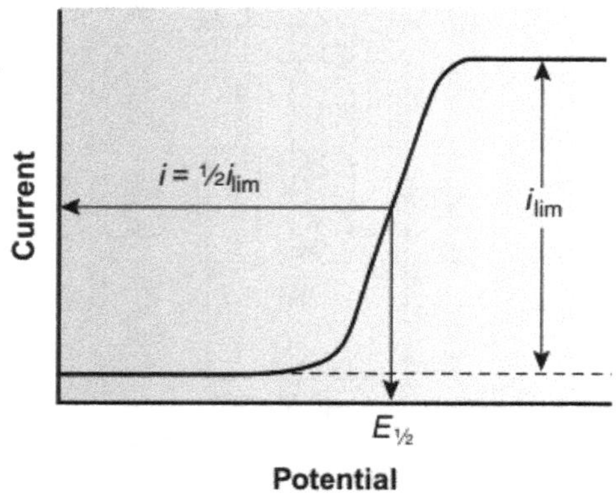

Typical voltammogram

10.9.1 Types of Voltammetric Techniques

1. Polarography

The earliest voltammetric experiment was normal **polarography** at a dropping mercury electrode. Although polarography takes place in an unstirred solution, a limiting current is obtained because the falling Hg drops mix the solution. Each new Hg drop, therefore, grows in a solution whose composition is identical to that of the initial bulk solution.

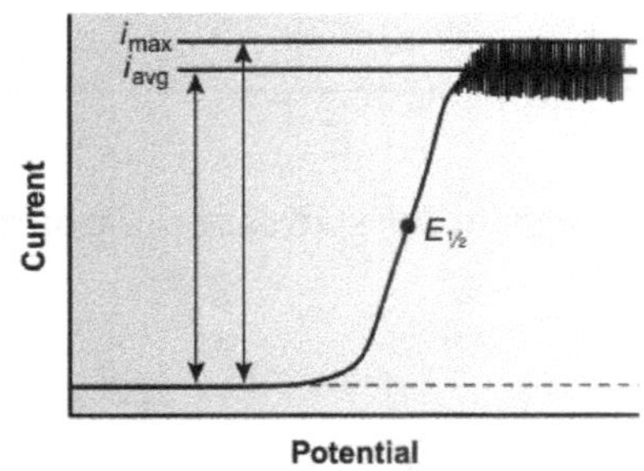

Voltammogram for normal polarography

Oscillations in the current are due to the growth of the Hg drop, which leads to a time-dependent change in the area of the working electrode. The limiting current, which is also called the diffusion current, may be measured from the maximum current, *i*max, or from the average current, *i*avg. The half-wave potential, *E*1/2, provides qualitative information about the redox reaction.

The relationship between the concentration of analyte, CA, and the limiting current is given by the Ilikovic equation:

$$(i_{lim})_{max} = 706nD^{\frac{1}{2}}m^{\frac{2}{3}}t^{\frac{1}{6}}C_A$$

$$(i_{lim})_{avg} = 607nD^{\frac{1}{2}}m^{\frac{2}{3}}t^{\frac{1}{6}}C_A$$

where, n = number of electrons transferred in the redox reaction, D = analyte's diffusion coefficient, m = flow rate of the Hg, and t = drop time.

There are other forms of polarography, in addition to normal, which include differential pulse polarography, staircase polarography and square-wave polarography.

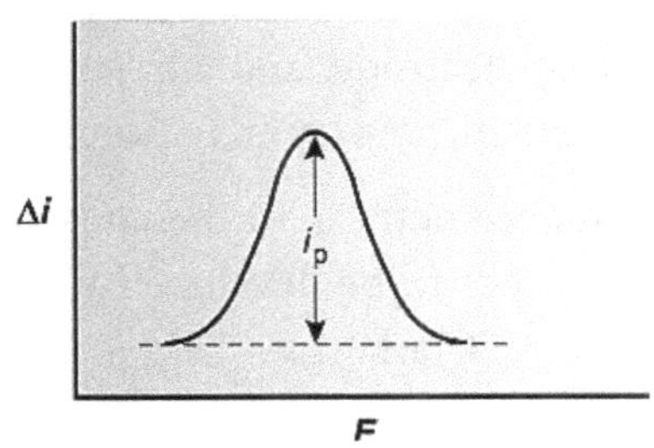

Voltammogram for differential pulse polarography

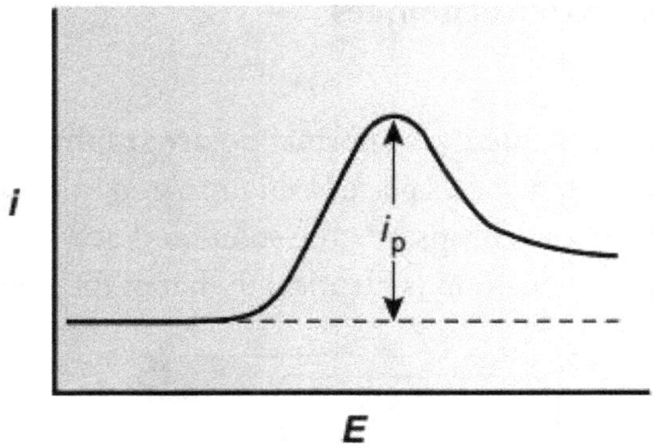

Voltammogram for staircase polarography

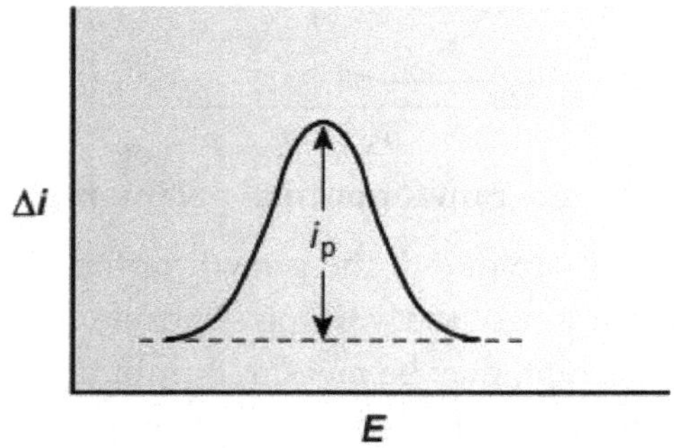

Voltammogram for square-wave polarography

2. Hydrodynamic Voltammetry

In hydrodynamic voltammetry, current is measured as a function of the potential applied to a solid working electrode. The same potential profiles used for polarography, such as a linear scan or a differential pulse, are used in hydrodynamic voltammetry.

3. Stripping Voltammetry

One of the most important quantitative voltammetric techniques is stripping voltammetry, which is composed of three related techniques: anodic, cathodic, and adsorptive stripping voltammetry.

4. Amperometry

The final voltammetric technique to be considered is amperometry, in which a constant potential is applied to the working electrode, and current is measured as a function of time. Since the potential is not scanned, amperometry does not lead to a voltammogram.

One important application of amperometry is in the construction of chemical sensors. One of the first amperometric sensors to be developed was for dissolved O2 in blood, which was developed in 1956 by L. C. Clark.

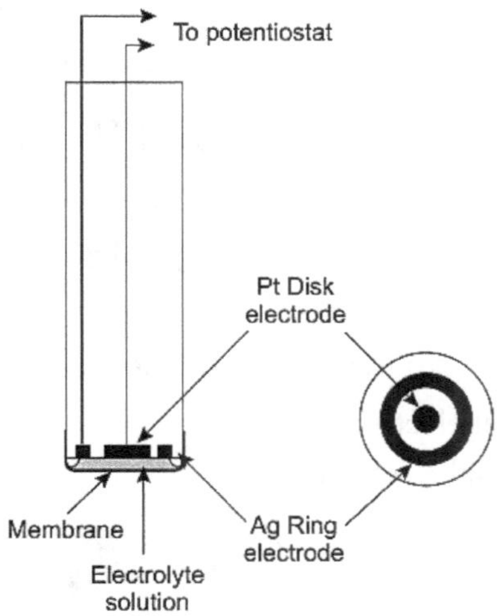

Clark amperometric sensor for the determination of dissolved O2

A gas-permeable membrane is stretched across the end of the sensor and is separated from the working and counter electrodes by a thin solution of KCl. The working electrode is a Pt disk cathode, and an Ag ring anode is the counter electrode.

10.10 APPLICATIONS OF VOLTAMMETRIC METHOD OF ANALYSIS

- **Multicomponent Analysis:** One advantage of voltammetry as a quantitative method of analysis is its capability for analyzing two or more analytes in a single sample.
- **Environmental Samples:** One area in which quantitative voltammetry has had a significant effect is in the analysis of trace metals in environmental samples. The most common samples are natural waters, including groundwater, lakes, rivers and streams, sea water, rain, and snow. Concentrations of trace metals at the parts-per billion (ppb) level can be determined using differential pulse polarography, whereas with anodic stripping voltammetry the determination of trace metals at the parts per-trillion (pptr) level is possible.
- **Clinical Samples:** Differential pulse polarography and stripping voltammetry have been used to determine the concentration of trace metals in a variety of matrices, including blood, urine, and tissue samples.
- **Miscellaneous Samples:** Besides environmental and clinical samples, differential pulse polarography and stripping voltammetry have been used for the analysis of trace metals in other samples, including food, steels and other alloys, gasoline, gunpowder residues, and pharmaceuticals.
- **Characterization Applications:** Voltammetry also can be used to obtain additional information, including verifying electrochemical reversibility,

determining the number of electrons transferred in a redox reaction, and determining equilibrium constants for coupled chemical reactions.

10.11 ELECTROGRAVIMETRIC METHOD OF ANALYSIS

Electrolytic precipitation or electrodeposition has been widely used for the determination of metals. In most applications, the metal is deposited on a weighed platinum cathode and the increase in mass is determined.

There are two types of electrogravimetric methods:

1. Constant current electrolysis
2. Constant potential electrolysis

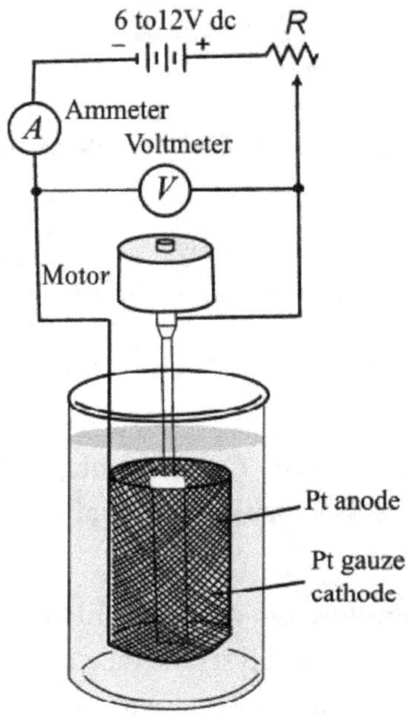

Apparatus for constant current electrolysis

10.11.1 Constant Current Electrolysis

It is important that the deposits produced by electrolysis is pure, strongly adherent, dense and smooth so that it can be washed, dried and weighed without any mechanical loss. Spongy, powdery or flaky deposits are likely to be less pure and less adherent.

Current density, temperature and the presence of complexing agents are the factors influencing the physical characteristics of the deposits. Ordinarily, the best deposits are formed at current densities that are less than 0.1A/cm^2.

Stirring generally improves the quality of a deposit. The time of deposition is reduced when the solution is stirred vigorously or if the electrode is rotated at a uniform speed. This type of stirring lowers the concentration overpotential and enables a higher current density without any adverse effect.

10.11.2 Constant Potential Electrolysis

By controlled potential electrolysis, it is possible to separate two elements whose deposition potentials differ sufficiently (by a few tenths of a volt). The potential of the cathode is controlled so that it never becomes sufficiently negative to allow the deposition of the next element.

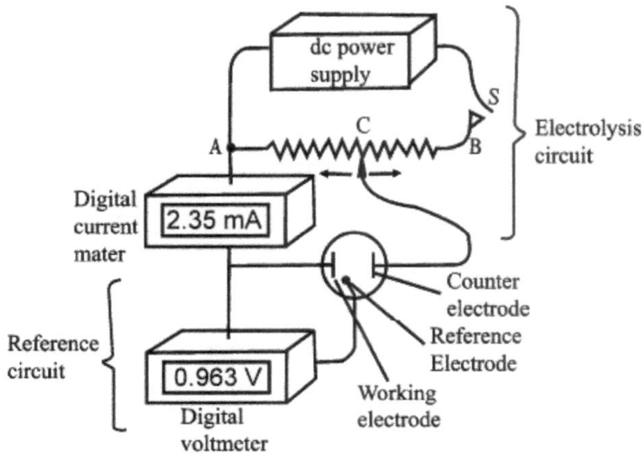

10.12 APPLICATIONS OF ELECTROGRAVIMETRIC METHOD OF ANALYSIS

- **Successive deposition of the metals:** A mixture of copper, bismuth, lead, cadmium, zinc and tin can be determined by successive deposition of the metals on a platinum cathode.
- **Simultaneous depositions:** A mixture of copper and lead can be determined by the deposition of copper in the presence of nitrate ions at the cathode. Lead is deposited as lead dioxide at the anode.
- **Preparative organic chemistry:** If an organic compound can undergo a series of reductions (or oxidations) each at a definite potential, it is then possible to reduce the starting material to the desired product by controlling the potential of the cathode during reduction.

11

TITRIMETRIC METHODS

11.1 INTRODUCTION: AN OVERVIEW

Titrimetry, in which volume serves as the analytical signal, first appears as an analytical method in the early eighteenth century. Titrimetric methods were not well received by the analytical chemists of that era because they could not duplicate the accuracy and precision of a gravimetric analysis. Not surprisingly, few standard texts from that era include titrimetric methods of analysis.

Precipitation gravimetry first developed as an analytical method without a general theory of precipitation. An empirical relationship between a precipitate's mass and the mass of analyte in a sample—what analytical chemists call a gravimetric factor—was determined experimentally by taking a known mass of analyte through the procedure. Today, we recognize this as an early example of an external standardization. Gravimetric factors were not calculated using the stoichiometry of a precipitation reaction because chemical formulas and atomic weights were not yet available! Unlike gravimetry, the development and acceptance of titrimetry required a deeper understanding of stoichiometry, of thermodynamics, and of chemical equilibria. By the 1900s, the accuracy and precision of titrimetric methods were comparable to that of gravimetric methods, establishing titrimetry as an accepted analytical technique.

In ***titrimetry*** we add a reagent, called the titrant, to a solution that contains another reagent, called the ***titrand***, and allow them to react. The type of reaction provides us with a simple way to divide titrimetry into four categories: acid–base titrations, in which an acidic or basic titrant reacts with a titrand that is a base or an acid; complexometric titrations , which are based on metal–ligand complexation; redox titrations, in which the titrant is an oxidizing or reducing agent; and precipitation titrations, in which the titrand and titrant form a precipitate.

We will deliberately avoid the term analyte at this point in our introduction to titrimetry. Although in most titrations the analyte is the titrand, there are circumstances where the analyte is the titrant. Later, when we discuss specific titrimetric methods, we will use the term analyte where appropriate.

Despite their difference in chemistry, all titrations share several common features. Before we consider individual titrimetric methods in greater detail, let's take a moment to consider some of these similarities. As you work through this chapter, this overview will help you focus on the similarities between different titrimetric methods. You will find it easier to understand a new analytical method when you can see its relationship to other similar methods.

11.1.1 Equivalence Points and End Points

If a titration is to give an accurate result we must combine the titrand and the titrant in stoichiometrically equivalent amounts. We call this stoichiometric mixture the equivalence point. Unlike precipitation gravimetry, where we add the precipitant in excess, an accurate titration requires that we know the exact volume of titrant at the equivalence point, Veq. The product of the titrant's equivalence point volume and its molarity, MT, is equal to the moles of titrant that react with the titrand.

$$\text{moles titrant} = M_T \times V_{eq}$$

if we know the stoichiometry of the titration reaction, then we can calculate the moles of titrand.

Unfortunately, for most titration reactions there is no obvious sign when we reach the equivalence point. Instead, we stop adding the titrant at an end point of our choosing. Often this end point is a change in the color of a substance, called an indicator, that we add to the titrand's solution. The difference between the end point's volume and the equivalence point's volume is a determinate titration error. If the end point and the equivalence point volumes coincide closely, then this error is insignificant and is safely ignored. Clearly, selecting an appropriate end point is of critical importance.

11.1.2 Volumes as a Signal

Instead of measuring the titrant's volume, we may choose to measure its mass. Although generally we can measure mass more precisely than we can measure volume, the simplicity of a volumetric titration makes it the more popular choice.

Almost any chemical reaction can serve as a titrimetric method provided that it meets the following four conditions. The first condition is that we must know the stoichiometry between the titrant and the titrand. If this is not the case, then we cannot convert the moles of titrant used to reach the end point to the moles of titrand in our sample. Second, the titration reaction effectively must proceed to completion; that is, the stoichiometric mixing of the titrant and the titrand must result in their complete reaction. Third, the titration reaction must occur rapidly. If we add the titrant faster than it can react with the titrand, then the end point and the equivalence point will differ significantly. Finally, we must have a suitable method for accurately determining the end point. These are significant limitations and, for this reason, there are several common titration strategies.

Depending on how we are detecting the endpoint, we may stop the titration too early or too late. If the end point is a function of the titrant's concentration, then adding the titrant too quickly leads to an early end point. On the other hand, if the end point is a function of the titrand's concentration, then the end point exceeds the equivalence point.

11.1.3 Titration Curves

To find a titration's end point, we need to monitor some property of the reaction that has a well-defined value at the equivalence point. For example, the equivalence point for a titration of HCl with NaOH occurs at a pH of 7.0. A simple method for finding the equivalence point is to monitor the titration mixture's pH using a pH electrode, stopping the titration when we reach a pH of 7.0. Alternatively, we can add an indicator to the titrand's solution that changes color at a pH of 7.0.

Why a pH of 7.0 is the equivalence point for this titration is a topic we will cover later in the section on acid–base titrations.

Suppose the only available indicator changes color at a pH of 6.8. Is the difference between this end point and the equivalence point small enough that we safely can ignore the titration error? To answer this question we need to know how the pH changes during the titration.

A titration curve provides a visual picture of how a property of the titration reaction changes as we add the titrant to the titrand. The titration curve in Figure 1 , for example, was obtained by suspending a pH electrode in a solution of 0.100 M HCl (the titrand) and monitoring the pH while adding 0.100 M NaOH (the titrant). A close examination of this titration curve should convince you that an end point pH of 6.8 produces a negligible titration error. Selecting a pH of 11.6 as the end point, however, produces an unacceptably large titration error.

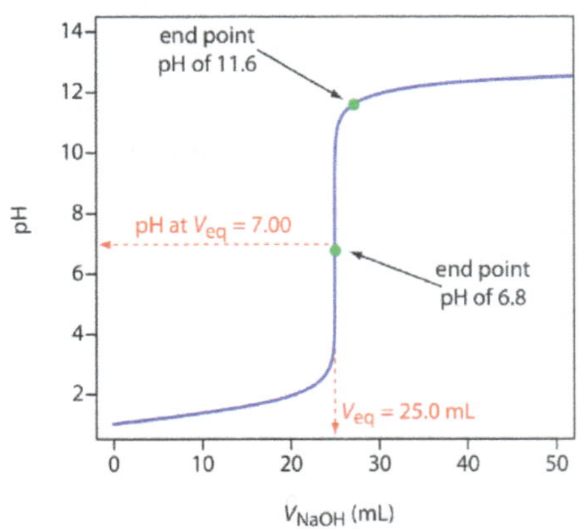

Fig. 1: Typical acid–base titration curve showing how the titrand's pH changes with the addition of titrant. The titrand is a 25.0 mL solution of 0.100 M HCl and the titrant is 0.100 M NaOH. The titration curve is the solid blue line, and the equivalence point

volume (25.0 mL) and pH (7.00) are shown by the dashed red lines. The green dots show two end points. The end point at a pH of 6.8 has a small titration error, and the end point at a pH of 11.6 has a larger titration error.

For the titration curve in Figure 1, the volume of titrant to reach a pH of 6.8 is 24.99995 mL, a titration error of -2.00×10^{-4}relative to the equivalence point of 25.00 mL. Typically, we can read the volume only to the nearest ±0.01 mL, which means this uncertainty is too small to affect our results. The volume of titrant to reach a pH of 11.6 is 27.07 mL, or a titration error of +8.28%. This is a significant error.

The shape of the titration curve in Figure 1 is not unique to an acid–base titration. Any titration curve that follows the change in concentration of a species in the titration reaction (plotted logarithmically) as a function of the titrant's volume has the same general sigmoidal shape. Several additional examples are shown in Figure 2 .

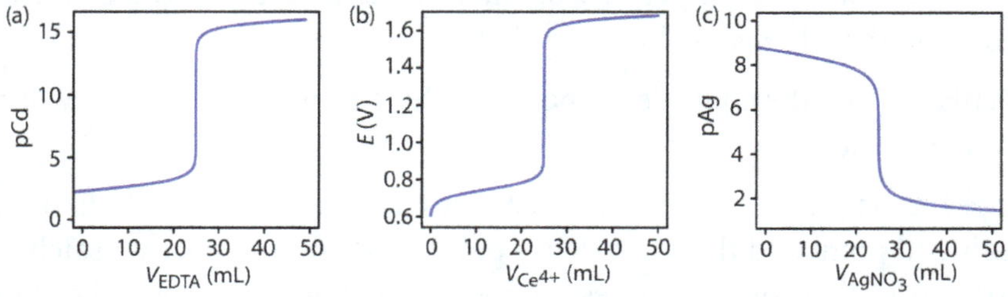

Fig. 2: Additional examples of titration curves. (a) Complexation titration of 25.0 mL of 1.0 mM Cd^{2+} with 1.0 mM EDTA at a pH of 10. The y-axis displays the titrand's equilibrium concentration as pCd. (b) Redox titration of 25.0 mL of 0.050 M Fe^{2+} with 0.050 M Ce^{4+} in 1 M $HClO_4$. The y-axis displays the titration mixture's electrochemical potential, E, which, through the Nernst equation is a logarithmic function of concentrations. (c) Precipitation titration of 25.0 mL of 0.10 M NaCl with 0.10 M $AgNO_3$. The y-axis displays the titrant's equilibrium concentration as pAg.

The titrand's or the titrant's concentration is not the only property we can use to record a titration curve. Other parameters, such as the temperature or absorbance of the titrand's solution, may provide a useful end point signal. Many acid–base titration reactions, for example, are exothermic. As the titrant and the titrand react, the temperature of the titrand's solution increases. Once we reach the equivalence point, further additions of titrant do not produce as exothermic a response. Figure 3 shows a typical ***thermometric*** titration curve where the intersection of the two linear segments indicates the equivalence point.

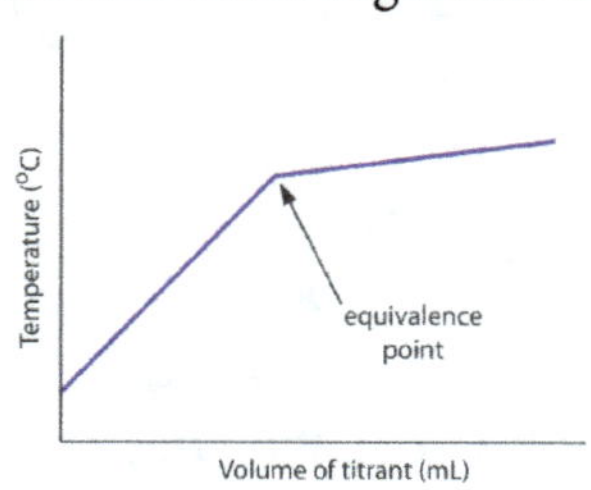

Fig. 3: Example of a thermometric titration curve showing the location of the equivalence point.

11.1.4 The Buret

The only essential equipment for an acid–base titration is a means for delivering the titrant to the titrand's solution. The most common method for delivering titrant is a ***buret***, which is a long, narrow tube with graduated markings and equipped with a stopcock for dispensing the titrant. The buret's small internal diameter provides a better defined meniscus, making it easier to read precisely the titrant's volume. Burets are available in a variety of sizes and tolerances, with the choice of buret determined by the needs of the analysis. Calibrating a buret corrects for variations in the buret's internal diameter.

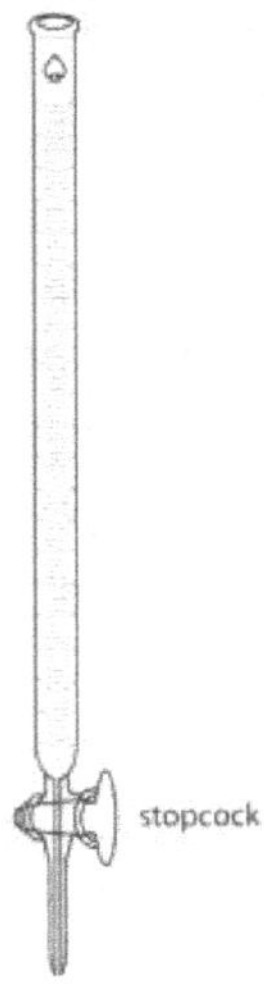

Fig. 4: A typical volumetric buret. The stopcock is shown here in the open position, which allows the titrant to flow into the titrand's solution. Rotating the stopcock controls the titrant's flow rate.

Table .1 . Specifications for Volumetric Burets

Volume (ml)	Class	Subdivision (ml)	Tolerance (±±)
5	A	0.01	±0.01
	B	0.01	±0.01
10	A	0.02	±0.02
	B	0.02	±0.04
25	A	0.1	±0.03
	B	0.1	±0.06
50	A	0.1	±0.05
	B	0.1	±0.10
100	A	0.2	±0.10
	B	0.2	±0.20

An automated titration uses a pump to deliver the titrant at a constant flow rate. Automated titrations offer the additional advantage of using a microcomputer for data storage and analysis.

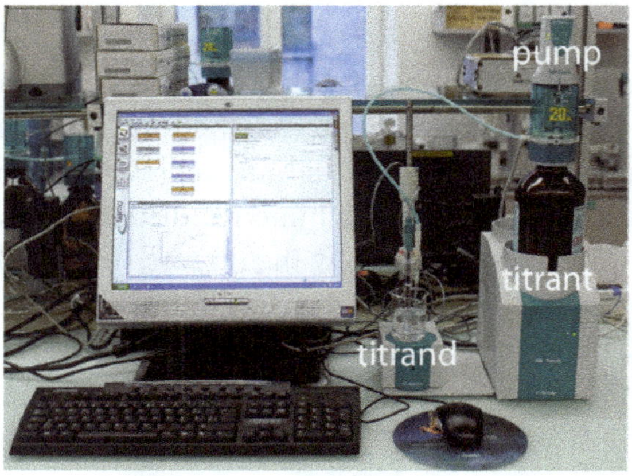

Fig. 5: Typical instrumentation for an automated acid–base titration showing the titrant, the pump, and the titrand. The pH electrode in the titrand's solution is used to monitor the titration's progress. You can see the titration curve in the lower-left quadrant of the computer's display. Modified from: Datamax.

11.2 ACID–BASE TITRATIONS

11.2.1 Introduction

Before 1800, most acid–base titrations used H_2SO_4, HCl, or HNO_3 as acidic titrants, and K_2CO_3 or Na_2CO_3 as basic titrants. A titration's end point was determined using litmus as an indicator, which is red in acidic solutions and blue in basic solutions, or by the cessation of CO_2 effervescence when neutralizing CO_2–$3CO_3^{2-}$. Early examples of acid–base titrimetry include determining the acidity or alkalinity of solutions, and determining the purity of carbonates and alkaline earth oxides.

Three limitations slowed the development of acid–base titrimetry: the lack of a strong base titrant for the analysis of weak acids, the lack of suitable indicators, and the absence of a theory of acid–base reactivity. The introduction, in 1846, of NaOH as a strong base titrant extended acid–base titrimetry to the determination of weak acids. The synthesis of organic dyes provided many new indicators. Phenolphthalein, for example, was first synthesized by Bayer in 1871 and used as an indicator for acid–base titrations in 1877.

Despite the increased availability of indicators, the absence of a theory of acid–base reactivity made it difficult to select an indicator. The development of equilibrium theory in the late 19th century led to significant improvements in the theoretical understanding of acid–base chemistry, and, in turn, of acid–base titrimetry. Sorenson's establishment of the pH scale in 1909 provided a rigorous means to compare indicators. The determination of acid–base dissociation constants made it possible to calculate a theoretical titration curve, as outlined by Bjerrum in 1914. For the first time analytical chemists had a rational method for selecting an indicator, making acid–base titrimetry a useful alternative to gravimetry.

11.2.2 Acid–Base Titration Curves

In the overview to this chapter we noted that a titration's end point should coincide with its equivalence point. To understand the relationship between an acid–base titration's end point and its equivalence point we must know how the titrand's pH changes during a titration. In this section we will learn how to calculate a titration curve using the equilibrium calculations from Chapter 6. We also will learn how to sketch a good approximation of any acid–base titration curve using a limited number of simple calculations.

Titrating Strong Acids and Strong Bases

For our first titration curve, let's consider the titration of 50.0 mL of 0.100 M HCl using a titrant of 0.200 M NaOH. When a strong base and a strong acid react the only reaction of importance is

$$H_3O+(aq)+OH-(aq) \rightarrow 2H_2O(l) \qquad 1$$

the first task is to calculate the volume of NaOH needed to reach the equivalence point, V_{eq}. At the equivalence point we know from reaction 1 that

$$\text{moles HCl} = \text{moles NaOH}$$

$$M_a \times V_a = M_b \times V_b$$

where the subscript '*a*' indicates the acid, HCl, and the subscript '*b*' indicates the base, NaOH. The volume of NaOH needed to reach the equivalence point is

$$V_{eq} = V_b = \frac{M_a V_a}{M_b} = \frac{(0.100\text{ M})(50.0\text{ mL})}{(0.200\text{ M})} = 25.0\text{ mL}$$

before the equivalence point, HCl is present in excess and the pH is determined by the concentration of unreacted HCl. At the start of the titration the solution is 0.100 M in HCl, which, because HCl is a strong acid, means the pH is

$$\text{pH} = -\log[H_3O^+] = -\log[\text{HCl}] = -\log(0.100) = 1.00$$

After adding 10.0 mL of NaOH the concentration of excess HCl is

$$[\text{HCl}] = \frac{(\text{mol HCl})_{\text{initial}} - (\text{mol NaOH})_{\text{added}}}{\text{total volume}} = \frac{M_a V_a - M_b V_b}{V_a + V_b}$$

$$[\text{HCl}] = \frac{(0.100\text{ M})(50.0\text{ mL}) - (0.200\text{ M})(10.0\text{ mL})}{50.0\text{ mL} + 10.0\text{ mL}} = 0.0500\text{ M}$$

and the pH increases to 1.30.

At the equivalence point the moles of HCl and the moles of NaOH are equal. Since neither the acid nor the base is in excess, the pH is determined by the dissociation of water.

'Thus, the pH at the equivalence point is 7.00.

$$K_w = 1.00 \times 10^{-14} = [H_3O^+][OH^-] = [H_3O^+]^2$$
$$[H_3O^+] = 1.00 \times 10^{-7}$$

For volumes of NaOH greater than the equivalence point, the pH is determined by the concentration of excess OH^-.

Titrating a Weak Acid with a Strong Base

For this example, let's consider the titration of 50.0 mL of 0.100 M acetic acid, CH_3COOH, with 0.200 M NaOH. Again, we start by calculating the volume of NaOH needed to reach the equivalence point; thus

$$\text{mol } CH_3COOH = \text{mol NaOH}$$

$$M_a \times V_a = M_b \times V_b$$

$$V_{eq} = V_b = \frac{M_a V_a}{M_b} = \frac{(0.100 \text{ M})(50.0 \text{ mL})}{(0.200 \text{ M})} = 25.0 \text{ mL}$$

$$CH_3COOH(aq) + H_2O(l) \rightleftharpoons H_3O^+(aq) + CH_3COO^-(aq)$$

$$K_a = \frac{[H_3O^+][CH_3COO^-]}{[CH_3COOH]} = \frac{(x)(x)}{0.100 - x} = 1.75 \times 10^{-5}$$

finding that the pH is 2.88.

Adding NaOH converts a portion of the acetic acid to its conjugate base, CH_3COO^-.

$$CH_3COOH(aq) + OH^-(aq) \longrightarrow H_2O(l) + CH_3COO^-(aq)$$ 2

$$K = K_a / K_w = 1.75 \times 10^9$$

We can treat the reaction as if it goes to completion.

Any solution that contains comparable amounts of a weak acid, HA, and its conjugate weak base, A^-, is a buffer.

$$pH = pK_a + \log\frac{[A^-]}{[HA]}$$

Before the equivalence point the concentration of unreacted acetic acid is

$$[CH_3COOH] = \frac{(\text{mol } CH_3COOH)_{\text{initial}} - (\text{mol NaOH})_{\text{added}}}{\text{total volume}} = \frac{M_a V_a - M_b V_b}{V_a + V_b}$$

and the concentration of acetate is

12.2.3 Selecting and Evaluating the End Point

Earlier we made an important distinction between a titration's end point and its equivalence point. The difference between these two terms is important and deserves repeating. An equivalence point, which occurs when we react stoichiometrically equal amounts of the analyte and the titrant, is a theoretical not an experimental value. A titration's end point is an experimental result that represents our best estimate of the equivalence point. Any difference between a titration's equivalence point and its corresponding end point is a source of determinate error.

Where is the Equivalence Point?

Earlier we learned how to calculate the pH at the equivalence point for the titration of a strong acid with a strong base, and for the titration of a weak acid with a strong base. We also learned how to sketch a titration curve with only a minimum of calculations. Can we also locate the equivalence point without performing any calculations. The answer, as you might guess, often is yes!

For most acid–base titrations the inflection point—the point on a titration curve that has the greatest slope—very nearly coincides with the titration's equivalence point.

Finding the End Point with an Indicator

One interesting group of weak acids and weak bases are organic dyes. Because an organic dye has at least one highly colored conjugate acid–base species, its titration results in a change in both its pH and its color. We can use this change in color to indicate the end point of a titration provided that it occurs at or near the titration's equivalence point.

As an example, let's consider an indicator for which the acid form, HIn, is yellow and the base form, In^-, is red. The color of the indicator's solution depends on the relative concentrations of HIn and In^-. To understand the relationship between pH and color we use the indicator's acid dissociation reaction

$$\mathrm{HIn}(aq) + \mathrm{H_2O}(l) \rightleftharpoons \mathrm{H_3O^+}(aq) + \mathrm{In^-}(aq)$$

and its equilibrium constant expression.

$$K_\mathrm{a} = \frac{[\mathrm{H_3O^+}][\mathrm{In^-}]}{[\mathrm{HIn}]}$$

Taking the negative log of each side of Equation 5, and rearranging to solve for pH leaves us with a equation that relates the solution's pH to the relative concentrations of HIn and In^-.

$$\mathrm{pH} = \mathrm{p}K_\mathrm{a} + \log\frac{[\mathrm{In^-}]}{[\mathrm{HIn}]}$$

If we can detect HIn and In– with equal ease, then the transition from yellow-to-red (or from red-to-yellow) reaches its midpoint, which is orange, when the concentrations of HIn and In– are equal, or when the pH is equal to the indicator's pK_a. If the indicator's pK_a and the pH at the equivalence point are identical, then titrating until the indicator turns orange is a suitable end point. Unfortunately, we rarely know the exact pH at the equivalence point. In addition, determining when the concentrations of HIn and In– are equal is difficult if the indicator's change in color is subtle.

We can establish the range of pHs over which the average analyst observes a change in the indicator's color by making two assumptions: that the indicator's color is yellow if the concentration of HIn is 10×10× greater than that of In– and that its color is red if the concentration of HIn is 10×10× smaller than that of In–. Substituting these inequalities into Equation 6

$$\mathrm{pH} = \mathrm{p}K_\mathrm{a} + \log\tfrac{1}{10} = \mathrm{p}K_\mathrm{a} - 1$$
$$\mathrm{pH} = \mathrm{p}K_\mathrm{a} + \log\tfrac{10}{1} = \mathrm{p}K_\mathrm{a} + 1$$

shows that the indicator changes color over a pH range that extends ±1 unit on either side of its pK_a. As shown in Figure 6, the indicator is yel-ow when the pH is less than pK_a - 1 and it is red when the pH is greater than pK_a + 1. For pH values between pK_a - 1 and pK_a + 1 the indicator's color passes through various shades of orange. The properties of several common acid–base indicators are listed in Table 2 .

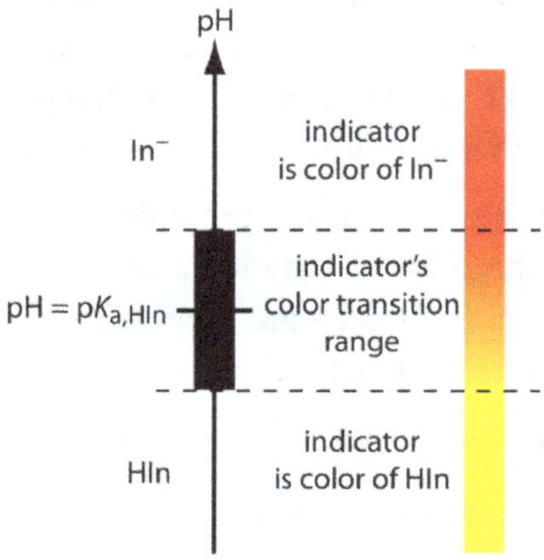

Fig. 6: Diagram showing the relationship between pH and an indicator's color. The ladder diagram defines pH values where HIn and In– are the predominate species. The indicator changes color when the pH is between pK_a^{-1} and pK_a^{+1}.

Table 2 . Properties of Selected Acid–Base Indicators

Indicator	Acid Color	Base Color	Ph Range	Pk_a
Cresol Red	Red	Yellow	0.2–1.8	—
Thymol Blue	Red	Yellow	1.2–2.8	1.7
Bromothymol Blue	Yellow	Blue	3.0–4.6	4.1
Methyl Orange	Red	Yellow	3.4–4.4	3.7
Congo Red	Blue	Red	3.0–5.0	—
Bromocresol Green	Yellow	Blue	3.8–5.4	4.7
Methyl Red	Red	Yellow	4.2–6.3	5.0
Bromocresol Purple	Yellow	Purple	5.2–6.8	6.1

Indicator	Acid Color	Base Color	Ph Range	Pk_a
Litmus	Red	Blue	5.0–8.0	—
Bromothymol Blue	Yellow	Blue	6.0–7.6	7.1
Phenol Red	Yellow	Blue	6.8–8.4	7.8
Cresol Red	Yellow	Red	7.2–8.8	8.2
Thymol Blue	Yellow	Red	8.0–9.6	8.9
Phenolphthalein	Colorless	Red	8.3–10.0	9.6
Alizarin Yellow R	Yellow	Orange-Red	10.1–12.0	—

You may wonder why an indicator's pH range, such as that for phenolphthalein, is not equally distributed around its pK_a value. The explanation is simple. Figure 5 presents an idealized view in which our sensitivity to the indicator's two colors is equal. For some indicators only the weak acid or the weak base is colored. For other indicators both the weak acid and the weak base are colored, but one form is easier to see. In either case, the indicator's pH range is skewed in the direction of the indicator's less colored form. Thus, phenolphthalein's pH range is skewed in the direction of its colorless form, shifting the pH range to values lower than those suggested by Figure.

The relatively broad range of pHs over which an indicator changes color places additional limitations on its ability to signal a titration's end point. To minimize a determinate titration error, the indicator's entire pH range must fall within the rapid change in pH near the equivalence point. For example, we see that phenolphthalein is an appropriate indicator for the titration of 50.0 mL of 0.050 M acetic acid with 0.10 M NaOH. Bromothymol blue, on the other hand, is an inappropriate indicator because its change in color begins well before the initial sharp rise in pH, and, as a result, spans a relatively large range of volumes. The early change in color increases the probability of obtaining an inaccurate result, and the range of possible end point volumes increases the probability of obtaining imprecise results.

12.2.4 Titrations in Nonaqueous Solvents

Thus far we have assumed that the titrant and the titrand are aqueous solutions. Although water is the most common solvent for acid–base titrimetry, switching to a nonaqueous solvent can improve a titration's feasibility.

For an amphoteric solvent, SH, the autoprotolysis constant, K_s, relates the concentration of its protonated form, SH+2SH2+, to its deprotonated form, S^-

$$2SH \rightleftharpoons SH^+_2 + S^-$$

$$Ks = [SH_2^+][S^-]$$

and the solvent's pH and pOH are

$$pH = -\log[SH^+_2]$$

$$pOh = -\log[S^-]$$

You should recognize that K_w is just specific form of K_s when the solvent is water.

The most important limitation imposed by K_s is the change in pH during a titration. To understand why this is true, let's consider the titration of 50.0 mL of 1.0×10−41.0×10−4 M HCl using 1.0×10−41.0×10−4 M NaOH as the titrant. Before the equivalence point, the pH is determined by the untitrated strong acid. For example, when the volume of NaOH is 90% of V_{eq}, the concentration of H_3O^+ is

$$[H_3O^+] = \frac{M_aV_a - M_bV_b}{V_a + V_b} = \frac{(1.0 \times 10^{-4}\ M)(50.0\ mL) - (1.0 \times 10^{-4}\ M)(45.0\ mL)}{50.0\ mL + 45.0\ mL} = 5.3 \times 10^{-6}\ M$$

and the pH is 5.3. When the volume of NaOH is 110% of V_{eq}, the concentration of OH− is

$$[OH^-] = \frac{M_bV_b - M_aV_a}{V_a + V_b} = \frac{(1.0 \times 10^{-4}\ M)(55.0\ mL) - (1.0 \times 10^{-4}\ M)(50.0\ mL)}{55.0\ mL + 50.0\ mL} = 4.8 \times 10^{-6}\ M$$

and the pOH is 5.3. The titrand's pH is

$$pH = pK_w - pOH = 14.0 - 5.3 = 8.7$$

and the change in the titrand's pH as the titration goes from 90% to 110% of V_{eq} is

$$\Delta pH = 8.7 - 5.3 = 3.4$$

If we carry out the same titration in a nonaqueous amphiprotic solvent that has a K_s of 1.0×10−201.0×10−20, the pH after adding 45.0 mL of NaOH is still 5.3. However, the pH after adding 55.0 mL of NaOH is

$$pH = pK_s - pOH = 20.0 - 5.3 = 14.7$$

In this case the change in pH

$$\Delta pH = 14.7 - 5.3 = 9.4$$

is significantly greater than that obtained when the titration is carried out in water. Figure 7 shows the titration curves in both the aqueous and the nonaqueous solvents.

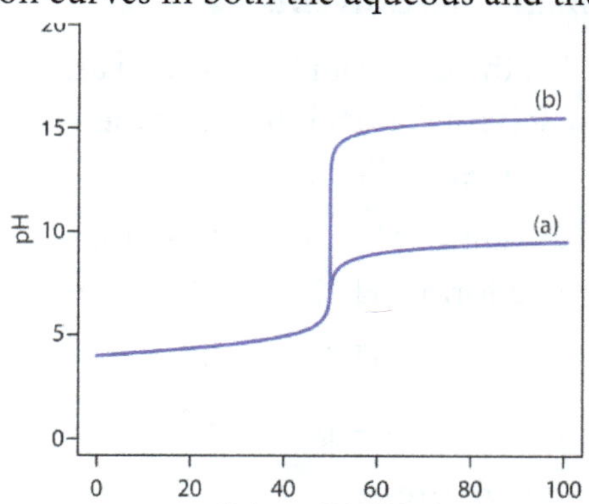

Fig. 7: Titration curves for 50.0 mL of 1.0×10−41.0×10−4 M HCl using 1.0×10−41.0×10−4 M NaOH in (a) water, K_w = 1.0×10−141.0×10−14, and (b) a nonaqueous amphiprotic solvent, K_s = 1.0×10−201.0×10−20.

Another parameter that affects the feasibility of an acid–base titration is the titrand's dissociation constant. Here, too, the solvent plays an important role. The strength of an acid or a base is a relative measure of how easy it is to transfer a proton from the acid to the solvent or from the solvent to the base.

For example, HF, with a Ka of 6.8×10−46.8×10−4, is a better proton donor than the strongest acid that can exist in water is the hydronium ion, H_3O+. HCl and HNO_3 are strong acids because they are better proton donors than H_3O+ and essentially donate all their protons to H_2O, leveling their acid strength to that of H_3O+. In a different solvent HCl and HNO_3 may not behave as strong acids. If we place acetic acid in water the dissociation reaction

$$CH_3COOH\ (aq) + H_2O(l) \rightleftharpoons H_3O+(aq)+CH_3COO-(aq)$$

does not proceed to a significant extent because CH_3COO- is a stronger base than H_2O and H_3O^+ is a stronger acid than CH_3COOH. If we place acetic acid in a solvent that is a stronger base than water, such as ammonia, then the reaction

$$CH_3COOH + NH_3 \rightleftharpoons NH^{+4}+CH_3COO^-$$

proceeds to a greater extent. In fact, both HCl and CH_3COOH are strong acids in ammonia.

All other things being equal, the strength of a weak acid increases if we place it in a solvent that is more basic than water, and the strength of a weak base increases if we place it in a solvent that is more acidic than water. In some cases, however, the opposite effect is observed. For example, the pK_b for NH_3 is 4.75 in water and it is 6.40 in the more acidic glacial acetic acid. In contradiction to our expectations, NH_3 is a weaker base in the more acidic solvent. A full description of the solvent's effect on the pK_a of weak acid or the pK_b of a weak base is beyond the scope of this text. You should be aware, however, that a titration that is not feasible in water may be feasible in a different solvent.

11.3 QUANTITATIVE APPLICATIONS

Although many quantitative applications of acid–base titrimetry have been replaced by other analytical methods, a few important applications continue to find use. In this section we review the general application of acid–base titrimetry to the analysis of inorganic and organic compounds, with an emphasis on applications in environmental and clinical analysis. First, however, we discuss the selection and standardization of acidic and basic titrants.

Selecting and Standardizing a Titrant

The most common strong acid titrants are HCl, $HClO_4$, and H_2SO_4. Solutions of these titrants usually are prepared by diluting a commercially available concentrated stock solution. Because the concentration of a concentrated acid is known only approximately, the titrant's concentration is determined by standardizing against one of the primary standard weak bases.

The nominal concentrations of the concentrated stock solutions are 12.1 M HCl, 11.7 M $HClO_4$, and 18.0 M H_2SO_4. The actual concentrations of these acids are given as %w/v and vary slightly from lot-to-lot.

Table 3 . Selected Primary Standards for Standardizing Strong Acid and Strong Base Titrants

Titrant type	Primary standard	Titration reaction	Comment
Strong acid	Na_2CO_3	Na_2CO_3+2H_3O+→H_2CO_3+2Na++2H_2O	a
Strong acid	($HOCH_2$) 3CNH_2	($HOCH_2$)3CNH2+H_3O+⟶($HOCH_2$)3CNH^+_3+H_2O	b
Strong acid	Na2B_4O_7	$Na_2B_4O_7$+2H_3O^++3H^2O→2Na_++4H_3BO_3	
Strong base	$KHC_8H_4O_4$	$KHC_8H_4O_4$+OH−→K^++C_8H_4O−4+H_2O	c
Strong base	C_6H_5COOH	C_6H_5COOH+OH^-→C_6H_5COO−+H_2O	d
Strong base	KH$(IO_3)_2$	KH(IO_3)2+OH−→K^++2IO^-_3+H_2O	

a. The end point for this titration is improved by titrating to the second equivalence point, boiling the solution to expel CO_2, and retitrating to the second equivalence point. The reaction in this case is

$$Na_2CO_3+2H_3O+\rightarrow CO^2+2Na^++3H_2O$$

Tris- (hydroxymethyl) aminomethane often goes by the shorter name of TRIS or THAM.

b. Potassium hydrogen phthalate often goes by the shorter name of KHP.

c. Because it is not very soluble in water, dissolve benzoic acid in a small amount of ethanol before diluting with water.

The most common strong base titrant is NaOH, which is available both as an impure solid and as an approximately 50% w/v solution. Solutions of NaOH are standardized against any of the primary weak acid standards listed in Table 2.

Using NaOH as a titrant is complicated by potential contamination from the following reaction between dissolved CO_2 and OH−.

$$CO_2(aq)+2OH^-(aq) \rightarrow CO_2-3(aq) + H_2O\ (l) \qquad 7$$

Any solution in contact with the atmosphere contains a small amount of CO_2*(aq)* from the equilibrium

$$CO_2(g) \rightleftharpoons CO^2(aq)$$

During the titration, NaOH reacts both with the titrand and with CO_2, which increases the volume of NaOH needed to reach the titration's end point. This is not a problem if the end point pH is less than 6. Below this pH the $CO_2^{-3}CO_3^{2-}$ from reaction 7 reacts with H_3O^+ to form carbonic acid.

$$CO_2^{-3}(aq)+2H_3O+(aq)\rightarrow 2H_2O(l)+H_2CO^3(aq). \qquad (8)$$

Combining reaction 7 and reaction 8 gives an overall reaction that does not include OH^-.

$$CO_2(aq)+H_2O(l)\longrightarrow H_2CO^3(aq)$$

Under these conditions the presence of CO_2 does not affect the quantity of OH− used in the titration and is not a source of determinate error.

If the end point pH is between 6 and 10, however, the neutralization of CO2−3CO32− requires one proton

$$CO^{2-}{}_3(aq)+H_3O+(aq)\rightarrow H_2O(l)+HCO^{-3}(aq)$$

and the net reaction between CO_2 and OH^- is

$$CO_2\ (aq)+OH^-(aq)\rightarrow HCO^{-3}(aq)$$

Under these conditions some OH− is consumed in neutralizing CO_2, which results in a determinate error. We can avoid the determinate error if we use the same end point pH for both the standardization of NaOH and the analysis of our analyte, although this is not always practical.

Solid NaOH is always contaminated with carbonate due to its contact with the atmosphere, and we cannot use it to prepare a carbonate-free solution of NaOH. Solutions of carbonate-free NaOH are prepared from 50% w/v NaOH because Na_2CO_3 is insoluble in concentrated NaOH. When CO_2 is absorbed, Na_2CO_3 precipitates and settles to the bottom of the container, which allow access to the carbonate-free NaOH. When preparing a solution of NaOH, be sure to use water that is free from dissolved CO_2. Briefly boiling the water expels CO_2; after it cools, the water is used to prepare carbonate-free solutions of NaOH. A solution of carbonate-free NaOH is relatively stable if we limit its contact with the atmosphere. Standard solutions of sodium hydroxide are not stored in glass bottles as NaOH reacts with glass to form silicate; instead, store such solutions in polyethylene bottles.

11.4 END POINT VOLUME AND SOURCES OF ALKANITY

A mixture of OH− and CO_2^{-3} or a mixture of HCO−3 and CO_2^{-3} also is possible. Consider, for example, a mixture of OH− and $CO_2^{-3}CO_3^{2-}$. The volume of strong acid to titrate OH− is the same whether we titrate to a pH of 8.3 or a pH of 4.5. Titrating CO_2^{-3} to a pH of 4.5, however, requires twice as much strong acid as titrating to a pH of 8.3. Consequently, when we titrate a mixture of these two ions, the volume of strong acid needed to reach a pH of 4.5 is less than twice that needed to reach a pH of 8.3. For a mixture of HCO^{-3} and CO_2^{-3} the volume of strong acid needed to reach a pH of 4.5 is more than twice that needed to reach a pH of 8.3. Table 4 summarizes the relationship between the sources of alkalinity and the volumes of titrant needed to reach the two end points.

A mixture of OH− and HCO^{-3} is unstable with respect to the formation of CO_2^{-3}.

Table 4 . Relationship Between End Point Volumes and Sources of Alkalinity

Source of alkalinity	Relationship between end point volumes
OH^-	VpH 4.5=VpH 8.3VpH 4.5=VpH 8.3
$CO_2^{-3}CO_3^{2-}$	VpH 4.5=2×VpH 8.3VpH 4.5=2×VpH 8.3
$HCO^{-3}HCO^{3-}$	VpH 4.5>0;VpH 8.3=0VpH 4.5>0;VpH 8.3=0
OH^- and CO2−3CO32−	VpH 4.5<2×VpH 8.3VpH 4.5<2×VpH 8.3
$CO_2^{-3}CO_3^{2-}$ and $HCO^{-3}HCO^{3-}$	VpH 4.5>2×VpH 8.3VpH 4.5>2×VpH 8.3

Acidity is a measure of a water sample's capacity to neutralize base and is divided into strong acid and weak acid acidity. Strong acid acidity from inorganic acids such as HCl, HNO_3, and H_2SO_4 is common in industrial effluents and in acid mine drainage. Weak acid acidity usually is dominated by the formation of H_2CO_3 from dissolved CO_2, but also includes contributions from hydrolyzable metal ions such as Fe^{3+}, Al^{3+}, and Mn^{2+}. In addition, weak acid acidity may include a contribution from organic acids.

Acidity is determined by titrating with a standard solution of NaOH to a fixed pH of 7 (or the bromothymol blue end point) and to a fixed pH of 3 (or the phenolphthalein end point). Titrating to a pH of 3.7 provides a measure of strong acid acidity, and titrating to a pH of 8.3 provides a measure of total acidity. Weak acid acidity is the difference between the total acidity and the strong acid acidity. Results are expressed as the amount of $CaCO_3$ that can be neutralized by the sample's acidity. An alternative approach for determining strong acid and weak acid acidity is to obtain a potentiometric titration curve and use a Gran plot to determine the two equivalence points. This approach has been used, for example, to determine the forms of acidity in atmospheric aerosols.

As is the case with alkalinity, acidity is reported as mg $CaCO_3$/L.

Water in contact with either the atmosphere or with carbonate-bearing sediments contains free CO_2 in equilibrium with CO_2(g) and with aqueous H_2CO_3, $HCO^{-3}HCO^{3-}$ and $CO_2^{-3}CO_3^{2-}$. The concentration of free CO_2 is determined by titrating with a standard solution of NaOH to the phenolphthalein end point, or to a pH of 8.3, with results reported as mg CO_2/L. This analysis essentially is the same as that for the determination of total acidity and is used only for water samples that do not contain strong acid acidity.

Free CO_2 is the same thing as CO_2(aq).

Table 5 . Selected Acid–Base Titrimetric Methods for Organic Functional Groups Based on the Production or Consumption of Acid or Base

Functional group	Reaction producing titratable species	Titration details
Ester	RCOOR' (aq) + OH− (aq) →→ RCOO− (aq) + HOR' (aq)	Titrate oh− with hcl
Carbonyl	R_2CO (aq) + $NH_4OH•HCl$ (aq) →→ R2CNOH (aq) + HCl (aq) + H_2O (l)	Titrate hcl with naoh

Functional group	Reaction producing titratable species	Titration details
Alcohol	[1]:(CH3CO)2O+ROH →→ CH3COOR + CH3COOH [2]: $(CH_3CO)_2)$ + H_2O →→ $_2CH_3COOH$	Titrate ch3cooh with naoh; a blank titration of acetic anhydride, (ch3co)2o, corrects for the contribution of reaction [2]
The species that is titrated is shown in bold for alcohols, reaction [1] is carried out in pyridine to prevent the hydrolysis of acetic anhydride by water. After reaction [1] is complete, water is added to covert any unreacted acetic anhydride to acetic acid (reaction [2])		

Many pharmaceutical compounds are weak acids or weak bases that are analyzed by an aqueous or a nonaqueous acid–base titration; examples include salicylic acid, phenobarbital, caffeine, and sulfanilamide. Amino acids and proteins are analyzed in glacial acetic acid using $HClO_4$ as the titrant. For example, a procedure for determining the amount of nutritionally available protein uses an acid–base titration of lysine residues.

Quantitative Calculations

The quantitative relationship between the titrand and the titrant is determined by the titration reaction's stoichiometry. If the titrand is polyprotic, then we must know to which equivalence point we are titrating. The following example illustrates how we can use a ladder diagram to determine a titration reaction's stoichiometry.

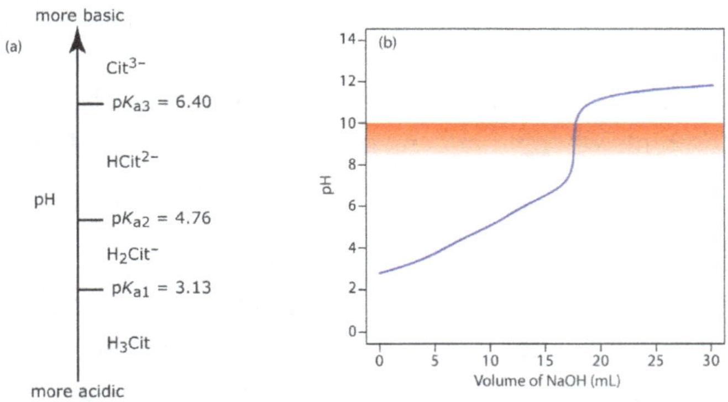

Fig. 8: (a) Ladder diagram for citric acid; (b) Titration curve for the sample showing phenolphthalein's pH transition region.

11.5 QUALITATIVE APPLICATIONS

Example shows how we can use an acid–base titration to determine the forms of alkalinity in waters and their concentrations. We can extend this approach to other systems. For example, if we titrate a sample to the methyl orange end point and the phenolphthalein end point using either a strong acid or a strong base, we can determine which of the following species are present and their concentrations: H_3PO_4, $H_2PO^{-4}H_2PO^{4-}$, $HPO_2{}^{-4}HPO_4{}^{2-}$, $PO_3{}^{-4}PO_4{}^{3-}$, HCl, and NaOH. As outlined in Table 5, each species or mixture of species has a unique relationship between the volumes of titrant needed to reach these two end points. Note that mixtures containing three or more these species are not possible.

Use a ladder diagram to convince yourself that mixtures containing three or more of these species are unstable.

Table 6 . Relationship Between End Point Volumes for Mixtures of Phosphate Species with HCl and NaOH

Solution composition	Relationship between end point volumes with strong base titrant	Relationship between end point volumes with strong acid titrant
H_3PO_4	VPH=2×VMOVPH=2×VMO	—
$H_2PO{-}4H_2PO_{4-}$	VPH>0;VMO=0VPH>0;VMO=0	—
$HPO_2{-}4HPO_4^{2-}$ — VMO>0;VPH=0VMO>0;VPH=0		
PO3−4PO43−	—	VMO=2×VPHVMO=2×VPH
HCl	VPH=VMOVPH=VMO	—
NaOH	—	VMO=VPHVMO=VPH
HCl and H_3PO_4	VPH<2×VMOVPH<2×VMO	—
H_3PO_4 and $H2PO^{-4}H2PO^{4-}$	VPH>2×VMOVPH>2×VMO	—
H2PO−4H2PO4− and HPO2−4HPO42−	VPH>0;VMO=0VPH>0;VMO=0	VMO>0;VPH=0VMO>0;VPH=0
HPO2−4HPO42− and PO3−4PO43−	—	VMO>2×VPHVMO>2×VPH
PO3−4PO43− and NaOH	—	VMO<2×VPHVMO<2×VPH

VPH and VMO are, respectively, the volume of titrant at the phenolphthalein and methyl orange end points when no information is provided, the volume at each end point is zero

11.7.1. Characterization Applications

In addition to a quantitative analysis and a qualitative analysis, we also can use an acid–base titration to characterize the chemical and physical properties of matter. Two useful characterization applications are the determination of a compound's equivalent weight and the determination of its acid dissociation constant or its base dissociation constant

11.7.2 Equivalent Weights

Suppose we titrate a sample of an impure weak acid to a well-defined end point using a monoprotic strong base as the titrant. If we assume the titration involves the transfer of *n* protons, then the moles of titrant needed to reach the end point is

$$\text{moles titrant} = \frac{n \text{ moles titrant}}{\text{moles analyte}} \times \text{moles analyte}$$

If we know the analyte's identity, we can use this equation to determine the amount of analyte in the sample

$$\text{grams analyte} = \text{moles titrant} \times \frac{1 \text{ mole analyte}}{n \text{ moles analyte}} \times FW \text{ analyte}$$

where *FW* is the analyte's formula weight.

But what if we do not know the analyte's identify? If we titrate a pure sample of the analyte, we can obtain some useful information that may help us establish its identity. Because we do not know the number of protons that are titrated, we let $n = 1$ and replace the analyte's formula weight with its equivalent weight (*EW*)

$$\text{grams analyte} = \text{moles titrant} \times \frac{1 \text{ equivalent analyte}}{1 \text{ mole analyte}} = EW \text{ analyte}$$

Where

$$FW = n \times EW$$

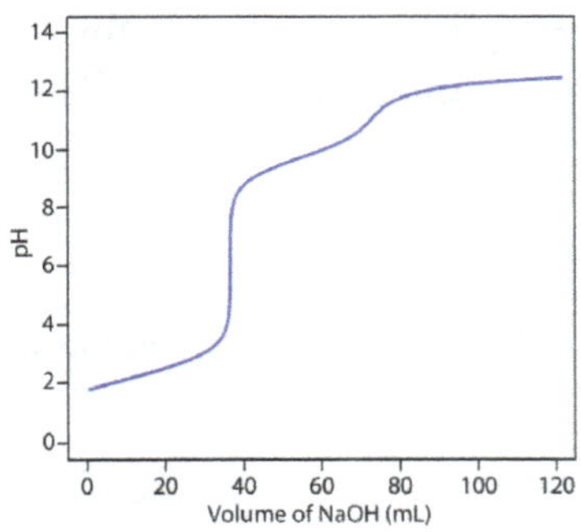

Fig. 9: Titration curve

Equilibrium Constants

Another application of acid–base titrimetry is the determination of a weak acid's or a weak base's dissociation constant. Consider, for example, a solution of acetic acid, CH_3COOH, for which the dissociation constant is

$$K_a = \frac{[H_3O^+][CH_3COO^-]}{[CH_3COOH]}$$

When the concentrations of CH_3COOH and CH_3COO- are equal, the Ka expression reduces to $K_a = [H_3O^+]$, or $pH = pK_a$. If we titrate a solution of acetic acid with NaOH, the pH equals the pK_a when the volume of NaOH is approximately $1/2V_{eq}$.

Recall that pH = pKa is a step on a ladder diagram, which divides the pH axis into two regions, one where the weak acid is the predominate species, and one where its conjugate weak base is the predominate species.

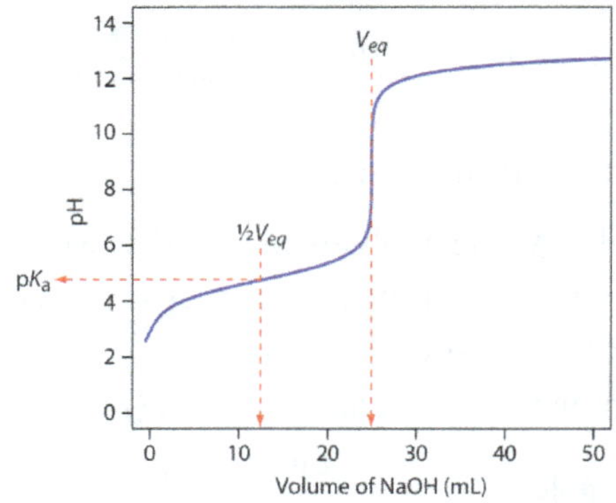

Fig. 10: Estimating acetic acid's pK_a using its potentiometric titration curve.

This method provides a reasonable estimate for a weak acid's pKa if the acid is neither too strong nor too weak. These limitations are easy to appreciate if we consider two limiting cases. For the first limiting case, let's assume the weak acid, HA, is more than 50% dissociated before the titration begins (a relatively large Ka value); in this case the concentration of HA before the equivalence point is always less than the concentration of A– and there is no point on the titration curve where [HA] = [A–]. At the other extreme, if the acid is too weak, then less than 50% of the weak acid reacts with the titrant at the equivalence point. In this case the concentration of HA before the equivalence point is always greater than that of A–. Determining the pKa by the half-equivalence point method overestimates its value if the acid is too strong and underestimates its value if the acid is too weak.

A second approach for determining a weak acid's pKa is to use a Gran plot. For example, earlier in this chapter we derived the following equation for the titration of a weak acid with a strong base.

$$[H_3O^+] \times Vb = K_aV_{eq} - K_aV_b$$

A plot of $[H_3O+] \times\times V_b$ versus Vb for volumes less than the equivalence point yields a straight line with a slope of $-K_a$. Other linearizations have been developed that use the entire titration curve or that require no assumptions. This approach to determining an acidity constant has been used to study the acid–base properties of humic acids, which are naturally occurring, large molecular weight organic acids with multiple acidic sites. In one study a humic acid was found to have six titratable sites, three which were identified as carboxylic acids, two which were believed to be secondary or tertiary amines, and one which was identified as a phenolic group.

11.7.3 Evaluation of Acid–Base Titrimetry

Scale of Operation

In an acid–base titration, the volume of titrant needed to reach the equivalence point is proportional to the moles of titrand. Because the pH of the titrand or the titrant is a function of its concentration, the change in pH at the equivalence point—and thus the feasibility of an acid–base titration—depends on their respective concentrations. Figure 11 , for example, shows a series of titration curves for the titration of several concentrations of HCl with equimolar solutions NaOH. For titrand and titrant concentrations smaller than 10^{-3} M, the change in pH at the end point is too small to provide an accurate and a precise result.

Acid–base titrimetry is an example of a total analysis technique in which the signal is proportional to the absolute amount of analyte.

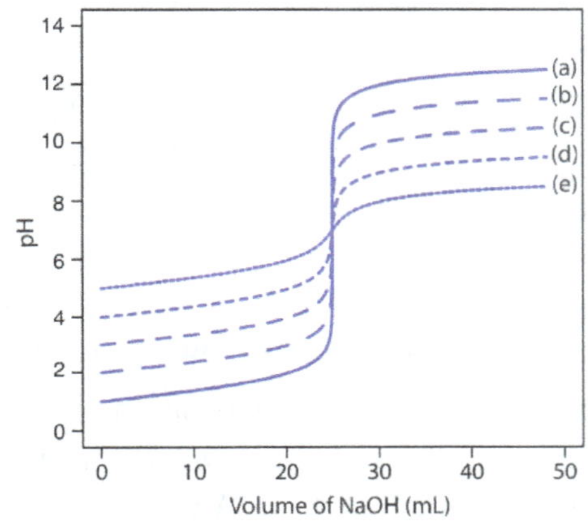

Fig. 11: Titration curves for 25.0 mL of (a) 10–1 M HCl, (b) 10–2 M HCl, (c) 10–3 M HCl, (d) 10–4 M HCl, and (e) 10–5 M HCl. In each case the titrant is an equimolar solution of NaOH.

A minimum concentration of 10^{-3} M places limits on the smallest amount of analyte we can analyze successfully. For example, suppose our analyte has a formula weight of 120 g/mol. To successfully monitor the titration's end point using an indicator or a pH probe, the titrand needs an initial volume of approximately 25 mL. If we assume the analyte's formula weight is 120 g/mol, then each sample must contain at least 3 mg of analyte. For this reason, acid–base titrations generally are limited to major and minor analytes. We can extend the analysis of gases to trace analytes by pulling a large volume of the gas through a suitable collection solution.

One goal of analytical chemistry is to extend analyses to smaller samples. Here we describe two interesting approaches to titrating μL and pL samples. In one experimental design (Figure 9), samples of 20–100 μL are held by capillary action between a flat-surface pH electrode and a stainless steel sample stage . The titrant is added using the oscillations of a piezoelectric ceramic device to move an angled glass rod in and out of a tube connected to a reservoir that contains the titrant. Each time the glass tube is withdrawn an approximately 2 nL microdroplet of titrant is released. The microdroplets are allowed to fall onto the sample, with mixing accomplished by spinning the sample stage at 120 rpm. A total of 450 microdroplets, with a combined volume of 0.81–0.84 μL, is dispensed between each pH measurement. In this fashion a titration curve is constructed. This method has been used to titrate solutions of 0.1 M HCl and 0.1 M CH3COOH with 0.1 M NaOH. Absolute errors ranged from a minimum of +0.1% to a maximum of –4.1%, with relative standard deviations from 0.15% to 4.7%. Samples as small as 20 μL were titrated successfully.

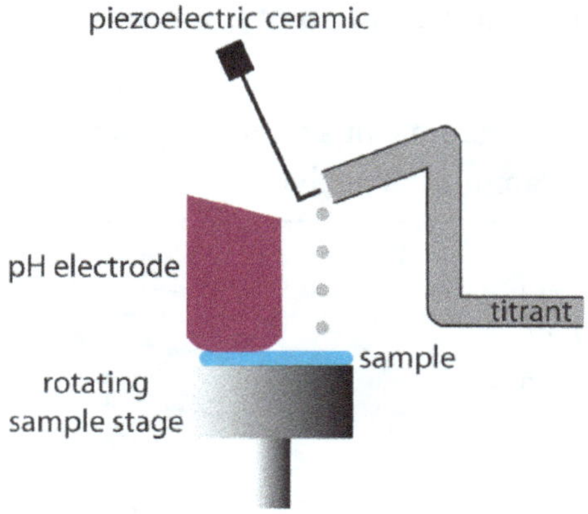

Fig. 12: Experimental design for a microdroplet titration apparatus.

Another approach carries out the acid–base titration in a single drop of solution [(a) Gratzl, M.; Yi, C. Anal. Chem. 1993, 65, 2085–2088; (b) Yi, C.; Gratzl, M. Anal. Chem. 1994, 66, 1976–1982; (c) Hui, K. Y.; Gratzl, M. Anal. Chem. 1997, 69, 695–698; (d) Yi, C.; Huang, D.; Gratzl, M. Anal. Chem. 1996, 68, 1580–1584; (e) Xie, H.; Gratzl, M. Anal. Chem. 1996, 68, 3665–3669]. The titrant is delivered using a microburet fashioned from a glass capillary micropipet (Figure 20). The microburet has a 1-2 μm tip filled with an agar gel membrane. The tip of the microburet is placed within a drop of the sample solution, which is suspended in heptane, and the titrant is allowed to diffuse into the sample. The titration's progress is monitored using an acid–base indicator and the time needed to reach the end point is measured. The rate of the titrant's diffusion from the microburet is determined by a prior calibration. Once calibrated the end point time is converted to an end point volume. Samples usually consist of picoliter volumes (10–12 liters), with the smallest sample being 0.7 pL. The precision of the titrations is about 2%.

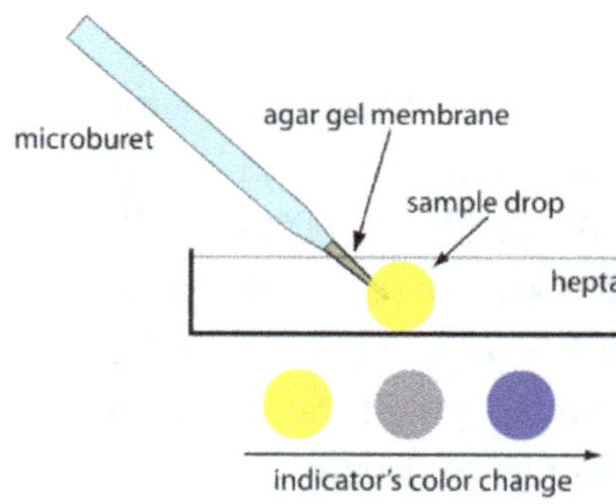

Fig. 13: Experimental set-up for a diffusional microtitration. The indicator is a mixture of bromothymol blue and bromocresol purple.

Titrations conducted with microliter or picoliter sample volumes require a smaller absolute amount of analyte. For example, diffusional titrations have been conducted on as little as 29 femtomoles (10^{-15}moles) of nitric acid. Nevertheless, the analyte must be present in the sample at a major or minor level for the titration to give accurate and precise results.

Accuracy

When working with a macro–major or a macro–minor sample, an acid–base titration can achieve a relative error of 0.1–0.2%. The principal limitation to accuracy is the difference between the end point and the equivalence point.

Precision

An acid–base titration's relative precision depends primarily on the precision with which we can measure the end point volume and the precision in detecting the end point. Under optimum conditions, an acid–base titration has a relative precision of 0.1–0.2%. We can improve the relative precision by using the largest possible buret and by ensuring we use most of its capacity in reaching the end point. A smaller volume buret is a better choice when using costly reagents, when waste disposal is a concern, or when we must complete the titration quickly to avoid competing chemical reactions. An automatic titrator is particularly useful for titrations that require small volumes of titrant because it provides significantly better precision (typically about ±0.05% of the buret's volume).

The precision of detecting the end point depends on how it is measured and the slope of the titration curve at the end point. With an indicator the precision of the end point signal usually is ±0.03–0.10 mL. Potentiometric end points usually are more precise.

Sensitivity

For an acid–base titration we can write the following general analytical equation to express the titrant's volume in terms of the amount of titrand

$$\text{volume of titrant} = k \times \text{moles of titrand}$$

where *k*, the sensitivity, is determined by the stoichiometry between the titrand and the titrant. Consider, for example, the determination of sulfurous acid, H_2SO_3, by titrating with NaOH to the first equivalence point

$$H_2SO_3(aq) + OH^-(aq) \rightarrow H_2O(l) + HSO^{-3}(aq)$$

At the equivalence point the relationship between the moles of NaOH and the moles of H_2SO_3 is

$$\text{mol NaOH} = \text{mol } H_2SO_3$$

Substituting the titrant's molarity and volume for the moles of NaOH and rearranging

$$M_{NaOH} \times V_{NNOH} = \text{mol } H_2SO_3$$

$$V_{NaOH} = \frac{1}{M_{NaOH}} \times \text{mol } H_2SO_3$$

we find that *k* is

$$k = \frac{1}{M_{NaOH}}$$

There are two ways in which we can improve a titration's sensitivity. The first, and most obvious, is to decrease the titrant's concentration because it is inversely proportional to the sensitivity, k. The second approach, which applies only if the titrand is multiprotic, is to titrate to a later equivalence point. If we titrate H_2SO^3 to its second equivalence point

$$H_2SO_3(aq) + 2OH^-(aq) \rightarrow 2H_2O(l) + SO_3^{2-}(aq)$$

then each mole of H_2SO_3 consumes two moles of NaOH

$$\text{mol NaOH} = 2\times\text{mol } H_2SO_3$$

and the sensitivity becomes

$$k = \frac{2}{M_{NaOH}}$$

In practice, however, any improvement in sensitivity is offset by a decrease in the end point's precision if a larger volume of titrant requires us to refill the buret. For this reason, standard acid–base titrimetric procedures are written to ensure that a titration uses 60–100% of the buret's volume.

Selectivity

Acid–base titrants are not selective. A strong base titrant, for example, reacts with all acids in a sample, regardless of their individual strengths. If the titrand contains an analyte and an interferent, then selectivity depends on their relative acid strengths. Let's consider two limiting situations.

If the analyte is a stronger acid than the interferent, then the titrant will react with the analyte before it begins reacting with the interferent. The feasibility of the analysis depends on whether the titrant's reaction with the interferent affects the accurate location of the analyte's equivalence point. If the acid dissociation constants are substantially different, the end point for the analyte can be determined accurately. Conversely, if the acid dissociation constants for the analyte and interferent are similar, then there may not be an accurate end point for the analyte. In the latter case a quantitative analysis for the analyte is not possible.

In the second limiting situation the analyte is a weaker acid than the interferent. In this case the volume of titrant needed to reach the analyte's equivalence point is determined by the concentration of both the analyte and the interferent. To account for the interferent's contribution to the end point, an end point for the interferent must be available. Again, if the acid dissociation constants for the analyte and interferent are significantly different, then the analyte's determination is possible. If the acid dissociation constants are similar, however, there is only a single equivalence point and we cannot separate the analyte's and the interferent's contributions to the equivalence point volume.

Time, Cost, and Equipment

Acid–base titrations require less time than most gravimetric procedures, but more time than many instrumental methods of analysis, particularly when analyzing many samples. With an automatic titrator, however, concerns about analysis time are less significant. When performing a titration manually our equipment needs—a buret and, perhaps, a pH meter—are few in number, inexpensive, routinely available, and easy to maintain. Automatic titrators are available for between \$3000 and \$10 000.

11.6 COMPLEXATION TITRATIONS

The earliest examples of metal–ligand complexation titrations are Liebig's determinations, in the 1850s, of cyanide and chloride using, respectively, Ag+ and Hg^{2+} as the titrant. Practical analytical applications of complexation titrimetry were slow to develop because many metals and ligands form a series of metal–ligand complexes. Liebig's titration of CN– with Ag+ was successful because they form a single, stable complex of $Ag(CN)^{-2}Ag(CN)^{2-}$, which results in a single, easily identified end point. Other metal–ligand complexes, such as $CdI_2^{-4}CdI4^{2-}$, are not analytically useful because they form a series of metal–ligand complexes (CdI+, CdI2(aq), $CdI^{-3}CdI_3$– and CdI2–4CdI42–) that produce a sequence of poorly defined end points.

Recall that an acid–base titration curve for a diprotic weak acid has a single end point if its two K_a values are not sufficiently different.

In 1945, Schwarzenbach introduced aminocarboxylic acids as multidentate ligands. The most widely used of these new ligands—ethylenediaminetetraacetic acid, or EDTA—forms a strong 1:1 complex with many metal ions. The availability of a ligand that gives a single, easily identified end point made complexation titrimetry a practical analytical method.

11.8.1 Chemistry and Properties of EDTA

Ethylenediaminetetraacetic acid, or EDTA, is an aminocarboxylic acid. EDTA, the structure of which is shown in Figure 1 a in its fully deprotonated form, is a Lewis acid with six binding sites—the four negatively charged carboxylate groups and the two tertiary amino groups—that can donate up to six pairs of electrons to a metal ion. The resulting metal–ligand complex, in which EDTA forms a cage-like structure around the metal ion , is very stable. The actual number of coordination sites depends on the size of the metal ion, however, all metal–EDTA complexes have a 1:1 stoichiometry.

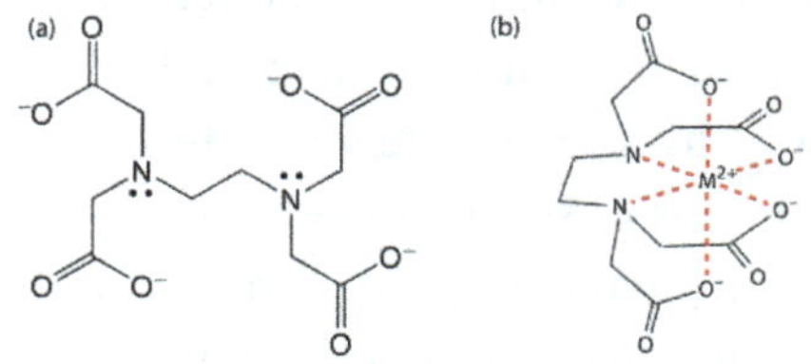

Fig. 14: Structures of (a) EDTA, in its fully deprotonated form, and (b) in a six-coordinate metal–EDTA complex with a divalent metal ion.

Metal–EDTA Formation Constants

To illustrate the formation of a metal–EDTA complex, let's consider the reaction between Cd^{2+} and EDTA

$$Cd^{2+}(aq)+Y^{4-}(aq) \rightleftharpoons CdY^{2-}(aq) \qquad 1$$

where Y^{4-} is a shorthand notation for the fully deprotonated form of EDTA shown in Figure 1 a. Because the reaction's formation constant

$$K_f = \frac{[CdY^{2-}]}{[Cd^{2+}][Y^{4-}]} = 2.9 \times 10^{16} \qquad 2$$

is large, its equilibrium position lies far to the right.

EDTA is a Weak Acid

In addition to its properties as a ligand, EDTA is also a weak acid. The fully protonated form of EDTA, H_6Y^{2+}, is a hexaprotic weak acid with successive pK_a values of

$$pK_{a1} = 0.0 \quad pK_{a2} = 1.5 \quad pK_{a3} = 2.0$$

$$pK_{a4} = 2.66 \quad pK_{a5} = 6.16 \quad pK_{a6} = 10.24$$

The first four values are for the carboxylic acid protons and the last two values are for the ammonium protons. Figure 15 shows a ladder diagram for EDTA. The specific form of EDTA in reaction 1 is the predominate species only when the pH is more basic than 10.24.

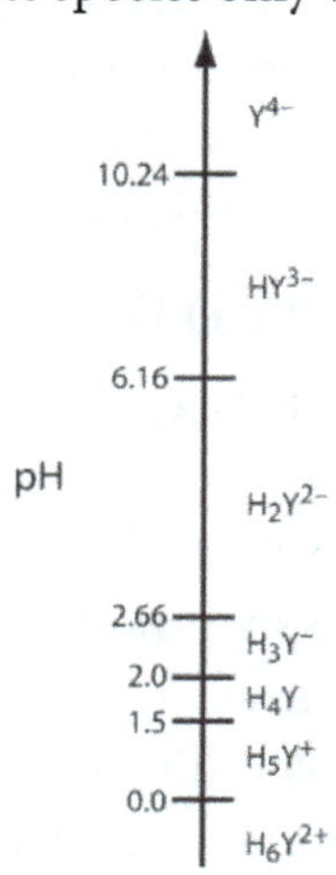

Fig. 15: Ladder diagram for EDTA.

11.8.2 Complexometric EDTA Titration Curves

Now that we know something about EDTA's chemical properties, we are ready to evaluate its usefulness as a titrant. To do so we need to know the shape of a complexometric titration curve.. The analogous result for a complexation titration shows the change in pM, where M is the metal ion's concentration, as a function of the volume of EDTA. In this section we will learn how to calculate a titration curve using the equilibrium calculations. We also will learn how to sketch a good approximation of any complexation titration curve using a limited number of simple calculations.

$$pM = -\log[M^{2+}]$$

11.8.3 Selecting and Evaluating the Endpoint

The equivalence point of a complexation titration occurs when we react stoichiometrically equivalent amounts of the titrand and titrant. As is the case for an acid–base titration, we estimate the equivalence point for a complexation titration using an experimental end point. A variety of methods are available for locating the end point, including indicators and sensors that respond to a change in the solution conditions.

Finding the End Point with an Indicator

Most indicators for complexation titrations are organic dyes—known as ***metallochromic indicators***—that form stable complexes with metal ions. The indicator, In^{m-}, is added to the titrand's solution where it forms a stable complex with the metal ion, MIn^{n-}. As we add EDTA it reacts first with free metal ions, and then displaces the indicator from MIn^{n-}.

$$MIn^{n-}(aq) + Y4^{-}(aq) \rightarrow MY^{2} - (aq) + Inm^{-}(aq)$$

If MInn– and Inm– have different colors, then the change in color signals the end point.

The accuracy of an indicator's end point depends on the strength of the metal–indicator complex relative to the strength of the metal–EDTA complex. If the metal–indicator complex is too strong, the change in color occurs after the equivalence point. If the metal–indicator complex is too weak, however, the end point occurs before we reach the equivalence point.

Most metallochromic indicators also are weak acids. One consequence of this is that the conditional formation constant for the metal–indicator complex depends on the titrand's pH. This provides some control over an indicator's titration error because we can adjust the strength of a metal–indicator complex by adjusted the pH at which we carry out the titration. Unfortunately, because the indicator is a weak acid, the color of the uncomplexed indicator also may change with pH. Figure .16 , for example, shows the color of the indicator calmagite as a function of pH and pMg, where H_2In^{-}, HIn^{2-}, and $In3^{-}$ are different forms of the uncomplexed indicator, and MgIn– is the Mg^{2+}–calmagite complex. Because the color of calmagite's metal–indicator complex is red, its use as a metallochromic indicator has a practical pH range of approximately 8.5–11 where the uncomplexed indicator, HIn^{2-}, has a blue color.

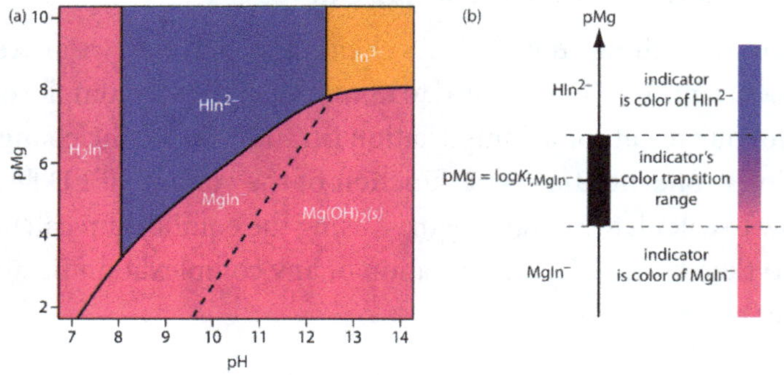

Fig. 16: (a) Predominance diagram for the metallochromic indicator calmagite showing the most important forms and colors of calmagite as a function of pH and pMg, where H_2In^-, HIn^{2-}, and In^{3-} are uncomplexed forms of calmagite, and MgIn– is its complex with Mg^{2+}. Conditions to the right of the dashed line, where Mg^{2+} precipitates as $Mg(OH)_2$, are not analytically useful for a complexation titration. A red to blue end point is possible if we maintain the titrand's pH in the range 8.5–11. (b) Diagram showing the relationship between the concentration of Mg^{2+} (as pMg) and the indicator's color. The ladder diagram defines pMg values where MgIn– and HIn2– are predominate species. The indicator changes color when pMg is between $\log K_f - 1$ and $\log K_f + 1$. This figure is essentially a two-variable ladder diagram. The solid lines are equivalent to a step on a conventional ladder diagram, indicating conditions where two (or three) species are equal in concentration.

Table 6 provides examples of metallochromic indicators and the metal ions and pH conditions for which they are useful. Even if a suitable indicator does not exist, it often is possible to complete an EDTA titration by introducing a small amount of a secondary metal–EDTA complex if the secondary metal ion forms a stronger complex with the indicator and a weaker complex with EDTA than the analyte. For example, calmagite has a poor end point when titrating Ca^{2+} with EDTA. Adding a small amount of Mg^{2+}–EDTA to the titrand gives a sharper end point. Because Ca2+ forms a stronger complex with EDTA, it displaces Mg^{2+}, which then forms the red-colored Mg^{2+}–calmagite complex. At the titration's end point, EDTA displaces Mg^{2+} from the Mg^{2+}–calmagite complex, signaling the end point by the presence of the uncomplexed indicator's blue form.

Table 7 . Selected Metallochromic Indicators

Indicator	Ph range	Metal ions	Indicator	Ph range	Metal ions
Calmagite	8.5–11	Ba, Ca, Mg, Zn	Eriochrome black t	7.5–10.5	Ba, Ca, Mg, Zn
Eriochrome blue black r	8–12	Ca, Mg, Zn, Cu	Pan	2–11	Cd, Cu, Zn
Murexide	6–13	Ca, Ni, Cu	Salicylic acid	2–3	Fe
All metal ions carry a +2 charge expect for iron, which is +3 Metal ions in italic font have poor end points.					

Finding the End Point By Monitoring Absorbance

An important limitation when using a metallochromic indicator is that we must be able to see the indicator's change in color at the end point. This may be difficult if the solution is already colored. For example, when titrating Cu^{2+} with EDTA, ammonia is used to adjust the titrand's pH. The intensely colored $Cu(NH_3)^{4+2}Cu\ (NH_3)24+$ complex obscures the indicator's color, making an accurate determination of the end point difficult. Other absorbing species present within the sample matrix may also interfere. This often is a problem when analyzing clinical samples, such as blood, or environmental samples, such as natural waters.

If at least one species in a complexation titration absorbs electromagnet- ic radiation, then we can identify the end point by monitoring the titrand's absorbance at a carefully selected wavelength. For example, we can identify the end point for a titration of Cu^{2+} with EDTA in the presence of NH_3 by monitoring the titrand's absorbance at a wavelength of 745 nm, where the $Cu(NH_3)4^{+2}Cu\ (NH_3)_2^{4+}$ complex absorbs strongly. At the beginning of the titration the absorbance is at a maximum. As we add EDTA, however, the reaction

$$Cu(NH_3)\ 2+\ 4(aq) + Y4^- \rightleftharpoons CuY^2-(aq) + 4NH_3(aq)$$

decreases the concentration of $Cu\ (NH_3)^4+2Cu(NH_3)24+$ and decreases the absorbance until we reach the equivalence point. After the equivalence point the absorbance essentially remains unchanged. Note that the titration curve's *y*-axis is not the measured absorbance, A_{meas}, but a corrected absorbance, A_{corr}

$$A_{corr} = A_{meas} \times \frac{V_{EDTA} + V_{Cu}}{V_{Cu}}$$

where V_{EDTA} and V_{Cu} are, respectively, the volumes of EDTA and Cu. Correcting the absorbance for the titrand's dilution ensures that the spectrophotometric titration curve consists of linear segments that we can extrapolate to find the end point. Other common spectrophotometric titration curves are shown in Figures 17 b-f.

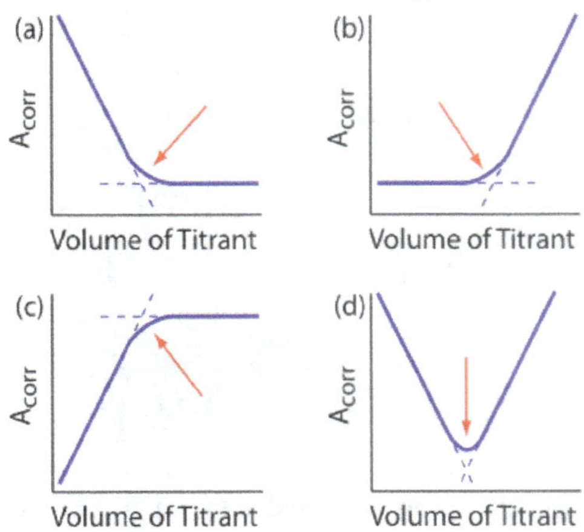

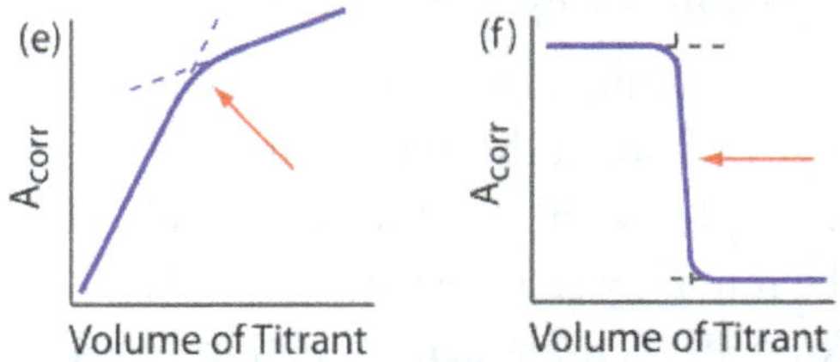

Fig. 17: Examples of spectrophotometric titration curves: (a) only the titrand absorbs; (b) only the titrant absorbs; (c) only the product of the titration reaction absorbs; (d) both the titrand and the titrant absorb; (e) both the titration reaction's product and the titrant absorb; (f) only the indicator absorbs. The red arrows indicate the end points for each titration curve.

11.7 DETERMINATION OF HARDNESS OF WATER AND WASTEWATER

The best way to appreciate the theoretical and the practical details discussed in this section is to carefully examine a typical complexation titrimetric method. Although each method is unique, the following description of the determination of the hardness of water provides an instructive example of a typical procedure.

11.9.1 Description of the Method

The operational definition of water hardness is the total concentration of cations in a sample that can form an insoluble complex with soap. Although most divalent and trivalent metal ions contribute to hardness, the two most important metal ions are $Ca^{2+}$and Mg^{2+}. Hardness is determined by titrating with EDTA at a buffered pH of 10. Calmagite is used as an indicator. Hardness is reported as mg $CaCO_3$/L.

11.9.2 Procedure

Select a volume of sample that requires less than 15 mL of titrant to keep the analysis time under 5 minutes and, if necessary, dilute the sample to 50 mL with distilled water. Adjust the sample's pH by adding 1–2 mL of a pH 10 buffer that contains a small amount of Mg^{2+}–EDTA. Add 1–2 drops of indicator and titrate with a standard solution of EDTA until the red-to-blue end point is reached (Figure 18).

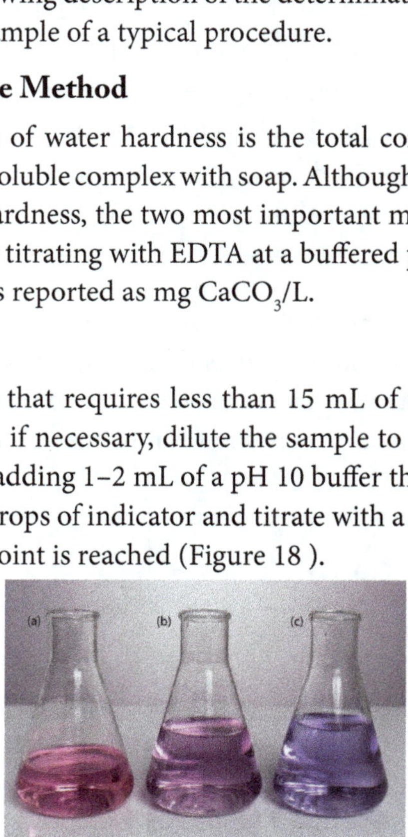

Fig. 18: End point for the titration of hardness with EDTA using calmagite as an indicator; the indicator is: (a) red prior to the end point due to the presence of the

Mg^{2+}–indicator complex; (b) purple at the titration's end point; and (c) blue after the end point due to the presence of uncomplexed indicator.

Of the two primary cations that contribute to hardness, Mg^{2+} forms the weaker complex with EDTA and is the last cation to react with the titrant. Calmagite is a useful indicator because it gives a distinct end point when titrating Mg^{2+} .Because of calmagite's acid–base properties, the range of pMg values over which the indicator changes color depends on the titrand's pH. Figure 19 shows the titration curve for a 50-mL solution of 10–3 M Mg^{2+} with 10^{-2} M EDTA at pHs of 9, 10, and 11. Superimposed on each titration curve is the range of conditions for which the average analyst will observe the end point. At a pH of 9 an early end point is possible, which results in a negative determinate error. A late end point and a positive determinate error are possible if the pH is 11.

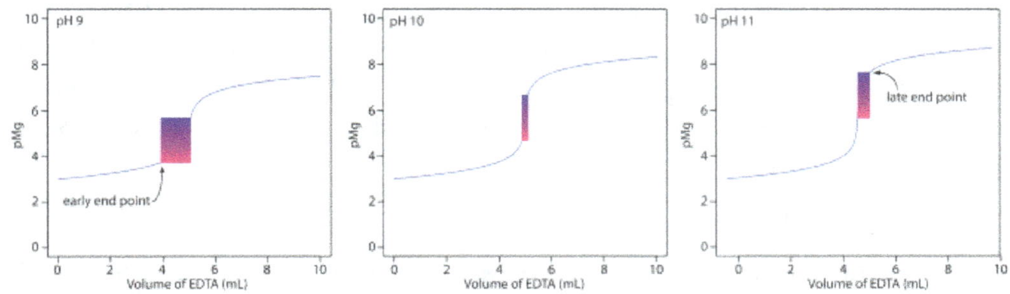

Fig. 19: Titration curves for 50 mL of 10^{-3} M Mg^{2+} with 10^{-3} M EDTA at pHs 9, 10, and 11 using calmagite as an indicator. The range of pMg and volume of EDTA over which the indicator changes color is shown for each titration curve.

Why is a small amount of the Mg^{2+}–EDTA complex added to the buffer?

The titration's end point is signaled by the indicator calmagite. The indicator's end point with Mg^{2+} is distinct, but its change in color when titrating Ca^{2+} does not provide a good end point (see Table 5). If the sample does not contain any Mg^{2+} as a source of hardness, then the titration's end point is poorly defined, which leads to an inaccurate and imprecise result. Adding a small amount of Mg^{2+}–EDTA to the buffer ensures that the titrand includes at least some Mg^{2+}. Because Ca^{2+} forms a stronger complex with EDTA, it displaces Mg^{2+} from the Mg^{2+}–EDTA complex, freeing the Mg^{2+} to bind with the indicator. This displacement is stoichiometric, so the total concentration of hardness cations remains unchanged. The displacement by EDTA of Mg^{2+} from the Mg^{2+}–indicator complex signals the titration's end point.

Why does the procedure specify that the titration take no longer than 5 minutes?

A time limitation suggests there is a kinetically-controlled interference, possibly arising from a competing chemical reaction. In this case the interference is the possible precipitation of $CaCO_3$ at a pH of 10.

11.8 QUANTITATIVE APPLICATIONS

Although many quantitative applications of complexation titrimetry have been replaced by other analytical methods, a few important applications continue to find relevance. In the section we review the general application of complexation titrimetry with an emphasis on applications from the analysis of water and wastewater. First, however, we discuss the selection and standardization of complexation titrants.

11.10.1 Selection and Standardization of Titrants

EDTA is a versatile titrant that can be used to analyze virtually all metal ions. Although EDTA is the usual titrant when the titrand is a metal ion, it cannot be used to titrate anions, for which Ag^+ or Hg^{2+} are suitable titrants.

Solutions of EDTA are prepared from its soluble disodium salt, $Na_2H_2Y2H_2O$, and standardized by titrating against a solution made from the primary standard $CaCO_3$. Solutions of Ag^+ and Hg^{2+} are prepared using $AgNO_3$ and $Hg(NO_3)_2$, both of which are secondary standards. Standardization is accomplished by titrating against a solution prepared from primary standard grade NaCl.

11.10.2 Inorganic Analytes

Complexation titrimetry continues to be listed as a standard method for the determination of hardness, Ca^{2+}, CN^-, and Cl^- in waters and wastewaters. The evaluation of hardness was described earlier in Representative Method 1. The determination of Ca^{2+} is complicated by the presence of Mg^{2+}, which also reacts with EDTA. To prevent an interference the pH is adjusted to 12–13, which precipitates Mg^{2+} as $Mg(OH)_2$. Titrating with EDTA using murexide or Eriochrome Blue Black R as the indicator gives the concentration of Ca^{2+}.

Cyanide is determined at concentrations greater than 1 mg/L by making the sample alkaline with NaOH and titrating with a standard solution of $AgNO_3$ to form the soluble $Ag(CN)^{-2}Ag\ (CN)^{2-}$ complex. The end point is determined using p-dimethylaminobenzalrhodamine as an indicator, with the solution turning from a yellow to a salmon color in the presence of excess Ag^+.

Chloride is determined by titrating with $Hg(NO_3)_2$, forming $HgCl_2(aq)$. The sample is acidified to a pH of 2.3–3.8 and diphenylcarba zone, which forms a colored complex with excess Hg^{2+}, serves as the indicator. The pH indicator xylene cyanol FF is added to ensure that the pH is within the desired range. The initial solution is a greenish blue, and the titration is carried out to a purple end point.

11.10.3 Quantitative Calculations

The quantitative relationship between the titrand and the titrant is determined by the titration reaction's stoichiometry. For a titration using EDTA, the stoichiometry is always 1:1.

11.9 EVALUATION OF COMPLEXATION TITRIMETRY

The scale of operations, accuracy, precision, sensitivity, time, and cost of a complexation titration are similar to those described earlier for acid–base titrations. Complexation titrations, however, are more selective. Although EDTA forms strong complexes with most metal ion, by carefully controlling the titrand's pH we can analyze samples that contain two or more analytes. The reason we can use pH to provide selectivity is shown in Figure 9 a. A titration of Ca^{2+} at a pH of 9 has a distinct break in the titration curve because the conditional formation constant for CaY^{2-} of 2.6×1092.6×109 is large enough to ensure that the reaction of Ca^{2+} and EDTA goes to completion. At a pH of 3, however, the conditional formation constant of 1.23 is so small that very little Ca^{2+} reacts with the EDTA. Suppose we need to analyze a mixture of Ni^{2+} and Ca^{2+}. Both analytes react with EDTA, but their conditional formation constants differ significantly. If we adjust the pH to 3 we can titrate Ni^{2+} with EDTA without titrating Ca^{2+} (Figure 20b). When the titration is complete, we adjust the titrand's pH to 9 and titrate the Ca^{2+} with EDTA.

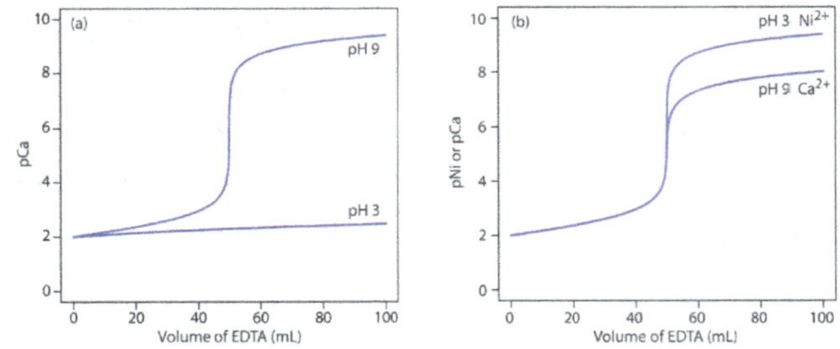

Fig. 20: Titration curves illustrating how we can use the titrand's pH to control EDTA's selectivity. (a) Titration of 50.0 mL of 0.010 M Ca^{2+} at a pH of 3 and a pH of 9 using 0.010 M EDTA. At a pH of 3 the CaY^{2-} complex is too weak to titrate successfully. (b) Titration of a 50.0 mL mixture of 0.010 M Ca^{2+} and 0.010 M Ni^{2+} at a pH of 3 and at a pH of 9 using 0.010 M EDTA. At a pH of 3 EDTA reacts only with Ni^{2+}. When the titration is complete, raising the pH to 9 allows for the titration of Ca^{2+}.

A spectrophotometric titration is a particularly useful approach for analyzing a mixture of analytes. For example, as shown in Figure 21 , we can determine the concentration of a two metal ions if there is a difference between the absorbance of the two metal-ligand complexes.

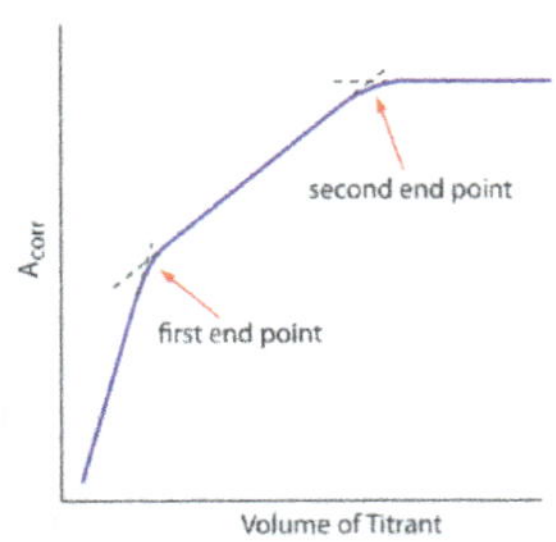

Fig. 21: Spectrophotometric titration curve for the complexation titration of a mixture of two analytes. The red arrows indicate the end points for each analyte.

11.10 REDOX TITRATIONS

Analytical titrations using oxidation–reduction reactions were introduced shortly after the development of acid–base titrimetry. The earliest redox titration took advantage of chlorine's oxidizing power. In 1787, Claude Berthollet introduced a method for the quantitative analysis of chlorine water (a mixture of Cl_2, HCl, and HOCl) based on its ability to oxidize indigo, a dye that is colorless in its oxidized state. In 1814, Joseph Gay-Lussac developed a similar method to determine chlorine in bleaching powder. In both methods the end point is a change in color. Before the equivalence point the solution is colorless due to the oxidation of indigo. After the equivalence point, however, unreacted indigo imparts a permanent color to the solution.

The number of redox titrimetric methods increased in the mid-1800s with the introduction of $MnO^{-4}MnO^{4-}$, $Cr_2O_{2\_}7Cr_2O7^{2-}$, and I2 as oxidizing titrants, and of Fe^{2+} and $S_2O_2^{-3}S_2O_3^{2-}$ as reducing titrants. Even with the availability of these new titrants, redox titrimetry was slow to develop due to the lack of suitable indicators. A titrant can serve as its own indicator if its oxidized and its reduced forms differ significantly in color For example, the intensely purple $MnO^{-4}MnO4^{-}$ ion serves as its own indicator since its reduced form, Mn^{2+}, is almost colorless. Other titrants require a separate indicator. The first such indicator, diphenylamine, was introduced in the 1920s. Other redox indicators soon followed, increasing the applicability of redox titrimetry.

11.12.1 Redox Titration Curves

To evaluate a redox titration we need to know the shape of its titration curve. In an acid–base titration or a complexation titration, the titration curve shows how the concentration of H_3O^+ (as pH) or Mn^+ (as pM) changes as we add titrant. For a redox titration it is convenient to monitor the titration reaction's potential instead of the concentration of one species.

You may recall from Chapter 6 that the Nernst equation relates a solution's potential to the concentrations of reactants and products that participate in the redox reaction. Consider, for example, a titration in which a titrand in a reduced state, Ared, reacts with a titrant in an oxidized state, Box.

$$A_{red} + B_{ox} \rightleftharpoons B_{red} + A_{ox}$$

where A_{ox} is the titrand's oxidized form, B_{red} is the titrant's reduced form, and the stoichiometry between the two is 1:1. The reaction's potential, E_{rxn}, is the difference between the reduction potentials for each half-reaction.

$$E_{rxn} = E_{B_{ox}/B_{red}} - E_{A_{ox}/A_{red}}$$

After each addition of titrant the reaction between the titrand and the titrant reaches a state of equilibrium. Because the potential at equilibrium is zero, the titrand's and the titrant's reduction potentials are identical.

$$E_{B_{ox}/B_{red}} = E_{A_{ox}/A_{red}}$$

This is an important observation as it allows us to use either half-reaction to monitor the titration's progress.

Before the equivalence point the titration mixture consists of appreciable quantities of the titrand's oxidized and reduced forms. The concentration of unreacted titrant, however, is very small. The potential, therefore, is easier to calculate if we use the Nernst equation for the titrand's half-reaction

$$E_{rxn} = E^{\circ}_{A_{ox}/A_{red}} - \frac{RT}{nF}\ln\frac{[A_{red}]}{[A_{ox}]}$$

After the equivalence point it is easier to calculate the potential using the Nernst equation for the titrant's half-reaction.

$$E_{rxn} = E^{\circ}_{B_{ox}/B_{red}} - \frac{RT}{nF}\ln\frac{[B_{red}]}{[B_{ox}]}$$

Although the Nernst equation is written in terms of the half-reaction's standard state potential, a matrix-dependent formal potential often is used in its place. See Appendix 13 for the standard state potentials and formal potentials for selected half-reactions.

11.12.2 Calculating the Titration Curve

Let's calculate the titration curve for the titration of 50.0 mL of 0.100 M Fe^{2+} with 0.100 M Ce^{4+} in a matrix of 1 M $HClO_4$. The reaction in this case is

$$Fe^{2+}(aq) + Ce^{4+}(aq) \rightleftharpoons Ce^{3+}(aq) + Fe^{3+}(aq) \qquad (1)$$

Because the equilibrium constant for reaction 9.4.19.4.1 is very large—it is approximately 6×10156×1015—we may assume that the analyte and titrant react completely.

In 1 M HClO4, the formal potential for the reduction of Fe^{3+} to Fe^{2+} is +0.767 V, and the formal potential for the reduction of Ce^{4+} to Ce^{3+} is +1.70 V.

The first task is to calculate the volume of Ce needed to reach the titration's equivalence point. From the reaction's stoichiometry we know that

$$\text{mol } Fe^{2+} = M_{Fe}V_{Fe} = M_{Ce}V_{Ce} = \text{mol } Ce^{4+}$$

Solving for the volume of Ce^{4+} gives the equivalence point volume as

$$V_{eq} = V_{Ce} = \frac{M_{Fe}V_{Fe}}{M_{Ce}} = \frac{(0.100\text{ M})(50.0\text{ mL})}{(0.100\text{ M})} = 50.0\text{ mL}$$

before the equivalence point, the concentration of unreacted Fe^{2+} and the concentration of Fe^{3+} are easy to calculate. For this reason we find the potential using the Nernst equation for the Fe^{3+}/Fe^{2+} half-reaction.

$$E = +0.767\text{ V} - 0.05916\log\frac{[\text{Fe}^{2+}]}{[\text{Fe}^{3+}]}$$

For example, the concentrations of Fe^{2+} and Fe^{3+} after adding 10.0 mL of titrant are

$$[\text{Fe}^{2+}] = \frac{(\text{mol Fe}^{2+})_{\text{initial}} - (\text{mol Ce}^{4+})_{\text{added}}}{\text{total volume}} = \frac{M_{\text{Fe}}V_{\text{Fe}} - M_{\text{Ce}}V_{\text{Ce}}}{V_{\text{Fe}} + V_{\text{Ce}}}$$

$$[\text{Fe}^{2+}] = \frac{(0.100\text{ M})(50.0\text{ mL}) - (0.100\text{ M})(10.0\text{ mL})}{50.0\text{ mL} + 10.0\text{ mL}} = 6.67 \times 10^{-2}\text{ M}$$

$$[\text{Fe}^{3+}] = \frac{(\text{mol Ce}^{4+})_{\text{added}}}{\text{total volume}} = \frac{M_{\text{Ce}}V_{\text{Ce}}}{V_{\text{Fe}} + V_{\text{Ce}}}$$

$$[\text{Fe}^{3+}] = \frac{(0.100\text{ M})(10.0\text{ mL})}{50.0\text{ mL} + 10.0\text{ mL}} = 1.67 \times 10^{-2}\text{ M}$$

Substituting these concentrations into Equation 9.4.29.4.2 gives the potential as

$$E = +0.767\text{ V} - 0.05916\log\frac{6.67 \times 10^{-2}}{1.67 \times 10^{-2}} = +0.731\text{ V}$$

after the equivalence point, the concentration of Ce^{3+} and the concentration of excess Ce^{4+} are easy to calculate. For this reason we find the potential using the Nernst equation for the Ce^{4+}/Ce^{3+} half-reaction in a manner similar to that used above to calculate potentials before the equivalence point.

$$E = +1.70\text{ V} - 0.05916\log\frac{[\text{Ce}^{3+}]}{[\text{Ce}^{4+}]}$$

For example, after adding 60.0 mL of titrant, the concentrations of Ce^{3+} and Ce^{4+} are

$$[\text{Ce}^{3+}] = \frac{(\text{mol Fe}^{2+})_{\text{initial}}}{\text{total volume}} = \frac{M_{\text{Fe}}V_{\text{Fe}}}{V_{\text{Fe}} + V_{\text{Ce}}}$$

$$[\text{Ce}^{3+}] = \frac{(0.100\text{ M})(50.0\text{ mL})}{50.0\text{ mL} + 60.0\text{ mL}} = 4.55 \times 10^{-2}\text{ M}$$

$$[\text{Ce}^{4+}] = \frac{(\text{mol Ce}^{4+})_{\text{added}} - (\text{mol Fe}^{2+})_{\text{initial}}}{\text{total volume}} = \frac{M_{\text{Ce}}V_{\text{Ce}} - M_{\text{Fe}}V_{\text{Fe}}}{V_{\text{Fe}} + V_{\text{Ce}}}$$

$$[\text{Ce}^{4+}] = \frac{(0.100\text{ M})(60.0\text{ mL}) - (0.100\text{ M})(50.0\text{ mL})}{50.0\text{ mL} + 60.0\text{ mL}} = 9.09 \times 10^{-3}\text{ M}$$

at the titration's equivalence point, the potential, E_{eq}, in Equation 2 and Equation .3 are identical. Adding the equations together to gives

$$E = +1.70\text{ V} - 0.05916\log\frac{4.55 \times 10^{-2}\text{ M}}{9.09 \times 10^{-3}\text{ M}} = +1.66\text{ V}$$

Because $[Fe^{2+}] = [Ce^{4+}]$ and $[Ce^{3+}] = [Fe^{3+}]$ at the equivalence point, the log term has a value of zero and the equivalence point's potential is

$$2E_{eq} = E^{\circ}_{\text{Fe}^{3+}/\text{Fe}^{2+}} + E^{\circ}_{\text{Ce}^{4+}/\text{Ce}^{3+}} - 0.05916\log\frac{[\text{Fe}^{2+}][\text{Ce}^{3+}]}{[\text{Fe}^{3+}][\text{Ce}^{4+}]}$$

11.12.3 Selecting and Evaluating the End Point

A redox titration's equivalence point occurs when we react stoichiometrically equivalent amounts of titrand and titrant. As is the case for acid–base titrations and complexation titrations, we estimate the equivalence point of a redox titration using an experimental end point. A variety of methods are available for locating a redox titration's end point, including indicators and sensors that respond to a change in the solution conditions.

Where is the Equivalence Point

For an acid–base titration or a complexometric titration the equivalence point is almost identical to the inflection point on the steeply rising part of the titration curve. If the stoichiometry of a redox titration is 1:1—that is, one mole of titrant reacts with each mole of titrand—then the equivalence point is symmetric. If the titration reaction's stoichiometry is not 1:1, then the equivalence point is closer to the top or to the bottom of the titration curve's sharp rise. In this case we have an ***asymmetric equivalence point***.

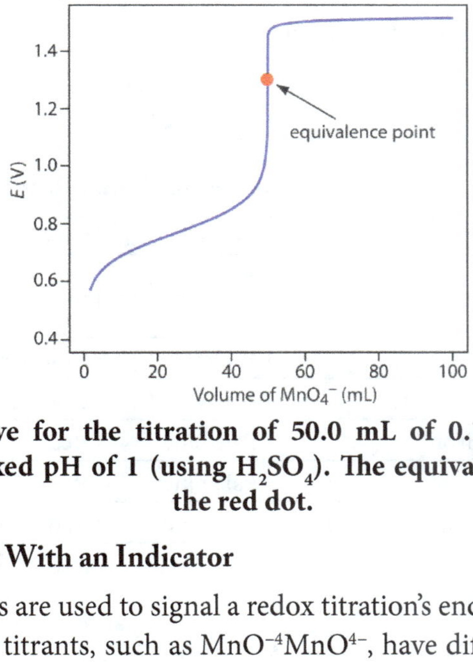

Fig. 22: Titration curve for the titration of 50.0 mL of 0.100 M Fe^{2+} with 0.0200 M $MnO^{-4}MnO^{4-}$ at a fixed pH of 1 (using H_2SO_4). The equivalence point is shown by the red dot.

Finding the End Point With an Indicator

Three types of indicators are used to signal a redox titration's end point. The oxidized and reduced forms of some titrants, such as $MnO^{-4}MnO^{4-}$, have different colors. A solution of $MnO^{-4}MnO^{4-}$ is intensely purple. In an acidic solution, however, permanganate's reduced form, Mn^{2+}, is nearly colorless. When using $MnO^{-4}MnO^{4-}$ as a titrant, the titrand's solution remains colorless until the equivalence point. The first drop of excess $MnO^{-4}MnO^{4-}$ produces a permanent tinge of purple, signaling the end point.

Some indicators form a colored compound with a specific oxidized or reduced form of the titrant or the titrand. Starch, for example, forms a dark purple complex with I–3I3–. We can use this distinct color to signal the presence of excess $I^{-3}I^{3-}$ as a titrant—a change in color from colorless to purple—or the completion of a reaction that consumes $I^{-3}I^{3-}$ as the titrand— a change in color from purple to colorless. Another example of a specific indicator is thiocyanate, SCN–, which forms the soluble red-colored complex of $Fe(SCN)^{2+}$ in the presence of Fe^{3+}.

The most important class of indicators are substances that do not participate in the redox titration, but whose oxidized and reduced forms differ in color. When we add a redox indicator to the titrand, the indicator imparts a color that depends on the solution's potential. As the solution's potential changes with the addition of titrant, the indicator eventually changes oxidation state and changes color, signaling the end point.

To understand the relationship between potential and an indicator's color, consider its reduction half-reaction

$$In_{ox} + ne^{-} \rightleftharpoons In_{red}$$

where Inox and In_{red} are, respectively, the indicator's oxidized and reduced forms.

For simplicity, Inox and In_{red} are shown without specific charges. Because there is a change in oxidation state, Inox and In_{red} cannot both be neutral.

The Nernst equation for this half-reaction is

$$E = E^{o}_{In_{ox}/In_{red}} - \frac{0.05916}{n} \log \frac{[In_{red}]}{[In_{ox}]}$$

Table 8 . Selected Examples of Redox Indicators

Indicator	Color in$_{ox}$	Color of in$_{red}$	E^o (v)
Indigo tetrasulfate	Blue	Colorless	0.36
Methylene blue	Blue	Colorless	0.53
Diphenylamine	Violet	Colorless	0.75
Diphenylamine sulfonic acid	Red-violet	Colorless	0.85
Tris(2,2'-bipyradine)iron	Pale blue	Red	1.120
Ferroin	Pale blue	Red	1.147
Tris(5-nitro-1,10-phenanthroline)iron	Pale blue	Red-violet	1.25

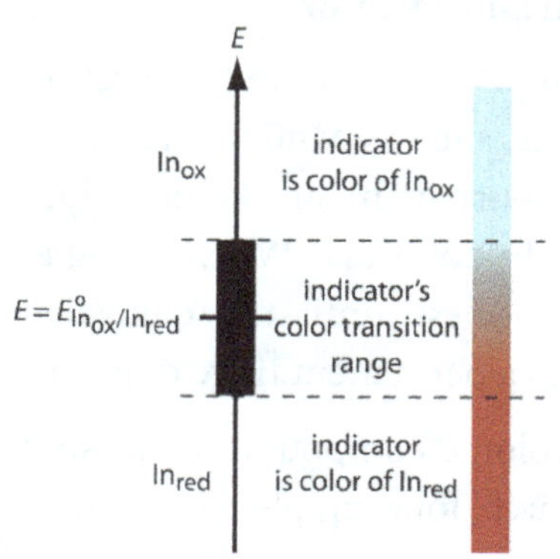

Fig. 23: Diagram showing the relationship between E and an indicator's color. The ladder diagram defines potentials where Inr$_{ed}$ and In$_{ox}$ are the predominate species.

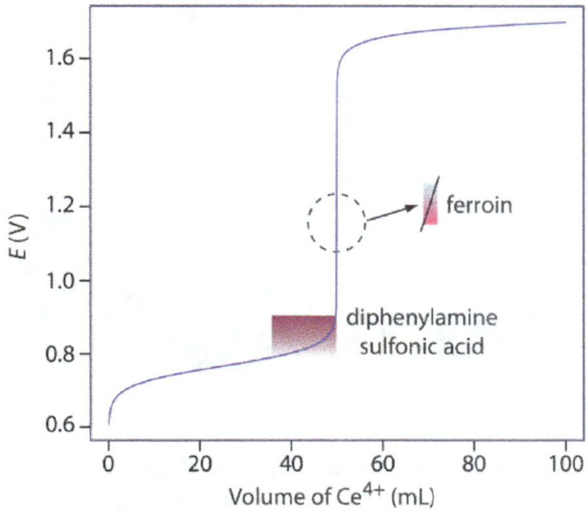

Fig. 24: Titration curve for titration of 50.0 mL of 0.100 M Fe^{2+} with 0.100 M Ce^{4+}. The end point transitions for the indicators diphenylamine sulfonic acid and ferroin are superimposed on the titration curve. Because the transition for ferroin is too small to see on the scale of the x-axis—it requires only 1-2 drops of titrant—the color change is expanded to the right.

11.11 DETERMINATION OF TOTAL CHLORINE RESIDUAL

The best way to appreciate the theoretical and the practical details discussed in this section is to carefully examine a typical redox titrimetric method. Although each method is unique, the following description of the determination of the total chlorine residual in water provides an instructive example of a typical procedure.

11.13.1 Description of the Method

The chlorination of a public water supply produces several chlorine-containing species, the combined concentration of which is called the total chlorine residual. Chlorine is present in a variety of chemical states, including the free residual chlorine, which consists of Cl_2, HOCl and OCl–, and the combined chlorine residual, which consists of NH_2Cl, $NHCl_2$, and NCl_3. The total chlorine residual is determined by using the oxidizing power of chlorine to convert I^- to I^{-3}I3–. The amount of $I^{-3}I^{3-}$ formed is then determined by titrating with $Na_2S_2O_3$ using starch as an indicator. Regardless of its form, the total chlorine residual is reported as if Cl_2 is the only source of chlorine, and is reported as mg Cl/L.

11.12.2 Procedure

Select a volume of sample that requires less than 20 mL of $Na_2S_2O_3$ to reach the end point. Using glacial acetic acid, acidify the sample to a pH between 3 and 4, and add about 1 gram of KI. Titrate with $Na_2S_2O_3$ until the yellow color of I–3I3– begins to disappear. Add 1 mL of a starch indicator solution and continue titrating until the blue color of the starch–I–3I3– complex disappears (Figure 25). Use a blank titration to correct the volume of titrant needed to reach the end point for reagent impurities.

Fig. 25: Endpoint for the determination of the total chlorine residual. (a) Acidifying the sample and adding KI forms a brown solution of I–3I3–. (b) Titrating with $Na_2S_2O_3$ converts I–3I3– to I– with the solution fading to a pale yellow color as we approach the end point. (c) Adding starch forms the deep purple starch–I–3I3– complex. (d) As the titration continues, the end point is a sharp transition from a purple to a colorless solution. The change in color from (c) to (d) typically takes 1–2 drops of titrant.

11.12.3 Quantitative Applications

Although many quantitative applications of redox titrimetry have been re- placed by other analytical methods, a few important applications continue to find relevance. In this section we review the general application of redox titrimetry with an emphasis on environmental, pharmaceutical, and industrial applications. We begin, however, with a brief discussion of selecting and characterizing redox titrants, and methods for controlling the titrand's oxidation state.

Adjusting the Titrand's Oxidation State

If a redox titration is to be used in a quantitative analysis, the titrand initially must be present in a single oxidation state. For example, iron is determined by a redox titration in which Ce^{4+} oxidizes Fe^{2+} to Fe^{3+}. Depending on the sample and the method of sample preparation, iron initially may be present in both the +2 and +3 oxidation states. Before titrating, we must reduce any Fe^{3+} to Fe^{2+} if we want to determine the total concentration of iron in the sample. This type of pretreatment is accomplished using an auxiliary reducing agent or oxidizing agent.

A metal that is easy to oxidize—such as Zn, Al, and Ag—can serve as an ***auxiliary reducing agent***. The metal, as a coiled wire or powder, is added to the sample where it reduces the titrand. Because any unreacted auxiliary reducing agent will react with the titrant, it is removed before we begin the titration by removing the coiled wire or by filtering.

An alternative method for using an auxiliary reducing agent is to immobilize it in a column. To prepare a reduction column an aqueous slurry of the finally divided metal is packed in a glass tube equipped with a porous plug at the bottom. The sample is placed at the top of the column and moves through the column under the influence of gravity or vacuum suction. The length of the reduction column and the flow rate are selected to ensure the analyte's complete reduction.

Two common reduction columns are used. In the ***Jones reductor*** the column is filled with amalgamated zinc, Zn(Hg), which is prepared by briefly placing Zn granules in a solution of HgCl2. Oxidation of zinc

$$Zn(Hg)\ (s) \rightarrow Zn^{2+}(aq) + Hg\ (l) + 2e^-$$

provides the electrons for reducing the titrand. In the ***Walden reductor*** the column is filled with granular Ag metal. The solution containing the titrand is acidified with HCl and passed through the column where the oxidation of silver

$$Ag(s) + Cl^-(aq) \rightarrow AgCl(s) + e^-$$

provides the necessary electrons for reducing the titrand.

Table 9. Examples of Reactions for Reducing a Titrand's Oxidation State Using a Reduction Column

Oxidized titrand	Walden reductor	Jones reducato
Cr^{3+}	—	C r 3+ (a q) + e − → C r 2+ (a q) $Cr^{3+}(aq)+e^- \rightarrow Cr^{2+}(aq)$
Cu^{2+}	C u 2+ (a q) + e − → C u + (a q) $Cu^{2+}(aq)+e^- \rightarrow Cu^+(aq)$	C u 2+ (a q) + 2 e − → C u (s) $Cu^{2+}(aq)+2e^- \rightarrow Cu(s)$
Fe^{3+}	F e 3+ (a q) + e − → F e 2+ (a q) $Fe^{3+}(aq)+e^- \rightarrow Fe^{2+}(aq)$	F e 3+ (a q) + e − → F e 2+ (a q) $Fe^{3+}(aq)+e^- \rightarrow Fe^{2+}(aq)$
TiO^{2+}	—	$TiO^{2+}(aq)^{+2}H+(aq)+e- \rightarrow Ti^{3+}(aq)+H2$ $O(l)TiO^{2+}(aq)^{+2}H+(aq)+e- \rightarrow Ti^{3-}(aq)$ $+H_2O(l)$
$MoO_2^{+4}MoO_4^{2+}$	$MoO_2^{+2}(aq)+e- \rightarrow MoO+2(aq)$ $MoO_2^{2+}(aq)+e- \rightarrow MoO^{2+}(aq)$	$MoO_2^{+2}(aq)+4H+(aq)^{+3}e- \rightarrow Mo^{3+}(aq)^{+}$ $^{2}H2O(l)MoO22+(aq)+4H+(aq)+3e-$ $\rightarrow Mo3+(aq)+2H2O(l)$
$VO^{+4}VO^{4+}$	$VO^{+2}(aq)^{+2}H+(aq)+e- \rightarrow VO^{2+}(aq)+$ $H_2O(l)VO^{2+}(aq)^{+2}H+(aq)+e- \rightarrow VO^{2+}$ $(aq)+H_2O(l)$	$VO^{+2}(aq)^{+4}H+(aq)^{+3}e- \rightarrow V2+(aq)^{+2}H_2$ $O(l)VO^{2+}(aq)^{+4}H+(aq)^{+3}e- \rightarrow V^{2+}(aq)^{+2}$ $H_2O(l)$

Several reagents are used as ***auxiliary oxidizing agents***, including ammonium peroxydisulfate, $(NH_4)_2S_2O_8$, and hydrogen peroxide, H_2O_2. Peroxydisulfate is a powerful oxidizing agent

$$S_2O^{2-}{}_8\,(aq) + 2e^- \rightarrow 2SO^{2-}{}_4(aq)$$

that is capable of oxidizing Mn^{2+} to MnO−4MnO4−, Cr^{3+} to $Cr_2O^{2-}7Cr_2O_7^{2-}$, and Ce^{3+} to Ce^{4+}. Excess peroxydisulfate is destroyed by briefly boiling the solution. The reduction of hydrogen peroxide in an acidic solution

$$H_2O^2(aq)^{+2}H+(aq)^{+2}e^- \rightarrow 2H_2O(l)$$

provides another method for oxidizing a titrand. Excess H_2O_2 is destroyed by briefly boiling the solution.

Inorganic Analysis

One of the most important applications of redox titrimetry is evaluating the chlorination of public water supplies. Representative Method 9.4.1, for example, describes an approach for determining the total chlorine residual using the oxidizing power of chlorine to oxidize I– to $I^{-3}I3^-$.The amount of $I^{-3}I3-$ is determined by back titrating with $S_2O_2^{-3}S_2O_32-$.

The efficiency of chlorination depends on the form of the chlorinating species. There are two contributions to the total chlorine residual—the free chlorine residual and the combined chlorine residual. The free chlorine residual includes forms of chlorine that are available for disinfecting the water supply. Examples of species that contribute to the free chlorine residual include Cl_2, HOCl and OCl^-. The combined chlorine residual includes those species in which chlorine is in its reduced form and, therefore, no longer capable of providing disinfection. Species that contribute to the combined chlorine residual are NH_2Cl, $NHCl_2$ and NCl_3.

When a sample of iodide-free chlorinated water is mixed with an excess of the indicator N,N-diethyl-p-phenylenediamine (DPD), the free chlorine oxidizes a stoichiometric portion of DPD to its red-colored form. The oxidized DPD is then back-titrated to its colorless form using ferrous ammonium sulfate as the titrant. The volume of titrant is proportional to the free residual chlorine.

Having determined the free chlorine residual in the water sample, a small amount of KI is added, which catalyzes the reduction of monochloramine, NH_2Cl, and oxidizes a portion of the DPD back to its red-colored form. Titrating the oxidized DPD with ferrous ammonium sulfate yields the amount of NH_2Cl in the sample. The amount of dichloramine and trichloramine are determined in a similar fashion.

The methods described above for determining the total, free, or combined chlorine residual also are used to establish a water supply's chlorine demand. Chlorine demand is defined as the quantity of chlorine needed to react completely with any substance that can be oxidized by chlorine, while also maintaining the desired chlorine residual. It is determined by adding progressively greater amounts of chlorine to a set of samples drawn from the water supply and determining the total, free, or combined chlorine residual.

Another important example of redox titrimetry, which finds applications in both public health and environmental analysis, is the determination of dissolved oxygen. In natural waters, such as lakes and rivers, the level of dissolved O_2 is important for two reasons: it is the most readily available oxidant for the biological oxidation of inorganic and organic pollutants; and it is necessary for the support of aquatic life. In a wastewater treatment plant dissolved O_2 is essential for the aerobic oxidation of waste materials. If the concentration of dissolved O_2 falls below a critical value, aerobic bacteria are replaced by anaerobic bacteria, and the oxidation of organic waste produces undesirable gases, such as CH_4 and H_2S.

One standard method for determining dissolved O_2 in natural waters and wastewaters is the Winkler method. A sample of water is collected without exposing it to the atmosphere, which might change the concentration of dissolved O_2. The sample first is treated with a solution of $MnSO_4$ and then with a solution of NaOH and KI. Under these alkaline conditions the dissolved oxygen oxidizes Mn^{2+} to MnO_2.

$$2Mn^{2+}(aq)+4OH-(aq)+O_2(g) \rightarrow 2MnO_2(s)+2H_2O(l)$$

After the reaction is complete, the solution is acidified with H_2SO_4. Under the now acidic conditions, I− is oxidized to I−3I3− by MnO_2.

$$MnO_2(s)+3I^-(aq)+4H^+(aq) \rightarrow Mn^{2+}(aq)+I^-_3(aq)+2H_2O(l)$$

The amount of I−3I3− that forms is determined by titrating with S2O2−3S2O32− using starch as an indicator. The Winkler method is subject to a variety of interferences and several modifications to the original procedure have been proposed. For example, NO−2NO2− interferes because it reduces I−3I3− to I^- under acidic conditions. This interference is eliminated by adding sodium azide, NaN_3, which reduces NO−2NO2− to N_2. Other reducing agents, such as Fe^{2+}, are eliminated by pretreating the sample with $KMnO_4$ and destroying any excess permanganate with $K_2C_2O_4$.

Another important example of redox titrimetry is the determination of water in nonaqueous solvents. The titrant for this analysis is known as the Karl Fischer reagent and consists of a mixture of iodine, sulfur dioxide, pyridine, and methanol. Because the concentration of pyridine is sufficiently large, I2 and SO_2 react with pyridine (py) to form the complexes py I2 and $pySO_2$. When added to a sample that contains water, I2 is reduced to I− and SO_2 is oxidized to SO_3.

$$py \cdot I2+py \cdot SO_2+H_2O + 2py \rightarrow 2py \cdot HI+py \cdot SO^3$$

Methanol is included to prevent the further reaction of $pySO_3$ with water. The titration's end point is signaled when the solution changes from the product's yellow color to the brown color of the Karl Fischer reagent.

11.12.4 Organic Analysis

Redox titrimetry also is used for the analysis of organic analytes. One important example is the determination of the chemical oxygen demand (COD) of natural waters and wastewaters. The COD is a measure of the quantity of oxygen necessary to oxidize completely all the organic matter in a sample to CO_2 and H_2O. Because no attempt is made to correct for organic matter that is decomposed biologically, or for slow decomposition kinetics, the COD always overestimates a sample's true oxygen demand. The determination of COD is particularly important in the management of industrial wastewater treatment facilities where it is used to monitor the release of organic-rich wastes into municipal sewer systems or into the environment.

A sample's COD is determined by refluxing it in the presence of excess K_2Cr_2O7, which serves as the oxidizing agent. The solution is acidified with H_2SO_4, using Ag2SO4 to catalyze the oxidation of low molecular weight fatty acids. Mercuric sulfate, $HgSO_4$, is added to complex any chloride that is present, which prevents the precipitation of the Ag^+ catalyst as AgCl. Under these conditions, the efficiency for oxidizing organic matter is 95–100%. After refluxing for two hours, the solution is cooled to room temperature and the excess Cr_2O_2–7Cr_2O7_2– determined by a back titration using ferrous ammonium sulfate as the titrant and ferroin as the indicator. Because it is difficult to remove completely all traces of organic matter from the reagents, a blank titration is performed. The difference in the amount of ferrous ammonium sulfate needed to titrate the sample and the blank is proportional to the COD.

Iodinehasbeenusedasanoxidizingtitrantforanumberofcompoundsofpharmaceutical interest. Earlier we noted that the reaction of $S_2O_2^{-3}S_2O_3$2– with I–3I3– produces the tetrathionate ion, $S_4O_2^{-6}S_4O_6^{2-}$. The tetrathionate ion is actually a dimer that consists of two thiosulfate ions connected through a disulfide (–S–S–) linkage. In the same fashion, I–3I3– is used to titrate mercaptans of the general formula RSH, forming the dimer RSSR as a product. The amino acid cysteine also can be titrated with I–3I3–. The product of this titration is cystine, which is a dimer of cysteine. Triiodide also is used for the analysis of ascorbic acid (vitamin C) by oxidizing the enediol functional group to an alpha diketone

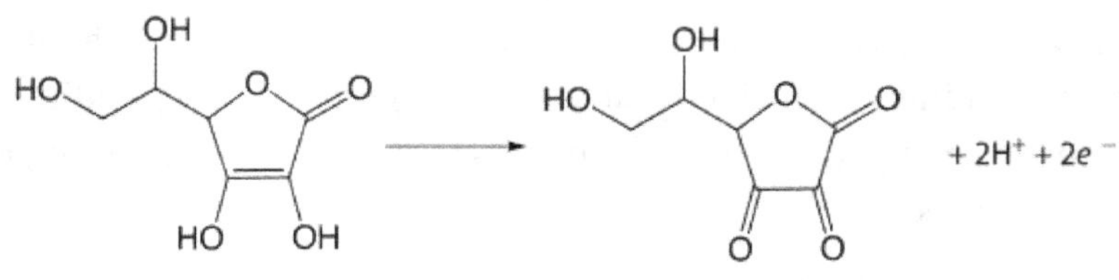

and for the analysis of reducing sugars, such as glucose, by oxidizing the aldehyde functional group to a carboxylate ion in a basic solution.

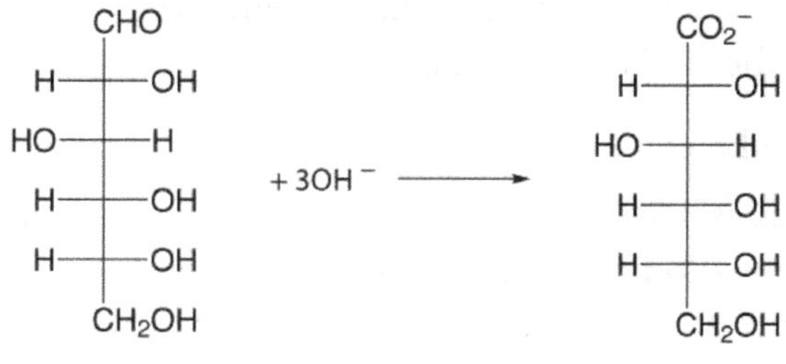

An organic compound that contains a hydroxyl, a carbonyl, or an amine functional group adjacent to an hydoxyl or a carbonyl group can be oxidized using metaperiodate, IO–4IO4–, as an oxidizing titrant.

$$IO^{-4}(aq)+H_2O(l)^{+2}e^- \rightleftharpoons IO^{-3}(aq)^{+2}OH^-(aq)IO4^-(aq)+H_2O(l)^{+2}e^- \rightleftharpoons IO3^-(aq)^{+2}OH^-(aq)$$

A two-electron oxidation cleaves the C–C bond between the two functional groups with hydroxyl groups oxidized to aldehydes or ketones, carbonyl groups oxidized to carboxylic acids, and amines oxidized to an aldehyde and an amine (ammonia if a primary amine). The analysis is conducted by adding a known excess of $IO^{-4}IO^{4-}$ to the solution that contains the analyte and allowing the oxidation to take place for approximately one hour at room temperature. When the oxidation is complete, an excess of KI is added, which converts any unreacted $IO^{-4}IO^{4-}$ to $IO^{-3}IO^{3-}$ and $I^{-3}I^{3-}$.

$$IO^{-4}(aq)^{+3}I-(aq)+H_2O(l)\rightarrow IO^{-3}(aq)+I^{-3}(aq)^{+2}OH-(aq)IO4-(aq)^{+3}I-(aq)+H_2O(l)\rightarrow IO3-(aq)+I3-(aq)^{+2}OH-(aq)$$

The $I^{-3}I3-$ is then determined by titrating with $S_2O_2^{-3}S_2O_32-$ using starch as an indicator.

11.12.5 Quantitative Calculations

The quantitative relationship between the titrand and the titrant is determined by the stoichiometry of the titration reaction. If you are unsure of the balanced reaction, you can deduce its stoichiometry by remembering that the electrons in a redox reaction are conserved.

11.12 EVALUATION OF REDOX TITRIMETRY

The scale of operations, accuracy, precision, sensitivity, time, and cost of a redox titration are similar to those described earlier in this chapter for an acid–base or a complexation titration. As with an acid–base titration, we can extend a redox titration to the analysis of a mixture of analytes if there is a significant difference in their oxidation or reduction potentials. Figure 26 shows an example of the titration curve for a mixture of Fe^{2+} and Sn^{2+} using Ce^{4+} as the titrant. A titration of a mixture of analytes is possible if their standard state potentials or formal potentials differ by at least 200 mV.

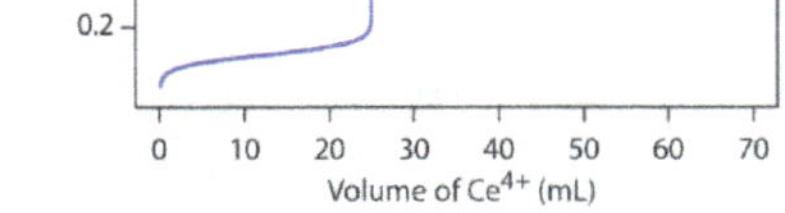

Fig. 26: Titration curve for the titration of 50.0 mL of 0.0125 M Sn^{2+} and 0.0250 M Fe^{2+} with 0.050 M Ce^{4+}. Both the titrand and the titrant are 1M in HCl.

11.13 PRECIPITATION TITRATIONS

Thus far we have examined titrimetric methods based on acid–base, complexation, and oxidation–reduction reactions. A reaction in which the analyte and titrant form an insoluble precipitate also can serve as the basis for a titration. We call this type of titration a ***precipitation titration***.

One of the earliest precipitation titrations—developed at the end of the eighteenth century—was the analysis of K_2CO_3 and K_2SO_4 in potash. Calcium nitrate, $Ca(NO_3)_2$, was used as the titrant, which forms a precipitate of $CaCO_3$ and $CaSO_4$. The titration's end point was signaled by noting when the addition of titrant ceased to generate additional precipitate. The importance of precipitation titrimetry as an analytical method reached its zenith in the nineteenth century when several methods were developed for determining Ag+ and halide ions.

11.14.1 Titration Curves

A precipitation titration curve follows the change in either the titrand's or the titrant's concentration as a function of the titrant's volume. As we did for other titrations, we first show how to calculate the titration curve and then demonstrate how we can sketch a reasonable approximation of the titration curve.

11.14.2 Calculating the Titration Curve

Let's calculate the titration curve for the titration of 50.0 mL of 0.0500 M NaCl with 0.100 M $AgNO_3$. The reaction in this case is

$$Ag^+(aq) + Cl^-(aq) \rightleftharpoons AgCl(s)$$

because the reaction's equilibrium constant is so large

$$K = (K_{sp})^{-1} = (1.8 \times 10^{-10})^{-1} = 5.6 \times 10^9$$

we may assume that Ag^+ and Cl– react completely.

By now you are familiar with our approach to calculating a titration curve. The first task is to calculate the volume of Ag^+ needed to reach the equivalence point. The stoichiometry of the reaction requires that

$$\text{mol Ag}^+ = M_{Ag}V_{Ag} = M_{Cl}V_{Cl} = \text{mol Cl}^-$$

Solving for the volume of Ag^+

$$V_{eq} = V_{Ag} = \frac{M_{Cl}V_{Cl}}{M_{Ag}} = \frac{(0.0500 \text{ M})(50.0 \text{ mL})}{0.100 \text{ M}} = 25.0 \text{ mL}$$

shows that we need 25.0 mL of Ag^+ to reach the equivalence point.

Before the equivalence point the titrand, Cl^-, is in excess. The concentration of unreacted Cl^- after we add 10.0 mL of Ag^+, for example, is

$$[Cl^-] = \frac{(\text{mol Cl}^-)_{\text{initial}} - (\text{mol Ag}^+)_{\text{added}}}{\text{total volume}} = \frac{M_{Cl}V_{Cl} - M_{Ag}V_{Ag}}{V_{Cl} + V_{Ag}}$$

$$[Cl^-] = \frac{(0.0500\text{ M})(50.0\text{ mL}) - (0.100\text{ M})(10.0\text{ mL})}{50.0\text{ mL} + 10.0\text{ mL}} = 2.50 \times 10^{-2}\text{ M}$$

which corresponds to a pCl of 1.60.

At the titration's equivalence point, we know that the concentrations of Ag+ and Cl^- are equal. To calculate the concentration of Cl^- we use the K_{sp} for AgCl; thus

$$K_{sp} = [Ag^+][Cl^-] = (x)(x) = 1.8 \times 10^{-10}$$

Solving for x gives [Cl–] as $1.3\times10^{-5}1.3\times10^{-5}$ M, or a pCl of 4.89.

After the equivalence point, the titrant is in excess. We first calculate the concentration of excess Ag^+ and then use the K_{sp} expression to calculate the concentration of Cl^-. For example, after adding 35.0 mL of titrant

$$[Ag^+] = \frac{(\text{mol Ag}^+)_{\text{added}} - (\text{mol Cl}^-)_{\text{initial}}}{\text{total volume}} = \frac{M_{Ag}V_{Ag} - M_{Cl}V_{Cl}}{V_{Ag} + V_{Cl}}$$

$$[Ag^+] = \frac{(0.100\text{ M})(35.0\text{ mL}) - (0.0500\text{ M})(50.0\text{ mL})}{35.0\text{ mL} + 50.0\text{ mL}} = 1.18 \times 10^{-2}\text{ M}$$

$$[Cl^-] = \frac{K_{sp}}{[Ag^+]} = \frac{1.8 \times 10^{-10}}{1.18 \times 10^{-2}} = 1.5 \times 10^{-8}\text{ M}$$

or a pCl of 7.81.

Table 10. Titration of 50.0 mL of 0.0500 M NaCl with 0.100 M AgNO3AgNO3

Volume of agno$_3$ (ml)	Pcl	Volume of agno$_3$ (ml)	Pcl
0.00	1.30	30.0	7.54
5.00	1.44	35.0	7.82
10.0	1.60	40.0	7.97
15.0	1.81	45.0	8.07
20.0	2.15	50.0	8.14
25.0	4.89		

11.14.3 Quantitative Applications

Although precipitation titrimetry rarely is listed as a standard method of analysis, it is useful as a secondary analytical method to verify other analytical methods. Most precipitation titrations use Ag^+ as either the titrand or the titrant. A titration in which Ag^+ is the titrant is called an ***argentometric titration***. Table 11 provides a list of several typical precipitation titrations.

Table 11 . Representative Examples of Precipitation Titrations

Titrand	Titrant	End point
$AsO_3^{-4}AsO_4 3-$	$AgNO_3$ and KSCN	Volhard
Br–	$AgNO_3$ $AgNO_3$ and KSCN	Mohr or Fajans Volhard
Cl–	$AgNO_3$ $AgNO_3$ and KSCN	Mohr or Fajarns Volhard*
$CO^{-3}CO3-$	$AgNO_3$ and KSCN	Volhard*
$C_2O_2^{-4}C_2O_4 2-$	$AgNO_3$ and KSCN	Volhard*
$CrO_2^{-4}CrO_4 2-$	$AgNO_3$ and KSCN	Volhard*
I–	$AgNO_3$ $AgNO_3$ and KSCN	Fajans Volhard
$PO_3^{-4}PO_4 3-$	$AgNO_3$ and KSCN	Volhard*
S2–	$AgNO_3$ and KSCN	Volhard*
SCN–	$AgNO_3$ and KSCN	Volhard*

When two titrants are listed ($AgNO_3$ and KSCN), the analysis is by a back titration; the first titrant, $AgNO_3$, is added in excess and the excess is titrated using the second titrant, KSCN. For those Volhard methods identified with an asterisk (*), the precipitated silver salt is removed before carrying out the back titration.

11.14.4 Quantitative Calculations

The quantitative relationship between the titrand and the titrant is determined by the stoichiometry of the titration reaction. If you are unsure of the balanced reaction, you can deduce the stoichiometry from the precipitate's formula.

12

SPECTROSCOPIC METHODS: GENERAL

12.1. INTRODUCTION

The focus of this chapter is on the interaction of ultraviolet, visible, and infrared radiation with matter. Because these techniques use optical materials to disperse and focus the radiation, they often are identified as optical spectroscopies. For convenience we will use the simpler term spectroscopy in place of optical *spectroscopy*; however, you should understand we will consider only a limited piece of what is a much broader area of analytical techniques.

Despite the difference in instrumentation, all spectroscopic techniques share several common features. Before we consider individual examples in greater detail, let's take a moment to consider some of these similarities.

12.2 ELECTROMAGNETIC RADIATION

Electromagnetic radiation—light—is a form of energy whose behavior is described by the properties of both waves and particles. Some properties of electromagnetic radiation, such as its refraction when it passes from one medium to another , are explained best when we describe light as a wave. Other properties, such as absorption and emission, are better described by treating light as a particle. The exact nature of electromagnetic radiation remains unclear, as it has since the development of quantum mechanics in the first quarter of the 20th century. Nevertheless, this dual model of wave and particle behavior provide a useful description for electromagnetic radiation.

The focus of this chapter is on the interaction of ultraviolet, visible, and infrared radiation with matter. Because these techniques use optical materials to disperse and focus the radiation, they often are identified as optical spectroscopies. For convenience we will use the simpler term spectroscopy in place of optical *spectroscopy*; however, you should understand we will consider only a limited piece of what is a much broader area of analytical techniques.

Fig. 1: The Golden Gate bridge as seen through rain drops. Refraction of light by the rain drops produces the distorted images.

12.2.1 Wave Properties of Electromagnetic Radiation

Electromagnetic radiation consists of oscillating electric and magnetic fields that propagate through space along a linear path and with a constant velocity. In a vacuum, electromagnetic radiation travels at the speed of light, c, which is 2.99792×1082.99792×108 m/s. When electromagnetic radiation moves through a medium other than a vacuum, its velocity, v, is less than the speed of light in a vacuum. The difference between v and c is sufficiently small (<0.1%) that the speed of light to three significant figures, 3.00×1083.00×108 m/s, is accurate enough for most purposes.

The oscillations in the electric field and the magnetic field are perpendicular to each other and to the direction of the wave's propagation. Figure 2 shows an example of plane-polarized electromagnetic radiation, which consists of a single oscillating electric field and a single oscillating magnetic field.

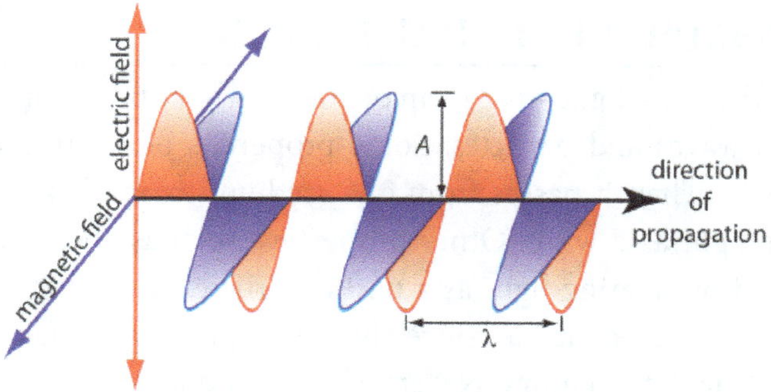

Fig. 2: Plane-polarized electromagnetic radiation showing the oscillating electric field in blue and the oscillating magnetic field in red. The radiation's amplitude, A, and its wavelength, λ, are shown. Normally, electromagnetic radiation is unpolarized, with oscillating electric and magnetic fields present in all possible planes perpendicular to the direction of propagation.

An electromagnetic wave is characterized by several fundamental properties, including its velocity, amplitude, frequency, phase angle, polarization, and direction of propagation. For example, the amplitude of the oscillating electric field at any point along the propagating wave is

$$A_t = A_e \sin(2\pi\nu t + \Phi)$$

where At is the magnitude of the electric field at time t, Ae is the electric field's maximum amplitude, νν is the wave's frequency—the number of oscillations in the electric field per unit time—and ΦΦ is a phase angle that accounts for the fact that At need not have a value of zero at t = 0. The identical equation for the magnetic field is where Am is the magnetic field's maximum amplitude.

$$A_t = A_m \sin(2\pi\nu t + \Phi)$$

Other properties also are useful for characterizing the wave behavior of electromagnetic radiation. The wavelength, λλ, is defined as the distance between successive maxima (see Figure 10.1.2). For ultraviolet and visible electromagnetic radiation the wavelength usually is expressed in nanometers (1 nm = 10–9 m), and for infrared radiation it is expressed in microns (1 mm = 10–6 m). The relationship between wavelength and frequency is

$$\lambda = \frac{c}{\nu}$$

another unit useful unit is the wavenumber, ν¯, which is the reciprocal of wavelength

$$\overline{\nu} = \frac{1}{\lambda}$$

wavenumbers frequently are used to characterize infrared radiation, with the units given in cm^{-1}.

12.2.2 Particle Properties of Electromagnetic Radiation

When matter absorbs electromagnetic radiation it undergoes a change in energy. The interaction between matter and electromagnetic radiation is easiest to understand if we assume that radiation consists of a beam of energetic particles called photons. When a photon is absorbed by a sample it is "destroyed" and its energy acquired by the sample. The energy of a photon, in joules, is related to its frequency, wavelength, and wave number by the following equalities where h is Planck's constant, which has a value of 6.626 × 10–34 Js.

$$E = h\nu = \frac{hc}{\lambda} = hc\overline{\nu}$$

12.2.3 The Electromagnetic Spectrum

The frequency and the wavelength of electromagnetic radiation vary over many orders of magnitude. For convenience, we divide electromagnetic radiation into different regions—the *electromagnetic spectrum*—based on the type of atomic or molecular transitions that gives rise to the absorption or emission of photons (Figure 3). The boundaries between the regions of the electromagnetic spectrum are not rigid and overlap between spectral regions is possible.

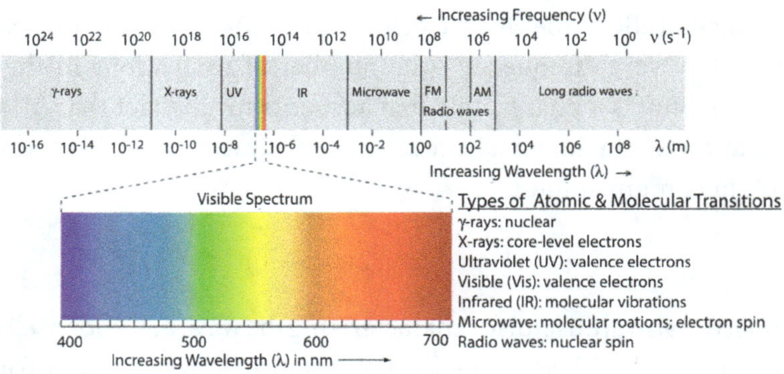

Fig. 3: The electromagnetic spectrum showing the boundaries between different regions and the type of atomic or molecular transitions responsible for the change in energy. The colored inset shows the visible spectrum.

12.2.4 Photons as a Signal Source

In the previous section we defined several characteristic properties of electromagnetic radiation, including its energy, velocity, amplitude, frequency, phase angle, polarization, and direction of propagation. A spectroscopic measurement is possible only if the photon's interaction with the sample leads to a change in one or more of these characteristic properties.

We will divide spectroscopy into two broad classes of techniques. In one class of techniques there is a transfer of energy between the photon and the sample. Table 1 provides a list of several representative examples.

Table 1 . Examples of Spectroscopic Techniques That Involve an Exchange of Energy Between a Photon and the Sample

Type of energy transfer	Region of electromagnetic spectrum	Spectroscopic technique
Absorption	γγ-ray	Mossbauer spectroscopy
	X-ray	X-ray absorption spectroscopy
	UV/Vis	UV/Vis spectroscopy
	IR	Infrared spectroscopy
	Microwave	Raman spectroscopy
	Radio wave	Electron spin resonance Nuclear magnetic resonance
Emission (thermal excitation)	UV/Vis	Atomic emission spectroscopy

Type of energy transfer	Region of electromagnetic spectrum	Spectroscopic technique
Photoluminescence	X-ray	X-ray fluorescence
	uV/Vis	Fluorescence spectroscopy Phosphorescence spectroscopy Atomic fluorescence spectroscopy
Chemiluminescence	UV/Vis	Chemiluminescence spectroscopy

In absorption spectroscopy a photon is absorbed by an atom or molecule, which undergoes a transition from a lower-energy state to a higher-energy, or excited state (Figure 4). The type of transition depends on the photon's energy. The electromagnetic spectrum in Figure 3 , for example, shows that absorbing a photon of visible light promotes one of the atom's or molecule's valence electrons to a higher-energy level. When an molecule absorbs infrared radiation, on the other hand, one of its chemical bonds experiences a change in vibrational energy.

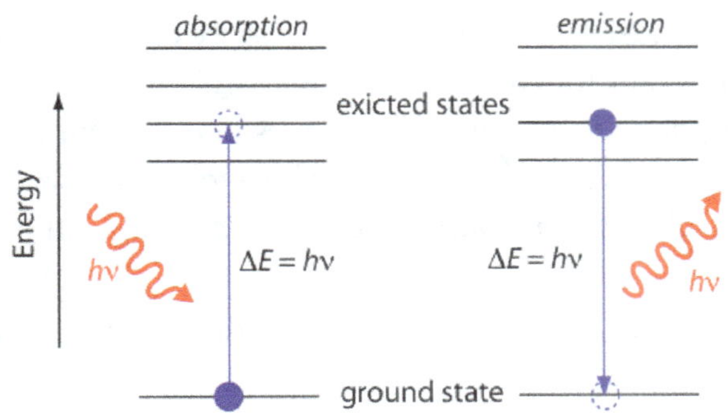

Fig. 4: A simplified energy diagram that shows the absorption and emission of a photon by an atom or a molecule. When a photon of energy hνhν strikes the atom or molecule, absorption may occur if the difference in energy, ΔEΔE, between the ground state and the excited state is equal to the photon's energy. An atom or molecule in an excited state may emit a photon and return to the ground state. The photon's energy, hνhν, equals the difference in energy, ΔEΔE, between the two states.

When it absorbs electromagnetic radiation the number of photons passing through a sample decreases. The measurement of this decrease in photons, which we call *absorbance*, is a useful analytical signal. Note that each energy level in Figure 4 has a well-defined value because each is quantized. Absorption occurs only when the photon's energy, hνhν, matches the difference in energy, ΔEΔE, between two energy levels. A plot of absorbance as a function of the photon's energy is called an *absorbance spectrum*. Figure.5 , for example, shows the absorbance spectrum of cranberry juice.

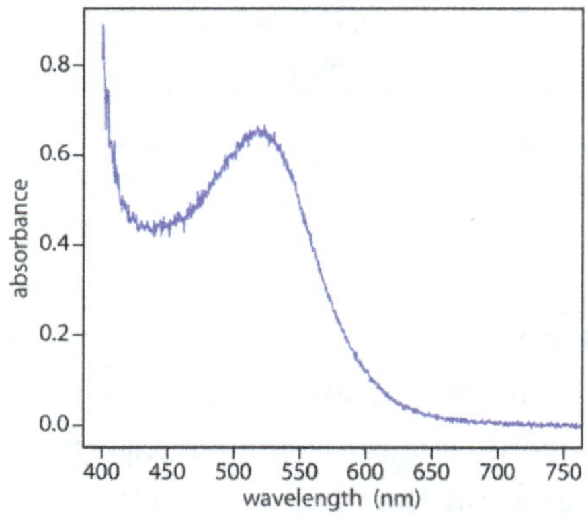

Fig. 5: Visible absorbance spectrum for cranberry juice. The anthocyanin dyes in cranberry juice absorb visible light with blue, green, and yellow wavelengths (Figure 3). As a result, the juice appears red.

When an atom or molecule in an excited state returns to a lower energy state, the excess energy often is released as a photon, a process we call *emission* (Figure 4). There are several ways in which an atom or a molecule may end up in an excited state, including thermal energy, absorption of a photon, or as the result of a chemical reaction. Emission following the absorption of a photon is also called *photoluminescence*, and that following a chemical reaction is called *chemiluminescence*. A typical emission spectrum is shown in Figure 6 .

Molecules also can release energy in the form of heat. We will return to this point later in the chapter.

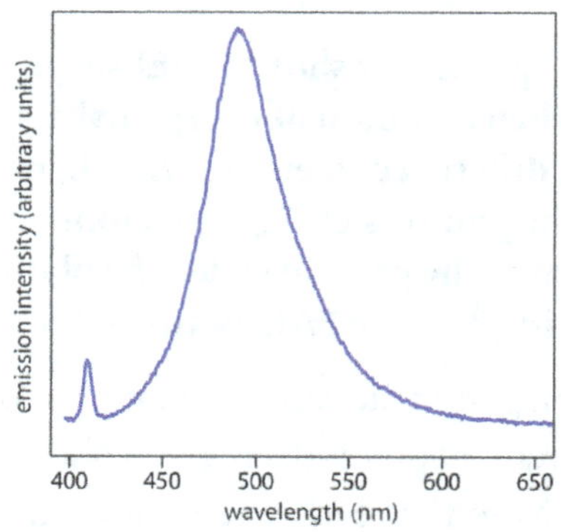

Fig. 6: Photoluminescence spectrum of the dye coumarin 343, which is incorporated in a reverse micelle suspended in cyclohexanol. The dye's absorbance spectrum (not shown) has a broad peak around 400 nm. The sharp peak at 409 nm is from the laser source used to excite coumarin 343. The broad band centered at approximately 500 nm is the dye's emission band. Because the dye absorbs blue light, a solution of coumarin 343 appears yellow in the absence of photoluminescence. Its photoluminescent emission is blue-green. Source: data from Bridget Gourley, Department of Chemistry & Biochemistry, DePauw University.

In the second broad class of spectroscopic techniques, the electromagnetic radiation undergoes a change in amplitude, phase angle, polarization, or direction of propagation as a result of its refraction, reflection, scattering, diffraction, or dispersion by the sample. Several representative spectroscopic techniques are listed in Table 2 .

Table 2 . Examples of Spectroscopic Techniques That Do Not Involve an Exchange of Energy Between a Photon and the Sample

Region of electromagnetic spectrum	Type of interaction	Spectroscopic technique
X-ray	Diffraction	X-ray diffraction
UV/Vis	Refraction	Refractrometry
	Scattering	*Nephelometry* *Turbidimetry*
	Dispersion	Optical rotary dispersion

12.3 BASIC COMPONENTS OF SPECTROSCOPIC INSTRUMENTS

The spectroscopic techniques in Table1 and Table 2 use instruments that share several common basic components, including a source of energy, a means for isolating a narrow range of wavelengths, a detector for measuring the signal, and a signal processor that displays the signal in a form convenient for the analyst. In this section we introduce these basic components.

12.3.1 Sources of Energy

All forms of spectroscopy require a source of energy. In absorption and scattering spectroscopy this energy is supplied by photons. Emission and photoluminescence spectroscopy use thermal, radiant (photon), or chemical energy to promote the analyte to a suitable excited state.

12.3.2 Sources of Electromagnetic Radiation

A source of electromagnetic radiation must provide an output that is both intense and stable. Sources of electromagnetic radiation are classified as either continuum or line sources. A continuum source emits radiation over a broad range of wavelengths, with a relatively smooth variation in intensity (Figure 7). A line source, on the other hand, emits radiation at selected wavelengths (Figure 8). Table 3 provides a list of the most common sources of electromagnetic radiation.

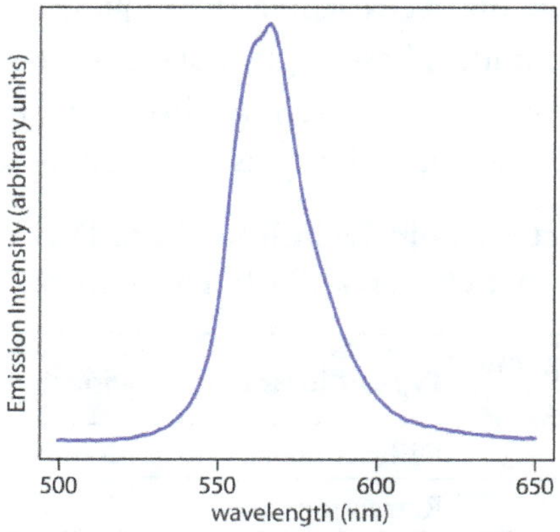

Figure 7 . Spectrum showing the emission from a green LED, which provides continuous emission over a wavelength range of approximately 530–640 nm.

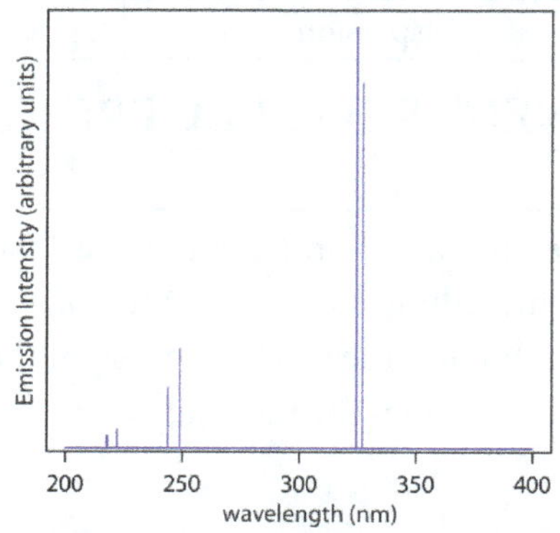

Figure 8 . Emission spectrum from a Cu hollow cathode lamp. This spectrum consists of seven distinct emission lines (the first two differ by only 0.4 nm and are not resolved at the scale shown in this spectrum). Each emission line has a width of approximately 0.01 nm at 1/2 of its maximum intensity.

12.3.3 Sources of Thermal Radiation

The most common sources of thermal energy are flames and plasmas. A flame source uses a combustion of a fuel and an oxidant to achieve temperatures of 2000–3400 K. Plasmas, which are hot, ionized gases, provide temperatures of 6000–10000 K.

12.3.4 Chemical Sources of Energy

Exothermic reactions also may serve as a source of energy. In chemiluminescence the analyte is raised to a higher-energy state by means of a chemical reaction, emitting characteristic radiation when it returns to a lower-energy state. When the chemical reaction results from a biological or enzymatic reaction, the emission of radiation is called bioluminescence. Commercially available “light sticks” and the flash of light from a firefly are examples of chemiluminescence and bioluminescence.

Table 3 . Common Source of Electromagnetic Radiation

Source	Wavelength region	Useful for...
H_2 and D_2 lamp	Continuum source from 160–380 nm	Molecular absorption
Tungsten lamp	Continuum source from 320–2400 nm	Molecular absorption
Xe arc lamp	Continuum source from 200–1000 nm	Molecular fluorescence
Nernst glower	Continuum source from 0.4–20 Mm	Molecular absorption
Nichrome wire	Continuum source from 0.75–20 Mm	Molecular absorption
Globar	Continuum source from 1–40 μm	Molecular absorption
Hollow cathode lamp	Line source in uv/vis	Atomic absorption
Hg vapor lamp	Line source in uv/vis	Molecular fluorescence
Laser	Line source in uv/vis/ir	Atomic and molecular absorption, fluorescence, and scattering

12.4 WAVELENGTH SELECTION

In Nessler's original colorimetric method for ammonia, which was described at the beginning of the chapter, the sample and several standard solutions of ammonia are placed in separate tall, flat-bottomed tubes. As shown in Figure 9 , after adding the reagents and allowing the color to develop, the a nalyst evaluates the color by passing ambient light through the bottom of the tubes and looking down through the solutions. By matching the sample's color to that of a standard, the analyst is able to determine the concentration of ammonia in the sample.

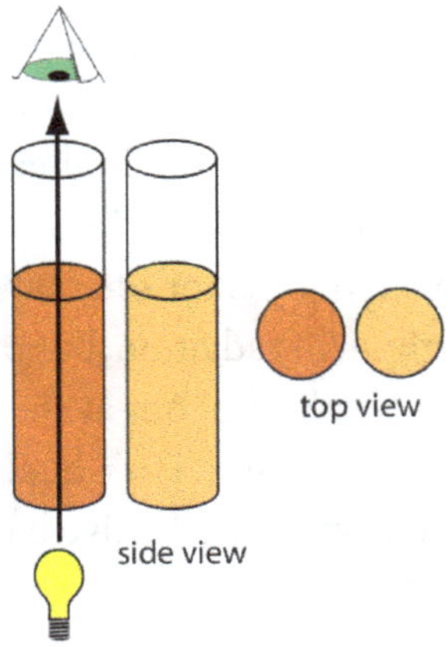

Fig. 9: Nessler's original method for comparing the color of two solutions. Natural light passes upwards through the samples and standards and the analyst views the solutions by looking down toward the light source. The top view, shown on the right, is what the analyst sees. To determine the analyte's concentration, the analyst exchanges standards until the two colors match.

In Figure 9 every wavelength of light from the source passes through the sample. This is not a problem if there is only one absorbing species in the sample. If the sample contains two components, then a quantitative analysis using Nessler's original method is impossible unless the standards contains the second component at the same concentration it has in the sample.

To overcome this problem, we want to select a wavelength that only the analyte absorbs. Unfortunately, we can not isolate a single wavelength of radiation from a continuum source, although we can narrow the range of wavelengths that reach the sample. As seen in Figure 10 , a wavelength selector always passes a narrow band of radiation characterized by a nominal wavelength, an effective bandwidth, and a maximum throughput of radiation. The effective bandwidth is defined as the width of the radiation at half of its maximum throughput.

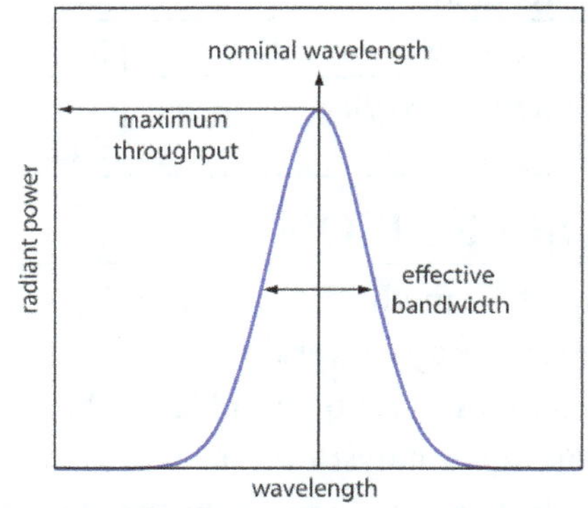

Fig. 10: Radiation exiting a wavelength selector showing the band's nominal wavelength and its effective bandwidth.

The ideal wavelength selector has a high throughput of radiation and a narrow effective bandwidth. A high throughput is desirable because the more photons that pass through the wavelength selector, the stronger the signal and the smaller the background noise. A narrow effective bandwidth provides a higher resolution, with spectral features separated by more than twice the effective bandwidth being resolved. As shown in Figure 11 , these two features of a wavelength selector often are in opposition. A larger effective bandwidth favors a higher throughput of radiation, but provide less resolution. Decreasing the effective bandwidth improves resolution, but at the cost of a noisier signal [Jiang, S.; Parker, G. A. Am. Lab. 1981, October, 38–43]. For a qualitative analysis, resolution usually is more important than noise and a smaller effective bandwidth is desirable; however, in a quantitative analysis less noise usually is desirable.

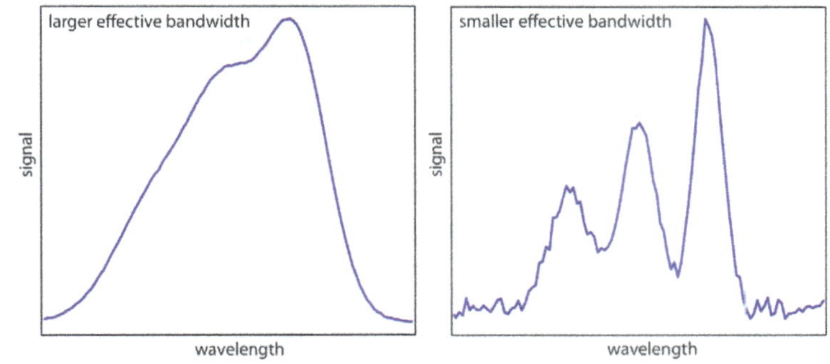

Fig. 11: Example showing the effect of the wavelength selector's effective bandwidth on resolution and noise. The spectrum with the smaller effective bandwidth (on the right) has a better resolution, allowing us to see the presence of three peaks, but at the expense of a noisier signal. The spectrum with the larger effective bandwidth (on the left) has less noise, but at the expense of less resolution between the three peaks.

12.4.1 Wavelength Selection Using Filters

The simplest method for isolating a narrow band of radiation is to use an absorption or interference filter. Absorption filters work by selectively absorbing radiation from a narrow region of the electromagnetic spectrum. Interference filters use constructive and destructive interference to isolate a narrow range of wavelengths. A simple example of an absorption filter is a piece of colored glass. A purple filter, for example, removes the complementary color green from 500–560 nm.

Commercially available absorption filters provide effective bandwidths of 30–250 nm, although the throughput at the low end of this range often is only 10% of the source's emission intensity. Interference filters are more expensive than absorption filters, but have narrower effective bandwidths, typically 10–20 nm, with maximum throughputs of at least 40%.

12.4.2 Wavelength Selection Using Monochromators

A filter has one significant limitation—because a filter has a fixed nominal wavelength, if we need to make measurements at two different wavelengths, then we must use two different filters. A monochromator is an alternative method for selecting a narrow band of radiation that also allows us to continuously adjust the band's nominal wavelength.

The construction of a typical monochromator is shown in Figure 12 . Radiation from the source enters the monochromator through an entrance slit. The radiation is collected by a collimating mirror, which reflects a parallel beam of radiation to a diffraction grating. The diffraction grating is an optically reflecting surface with a large number of parallel grooves (see insert to Figure 12). The diffraction grating disperses the radiation and a second mirror focuses the radiation onto a planar surface that contains an exit slit. In some monochromators a prism is used in place of the diffraction grating.

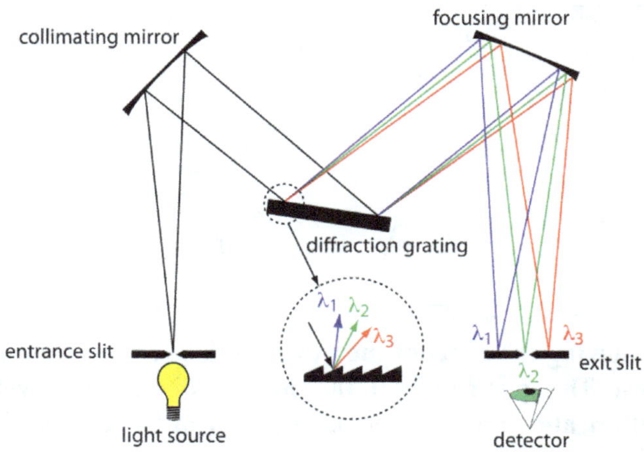

Fig. 12: Schematic diagram of a monochromator that uses a diffraction grating to disperse the radiation.

Radiation exits the monochromator and passes to the detector. As shown in Figure 10.1.12 , a monochromator converts a polychromatic source of radiation at the entrance slit to a monochromatic source of finite effective bandwidth at the exit slit. The choice of which wavelength exits the monochromator is determined by rotating the diffraction grating. A narrower exit slit provides a smaller effective bandwidth and better resolution than does a wider exit slit, but at the cost of a smaller throughput of radiation.

Polychromatic means many colored. Polychromatic radiation contains many different wavelengths of light. Monochromatic means one color, or one wavelength. Although the light exiting a monochromator is not strictly of a single wavelength, its narrow effective bandwidth allows us to think of it as monochromatic.

Monochromators are classified as either fixed-wavelength or scanning. In a fixed-wavelength monochromator we manually select the wavelength by rotating the grating. Normally a fixed-wavelength monochromator is used for a quantitative analysis where measurements are made at one or two wavelengths. A scanning monochromator includes a drive mechanism that continuously rotates the grating, which allows successive wavelengths of light to exit from the monochromator. A scanning monochromator is used to acquire spectra, and, when operated in a fixed-wavelength mode, for a quantitative analysis.

12.4.3 Interferometers

An interferometer provides an alternative approach for wavelength selection. Instead of filtering or dispersing the electromagnetic radiation, an interferometer allows source radiation of all wavelengths to reach the detector simultaneously (Figure 13). Radiation from the source is focused on a beam splitter that reflects half of the radiation to a fixed mirror and transmits the other half to a moving mirror. The radiation recombines at the beam splitter, where constructive and destructive interference determines, for each wavelength, the intensity of light that reaches the detector. As the moving mirror changes position, the wavelength of light that experiences maximum constructive interference and maximum destructive interference also changes. The signal at the detector shows intensity as a function of the moving mirror's position, expressed in units of distance or time. The result is called an interferogram or a time domain spectrum. The time domain spectrum is converted mathematically, by a process called a Fourier transform, to a spectrum (a frequency domain spectrum) that shows intensity as a function of the radiation's energy.

The mathematical details of the Fourier transform are beyond the level of this textbook. You can consult the chapter's additional resources for additional information.

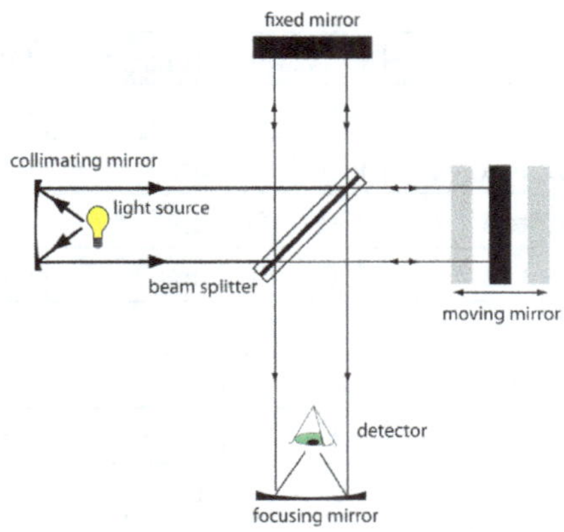

Fig. 13: Schematic diagram of an interferometers.

In comparison to a monochromator, an interferometer has two significant advantages. The first advantage, which is termed *Jacquinot's advantage*, is the greater throughput of source radiation. Because an interferometer does not use slits and has fewer optical components from which radiation is scattered and lost, the throughput of radiation reaching the detector is 80–200×80–200× greater than that for a monochromator. The result is less noise. The second advantage, which is called *Fellgett's advantage*, is a savings in the time needed to obtain a spectrum. Because the detector monitors all frequencies simultaneously, a spectrum takes approximately one second to record, as compared to 10–15 minutes when using a scanning monochromator.

12.4.4 Detectors

In Nessler's original method for determining ammonia (Figure 9) the analyst's eye serves as the detector, matching the sample's color to that of a standard. The human eye, of course, has a poor range—it responds only to visible light—and it is not particularly sensitive or accurate. Modern detectors use a sensitive transducer to convert a signal consisting of photons into an easily measured electrical signal. Ideally the detector's signal, S, is a linear function of the electromagnetic radiation's power, P,

$$S=kP + D$$

where k is the detector's sensitivity, and D is the detector's dark current, or the background current when we prevent the source's radiation from reaching the detector.

There are two broad classes of spectroscopic transducers: thermal transducers and photon transducers. Table. 4 provides several representative examples of each class of transducers.

Transducer is a general term that refers to any device that converts a chemical or a physical property into an easily measured electrical signal. The retina in your eye, for example, is a transducer that converts photons into an electrical nerve impulse; your eardrum is a transducer that converts sound waves into a different electrical nerve impulse.

Table 4 . Examples of Transducers for Spectroscopy

Transducer	Class	Wavelength range	Output signal
Phototube	Photon	200–1000 Nm	Current
Photomultiplier	Photon	110–1000 Nm	Current
Si photodiode	Photon	250–1100 Nm	Current
Photoconductor	Photon	750–6000 Nm	Change in resistance
Photovoltaic cell	Photon	400–5000 Nm	Current or voltage
Thermocouple	Thermal	0.8–40 Mm	Voltage
Thermistor	Thermal	0.8–40 Mm	Change in resistance
Pneumatic	Thermal	0.8–1000 Mm	Membrane displacement
Pyroelectric	Thermal	0.3–1000 Mm	Current

12.4.5 Photon Transducers

Phototubes and photomultipliers use a photosensitive surface that absorbs radiation in the ultraviolet, visible, or near IR to produce an electrical current that is proportional to the number of photons reaching the transducer (Figure 14). Other photon detectors use a semiconductor as the photosensitive surface. When the semiconductor absorbs photons, valence electrons move to the semiconductor's conduction band, producing a measurable current. One advantage of the Si photodiode is that it is easy to miniaturize. Groups of photodiodes are gathered together in a linear array that contains 64–4096 individual photodiodes. With a width of 25 μm per diode, a linear array of 2048 photodiodes requires only 51.2 mm of linear space. By placing a photodiode array along the monochromator's focal plane, it is possible to monitor simultaneously an entire range of wavelengths.

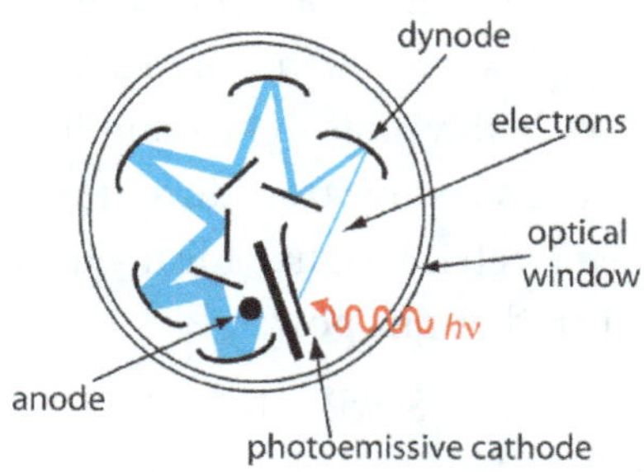

Fig. :14 Schematic of a photomultiplier. A photon strikes the photoemissive cathode producing electrons, which accelerate toward a positively charged dynode. Collision of

these electrons with the dynode generates additional electrons, which accelerate toward the next dynode. A total of 10^6–10^7 electrons per photon eventually reach the anode, generating an electrical current.

12.4.6 Thermal Transducers

Infrared photons do not have enough energy to produce a measurable current with a photon transducer. A thermal transducer, therefore, is used for infrared spectroscopy. The absorption of infrared photons increases a thermal transducer's temperature, changing one or more of its characteristic properties. A pneumatic transducer, for example, is a small tube of xenon gas with an IR transparent window at one end and a flexible membrane at the other end. Photons enter the tube and are absorbed by a blackened surface, increasing the temperature of the gas. As the temperature inside the tube fluctuates, the gas expands and contracts and the flexible membrane moves in and out. Monitoring the membrane's displacement produces an electrical signal.

12.6.7 Signal Processors

A transducer's electrical signal is sent to a signal processor where it is displayed in a form that is more convenient for the analyst. Examples of signal processors include analog or digital meters, recorders, and computers equipped with digital acquisition boards. A signal processor also is used to calibrate the detector's response, to amplify the transducer's signal, to remove noise by filtering, or to mathematically transform the signal.

13

SPECTROSCOPY BASED ON ABSORPTION

13.1 INTRODUCTION

In absorption spectroscopy a beam of electromagnetic radiation passes through a sample. Much of the radiation passes through the sample without a loss in intensity. At selected wavelengths, however, the radiation's intensity is attenuated. This process of attenuation is called absorption.

13.2 ABSORPTION SPECTRA

There are two general requirements for an analyte's absorption of electromagnetic radiation. First, there must be a mechanism by which the radiation's electric field or magnetic field interacts with the analyte. For ultraviolet and visible radiation, absorption of a photon changes the energy of the analyte's valence electrons. A bond's vibrational energy is altered by the absorption of infrared radiation. Figure 3 provides a list of the types of atomic and molecular transitions associated with different types of electromagnetic radiation.

The second requirement is that the photon's energy, hvhv, must exactly equal the difference in energy, ΔEΔE, between two of the analyte's quantized energy states. Figure 4 shows a simplified view of a photon's absorption, which is useful because it emphasizes that the photon's energy must match the difference in energy between a lower-energy state and a higher-energy state. What is missing, however, is information about what types of energy states are involved, which transitions between energy states are likely to occur, and the appearance of the resulting spectrum.

We can use the energy level diagram in Figure 1 to explain an absorbance spectrum. The lines labeled E0 and E1 represent the analyte's ground (lowest) electronic state and its first electronic excited state. Superimposed on each electronic energy level is a series of lines representing vibrational energy levels.

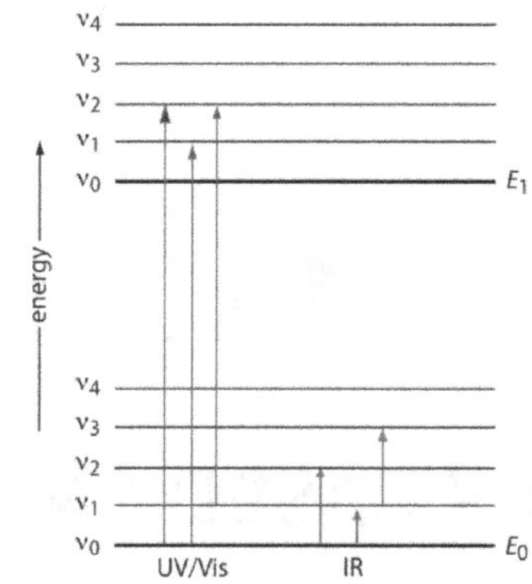

Fig. 1: Diagram showing two electronic energy levels (E_0 and E_1), each with five vibrational energy levels v0–4v0–4. Absorption of ultraviolet and visible radiation (shown by the blue arrows) leads to a change in the analyte's electronic energy levels and, possibly, a change in vibrational energy as well. A change in vibrational energy without a change in electronic energy levels occurs with the absorption of infrared radiation (shown by the red arrows).

13.2.1 Infrared Spectra for Molecules and Polyatomic Ions

The energy of infrared radiation produces a change in a molecule's or a polyatomic ion's vibrational energy, but is not sufficient to effect a change in its electronic energy. As shown in Figure 1 , vibrational energy levels are quantized; that is, a molecule or polyatomic ion has only certain, discrete vibrational energies. The energy for an allowed vibrational mode, E_v, is:

$$E_\nu = \nu + \frac{1}{2} h\nu_0$$

where ν is the vibrational quantum number, which has values of 0, 1, 2, ..., and ν0ν0 is the bond's fundamental vibrational frequency. The value of ν0ν0, which is determined by the bond's strength and by the mass at each end of the bond, is a characteristic property of a bond. For example, a carbon-carbon single bond (C–C) absorbs infrared radiation at a lower energy than a carbon-carbon double bond (C=C) because a single bond is weaker than a double bond.

At room temperature most molecules are in their ground vibrational state (ν=0) . A transition from the ground vibrational state to the first vibrational excited state (ν=1) requires absorption of a photon with an energy of hν0hν0. Transitions in which Δν=±1 give rise to the fundamental absorption lines. Weaker absorption lines, called overtones, result from transitions in which Δν is ±2 or ±3. The number of possible normal vibrational modes for a linear molecule is 3N – 5, and for a non-linear molecule is 3N – 6, where N is the number of atoms in the molecule. Not surprisingly, infrared spectra often show a considerable number of absorption bands. Even a relatively simple molecule, such as ethanol (C_2H_6O), for example, has 3×9–6, or 21 possible normal modes of vibration, although not all of these vibrational modes give rise to an absorption. The IR spectrum for ethanol is shown in Figure 2 .

Why does a non-linear molecule have 3N – 6 vibrational modes? Consider a molecule of methane, CH_4. Each of methane's five atoms can move in one of three directions (x, y, and z) for a total of 5×3=15 different ways in which the molecule's atoms can move. A molecule can move in three ways: it can move from one place to another, which we call translational motion; it can rotate around an axis, which we call rotational motion; and its bonds can stretch and bend, which we call vibrational motion. Because the entire molecule can move in the x, y, and z directions, three of methane's 15 different motions are translational. In addition, the molecule can rotate about its x, y, and z axes, accounting for three additional forms of motion. This leaves 15 – 3 – 3 = 9 vibrational modes. A linear molecule, such as CO_2, has 3N – 5 vibrational modes because it can rotate around only two axes.

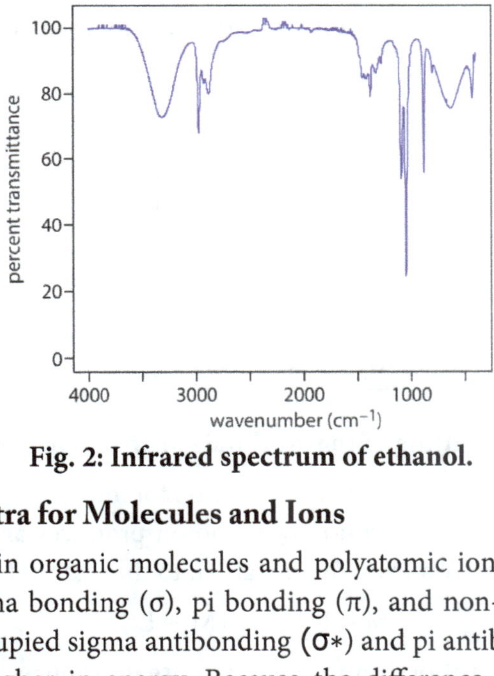

Fig. 2: Infrared spectrum of ethanol.

13.2.2 UV/VIS Spectra for Molecules and Ions

The valence electrons in organic molecules and polyatomic ions, such as $CO_2^{-3}CO_3 2-$, occupy quantized sigma bonding (σ), pi bonding (π), and non-bonding (*n*) molecular orbitals (MOs). Unoccupied sigma antibonding (σ*) and pi antibonding (п*) molecular orbitals are slightly higher in energy. Because the difference in energy between the highest-energy occupied MOs and the lowest-energy unoccupied MOs corresponds to ultraviolet and visible radiation, absorption of a photon is possible.

Four types of transitions between quantized energy levels account for most molecular U_V/V_{is} spectra. Table 1 lists the approximate wavelength ranges for these transitions, as well as a partial list of bonds, functional groups, or molecules responsible for these transitions. Of these transitions, the most important are n→π* and п→п* because they involve important functional groups that are characteristic of many analytes and because the wavelengths are easily accessible. The bonds and functional groups that give rise to the absorption of ultraviolet and visible radiation are called *chromophores.*

Table 1 . Electronic Transitions Involving n, σσ, and ππ Molecular Orbitals

Transition	Wavelength range	Examples
$\sigma \rightarrow \sigma*$	<200 nm	C—C, C—H
$n \rightarrow \sigma*$	160–260 nm	H_2O, CH_3OH, CH_3Cl
$\pi \rightarrow \pi*$	200–500 nm	C=C, C=O, C=N, C≡C
$n \rightarrow \pi*$	250–600 nm	C=O, C=N, N=N, N=O

Many transition metal ions, such as Cu^{2+} and Co^{2+}, form colorful solutions because the metal ion absorbs visible light. The transitions that give rise to this absorption are valence electrons in the metal ion's d-orbitals. For a free metal ion, the five d-orbitals are of equal energy. In the presence of a complexing ligand or solvent molecule, however, the d-orbitals split into two or more groups that differ in energy. For example, in an octahedral complex of $Cu(H_2O)_2^{+6}Cu(H_2O)_6 2+$ the six water molecules perturb the d-orbitals into the two groups shown in Figure 3 . The resulting d→d transitions for transition metal ions are relatively weak.

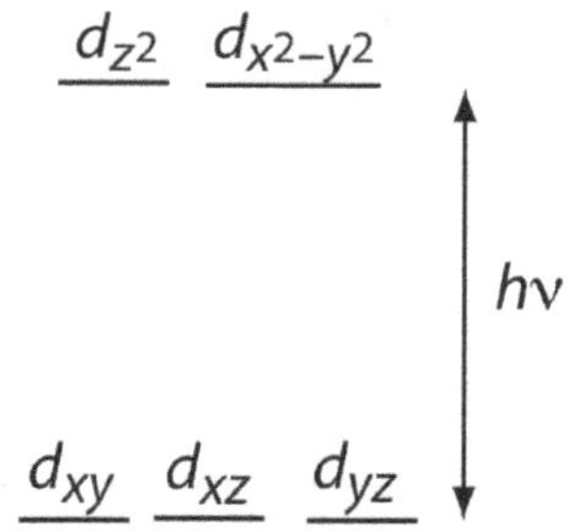

Fig. 3: Splitting of the *d*-orbitals in an octahedral field.

A more important source of UV/Vis absorption for inorganic metal–ligand complexes is charge transfer, in which absorption of a photon produces an excited state in which there is transfer of an electron from the metal, *M*, to the ligand, *L*.

$$M - L + h_v \rightarrow (M^+ - L^-)^*$$

Charge-transfer absorption is important because it produces very large absorbances. One important example of a charge-transfer complex is that of *o*-phenanthroline with Fe^{2+}, the UV/Vis spectrum for which is shown in Figure 4 . Charge - transfer absorption in which an electron moves from the ligand to the metal also is possible.

Why is a larger absorbance desirable? An analytical method is more sensitive if a smaller concentration of analyte gives a larger signal.

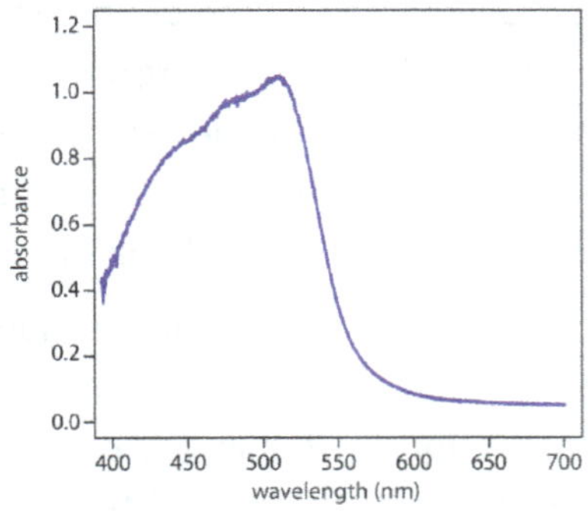

Fig. 4: UV/Vis spectrum for the metal–ligand complex Fe (phen) 2 + 3 Fe (phen)32⁺, where phen is the ligand *o*-phenanthroline.

Comparing the IR spectrum in Figure 2 to the UV/Vis spectrum in Figure 4 shows us that UV/Vis absorption bands are often significantly broader than those for IR absorption. We can use Figure 1 to explain why this is true. When a species absorbs UV/Vis radiation, the transition between electronic energy levels may also include a transition between vibrational energy levels. The result is a number of closely spaced absorption bands that merge together to form a single broad absorption band.

13.2.3 UV/VIS Spectra for Atoms

The energy of ultraviolet and visible electromagnetic radiation is sufficient to cause a change in an atom's valence electron configuration. Sodium, for example, has a single valence electron in its 3*s* atomic orbital. As shown in Figure 5 , unoccupied, higher energy atomic orbitals also exist.

The valence shell energy level diagram in Figure 5 might strike you as odd because it shows that the 3*p* orbitals are split into two groups of slightly different energy. The reasons for this splitting are unimportant in the context of our treatment of atomic absorption. For further information about the reasons for this splitting, consult the chapter's additional resources.

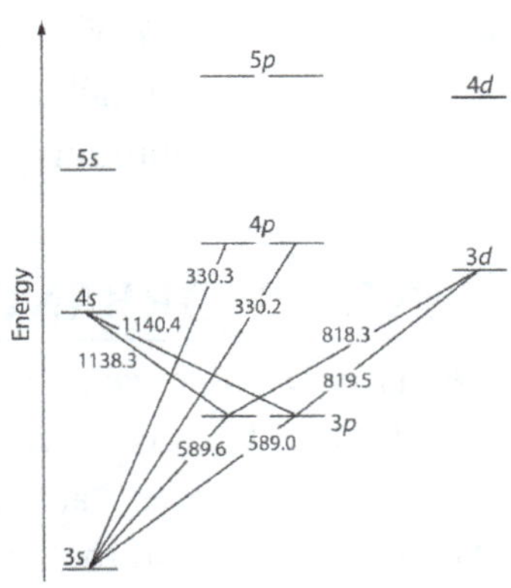

Fig. 5: Valence shell energy level diagram for sodium. The wavelengths (in wavenumbers) corresponding to several transitions are shown.

Absorption of a photon is accompanied by the excitation of an electron from a lower-energy atomic orbital to an atomic orbital of higher energy. Not all possible transitions between atomic orbitals are allowed. For sodium the only allowed transitions are those in which there is a change of ±1 in the orbital quantum number (l); thus transitions from s→p orbitals are allowed, but transitions from s→s and from s→d orbitals are forbidden.

The atomic absorption spectrum for Na is shown in Figure 6, and is typical of that found for most atoms. The most obvious feature of this spectrum is that it consists of a small number of discrete absorption lines that correspond to transitions between the ground state (the 3*s* atomic orbital) and the 3*p* and the 4*p* atomic orbitals. Absorption from excited states, such as the 3p→4s and the 3p→3d transitions included in Figure 5 , are too weak to detect. Because an excited state's lifetime is short—an excited state atom typically returns to a lower energy state in 10^{-7} to 10^{-8} seconds—an atom in the exited state is likely to return to the ground state before it has an opportunity to absorb a photon.

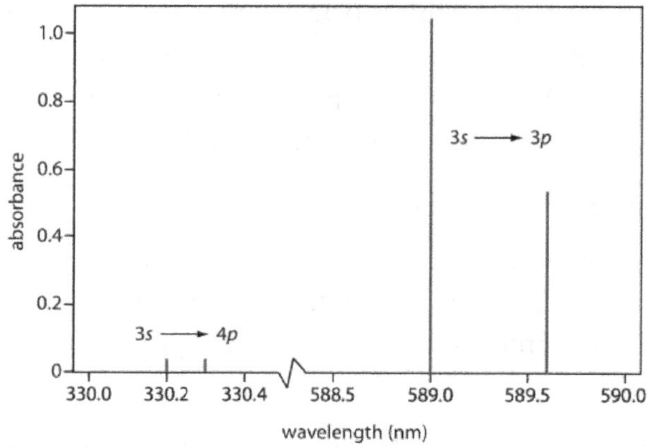

Fig. 6: Atomic absorption spectrum for sodium. Note that the scale on the *x*-axis includes a break.

Another feature of the atomic absorption spectrum in Figure 6 is the narrow width of the absorption lines, which is a consequence of the fixed difference in energy between the ground state and the excited state, and the lack of vibrational and rotational energy levels. Natural line widths for atomic absorption, which are governed by the uncertainty principle, are approximately 10^{-5} nm. Other contributions to broadening increase this line width to approximately 10^{-3} nm.

13.3 TRANSMITTANCE AND ABSORBANCE

As light passes through a sample, its power decreases as some of it is absorbed. This attenuation of radiation is described quantitatively by two separate, but related terms: transmittance and absorbance. As shown in Figure 7 a, transmittance is the ratio of the source radiation's power as it exits the sample, P_T, to that incident on the sample, P_0.

$$T = \frac{P_T}{P_0} \qquad 1$$

Multiplying the transmittance by 100 gives the percent transmittance, $\%T$, which varies between 100% (no absorption) and 0% (complete absorption). All methods of detecting photons—including the human eye and modern photoelectric transducers—measure the transmittance of electromagnetic radiation.

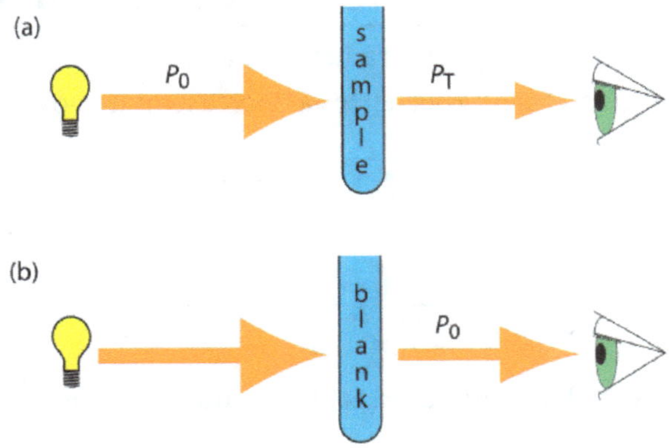

Fig. 7: (a) Schematic diagram showing the attenuation of radiation passing through a sample; P0 is the source's radiant power and PT is the radiant power transmitted by the sample. (b) Schematic diagram showing how we redefine P0 as the radiant power transmitted by the blank. Redefining P0 in this way corrects the transmittance in (a) for the loss of radiation due to scattering, reflection, absorption by the sample's container, and absorption by the sample's matrix.

Equation1 does not distinguish between different mechanisms that prevent a photon emitted by the source from reaching the detector. In addition to absorption by the analyte, several additional phenomena contribute to the attenuation of radiation, including reflection and absorption by the sample's container, absorption by other components in the sample's matrix, and the scattering of radiation. To compensate for this loss of the radiation's power, we use a method blank. As shown in Figure 7 b, we redefine P_0 as the power exiting the method blank.

An alternative method for expressing the attenuation of electromagnetic radiation is absorbance, A, which we define as

$$A = -\log T = -\log \frac{P_T}{P_0} \qquad 2$$

absorbance is the more common unit for expressing the attenuation of radiation because it is a linear function of the analyte's concentration.

Equation 1 has an important consequence for atomic absorption. As we learned from Figure 6, atomic absorption lines are very narrow. Even with a high quality monochromator, the effective bandwidth for a continuum source is 100 – 1000× greater than the width of an atomic absorption line. As a result, little radiation from a continuum source is absorbed when it passes through a sample of atoms; because $P_0 \approx P_T$ the measured absorbance effectively is zero. For this reason, atomic absorption requires that we use a line source instead of a continuum source.

13.4 ABSORBANCE AND CONCENTRATION: BEER'S LAW

When monochromatic electromagnetic radiation passes through an infinitesimally thin layer of sample of thickness *dx*, it experiences a decrease in its power of *dP* (Figure 8).

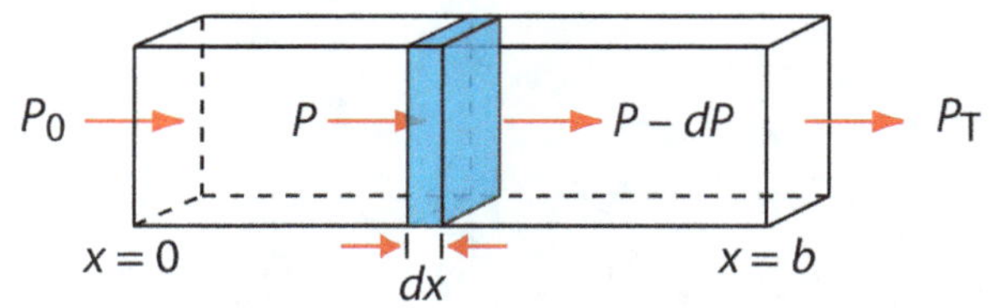

Fig. 8: Factors used to derive the Beer's law.

This fractional decrease in power is proportional to the sample's thickness and to the analyte's concentration, *C*; thus

$$-\frac{dP}{P} = \alpha C dx \qquad 3$$

where P is the power incident on the thin layer of sample and αα is a proportionality constant. Integrating the left side of Equation 3 over the sample's full thickness

$$-\int_{P_0}^{P_T} \frac{dP}{P} = \alpha C \int_0^b dx$$

$$\ln \frac{P_0}{P_T} = \alpha b C$$

converting from ln to log, and substituting into Equation 2, gives

$$A = abC \qquad 4$$

where a is the analyte's absorptivity with units of cm−1 conc−1. If we express the concentration using molarity, then we replace a with the molar absorptivity, εε, which has units of cm−1 M−1.

$$A = \varepsilon bC \qquad (5)$$

The absorptivity and the molar absorptivity are proportional to the probability that the analyte absorbs a photon of a given energy. As a result, values for both a and εε depend on the wavelength of the absorbed photon.

Equation 4 and Equation 5, which establish the linear relationship between absorbance and concentration, are known as Beer's law. Calibration curves based on Beer's law are common in quantitative analyses.

13.5 BEER'S LAW AND MULTICOMPONENT SAMPLES

We can extend Beer's law to a sample that contains several absorbing components. If there are no interactions between the components, then the individual absorbances, A_i, are additive. For a two-component mixture of analyte's *X* and *Y*, the total absorbance, A_{tot}, is

$$A_{tot}=A_X+A_Y=\varepsilon_X bC_X+\varepsilon_Y bC_Y$$

Generalizing, the absorbance for a mixture of *n* components, A_{mix}, is

$$A_{mix}=\sum_{i=1}^{n} A_i=\sum_{i=1}^{n}\varepsilon_i bC_i \qquad 6$$

13.6 LIMITATIONS TO BEER'S LAW

Beer's law suggests that a plot of absorbance vs. concentration—we will call this a Beer's law plot—is a straight line with a y-intercept of zero and a slope of ab or εbεb. In some cases a Beer's law plot deviates from this ideal behavior (see Figure 9), and such deviations from linearity are divided into three categories: fundamental, chemical, and instrumental.

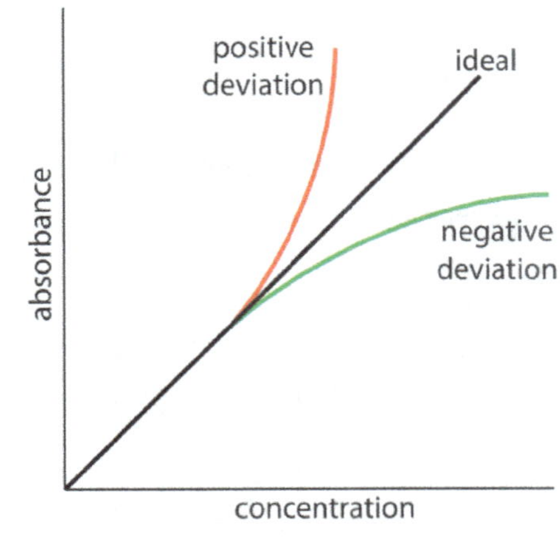

Fig. 9: Plots of absorbance vs. concentration showing positive and negative deviations from the ideal Beer's law relationship, which is a straight line.

13.6.1 Fundamental Limitations to Beer's Law

Beer's law is a limiting law that is valid only for low concentrations of analyte. There are two contributions to this fundamental limitation to Beer's law. At higher concentrations the individual particles of analyte no longer are independent of each other. The resulting interaction between particles of analyte may change the analyte's absorptivity. A second contribution is that an analyte's absorptivity depends on the solution's refractive index.

Because a solution's refractive index varies with the analyte's concentration, values of a and εε may change. For sufficiently low concentrations of analyte, the refractive index essentially is constant and a Beer's law plot is linear.

13.6.2 Chemical Limitations to Beer's Law

A chemical deviation from Beer's law may occur if the analyte is involved in an equilibrium reaction. Consider, for example, the weak acid, HA. To construct a Beer's law plot we prepare a series of standard solutions—each of which contains a known total concentration of HA—and then measure each solution's absorbance at the same wavelength. Because HA is a weak acid, it is in equilibrium with its conjugate weak base, A^-.

In the equations that follow, the conjugate weak base A– is written as A as it is easy to mistake the symbol for anionic charge as a minus sign.

$$HA(aq)+H_2O(l) \rightleftharpoons H_3O+(aq)+A^-(aq)$$

If both HA and A– absorb at the selected wavelength, then Beer's law is

$$A = \varepsilon_{HA} bC_{HA} + \varepsilon_A bC_A \qquad 7$$

because the weak acid's total concentration, C_{total}, is

$$C_{total} = C_{HA} + C_A$$

we can write the concentrations of HA and A^- as

$$C_{HA} = \alpha_{HA} C_{total} \qquad 8$$

$$CA=(1 - \alpha_{HA})C_{total} \qquad 9$$

where αHAαHA is the fraction of weak acid present as HA. Substituting Equation 8 and Equation 9 into Equation 7 and rearranging, gives

$$A=(\varepsilon_{HA}\alpha\ H_A+\varepsilon_A-\varepsilon_A\alpha_A)bC_{total} \qquad 10$$

To obtain a linear Beer's law plot, we must satisfy one of two conditions. If εHAεHA and εAεA have the same value at the selected wavelength, then Equation 10 simplifies to

$$A = \varepsilon_A bC_{total}$$

Alternatively, if αHAαHA has the same value for all standard solutions, then each term within the parentheses of Equation 10 is constant—which we replace with k—and a linear calibration curve is obtained at any wavelength.

$$A = kbC_{total}$$

Because HA is a weak acid, the value of αHA varies with pH. To hold αHA constant we buffer each standard solution to the same pH. Depending on the relative values of αHA and αA, the calibration curve has a positive or a negative deviation from Beer's law if we do not buffer the standards to the same pH.

13.6.3 Instrumental Limitations to Beer's Law

There are two principal instrumental limitations to Beer's law. The first limitation is that Beer's law assumes that radiation reaching the sample is of a single wavelength—that is, it assumes a purely monochromatic source of radiation. As shown in Figure 10, even the best wavelength selector passes radiation with a small, but finite effective bandwidth. Polychromatic radiation always gives a negative deviation from Beer's law, but the effect is smaller if the value of εε essentially is constant over the wavelength range passed by the wavelength selector. For this reason, as shown in Figure 10 , it is better to make absorbance measurements at the top of a broad absorption peak. In addition, the deviation from Beer's law is less serious if the source's effective bandwidth is less than one-tenth of the absorbing species' natural bandwidth [(a) Strong, F. C., III Anal. Chem. 1984, 56, 16A–34A; Gilbert, D. D. J. Chem. Educ. 1991, 68, A278–A281]. When measurements must be made on a slope, linearity is improved by using a narrower effective bandwidth.

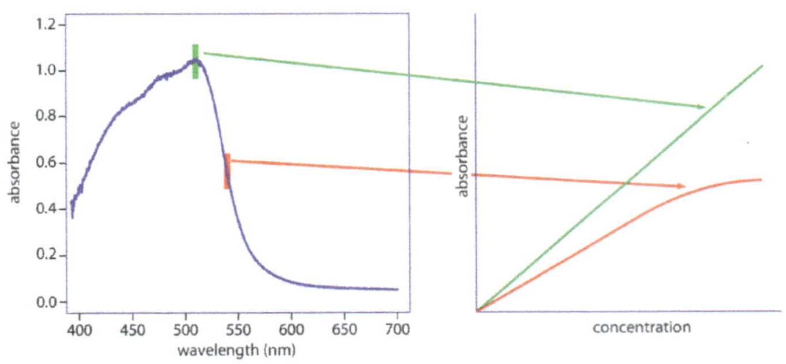

Fig. 10: Effect of wavelength selection on the linearity of a Beer's law plot. Another reason for measuring absorbance at the top of an absorbance peak is that it provides for a more sensitive analysis. Note that the green Beer's law plot has a steeper slope—and, therefore, a greater sensitivity—than the red Beer's law plot. A Beer's law plot, of course, is equivalent to a calibration curve.

Stray radiation is the second contribution to instrumental deviations from Beer's law. Stray radiation arises from imperfections in the wavelength selector that allow light to enter the instrument and to reach the detector without passing through the sample. Stray radiation adds an additional contribution, Pstray, to the radiant power that reaches the detector; thus

$$A = -\log \frac{P_T + P_{stray}}{P_0 + P_{stray}}$$

For a small concentration of analyte, Pstray is significantly smaller than P0 and PT, and the absorbance is unaffected by the stray radiation. For higher concentrations of analyte, less light passes through the sample and PT and Pstray become similar in magnitude. This results is an absorbance that is smaller than expected, and a negative deviation from Beer's law.

14

UV/VIS AND IR SPECTROSCOPY

14.1 INSTRUMENTATION

Frequently an analyst must select from among several instruments of different design, the one instrument best suited for a particular analysis. In this section we examine several different instruments for molecular absorption spectroscopy, with an emphasis on their advantages and limitations. Methods of sample introduction also are covered in this section.

14.1.1 Instrument Designs for Molecular UV/VIS Absorption

Filter Photometer: The simplest instrument for molecular UV/Vis absorption is a filter photometer (Figure 1), which uses an absorption or interference filter to isolate a band of radiation. The filter is placed between the source and the sample to prevent the sample from decomposing when exposed to higher energy radiation. A filter photometer has a single optical path between the source and detector, and is called a single-beam instrument. The instrument is calibrated to 0% T while using a shutter to block the source radiation from the detector. After opening the shutter, the instrument is calibrated to 100% T using an appropriate blank. The blank is then replaced with the sample and its transmittance measured. Because the source's incident power and the sensitivity of the detector vary with wavelength, the photometer is recalibrated whenever the filter is changed. Photometers have the advantage of being relatively inexpensive, rugged, and easy to maintain. Another advantage of a photometer is its portability, making it easy to take into the field. Disadvantages of a photometer include the inability to record an absorption spectrum and the source's relatively large effective bandwidth, which limits the calibration curve's linearity.

The percent transmittance varies between 0% and 100%. As we learned from Figure, we use a blank to determine P_0, which corresponds to 100%T. Even in the absence of light the detector records a signal. Closing the shutter allows us to assign 0%T to this signal. Together, setting 0% T and 100%T calibrates the instrument. The amount of light that passes through a sample produces a signal that is greater than or equal to 0%T and smaller than or equal to 100%T.

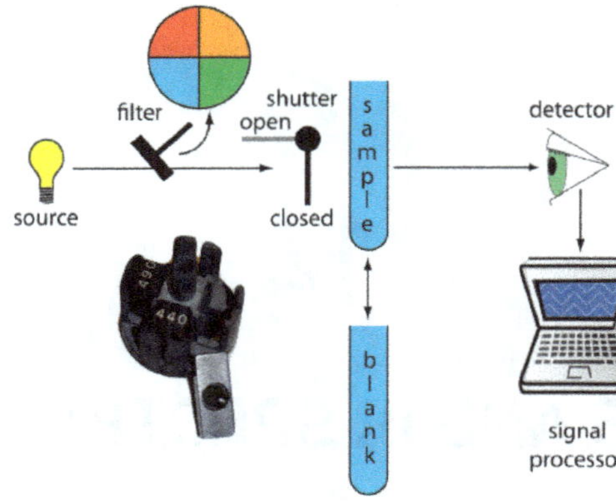

Fig. 1: Schematic diagram of a filter photometer. The analyst either inserts a removable filter or the filters are placed in a carousel, an example of which is shown in the photographic inset. The analyst selects a filter by rotating it into place.

14.1.2 Single-Beam Spectrophotometer.

An instrument that uses a monochromator for wavelength selection is called a spectrophotometer. The simplest spectrophotometer is a single-beam instrument equipped with a fixed-wavelength monochromator (Figure 2). Single-beam spectrophotometers are calibrated and used in the same manner as a photometer. One example of a single-beam spectrophotometer is Thermo Scientific's Spectronic 20D+, which is shown in the photographic insert to Figure 2 . The Spectronic 20D+ has a wavelength range of 340–625 nm (950 nm when using a red-sensitive detector), and a fixed effective bandwidth of 20 nm. Battery-operated, hand-held single-beam spectrophotometers are available, which are easy to transport into the field. Other single-beam spectrophotometers also are available with effective bandwidths of 2–8 nm. Fixed wavelength single-beam spectrophotometers are not practical for recording spectra because manually adjusting the wavelength and recalibrating the spectrophotometer is awkward and time-consuming. The accuracy of a single-beam spectrophotometer is limited by the stability of its source and detector over time.

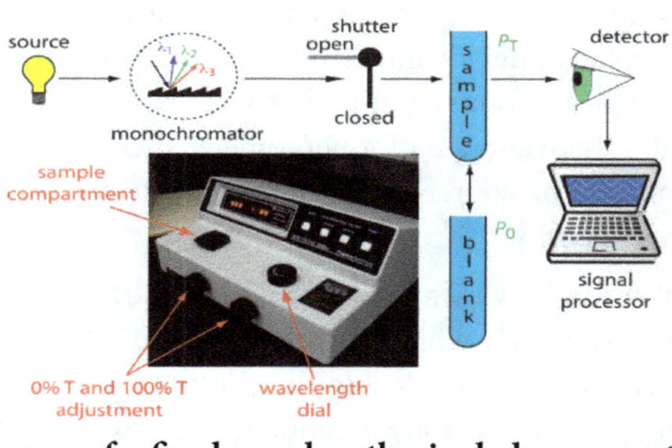

Fig. 2: Schematic diagram of a fixed-wavelength, single-beam spectrophotometer. The photographic inset shows a typical instrument. The shutter remains closed until the

sample or blank is placed in the sample compartment. The analyst manually selects the wavelength by adjusting the wavelength dial. Inset photo modified from.

14.1.3 Double-Beam Spectrophotometer.

The limitations of a fixed-wavelength, single-beam spectrophotometer is minimized by using a double-beam spectrophotometer (Figure 3). A chopper controls the radiation's path, alternating it between the sample, the blank, and a shutter. The signal processor uses the chopper's speed of rotation to resolve the signal that reaches the detector into the transmission of the blank, P0, and the sample, PT. By including an opaque surface as a shutter, it also is possible to continuously adjust 0%T. The effective bandwidth of a double-beam spectrophotometer is controlled by adjusting the monochromator's entrance and exit slits. Effective bandwidths of 0.2–3.0 nm are common. A scanning monochromator allows for the automated recording of spectra. Double-beam instruments are more versatile than single-beam instruments, being useful for both quantitative and qualitative analyse s, but also are more expensive and not particularly portable.

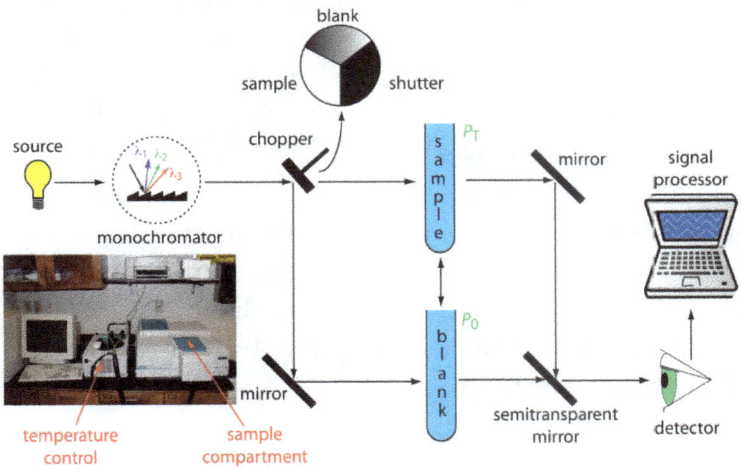

Fig. 3: Schematic diagram of a scanning, double-beam spectrophotometer. A chopper directs the source's radiation, using a transparent window to pass radiation to the sample and a mirror to reflect radiation to the blank. The chopper's opaque surface serves as a shutter, which allows for a constant adjustment of the spectrophotometer's 0%T. The photographic insert shows a typical instrument. The module in the middle of the photo is a temperature control unit that makes it possible to heat or cool the sample to a constant temperature.

14.1.4 Diode Array Spectrometer

An instrument with a single detector can monitor only one wavelength at a time. If we replace a single photomultiplier with an array of photodiodes, we can use the resulting detector to record a full spectrum in as little as 0.1 s. In a diode array spectrometer the source radiation passes through the sample and is dispersed by a grating. The photodiode array detector is situated at the grating's focal plane, with each diode recording the radiant power over a narrow range of wavelengths. Because we replace a full monochromatic with just a grating, a diode array spectrometer is small and compact.

One advantage of a diode array spectrometer is the speed of data acquisition, which allows us to collect multiple spectra for a single sample. Individual spectra are added and averaged to obtain the final spectrum. This signal averaging improves a spectrum's signal-to-noise ratio. If we add together n spectra, the sum of the signal at any point, x, increases as nSx, where Sx is the signal. The noise at any point, Nx, is a random event, which increases as n−√NxnNx when we add together n spectra. The signal-to-noise ratio after n scans, (S/N)n is:

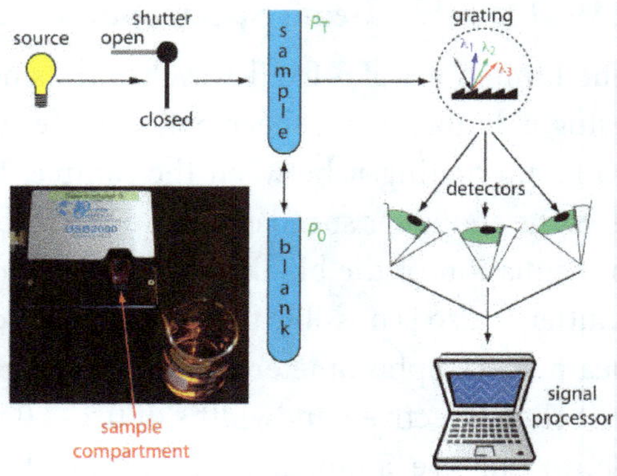

Fig. 5: Schematic diagram of a diode array spectrophotometer. The photographic insert shows a typical instrument. Note that the 50-mL beaker provides a sense of scale. Because the spectrometer is small and compact, it is easy to transport into the field.

$$\frac{S}{N} = \frac{nS_x}{\sqrt{n}N_x} = \sqrt{n}\frac{S_x}{N_x}$$

where S_x/N_x is the signal-to-noise ratio for a single scan. The impact of signal averaging is shown in Figure 10.3.5 . The first spectrum shows the signal after one scan, which consists of a single, noisy peak. Signal averaging using 4 scans and 16 scans decreases the noise and improves the signal-to-noise ratio. One disadvantage of a photodiode array is that the effective bandwidth per diode is roughly an order of magnitude larger than that for a high quality monochromator.

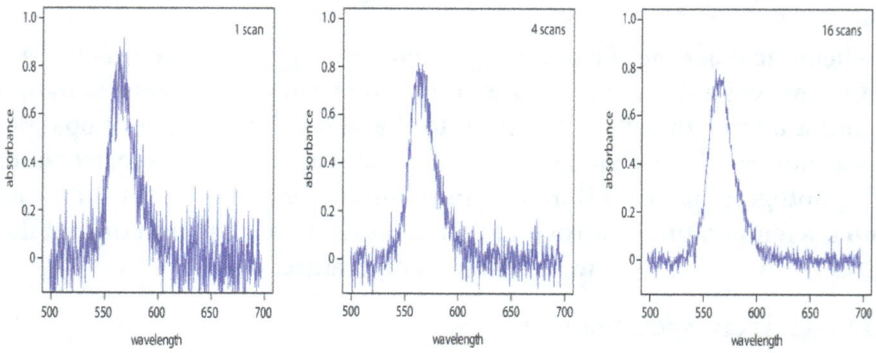

Fig. 5: Effect of signal averaging on a spectrum's signal-to-noise ratio. From top to bottom: spectrum for a single scan; average spectrum after four scans; and average spectrum after adding 16 scans.

For more details on signals and noise, see Introduction to Signals and Noise by Steven Petrovic, an on-line resource that is part of the Analytical Sciences Digital Library.

14.1.5 Sample Cells

The sample compartment provides a light-tight environment that limits stray radiation. Samples normally are in a liquid or solution state, and are placed in cells constructed with UV/Vis transparent materials, such as quartz, glass, and plastic (Figure 6). A quartz or fused-silica cell is required when working at a wavelength <300 nm where other materials show a significant absorption. The most common pathlength is 1 cm (10 mm), although cells with shorter (as little as 0.1 cm) and longer pathlengths (up to 10 cm) are available. Longer pathlength cells are useful when analyzing a very dilute solution or for gas samples. The highest quality cells allow the radiation to strike a flat surface at a 90o angle, minimizing the loss of radiation to reflection. A test tube often is used as a sample cell with simple, single-beam instruments, although differences in the cell's pathlength and optical properties add an additional source of error to the analysis.

Fig. 6: Examples of sample cells for UV/Vis spectroscopy. From left to right (with path lengths in parentheses): rectangular plastic cuvette (10.0 mm), rectangular quartz cuvette (5.000 mm), rectangular quartz cuvette (1.000 mm), cylindrical quartz cuvette (10.00 mm), cylindrical glass cuvette with quartz windows (100.0 mm). Cells often are available as a matched pair, which is important when using a double-beam instrument.

If we need to monitor an analyte's concentration over time, it may not be possible to remove samples for analysis. This often is the case, for example, when monitoring an industrial production line or waste line, when monitoring a patient's blood, or when monitoring an environmental system, such as stream. With a fiber-optic probe we can analyze samples in situ. An example of a remote sensing fiber-optic probe is shown in Figure 7 . The probe consists of two bundles of fiber-optic cable. One bundle transmits radiation from the source to the probe's tip, which is designed to allow the sample to flow through the sample cell. Radiation from the source passes through the solution and is reflected back by a mirror. The second bundle of fiber-optic cable transmits the nonabsorbed radiation to the wavelength selector. Another design replaces the flow cell shown in Figure 7 with a membrane that contains a reagent that reacts with the analyte. When the analyte diffuses into the membrane it reacts with the reagent, producing a product that absorbs UV or visible radiation. The nonabsorbed radiation from the source is reflected or scattered back to the detector. Fiber optic probes that show chemical selectivity are called optrodes [(a) Seitz, W. R. Anal. Chem. 1984, 56, 16A–34A; (b) Angel, S. M. Spectroscopy 1987, 2(2), 38–48].

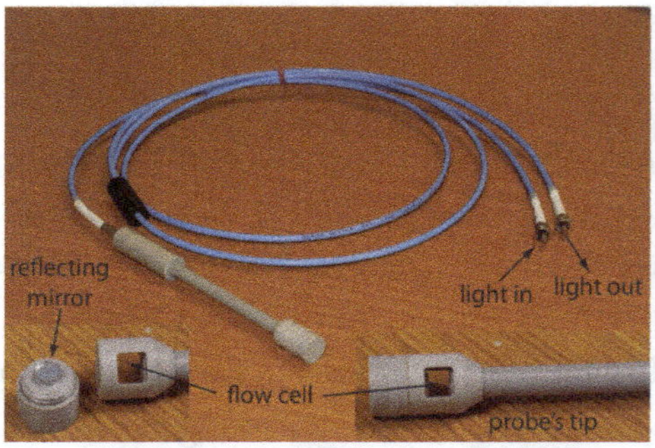

Figure 7. Example of a fiber-optic probe. The inset photographs at the bottom of the figure provide close-up views of the probe's flow cell and the reflecting mirror.

14.1.6 Instrument Designs for Infrared Adsorption

Filter Photometer

The simplest instrument for IR absorption spectroscopy is a filter photometer similar to that shown in Figure 1 for UV/Vis absorption. These instruments have the advantage of portability and typically are used as dedicated analyzers for gases such as HCN and CO.

Double-beam spectrophotometer

Infrared instruments using a monochromator for wavelength selection use double-beam optics similar to that shown in Figure 3 . Double-beam optics are preferred over single-beam optics because the sources and detectors for infrared radiation are less stable than those for UV/Vis radiation. In addition, it is easier to correct for the absorption of infrared radiation by atmospheric CO_2 and H_2O vapor when using double-beam optics. Resolutions of 1–3 cm^{-1} are typical for most instruments.

Fourier transform spectrometer

In a Fourier transform infrared spectrometer, or FT–IR, the monochromator is replaced with an interferometer (Figure 13). Because an FT-IR includes only a single optical path, it is necessary to collect a separate spectrum to compensate for the absorbance of atmospheric CO_2 and H_2O vapor. This is done by collecting a background spectrum without the sample and storing the result in the instrument's computer memory. The background spectrum is removed from the sample's spectrum by taking the ratio the two signals. In comparison to other instrument designs, an FT–IR provides for rapid data acquisition, which allows for an enhancement in signal-to-noise ratio through signal-averaging.

Sample Cells

Infrared spectroscopy routinely is used to analyze gas, liquid, and solid samples. Sample cells are made from materials, such as NaCl and KBr, that are transparent to infrared radiation. Gases are analyzed using a cell with a path length of approximately 10 cm. Longer path lengths are obtained by using mirrors to pass the beam of radiation through the sample several times.

A liquid sample may be analyzed using a variety of different sample cells (Figure 8). For non-volatile liquids a suitable sample is prepared by placing a drop of the liquid between two NaCl plates, forming a thin film that typically is less than 0.01 mm thick. Volatile liquids are placed in a sealed cell to prevent their evaporation.

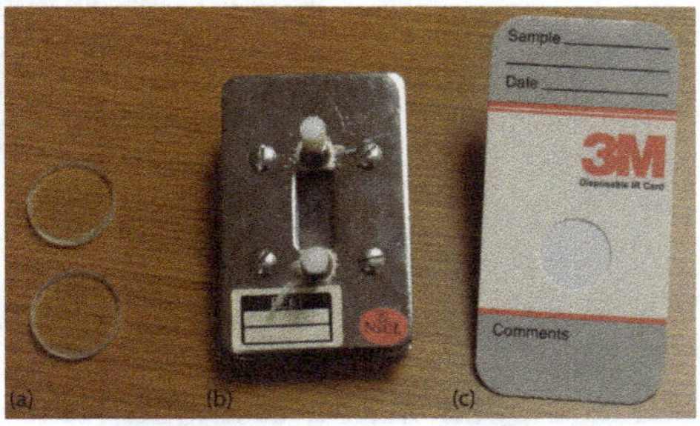

Fig. 8: Three examples of IR sample cells: (a) NaCl salts plates; (b) fixed pathlength (0.5 mm) sample cell with NaCl windows; (c) disposable card with a polyethylene window that is IR transparent with the exception of strong absorption bands at 2918 cm−1 and 2849 cm−1.

The analysis of solution samples is limited by the solvent's IR absorbing properties, with CCl4, CS2, and CHCl3 being the most common solvents. Solutions are placed in cells that contain two NaCl windows separated by a Teflon spacer. By changing the Teflon spacer, pathlengths from 0.015–1.0 mm are obtained.

Transparent solid samples are analyzed by placing them directly in the IR beam. Most solid samples, however, are opaque, and are first dispersed in a more transparent medium before recording the IR spectrum. If a suitable solvent is available, then the solid is analyzed by preparing a solution and analyzing as described above. When a suitable solvent is not available, solid samples are analyzed by preparing a mull of the finely powdered sample with a suitable oil. Alternatively, the powdered sample is mixed with KBr and pressed into an optically transparent pellet.

The analysis of an aqueous sample is complicated by the solubility of the NaCl cell window in water. One approach to obtaining an infrared spectrum of an aqueous solution is to use attenuated total reflectance instead of transmission. Figure 9 shows a diagram of a typical attenuated total reflectance (ATR) FT-IR instrument. The ATR cell consists of a high refractive index material, such as ZnSe or diamond, sandwiched between a low refractive index substrate and a lower refractive index sample. Radiation from the source enters the ATR crystal where it undergoes a series of internal reflections before exiting the crystal. During each reflection the radiation penetrates into the sample to a depth of a few microns, which results in a selective attenuation of the radiation at those wavelengths where the sample absorbs. ATR spectra are similar, but not identical, to those obtained by measuring the transmission of radiation.

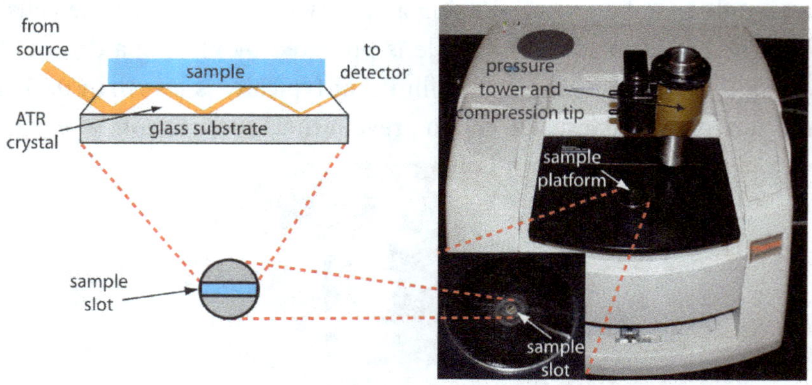

S

Fig. 9: FT-IR spectrometer equipped with a diamond ATR sample cell. The inserts show a close-up photo of the sample platform, a sketch of the ATR's sample slot, and a schematic showing how the source's radiation interacts with the sample. The pressure tower is used to ensure proper contact of a solid sample with the ATR crystal.

Solid samples also can be analyzed using an ATR sample cell. After placing the solid in the sample slot, a compression tip ensures that it is in contact with the ATR crystal. Examples of solids analyzed by ATR include polymers, fibers, fabrics, powders, and biological tissue samples. Another reflectance method is diffuse reflectance, in which radiation is reflected from a rough surface, such as a powder. Powdered samples are mixed with a non-absorbing material, such as powdered KBr, and the reflected light is collected and analyzed. As with ATR, the resulting spectrum is similar to that obtained by conventional transmission methods..

14.2 QUANTITATIVE APPLICATIONS

The determination of an analyte's concentration based on its absorption of ultraviolet or visible radiation is one of the most frequently encountered quantitative analytical methods. One reason for its popularity is that many organic and inorganic compounds have strong absorption bands in the UV/Vis region of the electromagnetic spectrum. In addition, if an analyte does not absorb UV/Vis radiation—or if its absorbance is too weak—we often can react it with another species that is strongly absorbing. For example, a dilute solution of Fe2+ does not absorb visible light. Reacting Fe2+ with o-phenanthroline, however, forms an orange-red complex of Fe(phen) 2+3Fe (phen) 32 + that has a strong, broad absorbance band near 500 nm. An additional advantage to UV/Vis absorption is that in most cases it is relatively easy to adjust experimental and instrumental conditions so that Beer's law is obeyed.

A quantitative analysis based on the absorption of infrared radiation, although important, is encountered less frequently than with UV/Vis absorption. One reason is the greater tendency for instrumental deviations from Beer's law when using infrared radiation. Because an infrared absorption band is relatively narrow, any deviation due to the lack of monochromatic radiation is more pronounced. In addition, infrared sources are less intense than UV/Vis sources, which makes stray radiation more of a problem. Differences between the pathlengths for samples and for standards when using thin liquid films or KBr pellets are a problem, although an internal standard can correct for any difference in pathlength. Finally, establishing a 100%T (A = 0) baseline often is difficult because the optical properties of NaCl sample cells may change significantly with wavelength due to contamination and degradation. We can minimize this problem by measuring absorbance relative to a baseline established for the absorption band. Figure 10.3.10 shows how this is accomplished.

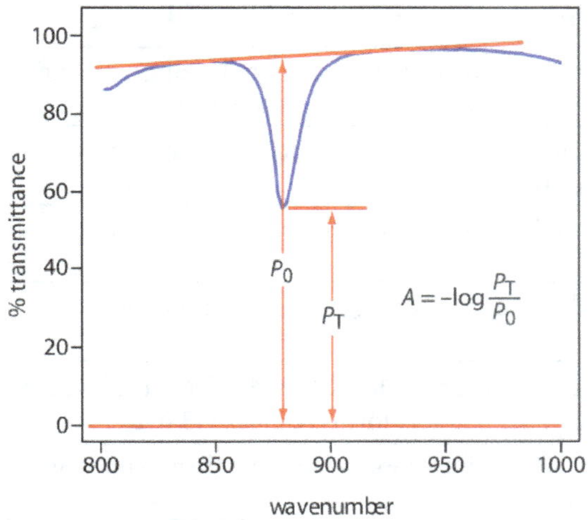

Fig. 10: Method for determining absorbance from an IR spectrum.

Another approach is to use a cell with a fixed pathlength, such as that shown in Figure 8 b.

14.2.1 Environmental Applications

The analysis of waters and wastewaters often relies on the absorption of ultraviolet and visible radiation. Many of these methods are outlined in Table 1 . Several of these methods are described here in more detail.

Table 1 . Examples of Molecular UV/Vis Analysis of Waters and Wastewaters

Analyte	Method	Λλ (nm)
Trace metals		
Aluminum	React with eriochrome cyanide r dye at ph 6; forms red to pink complex	535
Arsenic	Reduce to ash_3 using zn and react with silver diethyldithiocarbamate; forms red complex	535
Cadmium	Extract into $chcl_3$ containing dithizone from a sample made basic with naoh; forms pink to red complex	518
Chromium	Oxidize to cr(vi) and react with diphenylcarbazide; forms 540 red-violet product	540

Copper	React with neocuprine in neutral to slightly acid solution and extract into $chcl_3/ch_3oh$; forms yellow complex	457
Iron	Reduce to fe^{2+} and react with *o*-phenanthroline; forms orange-red complex	510
Lead	Extract into $chcl_3$ containing dithizone from sample made basic with nh_3/ nh_4^+ buffer; forms cherry red complex	510
Manganese	Oxidize to mno_4^- with persulfate; forms purple solution	525
Mercury	Extract into $chcl_3$ containing dithizone from acidic sample; forms orange complex	492
Zinc	React with zincon at ph 9; forms blue complex	620
Inorganic nonmetals		
Ammonia	Reaction with hypochlorite and phenol using a manganous 630 salt catalyst; forms blue indophenol as product	630
Cyanide	React with chloroamine-t to form cncl and then with a pyridine-barbituric acid; forms a red-blue dye	578
Fluoride	React with red zr-spadns lake; formation of zrf_6^{2-} decreases color of the red lake	570
Chlorine (residual)	React with leuco crystal violet; forms blue product	592
Nitrate	React with cd to form no_2^- and then react with sulfanilamide and *n*-(1-napthyl)-ethylenediamine; forms red azo 543 dye	543
Phosphate	React with ammonium molybdate and then reduce with $sncl_2$; forms molybdenum blue	690
Organics		
Phenol	React with 4-aminoantipyrine and $k_3fe(cn)_6$; forms yellow antipyrine dye	460
Anionic surfactants	React with cationic methylene blue dye and extract into $chcl_3$; forms blue ion pair	652

Although the quantitative analysis of metals in waters and wastewaters is accomplished primarily by atomic absorption or atomic emission spectroscopy, many metals also can be analyzed following the formation of a colored metal–ligand complex. One advantage to these spectroscopic methods is that they easily are adapted to the analysis of samples in the field using a filter photometer. One ligand used for the analysis of several metals is diphenylthiocarbazone, also known as dithizone. Dithizone is not soluble in water, but when a solution of dithizone in $CHCl_3$ is shaken with an aqueous solution that contains an appropriate metal ion, a colored metal–dithizonate complex forms that is soluble in $CHCl_3$. The selectivity of dithizone is controlled by adjusting the sample's pH. For example, Cd^{2+} is extracted from solutions made strongly basic with NaOH, Pb^{2+} from solutions made basic with an NH_3/ NH^{4+} buffer, and Hg^{2+} from solutions that are slightly acidic.

The structure of dithizone is shown below.

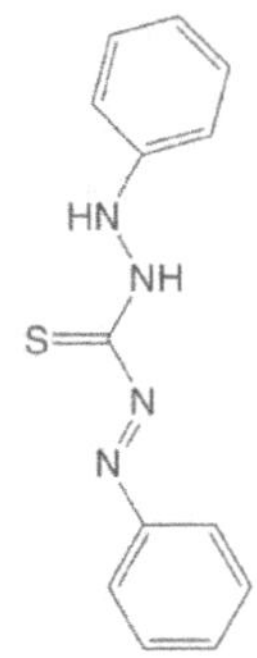

When chlorine is added to water the portion available for disinfection is called the chlorine residual. There are two forms of chlorine residual. The free chlorine residual includes Cl_2, HOCl, and OCl^-. The combined chlorine residual, which forms from the reaction of NH_3 with HOCl, consists of monochloramine, NH_2Cl, dichloramine, $NHCl_2$, and trichloramine, NCl_3. Because the free chlorine residual is more efficient as a disinfectant, there is an interest in methods that can distinguish between the total chlorine residual's different forms. One such method is the leuco crystal violet method. The free residual chlorine is determined by adding leuco crystal violet to the sample, which instantaneously oxidizes to give a blue-colored compound that is monitored at 592 nm. Completing the analysis in less than five minutes prevents a possible interference from the combined chlorine residual. The total chlorine residual (free + combined) is determined by reacting a separate sample with iodide, which reacts with both chlorine residuals to form HOI. When the reaction is complete, leuco crystal violet is added and oxidized by HOI, giving the same blue-colored product. The combined chlorine residual is determined by difference.

The concentration of fluoride in drinking water is determined indirectly by its ability to form a complex with zirconium. In the presence of the dye SPADNS, a solution of zirconium forms a red colored compound, called a lake, that absorbs at 570 nm. When fluoride is added, the formation of the stable $ZrF2^{-6}ZrF6^{2-}$ complex causes a portion of the lake to dissociate, decreasing the absorbance. A plot of absorbance versus the concentration of fluoride, therefore, has a negative slope.

SPADNS, the structure of which is shown below, is an abbreviation for the sodium salt of 2-(4-sulfophenylazo)-1,8-dihydroxy-3,6-napthalenedisulfonic acid, which is a mouthful to say.

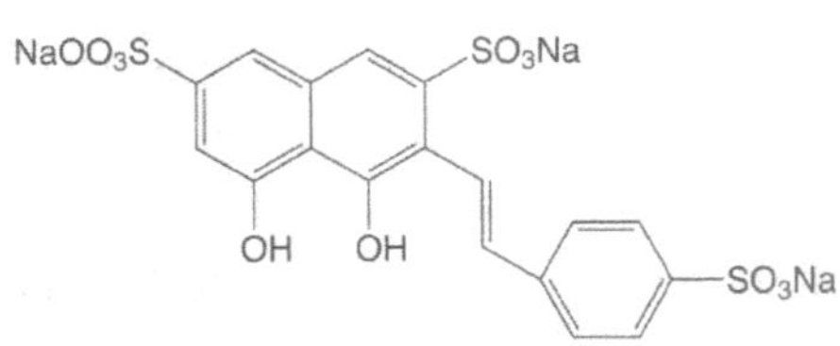

Spectroscopic methods also are used to determine organic constituents in water. For example, the combined concentrations of phenol and ortho- and meta-substituted phenols are determined by using steam distillation to separate the phenols from nonvolatile impurities. The distillate reacts with 4-aminoantipyrine at pH 7.9 ± 0.1 in the presence of $K_3Fe(CN)_6$ to a yellow colored antipyrine dye. After extracting the dye into $CHCl_3$, its absorbance is monitored at 460 nm. A calibration curve is prepared using only the unsubstituted phenol, C_6H5OH. Because the molar absorptivity of substituted phenols generally are less than that for phenol, the reported concentration represents the minimum concentration of phenolic compounds.

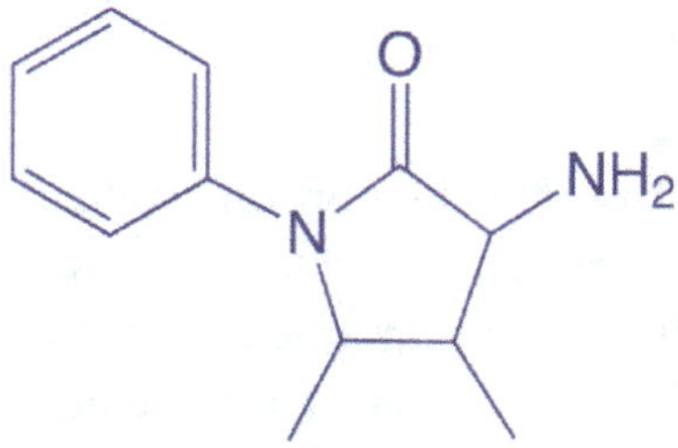

4-aminoantipyrene

Molecular absorption also is used for the analysis of environmentally significant airborne pollutants. In many cases the analysis is carried out by collecting the sample in water, converting the analyte to an aqueous form that can be analyzed by methods such as those described in Table 1 . For example, the concentration of NO_2 is determined by oxidizing NO_2 to NO^{-3}NO3−. The concentration of NO−3NO3− is then determined by first reducing it to NO^{-2}NO2− with Cd, and then reacting NO^{-2}NO2− with sulfanilamide and N-(1-naphthyl)-ethylenediamine to form a red azo dye. Another important application is the analysis for SO_2, which is determined by collecting the sample in an aqueous solution of $HgCl_2^{-4}$HgCl42− where it reacts to form $Hg(SO_3)_2^{-2}$$Hg(SO_3)_2$2−. Addition of pros-aniline and formaldehyde produces a purple complex that is monitored at 569 nm. Infrared absorption is useful for the analysis of organic vapors, including HCN, SO_2, nitrobenzene, methyl mercaptan, and vinyl chloride. Frequently, these analyses are accomplished using portable, dedicated infrared photometers.

14.2.2 Clinical Applications

The analysis of clinical samples often is complicated by the complexity of the sample's matrix, which may contribute a significant background absorption at the desired wavelength. The determination of serum barbiturates provides one example of how this problem is overcome. The barbiturates are first extracted from a sample of serum with $CHCl_3$ and then extracted from the $CHCl_3$ into 0.45 M NaOH (pH ≈ 13). The absorbance of the aqueous extract is measured at 260 nm, and includes contributions from the barbiturates as well as other components extracted from the serum sample. The pH of the sample is then lowered to approximately 10 by adding NH4Cl and the absorbance remeasured. Because the barbiturates do not absorb at this pH, we can use the absorbance at pH 10, ApH 10, to correct the absor-ance at pH 13, A_{pH} 13

A_{barb}= ApH 13−V_{samp}+V_{NH4Cl} V_{samp}×ApH) • A_{barb} = ApH 13−V_{samp}+V_{NH4Cl} V_{samp}×ApH) •

where Abarb is the absorbance due to the serum barbiturates and V_{samp} and VNH_4ClVNH_4Cl are the volumes of sample and NH_4Cl, respectively. Table 2 provides a summary of several other methods for analyzing clinical samples.

Table 2 . Examples of the Molecular UV/Vis Analysis of Clinical Samples

Analyte	Method	Λλ (nm)
Total serum protein	React with naoh and cu^{2+}; forms blue-violet complex	540
Serum cholesterol	React with fe^{3+} in presence of isopropanol, acetic acid, and h_2so_4; forms blue-violet complex	540
Uric acid	React with phosphotungstic acid; forms tungsten blue	710
Serum barbituates	Extract into $chcl_3$ to isolate from interferents and then extract into 0.45 M naoh	260
Glucose	React with *o*-toludine at 100°c; forms blue-green complex	630
Protein-bound iodine	Decompose protein to release iodide, which catalyzes redox reaction between ce^{3+} and as^{3+}; forms yellow colored ce^{4+}	420

14.2.3 Industrial Applications

UV/Vis molecular absorption is used for the analysis of a diverse array of industrial samples including pharmaceuticals, food, paint, glass, and metals. In many cases the methods are similar to those described in Table 1 and in Table 2 . For example, the amount of iron in food is determined by bringing the iron into solution and analyzing using the o-phenanthroline method listed in Table 1 .

Many pharmaceutical compounds contain chromophores that make them suitable for analysis by UV/Vis absorption. Products analyzed in this fashion include antibiotics, hormones, vitamins, and analgesics. One example of the use of UV absorption is in determining the purity of aspirin tablets, for which the active ingredient is acetylsalicylic acid. Salicylic acid, which is produced by the hydrolysis of acetylsalicylic acid, is an undesirable impurity in aspirin tablets, and should not be present at more than 0.01% w/w. Samples are screened for unacceptable levels of salicylic acid by monitoring the absorbance at a wavelength of 312 nm. Acetylsalicylic acid absorbs at 280 nm, but absorbs poorly at 312 nm. Conditions for preparing the sample are chosen such that an absorbance of greater than 0.02 signifies an unacceptable level of salicylic acid.

14.2.4 Forensic Applications

UV/Vis molecular absorption routinely is used for the analysis of narcotics and for drug testing. One interesting forensic application is the determination of blood alcohol using the Breathalyzer test. In this test a 52.5-mL breath sample is bubbled through an acidified solution of $K_2Cr_2O_7$, which oxidizes ethanol to acetic acid. The concentration of ethanol in the breath sample is determined by a decrease in the absorbance at 440 nm where the dichromate ion absorbs. A blood alcohol content of 0.10%, which is above the legal limit, corresponds to 0.025 mg of ethanol in the breath sample.

14.2.5 Developing a Quantitative Method for a Single Component

To develop a quantitative analytical method, the conditions under which Beer's law is obeyed must be established. First, the most appropriate wavelength for the analysis is determined from an absorption spectrum. In most cases the best wavelength corresponds to an absorption maximum because it provides greater sensitivity and is less susceptible to instrumental limitations. Second, if the instrument has adjustable slits, then an appropriate slit width is chosen. The absorption spectrum also aids in selecting a slit width by choosing a width that is narrow enough to avoid instrumental limita- tions to Beer's law, but wide enough to increase the throughput of source radiation. Finally, a calibration curve is constructed to determine the range of concentrations for which Beer's law is valid. Additional considerations that are important in any quantitative method are the effect of potential interferents and establishing an appropriate blank.

Representative Method 1: Determination of Iron in Water and Wastewater

The best way to appreciate the theoretical and the practical details discussed in this section is to carefully examine a typical analytical method. Although each method is unique, the following description of the determination of iron in water and waste- water provides an instructive example of a typical procedure.

14.2.6 Description of Method

Iron in the +2 oxidation state reacts with o-phenanthroline to form the orange-red $Fe(phen)_2^{+3}Fe(phen)_3^{2+}$ complex. The intensity of the complex's color is independent of the solution's acidity between a pH of 3 and 9. Because the complex forms more rapidly at lower pH levels, the reaction usually is carried out within a pH range of 3.0–3.5. Any iron present in the +3 oxidation state is reduced with hydroxylamine before adding o-phenanthroline. The most important interferents are strong oxidizing agents, polyphosphates, and metal ions such as Cu^{2+}, Zn2+, Ni^{2+}, and Cd^{2+}. An interference from oxidizing agents is minimized by adding an excess of hydroxylamine, and an interference from polyphosphate is minimized by boiling the sample in the presence of acid. The absorbance of samples and standards are measured at a wavelength of 510 nm using a 1-cm cell (longer pathlength cells also may be used). Beer's law is obeyed for concentrations of within the range of 0.2–4.0 mg Fe/L.

Procedure

For a sample that contains less than 2 mg Fe/L, directly transfer a 50-mL portion to a 125-mL Erlenmeyer flask. Samples that contain more than 2 mg Fe/L are diluted before acquiring the 50-mL portion. Add 2 mL of concentrated HCl and 1 mL of hydroxylamine to the sample. Bring the solution to a boil and continue boiling until the solution's volume is reduced to between 15 and 20 mL. After cooling to room temperature, transfer the solution to a 50-mL volumetric flask, add 10 mL of an ammonium acetate buffer, 2 mL of a 1000 ppm solution of o-phenanthroline, and dilute to volume. Allow 10–15 minutes for color development before measuring the absorbance, using distilled water to set 100% T. Calibration standards, including a blank, are prepared by the same procedure using a stock solution that contains a known concentration of Fe^{2+}.

14.2.7 Quantitative Analysis for a Single Sample

To determine the concentration of an analyte we measure its absorbance and apply Beer's law using any of the standardization methods described in Chapter 5. The most common methods are a normal calibration curve using external standards and the method of standard additions. A single point standardization also is possible, although we must first verify that Beer's law holds for the concentration of analyte in the samples and the standard.

14.2.8 Quantitative Analysis of Mixtures

Suppose we need to determine the concentration of two analytes, *X* and *Y*, in a sample. If each analyte has a wavelength where the other analyte does not absorb, then we can proceed using the approach in Example 10.3.5 . Unfortunately, UV/Vis absorption bands are so broad that frequently it is not possible to find suitable wavelengths. Because Beer's law is additive the mixture's absorbance, A_{mix}, is

$$(A_{mix})_{\lambda} = (\varepsilon_x)_{\lambda I} bC_X + (\varepsilon_Y)_{\lambda I} bC_Y \qquad (1)$$

where λ1 is the wavelength at which we measure the absorbance. Because Equation 1 includes terms for the concentration of both X and Y, the absorbance at one wavelength does not provide enough information to determine either C_X or C_Y. If we measure the absorbance at a second wavelength

$$(A_{mix})_{\lambda\Upsilon} = (\varepsilon_x)_{\lambda\Upsilon} bC_X + (\varepsilon_Y)_{\lambda\Upsilon} bC_Y \qquad (2)$$

then we can determine C_X and C_Y by solving simultaneously Equation 1 and Equation 2. Of course, we also must determine the value for εXεX and εYεY at each wavelength. For a mixture of n components, we must measure the absorbance at n different wavelengths.

When the analyte's spectra overlap severely, such that εX≈εYεX≈εY at all wavelengths, other computational methods may provide better accuracy and precision. In a multiwavelength linear regression analysis, for example, a mixture's absorbance is compared to that for a set of standard solutions at several wavelengths [Blanco, M.; Iturriaga, H.; Maspoch, S.; Tarin, P. J. Chem. Educ. 1989, 66, 178–180]. If ASX and ASY are the absorbance values for standard solutions of components X and Y at any wavelength, then

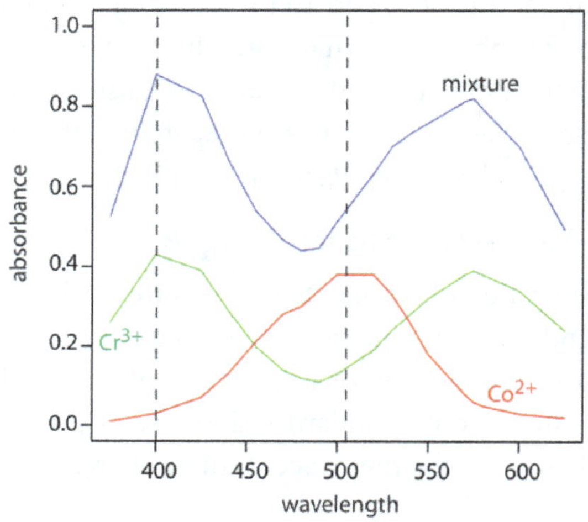

Fig. 11: Visible absorption spectra for 0.0250 M Cr^{3+}, 0.0750 M Co^{2+}, and for a mixture of Cr^{3+} and Co^{2+}. The two wavelengths used to analyze the mixture of Cr^{3+} and Co^{2+} are shown by the dashed lines.

$$A_{SX}=\varepsilon_X b C_{SX} \tag{3}$$

$$A_{SY}=\varepsilon_Y b C_{SY} \tag{4}$$

where C_{SX} and C_{SY} are the known concentrations of X and Y in the standard solutions. Solving Equation 3 and Equation 4 for ε_X and for ε_Y, substituting into Equation 1, and rearranging, gives

$$\frac{A_{mix}}{A_{SX}} = \frac{C_X}{C_{SX}} + \frac{C_Y}{C_{SY}} \times \frac{A_{SY}}{A_{SX}}$$

To determine C_X and C_Y the mixture's absorbance and the absorbances of the standard solutions are measured at several wavelengths. Graphing A_{mix}/A_{SX} versus A_{SY}/A_{SX} gives a straight line with a slope of C_Y/C_{SY} and a *y*-intercept of C_X/C_{SX}. This approach is particularly helpful when it is not possible to find wavelengths where $\varepsilon_X > \varepsilon_Y \varepsilon_X > \varepsilon_Y$ and $\varepsilon_X < \varepsilon_Y \varepsilon_X < \varepsilon_Y$.

The approach outlined here for a multiwavelength linear regression uses a single standard solution for each analyte. A more rigorous approach uses multiple standards for each analyte. The math behind the analysis of such data—which we call a multiple linear regression—is beyond the level of this text. For more details about multiple linear regression see Brereton, R. G. Chemometrics: Data Analysis for the Laboratory and Chemical Plant, Wiley: Chichester, England, 2003.

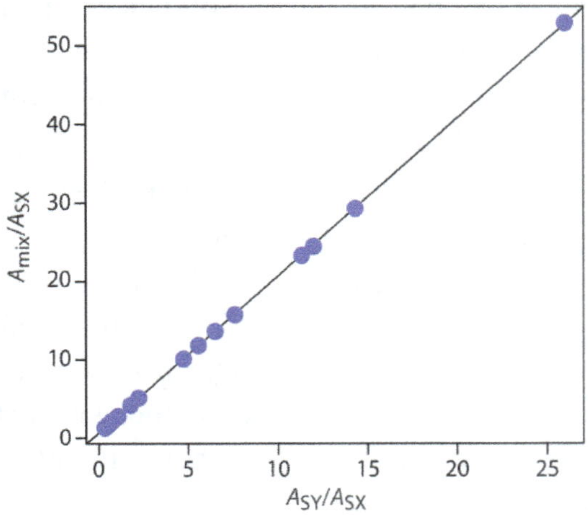

Fig. 12: Multiwavelength linear regression analysis for the data .

14.3 QUALITATIVE APPLICATIONS

As discussed ultraviolet, visible, and infrared absorption bands result from the absorption of electromagnetic radiation by specific valence electrons or bonds. The energy at which the absorption occurs, and the intensity of that absorption, is determined by the chemical environment of the absorbing moiety. For example, benzene has several ultraviolet absorption bands due to п→п*п→п* transitions. The position and intensity of two of these bands, 203.5 nm (εε = 7400 M^{-1} cm^{-1}) and 254 nm (εε = 204 M^{-1} cm^{-1}), are sensitive to substitution. For benzoic acid, in which a carboxylic acid group replaces one of the aromatic hydrogens, the two bands shift to 230 nm (εε = 11600 M^{-1} cm^{-1}) and 273 nm (εε = 970 M^{-1} cm^{-1}). A variety of rules have been developed to aid in correlating UV/Vis absorption bands to chemical structure. Similar correlations are available for infrared absorption bands. For example a carbonyl's C=O stretch is sensitive to adjacent functional groups, appearing at 1650 cm^{-1} for acids, 1700 cm^{-1} for ketones, and 1800 cm^{-1} for acid chlorides. The interpretation of UV/ Vis and IR spectra receives adequate coverage elsewhere in the chemistry curriculum, notably in organic chemistry, and is not considered further in this text.

With the availability of computerized data acquisition and storage it is possible to build digital libraries of standard reference spectra. The identity of an a unknown compound often can be determined by comparing its spectrum against a library of reference spectra, a process known as spectral searching. Comparisons are made using an algorithm that calculates the cumulative difference between the sample's spectrum and a reference spectrum. For example, one simple algorithm uses the following equation

$$D = \sum_{i=1}^{n} \left| (A_{sample})_i - (A_{reference})_i \right|$$

where D is the cumulative difference, A_{sample} is the sample's absorbance at wavelength or wavenumber i, $A_{reference}$ is the absorbance of the reference compound at the same wavelength or wavenumber, and *n* is the number of digitized points in the spectra. The cumulative difference is calculated for each reference spectrum. The reference compound with the smallest value of *D* is the closest match to the unknown compound. The accuracy of spectral searching is limited by the number and type of compounds included in the library, and by the effect of the sample's matrix on the spectrum.

Another advantage of computerized data acquisition is the ability to subtract one spectrum from another. When coupled with spectral searching it is possible to determine the identity of several components in a sample without the need of a prior separation step by repeatedly searching and sub- tracting reference spectra. An example is shown in Figure 13 in which the composition of a two-component mixture is determined by successive searching and subtraction. Figure 13 a shows the spectrum of the mixture. A search of the spectral library selects cocaine

- HCl (Figure 13 b) as a likely component of the mixture. Subtracting the reference spectrum for cocaine
- HCl from the mixture's spectrum leaves a result (Figure 13 c) that closely matches mannitol's reference spectrum (Figure 13 d). Subtracting the reference spectrum for mannitol leaves a small residual signal (Figure 13 e).

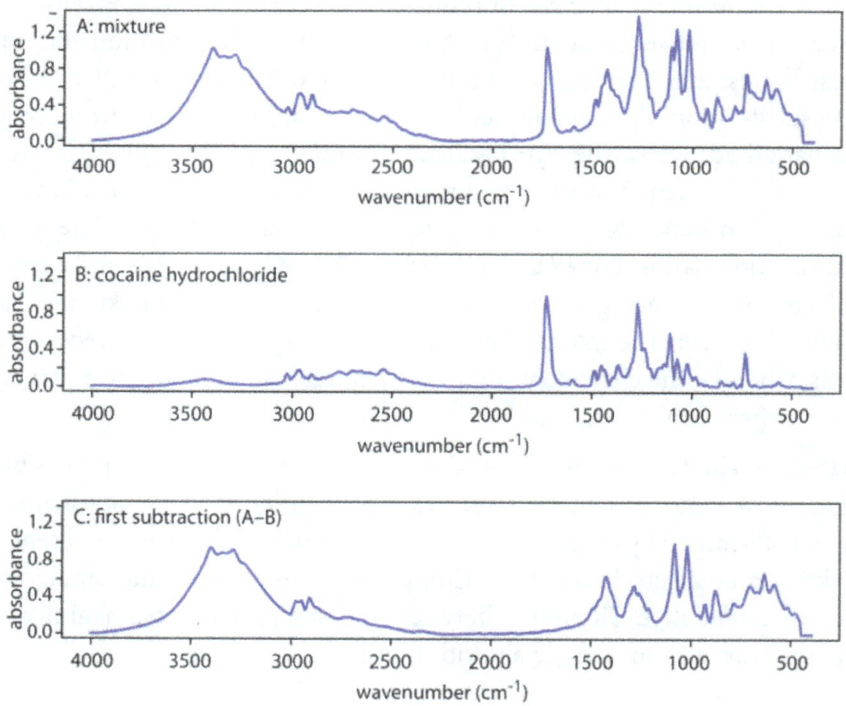

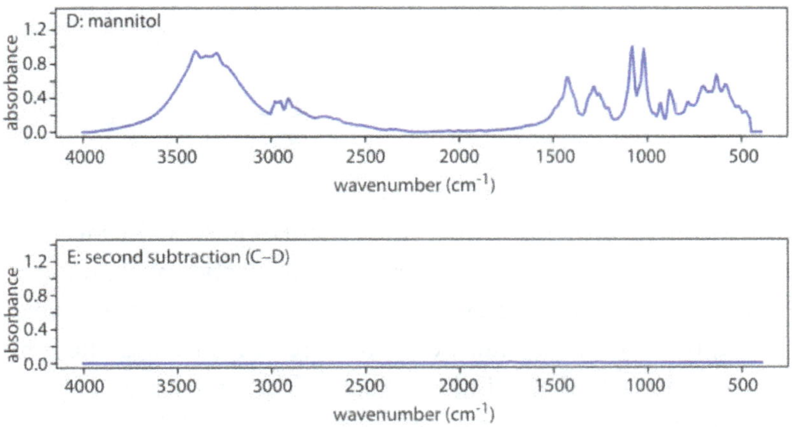

Fig. 13: Identifying the components of a mixture by spectral searching and subtracting. (a) IR spectrum of the mixture; (b) Reference IR spectrum of cocaine HCl; (c) Result of subtracting the spectrum of cocaine HCl from the mixture's spectrum; (d) Reference IR spectrum of mannitol; and (e) The residual spectrum after removing mannitol's contribution to the mixture's spectrum. IR spectra traditionally are displayed using percent transmittance, %T, along the y-axis (for example, see Figure 2). Because absorbance—not percent transmittance—is a linear function of concentration, spectral searching and spectral subtraction, is easier to do when displaying absorbance on the y-axis.

14.4 CHARACTERIZATION APPLICATIONS

Molecular absorption, particularly in the UV/Vis range, has been used for a variety of different characterization studies, including determining the stoichiometry of metal–ligand complexes and determining equilibrium constants. Both of these examples are examined in this section.

14.4.1 Stoichiometry of a Metal-Ligand Complex

We can determine the stoichiometry of the metal–ligand complexation reaction

$$M + yL \rightleftharpoons MLy$$

using one of three methods: the method of continuous variations, the mole-ratio method, and the slope-ratio method. Of these approaches, the *method of continuous variations*, also called Job's method, is the most popular. In this method a series of solutions is prepared such that the total moles of metal and of ligand, n_{total}, in each solution is the same. If $(n_M)_i$ and $(n_L)_i$ are, respectively, the moles of metal and ligand in solution i, then

$$n_{total} = (n_M)_i + (n_L)_i$$

The relative amount of ligand and metal in each solution is expressed as the mole fraction of ligand, $(X_L)i$, and the mole fraction of metal, $(X_M)_i$,

$$(X_L)_i = \frac{(n_L)_i}{n_{total}}$$

$$(X_M)_i = 1 - \frac{(n_L)_i}{n_{total}} = \frac{(n_M)_i}{n_{total}}$$

The concentration of the metal–ligand complex in any solution is determined by the limiting reagent, with the greatest concentration occurring when the metal and the ligand are mixed stoichiometrically. If we monitor the complexation reaction at a wavelength where only the metal–ligand complex absorbs, a graph of absorbance versus the mole fraction of ligand has two linear branches—one when the ligand is the limiting reagent and a second when the metal is the limiting reagent. The intersection of the two branches represents a stoichiometric mixing of the metal and the ligand. We use the mole fraction of ligand at the intersection to determine the value of y for the metal–ligand complex ML_y.

$$y = \frac{n_L}{n_M} = \frac{X_L}{X_M} = \frac{X_L}{1 - X_L}$$

You also can plot the data as absorbance versus the mole fraction of metal. In this case, y is equal to $(1 - X_M)/X_M$.

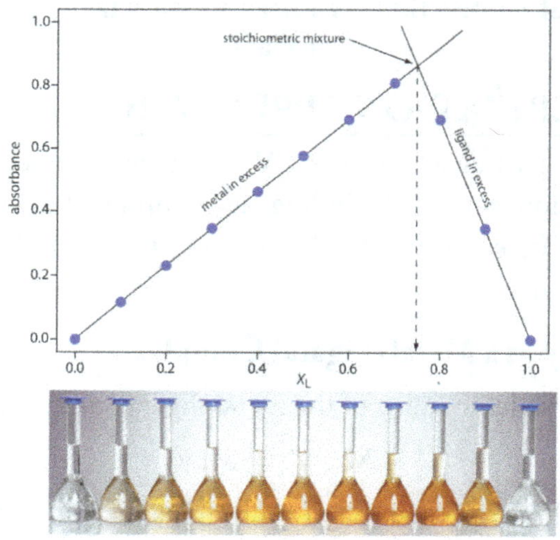

Fig. 14: Continuous variations plot for Example 4 . The photo shows the solutions used to gather the data. Each solution is displayed directly below its coresponding point on the continuous variations plot. To prepare these solutions I first prepared a solution of 3.15 ×10–4 M Fe^{2+} and a solution of 3.15×10–4 M o-phenanthroline. Because the two stock solutions have the same concentration, diluting a portion of one solution with the other solution gives a mixture in which the combined concentration of o-phenanthroline and Fe^{2+} is 3.15×10^{-4} M. Because each solution has the same volume, each solution also contains the same total moles of metal and ligand.

Several precautions are necessary when using the method of continuous variations. First, the metal and the ligand must form only one metal–ligand complex. To determine if this condition is true, plots of absorbance versus X_L are constructed at several different wavelengths and for several different values of n_{total}. If the maximum absorbance does not occur at the same value of X_L for each set of conditions, then more than one metal–ligand complex is present. A second precaution is that the metal–ligand complex's absorbance must obey Beer's law. Third, if the metal–ligand complex's formation constant is relatively small, a plot of absorbance versus X_L may show significant curvature. In this case it often is difficult to determine the stoichiometry by extrapolation. Finally, because the stability of a metal–ligand complex may be influenced by solution conditions, it is necessary to control carefully the composition of the solutions. When the ligand is a weak base, for example, each solutions must be buffered to the same pH.

In the mole-ratio method the moles of one reactant, usually the metal, is held constant, while the moles of the other reactant is varied. The absorbance is monitored at a wavelength where the metal–ligand complex absorbs. A plot of absorbance as a function of the ligand-to-metal mole ratio, nL/n_M, has two linear branches that intersect at a mole–ratio corresponding to the complex's formula. Figure 15 a shows a mole-ratio plot for the formation of a 1:1 complex in which the absorbance is monitored at a wavelength where only the complex absorbs. Figure 15 b shows a mole-ratio plot for a 1:2 complex in which all three species—the metal, the ligand, and the complex—absorb at the selected wavelength. Unlike the method of continuous variations, the mole-ratio method can be used for complexation reactions that occur in a stepwise fashion if there is a difference in the molar absorptivities of the metal–ligand complexes, and if the formation constants are sufficiently different. A typical mole-ratio plot for the step-wise formation of ML and ML_2 is shown in Figure 15 c.

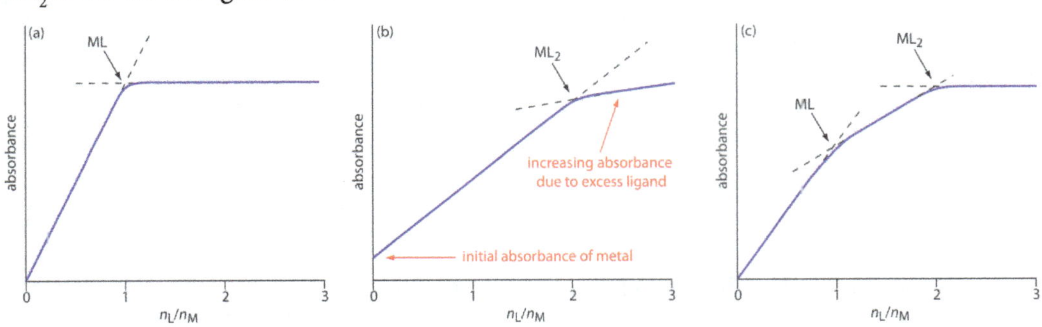

Fig. 15: Mole-ratio plots for: (a) a 1:1 metal–ligand complex in which only the complex absorbs; (b) a 1:2 metal–ligand complex in which the metal, the ligand, and the complex absorb; and (c) the stepwise formation of a 1:1 and a 1:2 metal–ligand complex.

For both the method of continuous variations and the mole-ratio method, we determine the complex's stoichiometry by extrapolating absorbance data from conditions in which there is a linear relationship between absorbance and the relative amounts of metal and ligand. If a metal–ligand complex is very weak, a plot of absorbance versus XL or nL/n_M becomes so curved that it is impossible to determine the stoichiometry by extrapolation. In this case the slope-ratio is used.

In the slope-ratio method two sets of solutions are prepared. The first set of solutions contains a constant amount of metal and a variable amount of ligand, chosen such that the total concentration of metal, C_M, is much larger than the total concentration of ligand, C_L. Under these conditions we may assume that essentially all the ligand reacts to form the metal–ligand complex. The concentration of the complex, which has the general form MxLy, is:

$$[M_xL_y] = \frac{C_L}{y}$$

If we monitor the absorbance at a wavelength where only M_xL_y absorbs, then

$$A = \varepsilon b[M_xL_y] = \frac{\varepsilon b C_L}{y}$$

and a plot of absorbance versus C_L is linear with a slope, s_L, of

$$s_L = \frac{\varepsilon b}{y}$$

A second set of solutions is prepared with a fixed concentration of ligand that is much greater than a variable concentration of metal; thus

$$[M_xL_y] = \frac{C_M}{x}$$

$$A = \varepsilon b[M_xL_y] = \frac{\varepsilon b C_M}{x}$$

$$s_M = \frac{\varepsilon b}{x}$$

A ratio of the slopes provides the relative values of x and y.

$$\frac{s_M}{s_L} = \frac{\varepsilon b / x}{\varepsilon b / y} = \frac{y}{x}$$

An important assumption in the slope-ratio method is that the complexation reaction continues to completion in the presence of a sufficiently large excess of metal or ligand. The slope-ratio method also is limited to systems in which only a single complex forms and for which Beer's law is obeyed.

14.4.2 Determination of Equilibrium Constants

Another important application of molecular absorption spectroscopy is the determination of equilibrium constants. Let's consider, as a simple example, an acid–base reaction of the general form

$$HIn(_{aq}) + H_2O(l) \rightleftharpoons H_3O^+(aq) + In^-(aq)$$

where HIn and In– are the conjugate weak acid and weak base forms of an acid–base indicator. The equilibrium constant for this reaction is :

$$K_a = \frac{[H_3O^+][In^-]}{[HIn]}$$

To determine the equilibrium constant's value, we prepare a solution in which the reaction is in a state of equilibrium and determine the equilibrium concentration for H3O+, HIn, and In–. The concentration of H_3O^+ is easy to determine by measuring the solution's pH. To determine the concentration of HIn and In– we can measure the solution's absorbance.

If both HIn and In– absorb at the selected wavelength, then, from Beer›s law, we know that

$$A = \varepsilon_{HIn} b[HIn] + \varepsilon_{In} b[In^-]$$

where εHInεHIn and εInεIn are the molar absorptivities for HIn and In–. The indicator's total concentration, C, is given by a mass balance equation

$$C = [HIn] + [In^-]$$

Solving Equation 6 for [HIn] and substituting into Equation 5 gives

$$A = \varepsilon_{HIn} b(C - [In^-]) + \varepsilon_{In} b[In^-]$$

which we simplify to

$$A = \varepsilon_{HIn} bC - \varepsilon_{HIn} b[In^-] + \varepsilon_{In} b[In^-]$$

$$A = A_{HIn} + b[In^-](\varepsilon_{In} - \varepsilon_{HIn})$$

where A_{HIn}, which is equal to εHInbCεHInbC, is the absorbance when the pH is acidic enough that essentially all the indicator is present as HIn. Solving Equation 7 for the concentration of In– gives

$$[In^-] = \frac{A - A_{HIn}}{b(\varepsilon_{In} - \varepsilon_{HIn})}$$

Proceeding in the same fashion, we derive a similar equation for the concentration of HIn

$$[\text{HIn}] = \frac{A_{In} - A}{b(\varepsilon_{In} - \varepsilon_{HIn})}$$

where AIn, which is equal to εInbCεInbC, is the absorbance when the pH is basic enough that only In– contributes to the absorbance. Substituting Equation 8 and Equation 9 into the equilibrium constant expression for HIn gives:

$$K_a = [\text{H}_3\text{O}^+]\frac{A - A_{HIn}}{A_{In} - A}$$

We can use Equation 10 to determine Ka in one of two ways. The simplest approach is to prepare three solutions, each of which contains the same amount, C, of indicator. The pH of one solution is made sufficiently acidic such that [HIn] >> [In–]. The absorbance of this solution gives A_{HIn}. The value of A_{In} is determined by adjusting the pH of the second solution such that [In–] >> [HIn]. Finally, the pH of the third solution is adjusted to an intermediate value, and the pH and absorbance, A, recorded. The value of Ka is calculated using.

14.5 EVALUATION OF UV/VIS AND IR SPECTROSCOPY

14.5.1 Scale of Operations

Molecular UV/Vis absorption routinely is used for the analysis of trace analytes in macro and meso samples. Major and minor analytes are determined by diluting the sample before analysis, and concentrating a sample may allow for the analysis of ultratrace analytes. The scale of operations for infrared absorption is generally poorer than that for UV/Vis absorption.

14.5.3 Accuracy

Under normal conditions a relative error of 1–5% is easy to obtained with UV/Vis absorption. Accuracy usually is limited by the quality of the blank. Examples of the type of problems that are encountered include the pres- ence of particulates in the sample that scatter radiation, and the presence of interferents that react with analytical reagents. In the latter case the interferent may react to form an absorbing species, which leads to a positive determinate error. Interferents also may prevent the analyte from reacting, which leads to a negative determinate error. With care, it is possible to improve the accuracy of an analysis by as much as an order of magnitude.

14.5.3 Precision

In absorption spectroscopy, precision is limited by indeterminate errors—primarily instrumental noise—which are introduced when we measure absorbance. Precision generally is worse for low absorbances where P0 ≈ PT, and for high absorbances where PT approaches 0. We might expect, therefore, that precision will vary with transmittance.

We can derive an expression between precision and transmittance by applying the propagation of uncertainty. To do so we rewrite Beer's law as

$$\log \frac{A - A_{\mathrm{HIn}}}{A_{\mathrm{In}} - A} = \mathrm{pH} - \mathrm{p}K_{\mathrm{a}}$$

the absolute uncertainty in the concentration, s_C, is

$$Sc = -0.4343\varepsilon_b \times s_{TT} \ (13)$$

where s_T is the absolute uncertainty in the transmittance. Dividing Equation 13 by Equation 12 gives the relative uncertainty in concentration, s_C/C, as

$$sc = 0.4343sTT_{\log} T\square T$$

If we know the transmittance's absolute uncertainty, then we can determine the relative uncertainty in concentration for any measured transmittance.

Determining the relative uncertainty in concentration is complicated because sT is a function of the transmittance. As shown in Table 3 , three categories of indeterminate instrumental error are observed [Rothman, L. D.; Crouch, S. R.; Ingle, J. D. Jr. Anal. Chem. 1975, 47, 1226–1233]. A constant sT is observed for the uncertainty associated with reading %T on a meter's analog or digital scale. Typical values are ±0.2–0.3% (a k_1 of ±0.002–0.003) for an analog scale and ±0.001% a (k_1 of ±0.00001) for a digital scale.

Table 3 . Effect of Indeterminate Errors on Relative Uncertainty in Concentration

Category	Sources of indeterminate error	Relative uncertainty in concentration
sT=k1sT=k1	%T readout resolution noise in thermal detectors	sCC=0.4343k1TlogTsCC=0.4343k1Tlog□T
sT=k_2T2+T−√sT =k_2T2+T	Noise in photon detectors	sCC=0.4343k2logT1+1T−−−√sCC=0.4343k2log□T1+1T
sT=k_3TsT=k_3T	Positioning of sample cell fluctuations in source intensity	sCC=0.4343k3logTsCC=0.4343k3log□T

A constant$s_T$also is observed for the thermal transducers used in infrared spectrophotometers. The effect of a constant$s_T$on the relative uncertainty in concentration is shown by curve A in Figure 16 . Note that the relative uncertainty is very large for both high absorbances and low absorbances, reaching a minimum when the absorbance is 0.4343. This source of indeterminate error is important for infrared spectrophotometers and for inexpensive UV/Vis spectrophotometers. To obtain a relative uncertainty in concentration of ±1–2%, the absorbance is kept within the range 0.1–1.

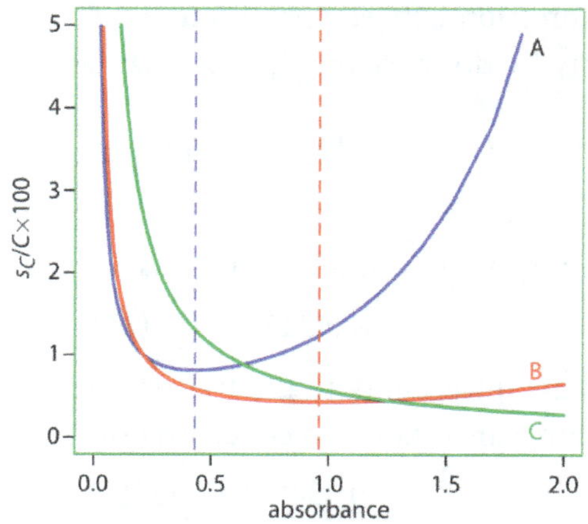

Fig. 16: Percent relative uncertainty in concentration as a function of absorbance for the categories of indeterminate errors in Table 10.3.10.310.3.10.3. A: k_1 = ±0.0030; B: k_2 = ±0.0030; and C: k_3= ±0.0130. The dashed lines correspond to the minimum uncertainty for curve A (absorbance of 0.4343) and for curve B (absorbance of 0.963).

Values of s_T are a complex function of transmittance when indeterminate errors are dominated by the noise associated with photon detectors. Curve B in Figure 10.3.16 shows that the relative uncertainty in concentration is very large for low absorbances, but is smaller at higher absorbances. Although the relative uncertainty reaches a minimum when the absorbance is 0.963, there is little change in the relative uncertainty for absorbances between 0.5 and 2. This source of indeterminate error generally limits the precision of high quality UV/Vis spectrophotometers for mid-to-high absorbances.

Finally, the value of s_T is directly proportional to transmittance for indeterminate errors that result from fluctuations in the source's intensity and from uncertainty in positioning the sample within the spectrometer. The latter is particularly important because the optical properties of a sample cell are not uniform. As a result, repositioning the sample cell may lead to a change in the intensity of transmitted radiation. As shown by curve C in Figure 16 , the effect is important only at low absorbances. This source of indeterminate errors usually is the limiting factor for high quality UV/Vis spectrophotometers when the absorbance is relatively small.

When the relative uncertainty in concentration is limited by the %*T* readout resolution, it is possible to improve the precision of the analysis by redefining 100% T and 0% T. Normally 100% T is established using a blank and 0% T is established while preventing the source's radiation from reaching the detector. If the absorbance is too high, precision is improved by resetting 100% T using a standard solution of analyte whose concentration is less than that of the sample). For a sample whose absorbance is too low, precision is improved by redefining 0% T using a standard solution of the analyte whose concentration is greater than that of the analyte. In this case a calibration curve is required because a linear relationship between absorbance and concentration no longer exists.

Precision is further increased by combining these two methods). Again, a calibration curve is necessary since the relation- ship between absorbance and concentration is no longer linear.

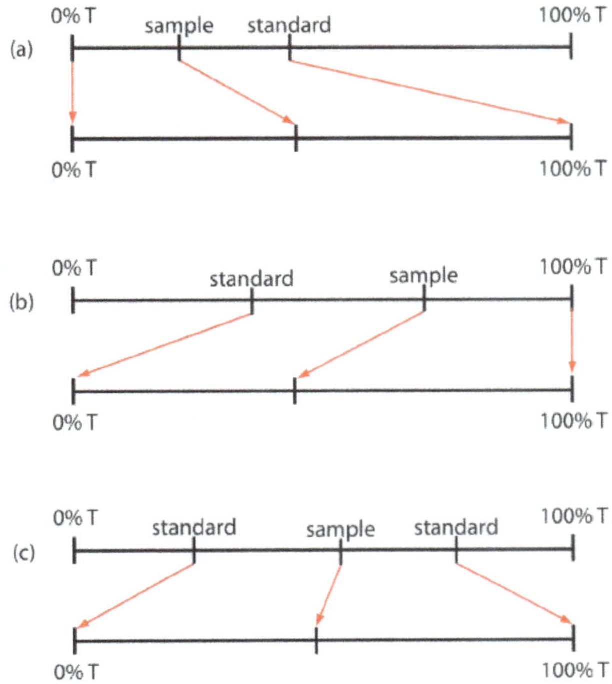

Fig. 17: Methods for improving the precision of absorption methods: (a) high-absorbance method; (b) low-absorbance method; (c) maximum precision method.

Sensitivity

The sensitivity of a molecular absorption method, which is the slope of a Beer's law calibration curve, is the product of the analyte's absorptivity and the pathlength of the sample cell (εbεb). You can improve a method's sensitivity by selecting a wavelength where absorbance is at a maximum or by increasing path length.

14.5.4 Selectivity

Selectivity rarely is a problem in molecular absorption spectrophotometry. In many cases it is possible to find a wavelength where only the analyte absorbs. When two or more species do contribute to the measured absorbance, a multicomponent analysis is still possible.

14.5.5 Time, Cost, and Equipment

The analysis of a sample by molecular absorption spectroscopy is relatively rapid, although additional time is required if we need to convert a non-absorbing analyte into an absorbing form. The cost of UV/Vis instrumentation ranges from several hundred dollars for a simple filter photometer, to more than $50,000 for a computer-controlled, high-resolution double-beam instrument equipped with variable slit widths, and operating over an extended range of wavelengths. Fourier transform infrared spectrometers can be obtained for as little as $15,000–$20,000, although more expensive models are available

15

ATOMIC ABSORPTION AND EMISSION SPECTROSCOPY

15.1 INSTRUMENTATION

Atomic absorption spectrophotometers use the same single-beam or double-beam optics described earlier for molecular absorption spectrophotometers (see Figure 2 and Figure 3). There is, however, an important additional need in atomic absorption spectroscopy: we first must covert the analyte into free atoms. In most cases the analyte is in solution form. If the sample is a solid, then we must bring the analyte into solution before the analysis. When analyzing a lake sediment for Cu, Zn, and Fe, for example, we bring the analytes into solution as Cu^{2+}, Zn^{2+}, and Fe^{3+} by extracting them with a suitable reagent. For this reason, only the introduction of solution samples is considered in this chapter.

What reagent we choose to use to bring an analyte into solution depends on our research goals. If we need to know the total amount of metal in the sediment, then we might try a microwave digestion using a mixture of concentrated acids, such as HNO_3, HCl, and HF. This destroys the sediment's matrix and brings everything into solution. On the other hand, if our interest is biologically available metals, we might extract the sample under milder conditions using, for example, a dilute solution of HCl or CH_3COOH at room temperature.

15.1.1 Atomization

The process of converting an analyte to a free gaseous atom is called atomization. Converting an aqueous analyte into a free atom requires that we strip away the solvent, volatilize the analyte, and, if necessary, dissociate the analyte into free atoms. Desolvating an aqueous solution of $CuCl_2$, for example, leaves us with solid particulates of $CuCl_2$. Converting the particulate $CuCl_2$ to gas phases atoms of Cu and Cl requires thermal energy.

$$CuCl_2(aq) \rightarrow CuCl_2\,(s) \rightarrow Cu(g) + 2Cl(g)$$

There are two common atomization methods: flame atomization and electro thermal atomization, although a few elements are atomized using other methods.

15.1.2 Flame Atomizer

Figure 1 shows a typical flame atomization assembly with close-up views of several key components. In the unit shown here, the aqueous sample is drawn into the assembly by passing a high-pressure stream of compressed air past the end of a capillary tube immersed in the sample. When the sample exits the nebulizer it strikes a glass impact bead, which converts it into a fine aerosol mist within the spray chamber. The aerosol mist is swept through the spray chamber by the combustion gases—compressed air and acetylene in this case—to the burner head where the flame's thermal energy desolvates the aerosol mist to a dry aerosol of small, solid particulates. The flame's thermal energy then volatilizes the particles, producing a vapor that consists of molecular species, ionic species, and free atoms.

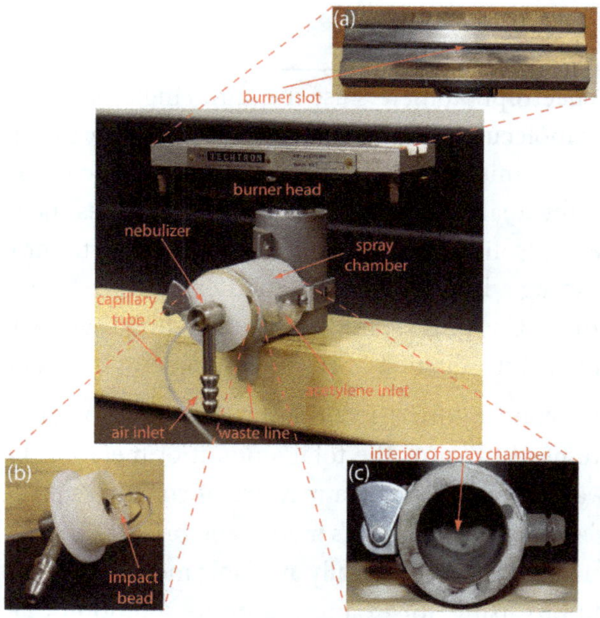

Fig. 1: Flame atomization assembly with expanded views of (a) the burner head showing the burner slot where the flame is located; (b) the nebulizer's impact bead; and (c) the interior of the spray chamber. Although the unit shown here is from an instrument dating to the 1970s, the basic components of a modern flame AA spectrometer are the same.

Burner: The slot burner in Figure 1 a provides a long optical pathlength and a stable flame. Because absorbance is directly proportional to pathlength, a long pathlength provides greater sensitivity. A stable flame minimizes uncertainty due to fluctuations in the flame.

The burner is mounted on an adjustable stage that allows the entire assembly to move horizontally and vertically. Horizontal adjustments ensure the flame is aligned with the instrument's optical path. Vertical adjustments change the height within the flame from which absorbance is monitored. This is important because two competing processes affect the concentration of free atoms in the flame. The more time an analyte spends in the flame the greater the atomization efficiency; thus, the production of free atoms increases with height. On the other hand, a longer residence time allows more opportunity for the free atoms to combine with oxygen to form a molecular oxide. As seen in Figure 2 , for a metal this is easy to oxidize, such as Cr, the concentration of free atoms is greatest just above the burner head. For a metal, such as Ag, which is difficult to oxidize, the concentration of free atoms increases steadily with height.

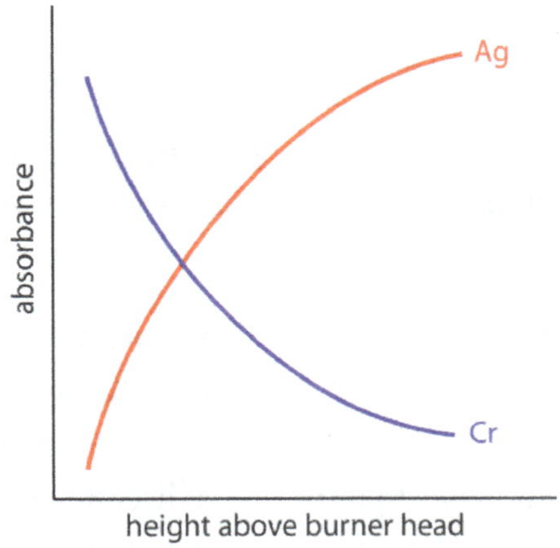

Fig. 2: Absorbance versus height profiles for Ag and Cr in flame atomic absorption spectroscopy.

Flame: The flame's temperature, which affects the efficiency of atomization, depends on the fuel–oxidant mixture, several examples of which are listed in Table 10.4.1 . Of these, the air–acetylene and the nitrous oxide–acetylene flames are the most popular. Normally the fuel and oxidant are mixed in an approximately stoichiometric ratio; however, a fuel-rich mixture may be necessary for easily oxidized analytes.

Table 1 . Fuels and Oxidants Used for Flame Combustion

Fuel	Oxidant	Temperature range (oc)
Natural gas	Air	1700–1900
Hydrogen	Air	2000–2100
Acetylene	Air	2100–2400
Acetylene	Nitrous oxide	2600–2800
Acetylene	Oxygen	3050–3150

Figure 3 shows a cross-section through the flame, looking down the source radiation's optical path. The primary combustion zone usually is rich in gas combustion products that emit radiation, limiting is useful- ness for atomic absorption. The interzonal region generally is rich in free atoms and provides the best location for measuring atomic absorption. The hottest part of the flame typically is 2–3 cm above the primary combustion zone. As atoms approach the flame's secondary combustion zone, the decrease in temperature allows for formation of stable molecular species.

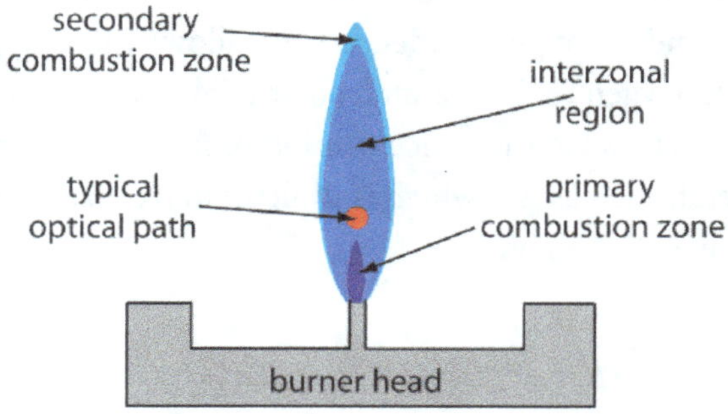

Fig. 3: Profile of typical flame using a slot burner. The relative size of each zone depends on many factors, including the choice of fuel and oxidant, and their relative proportions.

15.1.3 Sample Introduction

The most common means for introducing a sample into a flame atomizer is a continuous aspiration in which the sample flows through the burner while we monitor absorbance. Continuous aspiration is sample intensive, typically requiring from 2–5 mL of sample.

Flame microsampling allows us to introduce a discrete sample of fixed volume, and is useful if we have a limited amount of sample or when the sample's matrix is incompatible with the flame atomizer. For example, continuously aspirating a sample that has a high concentration of dissolved solids—sea water, for example, comes to mind—may build-up a solid de- posit on the burner head that obstructs the flame and that lowers the absorbance. Flame microsampling is accomplished using a micropipet to place 50–250 μL of sample in a Teflon funnel connected to the nebulizer, or by dipping the nebulizer tubing into the sample for a short time. Dip sampling usually is accomplished with an automatic sampler. The signal for flame microsampling is a transitory peak whose height or area is proportional to the amount of analyte that is injected.

15.1.4 Advantages and Disadvantages of Flame Atomization

The principal advantage of flame atomization is the reproducibility with which the sample is introduced into the spectrophotometer; a significant disadvantage is that the efficiency of atomization is quite poor. There are two reasons for poor atomization efficiency. First, the majority of the aerosol droplets produced during nebulization are too large to be carried to the flame by the combustion gases. Consequently, as much as 95% of the sample never reaches the flame, which is the reason for the waste line shown at the bottom of the spray chamber in Figure 1 . A second reason for poor atomization efficiency is that the large volume of combustion gases significantly dilutes the sample. Together, these contributions to the efficiency of atomization reduce sensitivity because the analyte's concentration in the flame may be a factor of 2.5×10–62.5×10–6 less than that in solution.

15.1.5 Electrothermal Atomizers

A significant improvement in sensitivity is achieved by using the resistive heating of a graphite tube in place of a flame. A typical electrothermal atomizer, also known as a *graphite furnace*, consists of a cylindrical graphite tube approximately 1–3 cm in length and 3–8 mm in diameter. As shown in Figure 4 , the graphite tube is housed in an sealed assembly that has an optically transparent window at each end. A continuous stream of inert gas is passed through the furnace, which protects the graphite tube from oxidation and removes the gaseous products produced during atomization. A power supply is used to pass a current through the graphite tube, resulting in resistive heating.

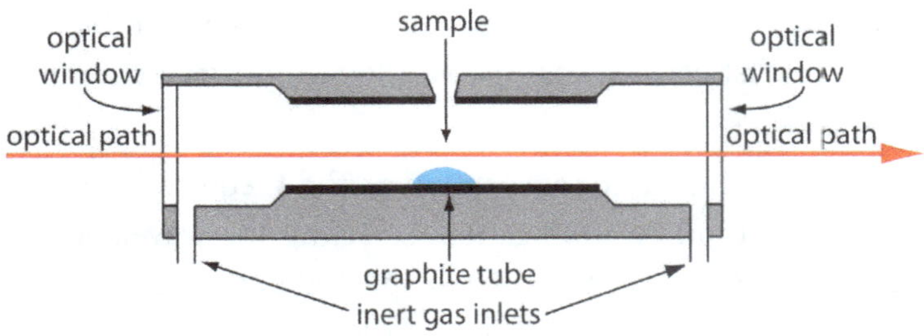

Fig. 4: Diagram showing a cross-section of an electrothermal analyzer.

Samples of between 5–50 µL are injected into the graphite tube through a small hole at the top of the tube. Atomization is achieved in three stages. In the first stage the sample is dried to a solid residue using a current that raises the temperature of the graphite tube to about 110°C. In the second stage, which is called ashing, the temperature is increased to between 350–1200°C. At these temperatures organic material in the sample is converted to CO_2 and H_2O, and volatile inorganic materials are vaporized. These gases are removed by the inert gas flow. In the final stage the sample is atomized by rapidly increasing the temperature to between 2000–3000°C. The result is a transient absorbance peak whose height or area is proportional to the absolute amount of analyte injected into the graphite tube. Together, the three stages take approximately 45–90 s, with most of this time used for drying and ashing the sample.

Electrothermal atomization provides a significant improvement in sensitivity by trapping the gaseous analyte in the small volume within the graphite tube. The analyte's concentration in the resulting vapor phase is as much as 1000×1000× greater than in a flame. This improvement in sensitivity—and the resulting improvement in detection limits—is offset by a significant decrease in precision. Atomization efficiency is influenced strongly by the sample's contact with the graphite tube, which is difficult to control reproducibly.

15.1.6 Miscellaneous Atomization Methods

A few elements are atomized by using a chemical reaction to produce a volatile product. Elements such as As, Se, Sb, Bi, Ge, Sn, Te, and Pb, for example, form volatile hydrides when they react with $NaBH_4$ in the presence of acid. An inert gas carries the volatile hydride to either a flame or to a heated quartz observation tube situated in the optical path. Mercury is determined by the cold-vapor method in which it is reduced to elemental mercury with $SnCl_2$. The volatile Hg is carried by an inert gas to an unheated observation tube situated in the instrument's optical path.

15.2 QUANTITATIVE APPLICATIONS

Atomic absorption is used widely for the analysis of trace metals in a variety of sample matrices. Using Zn as an example, there are standard atomic absorption methods for its determination in samples as diverse as water and wastewater, air, blood, urine, muscle tissue, hair, milk, breakfast cereals, shampoos, alloys, industrial plating baths, gasoline, oil, sediments, and rocks.

Developing a quantitative atomic absorption method requires several considerations, including choosing a method of atomization, selecting the wavelength and slit width, preparing the sample for analysis, minimizing spectral and chemical interferences, and selecting a method of standardization. Each of these topics is considered in this section.

15.2.1 Developing a Quantitative Method

15.2.1.1 Flame or Electrothermal Atomization?

The most important factor in choosing a method of atomization is the analyte's concentration. Because of its greater sensitivity, it takes less analyte to achieve a given absorbance when using electrothermal atomization. Table 2 , which compares the amount of analyte needed to achieve an absorbance of 0.20 when using flame atomization and electrothermal atomization, is useful when selecting an atomization method. For example, flame atomization is the method of choice if our samples contain 1–10 mg Zn^{2+}/L, but electrothermal atomization is the best choice for samples that contain 1–10 μg Zn^{2+}/L.

Table 2 . Concentration of Analyte (in mg/L) That Yields an Absorbance of 0.20

Element	Flame atomization	Electrothermal atomization
Ag	1.5	0.0035
Al	40	0.015
As	40	0.050
Ca	0.8	0.003
Cd	0.6	0.001
Co	2.5	0.021
Cr	2.5	0.0075
Cu	1.5	0.012

Element	Flame atomization	Electrothermal atomization
Fe	2.5	0.006
Hg	70	0.52
Mg	0.15	0.00075
Mn	1	0.003
Na	0.3	0.00023
Ni	2	0.024
Pb	5	0.080
Pt	70	0.29
Sn	50	0.023
Zn	0.3	0.00071
Source: Varian Cookbook, SpectraAA Software Version 4.00 Pro. As: 10 mg/L by hydride vaporization; Hg: 11.5 mg/L by cold-vapor; and Sn:18 mg/L by hydride vaporization		

15.2.1.2 Selecting the Wavelength and Slit Width

The source for atomic absorption is a hollow cathode lamp that consists of a cathode and anode enclosed within a glass tube filled with a low pressure of an inert gas, such as Ne or Ar (Figure.5). Applying a potential across the electrodes ionizes the filler gas. The positively charged gas ions collide with the negatively charged cathode, sputtering atoms from the cathode's surface. Some of the sputtered atoms are in the excited state and emit radiation characteristic of the metal(s) from which the cathode is manufactured. By fashioning the cathode from the metallic analyte, a hollow cathode lamp provides emission lines that correspond to the analyte's absorption spectrum.

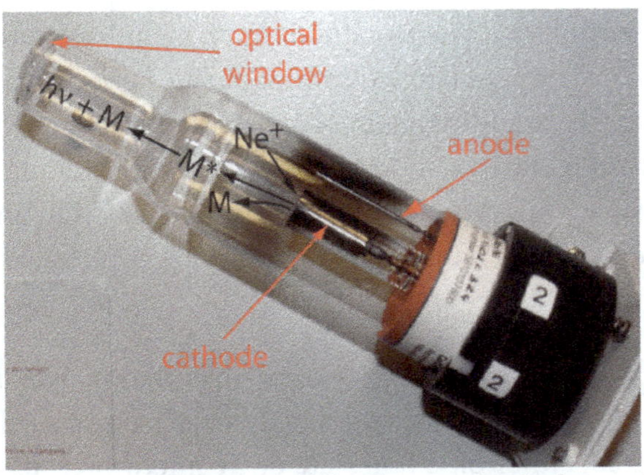

Fig. 5: Photo of a typical multi-elemental hollow cathode lamp. The cathode in this lamp is fashioned from an alloy of Co, Cr, Cu, Fe, Mn, and Ni, and is surrounded by a glass shield that isolates it from the anode. The lamp is filled with Ne gas. Also shown is the process that leads to atomic emission. Note the black deposit of sputtered metal on the outer wall of the hollow cathode lamp. See the text for additional details.

Because atomic absorption lines are narrow, we need to use a line source instead of a continuum source. The effective bandwidth when using a continuum source is roughly 1000×1000× larger than an atomic absorption line; thus, $P_T \approx P_0$, $\%T \approx 100$, and $A \approx 0$. Because a hollow cathode lamp is a line source, P_T and P_0 have different values giving a $\%T < 100$ and $A > 0$.

Each element in a hollow cathode lamp provides several atomic emission lines that we can use for atomic absorption. Usually the wavelength that provides the best sensitivity is the one we choose to use, although a less sensitive wavelength may be more appropriate for a sample that has higher concentration of analyte. For the Cr hollow cathode lamp in Table 3 , the best sensitivity is obtained using a wavelength of 357.9 nm.

Table 3 . Atomic Emission Lines for a Cr Hollow Cathode Lamp

Wavelength (nm)	Slit width (nm)	Mg cr/l giving a = 0.20	P0 (relative)
357.9	0.2	2.5	40
425.4	0.2	12	85
429.0	0.5	20	100
520.5	0.2	1500	15
520.8	0.2	500	20

Another consideration is the emission line's intensity. If several emission lines meet our requirements for sensitivity, we may wish to use the emission line with the largest relative P0 because there is less uncertainty in measuring P0 and PT. When analyzing a sample that is ≈10 mg Cr/L, for example, the first three wavelengths in Table 3 provide an appropriate sensitivity; the wavelengths of 425.4 nm and 429.0 nm, however, have a greater P0 and will provide less uncertainty in the measured absorbance.

The emission spectrum for a hollow cathode lamp includes, in addition to the analyte's emission lines, additional emission lines from impurities present in the metallic cathode and from the filler gas. These additional lines are a potential source of stray radiation that could result in an instrumental deviation from Beer's law. The monochromator's slit width is set as wide as possible to improve the throughput of radiation and narrow enough to eliminate these sources of stray radiation.

15.2.1.3 Preparing the Sample

Flame and electrothermal atomization require that the analyte is in solution. Solid samples are brought into solution by dissolving in an appropriate solvent. If the sample is not soluble it is digested, either on a hot-plate or by microwave, using HNO_3, H_2SO_4, or $HClO_4$. Alternatively, we can extract the analyte using a Soxhlet extractor. Liquid samples are analyzed directly or the analytes extracted if the matrix is in- compatible with the method of atomization. A serum sample, for instance, is difficult to aspirate when using flame atomization and may produce an unacceptably high background absorbance when using electrothermal atomization. A liquid–liquid extraction using an organic solvent and a chelating agent frequently is used to concentrate analytes. Dilute solutions of Cd^{2+}, Co^{2+}, Cu^{2+}, Fe^{3+}, Pb^{2+}, Ni^{2+}, and Zn^{2+}, for example, are concentrated by extracting with a solution of ammonium pyrrolidine dithiocarbamate in methyl isobutyl ketone.

15.2.1.4 Minimizing Spectral Interference.

A spectral interference occurs when an analyte's absorption line overlaps with an interferent's absorption line or band. Because they are so narrow, the overlap of two atomic absorption lines seldom is a problem. On the other hand, a molecule's broad absorption band or the scattering of source radiation is a potentially serious spectral interference.

An important consideration when using a flame as an atomization source is its effect on the measured absorbance. Among the products of combustion are molecular species that exhibit broad absorption bands and particulates that scatter radiation from the source. If we fail to compensate for these spectral interferences, then the intensity of transmitted radiation is smaller than expected. The result is an apparent increase in the sample's absorbance. Fortunately, absorption and scattering of radiation by the flame are corrected by analyzing a blank.

Spectral interferences also occur when components of the sample's matrix other than the analyte react to form molecular species, such as oxides and hydroxides. The resulting absorption and scattering constitutes the sample's background and may present a significant problem, particularly at wavelengths below 300 nm where the scattering of radiation becomes more important. If we know the composition of the sample's matrix, then we can prepare our samples using an identical matrix. In this case the background absorption is the same for both the samples and the standards. Alternatively, if the background is due to a known matrix component, then we can add that component in excess to all samples and standards so that the contribution of the naturally occurring interferent is insignificant. Finally, many interferences due to the sample's matrix are eliminated by increasing the atomization temperature. For example, switching to a higher temperature flame helps prevents the formation of interfering oxides and hydroxides.

If the identity of the matrix interference is unknown, or if it is not possible to adjust the flame or furnace conditions to eliminate the interference, then we must find another method to compensate for the background interference. Several methods have been developed to compensate for matrix interferences, and most atomic absorption spectrophotometers include one or more of these methods.

One of the most common methods for background correction is to use a continuum source, such as a D2 lamp. Because a D2 lamp is a continuum source, absorbance of its radiation by the analyte's narrow absorption line is negligible. Only the background, therefore, absorbs radiation from the D2 lamp. Both the analyte and the background, on the other hand, absorb the hollow cathode's radiation. Subtracting the absorbance for the D2 lamp from that for the hollow cathode lamp gives a corrected absorbance that compensates for the background interference. Although this method of background correction is effective, it does assume that the background absorbance is constant over the range of wavelengths passed by the monochromator. If this is not true, then subtracting the two absorbances underestimates or overestimates the background.

Other methods of background correction have been developed, including Zeeman effect background correction and Smith–Hieftje background correction, both of which are included in some commercially available atomic absorption spectrophotometers. Consult the chapter's additional resources for additional information.

15.2.1.5 Minimizing Chemical Interferences.

The quantitative analysis of some elements is complicated by chemical interferences that occur during atomization. The most common chemical interferences are the formation of nonvolatile compounds that contain the analyte and ionization of the analyte.

One example of the formation of a nonvolatile compound is the effect of $PO_3^{-4}PO_4 3-$ or Al^{3+} on the flame atomic absorption analysis of Ca^{2+}. In one study, for example, adding 100 ppm Al^{3+} to a solution of 5 ppm Ca^{2+} decreased calcium ions absorbance from 0.50 to 0.14, while adding 500 ppm $PO_3^{-4}PO_4^{3-}$ to a similar solution of Ca^{2+} decreased the absorbance from 0.50 to 0.38. These interferences are attributed to the formation of nonvolatile particles of $Ca_3(PO_4)_2$ and an Al–Ca–O oxide.

When using flame atomization, we can minimize the formation of non-volatile compounds by increasing the flame's temperature by changing the fuel-to-oxidant ratio or by switching to a different combination of fuel and oxidant. Another approach is to add a releasing agent or a protecting agent to the sample. A releasing agent is a species that reacts preferentially with the interferent, releasing the analyte during atomization. For example, Sr^{2+} and La^{3+} serve as releasing agents for the analysis of Ca^{2+} in the presence of $PO_3^{-4}PO_4^{3-}$ or Al^{3+}. Adding 2000 ppm $SrCl_2$ to the Ca^{2+}/ $PO_3^{-4}PO_4^{3-}$ and to the Ca^{2+}/ Al^{3+} mixtures described in the previous paragraph increased the absorbance to 0.48. A protecting agent reacts with the analyte to form a stable volatile complex. Adding 1% w/w EDTA to the Ca^{2+}/ $PO_3^{-4}PO_4^{3-}$ solution described in the previous paragraph increased the absorbance to 0.52.

An ionization interference occurs when thermal energy from the flame or the electrothermal atomizer is sufficient to ionize the analyte

$$M(s) \rightleftharpoons M+(aq)+e- \qquad (1)$$

where M is the analyte. Because the absorption spectra for M and M+ are different, the position of the equilibrium in reaction 1 affects the absorbance at wavelengths where M absorbs. To limit ionization we add a high concentration of an ionization suppressor, which is a species that ionizes more easily than the analyte. If the ionization suppressor's concentration is sufficient, then the increased concentration of electrons in the flame pushes reaction 1 to the left, preventing the analyte's ionization. Potassium and cesium frequently are used as an ionization suppressor because of their low ionization energy.

15.2.1.6 Standardizing the Method

Because Beer's law also applies to atomic absorption, we might expect atomic absorption calibration curves to be linear. In practice, however, most atomic absorption calibration curves are nonlinear or linear over a limited range of concentrations. Nonlinearity in atomic absorption is a consequence of instrumental limitations, including stray radiation from the hollow cathode lamp and the variation in molar absorptivity across the absorption line. Accurate quantitative work, therefore, requires a suitable means for computing the calibration curve from a set of standards.

When possible, a quantitative analysis is best conducted using external standards. Unfortunately, matrix interferences are a frequent problem, particularly when using electrothermal atomization. For this reason the method of standard additions often is used. One limitation to this method of standardization, however, is the requirement of a linear relationship between absorbance and concentration.

Most instruments include several different algorithms for computing the calibration curve. The instrument in my lab, for example, includes five algorithms. Three of the algorithms fit absorbance data using linear, quadratic, or cubic polynomial functions of the analyte's concentration. It also includes two algorithms that fit the concentrations of the standards to quadratic functions of the absorbance.

15.3 DETERMINATION OF CU AND ZN IN TISSUE SAMPLES

The best way to appreciate the theoretical and the practical details discussed in this section is to carefully examine a typical analytical method. Although each method is unique, the following description of the determination of Cu and Zn in biological tissues provides an instructive example of a typical procedure. The description here is based on Bhattacharya, S. K.; Goodwin, T. G.; Crawford, A. J. *Anal. Lett.* 1984, *17*, 1567–1593, and Crawford, A. J.; Bhattacharya, S. K. Varian Instruments at Work, Number AA-46, April 1985.

15.3.1 Description of Method

Copper and zinc are isolated from tissue samples by digesting the sample with HNO_3 after first removing any fatty tissue. The concentration of copper and zinc in the supernatant are determined by atomic absorption using an air-acetylene flame.

15.3.2 Procedure

Tissue samples are obtained by a muscle needle biopsy and dried for 24–30 h at 105°C to remove all traces of moisture. The fatty tissue in a dried sample is removed by extracting overnight with anhydrous ether. After removing the ether, the sample is dried to obtain the fat-free dry tissue weight (FFDT). The sample is digested at 68oC for 20–24 h using 3 mL of 0.75 M HNO_3. After centrifuging at 2500 rpm for 10 minutes, the supernatant is transferred to a 5-mL volumetric flask. The digestion is repeated two more times, for 2–4 hours each, using 0.9-mL aliquots of 0.75 M HNO_3. These supernatants are added to the 5-mL volumetric flask, which is diluted to volume with 0.75 M HNO_3. The concentrations of Cu and Zn in the diluted supernatant are determined by flame atomic absorption spectroscopy using an air-acetylene flame and external standards. Copper is analyzed at a wavelength of 324.8 nm with a slit width of 0.5 nm, and zinc is analyzed at 213.9 nm with a slit width of 1.0 nm. Background correction using a D_2 lamp is necessary for zinc. Results are reported as μg of Cu or Zn per gram of FFDT.

15.4 EVALUATION OF ATOMIC ABSORPTION SPECTROSCOPY

15.4.1 Scale of Operation

Atomic absorption spectroscopy is ideally suited for the analysis of trace and ultratrace analytes, particularly when using electrothermal atomization. For minor and major analytes, sample are diluted before the analysis. Most analyses use a macro or a meso sample. The small volume requirement for electrothermal atomization or for flame microsampling, however, makes practical the analysis of micro and ultramicro samples.

15.4.2 Accuracy

If spectral and chemical interferences are minimized, an accuracy of 0.5–5% is routinely attainable. When the calibration curve is nonlinear, accuracy is improved by using a pair of standards whose absorbances closely bracket the sample's absorbance and assuming that the change in absorbance is linear over this limited concentration range. Determinate errors for electrothermal atomization often are greater than those obtained with flame atomization due to more serious matrix interferences.

15.4.3 Precision

For an absorbance greater than 0.1–0.2, the relative standard deviation for atomic absorption is 0.3–1% for flame atomization and 1–5% for electrothermal atomization. The principle limitation is the uncertainty in the concentration of free analyte atoms that result from variations in the rate of aspiration, nebulization, and atomization for a flame atomizer, and the consistency of injecting samples for electrothermal atomization.

15.4.4 Sensitivity

The sensitivity of a flame atomic absorption analysis is influenced by the flame's composition and by the position in the flame from which we monitor the absorbance. Normally the sensitivity of an analysis is optimized by aspirating a standard solution of analyte and adjusting the fuel-to-oxidant ratio, the nebulizer flow rate, and the height of the burner, to give the greatest absorbance. With electrothermal atomization, sensitivity is influenced by the drying and ashing stages that precede atomization. The temperature and time at each stage is optimized for each type of sample.

Sensitivity also is influenced by the sample's matrix. We already noted, for example, that sensitivity is decreased by a chemical interference. An increase in sensitivity may be realized by adding a low molecular weight alcohol, ester, or ketone to the solution, or by using an organic solvent.

15.4.5 Selectivity

Due to the narrow width of absorption lines, atomic absorption provides excellent selectivity. Atomic absorption is used for the analysis of over 60 elements at concentrations at or below the level of μg/L.

15.4.6 Time, Cost, and Equipment

The analysis time when using flame atomization is short, with sample throughputs of 250–350 determinations per hour when using a fully automated system. Electrothermal atomization requires substantially more time per analysis, with maximum sample throughputs of 20–30 determinations per hour. The cost of a new instrument ranges from between $10,000– $50,000 for flame atomization, and from $18,000–$70,000 for electrothermal atomization. The more expensive instruments in each price range include double-beam optics, automatic samplers, and can be programmed for multielemental analysis by allowing the wavelength and hollow cathode lamp to be changed automatically.

15.5 EMISSION SPECTROSCOPY

An analyte in an excited state possesses an energy, E_2, that is greater than its energy when it is in a lower energy state, E_1. When the analyte returns to its lower energy state—a process we call relaxation—the excess energy, ΔEΔE

$$\Delta E = E_2 - E_1$$

is released. Figure 4 shows a simplified picture of this process.

The amount of time an analyte, A, spends in its excited state—what we call the excited state's lifetime—is short, typically 10^{-5}–10^{-9} s for an electronic excited state and 10–15 s for a vibrational excited state. Relaxation of the analyte's excited state, A*, occurs through several mechanisms, including collisions with other species in the sample, photochemical reactions, and the emission of photons. In the first process, which we call vibrational relaxation or nonradiative relaxation, the excess energy is released as heat.

$$A* \longrightarrow A + \text{heat}$$

Relaxation by a photochemical reaction may involve simple decomposition

$$A* \longrightarrow X + Y$$

or a reaction between A^* and another species

$$A^* + Z \longrightarrow X + Y$$

In both cases the excess energy is used up in the chemical reaction or released as heat.

In the third mechanism, the excess energy is released as a photon of electromagnetic radiation.

$$A^{*} \longrightarrow A + h\nu A^{*} \longrightarrow A + h\nu$$

The release of a photon following thermal excitation is called emission and that following the absorption of a photon is called photoluminescence. In chemiluminescence and bioluminescence, excitation results from a chemical or a biochemical reaction, respectively.

15.6. PHOTOLUMINESCENT SPECTROSCOPY

Photoluminescence is divided into two categories: fluorescence and phosphorescence. A pair of electrons that occupy the same electronic ground state have opposite spins and are in a singlet spin state.

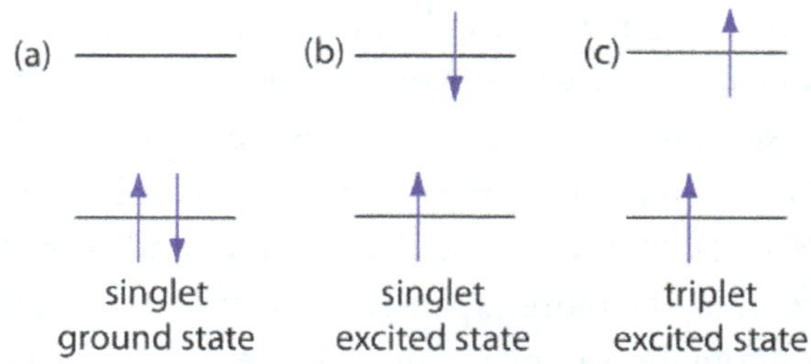

Fig. 1: Electron configurations for (a) a singlet ground state; (b) a singlet excited state; and (c) a triplet excited state.

When an analyte absorbs an ultraviolet or a visible photon, one of its valence electrons moves from the ground state to an excited state with a conservation of the electron's spin (Figure 10.6.1 b). Emission of a photon from a singlet excited state to the singlet ground state—or between any two energy levels with the same spin—is called fluorescence. The probability of fluorescence is very high and the average lifetime of an electron in the excited state is only 10^{-5}–10^{-8} s. Fluorescence, therefore, rapidly decays once the source of excitation is removed.

In some cases an electron in a singlet excited state is transformed to a triplet excited state (Figure 1 c) in which its spin no is longer paired with the ground state. Emission between a triplet excited state and a singlet ground state—or between any two energy levels that differ in their respective spin states–is called phosphorescence. Because the average lifetime for phosphorescence ranges from 10^{-4}–10^{4} s, phosphorescence may continue for some time after we remove the excitation source.

The use of molecular fluorescence for qualitative analysis and for semi-quantitative analysis dates to the early to mid 1800s, with more accurate quantitative methods appearing in the 1920s. Instrumentation for fluorescence spectroscopy using a filter or a monochromator for wavelength selection appeared in, respectively, the 1930s and 1950s. Although the discovery of phosphorescence preceded that of fluorescence by almost 200 years, qualitative and quantitative applications of molecular phosphorescence did not receive much attention until after the development of fluorescence instrumentation.

As you might expect, the persistence of long-lived phosphorescence made it more noticeable.

15.7 FLUORESCENCE AND PHOSPHORESCENCE SPECTRA

To appreciate the origin of fluorescence and phosphorescence we must consider what happens to a molecule following the absorption of a photon. Let's assume the molecule initially occupies the lowest vibrational energy level of its electronic ground state, which is the singlet state labeled S_0 in Figure 2 . Absorption of a photon excites the molecule to one of several vibrational energy levels in the first excited electronic state, S_1, or the second electronic excited state, S_2, both of which are singlet states. Relaxation to the ground state occurs by a number of mechanisms, some of which result in the emission of a photon and others that occur without the emission of a photon. These relaxation mechanisms are shown in Figure 2 . The most likely relaxation pathway from any excited state is the one with the shortest lifetime.

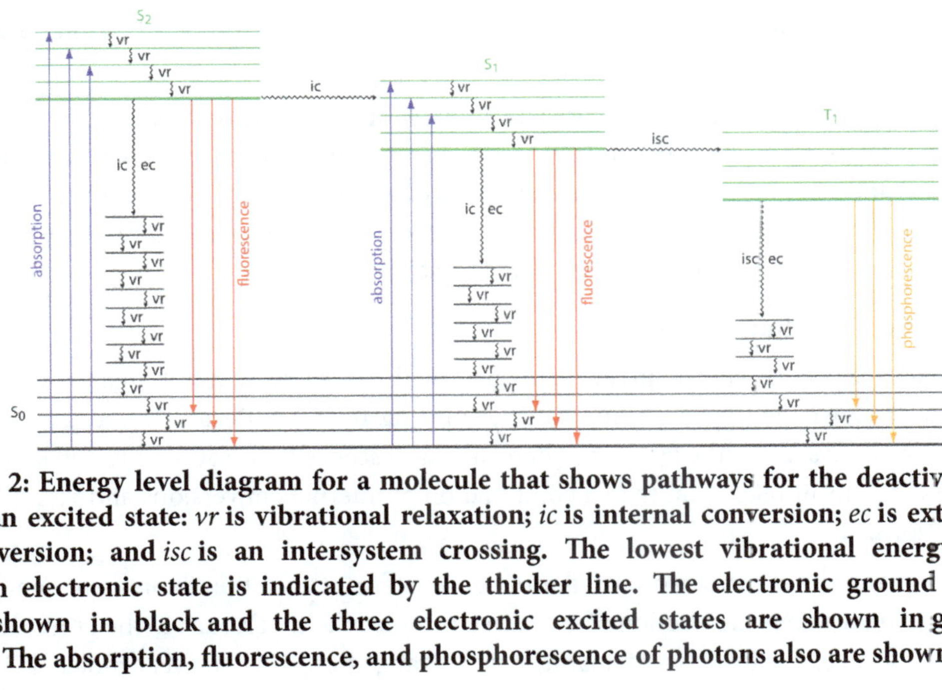

Fig. 2: Energy level diagram for a molecule that shows pathways for the deactivation of an excited state: *vr* is vibrational relaxation; *ic* is internal conversion; *ec* is external conversion; and *isc* is an intersystem crossing. The lowest vibrational energy for each electronic state is indicated by the thicker line. The electronic ground state is shown in black and the three electronic excited states are shown in green. The absorption, fluorescence, and phosphorescence of photons also are shown.

Radiationless Deactivation

When a molecule relaxes without emitting a photon we call the process radiationless deactivation. One example of radiationless deactivation is vibrational relaxation, in which a molecule in an excited vibrational energy level loses energy by moving to a lower vibrational energy level in the same electronic state. Vibrational relaxation is very rapid, with an average lifetime of $<10^{-12}$ s. Because vibrational relaxation is so efficient, a molecule in one of its excited state's higher vibrational energy levels quickly returns to the excited state's lowest vibrational energy level.

Another form of radiationless deactivation is an internal conversion in which a molecule in the ground vibrational level of an excited state passes directly into a higher vibrational energy level of a lower energy electronic state of the same spin state. By a combination of internal conversions and vibrational relaxations, a molecule in an excited electronic state may return to the ground electronic state without emitting a photon. A related form of radiationless deactivation is an external conversion in which excess energy is transferred to the solvent or to another component of the sample's matrix.

Let's use Figure 2 to illustrate how a molecule can relax back to its ground state without emitting a photon. Suppose our molecule is in the highest vibrational energy level of the second electronic excited state. After a series of vibrational relaxations brings the molecule to the lowest vibrational energy level of S_2, it undergoes an internal conversion into a higher vibrational energy level of the first excited electronic state. Vibrational relaxations bring the molecule to the lowest vibrational energy level of S_1. Following an internal conversion into a higher vibrational energy level of the ground state, the molecule continues to undergo vibrational relaxation until it reaches the lowest vibrational energy level of S_0.

A final form of radiationless deactivation is an intersystem crossing in which a molecule in the ground vibrational energy level of an excited electronic state passes into one of the higher vibrational energy levels of a lower energy electronic state with a different spin state. For example, an intersystem crossing is shown in Figure 2 between the singlet excited state S1 and the triplet excited state T_1.

Relaxation by Fluorescence

Fluorescence occurs when a molecule in an excited state's lowest vibrational energy level returns to a lower energy electronic state by emitting a photon. Because molecules return to their ground state by the fastest mechanism, fluorescence is observed only if it is a more efficient means of relaxation than a combination of internal conversions and vibrational relaxations.

A quantitative expression of fluorescence efficiency is the fluorescent quantum yield, ΦfΦf, which is the fraction of excited state molecules that return to the ground state by fluorescence. The fluorescent quantum yields range from 1 when every molecule in an excited state undergoes fluorescence, to 0 when fluorescence does not occur.

The intensity of fluorescence, I_f, is proportional to the amount of radiation absorbed by the sample, $P_0 - P_T$, and the fluorescent quantum yield

If=kΦf (PO−PT) (1)

where k is a constant that accounts for the efficiency of collecting and detecting the fluorescent emission. From Beer's law we know that

PTP0=10−εbC (2)

where C is the concentration of the fluorescing species. Solving Equation 2 for PT and substituting into Equation 1 gives, after simplifying

If=kΦfP0(1−10−εbC)(10.6.3)(10.6.3)If=kΦfP0(1−10−εbC)

when εbCεbC < 0.01, which often is the case when the analyte's concentration is small, Equation 10.6.310.6.3 simplifies to

If=2.303kΦfεbCP0=k′P0(4)

where k' is a collection of constants. The intensity of fluorescence, therefore, increases with an increase in the quantum efficiency, the source's incident power, and the molar absorptivity and the concentration of the fluorescing species.

Fluorescence generally is observed when the molecule's lowest energy absorption is a п□п* transition, although some n→π* transitions show weak fluorescence. Many unsubstituted, nonheterocyclic aromatic compounds have a favorable fluorescence quantum yield, although substitutions on the aromatic ring can effect ΦfΦf significantly. For example, the presence of an electron-withdrawing group, such as $-NO_2$, decreases ΦfΦf, while adding an electron-donating group, such as –OH, increases ΦfΦf. Fluorrescence also increases for aromatic ring systems and for aromatic molecules with rigid planar structures. Figure 3 shows the fluorescence of quinine under a UV lamp.

Fig. 3: Tonic water, which contains quinine, is fluorescent when placed under a UV lamp. Source: Splarka (commons.Wikipedia.org).

A molecule's fluorescent quantum yield also is influenced by external variables, such as temperature and solvent. Increasing the temperature generally decreases ΦfΦf because more frequent collisions between the molecule and the solvent increases external conversion. A decrease in the solvent's viscosity decreases ΦfΦf for similar reasons. For an analyte with acidic or basic functional groups, a change in pH may change the analyte's structure and its fluorescent properties.

As shown in Figure.2 , fluorescence may return the molecule to any of several vibrational energy levels in the ground electronic state. Fluorescence, therefore, occurs over a range of wavelengths. Because the change in energy for fluorescent emission generally is less than that for absorption, a molecule's fluorescence spectrum is shifted to higher wavelengths than its absorption spectrum.

Relaxation by Phosphorescence

A molecule in a triplet electronic excited state's lowest vibrational energy level normally relaxes to the ground state by an intersystem crossing to a singlet state or by an external conversion. Phosphorescence occurs when the molecule relaxes by emitting a photon. As shown in Figure.2 , phosphorescence occurs over a range of wavelengths, all of which are at lower energies than the molecule's absorption band. The intensity of phosphorescence, IpIp, is given by an equation similar to Equation 4 for fluorescence

IP=2.303kΦPεbCP0=k′P0(10.6.5)(10.6.6)(10.6.5)IP=2.303kΦPεbCP0(10.6.6)=k′P0

where ΦpΦp is the phosphorescent quantum yield.

Phosphorescence is most favorable for molecules with $n \rightarrow \pi*$ transitions, which have a higher probability for an intersystem crossing than $\pi \rightarrow \pi*$ transitions. For example, phosphorescence is observed with aromatic molecules that contain carbonyl groups or heteroatoms. Aromatic compounds that contain halide atoms also have a higher efficiency for phosphorescence. In general, an increase in phosphorescence corresponds to a decrease in fluorescence.

Because the average lifetime for phosphorescence can be quite long, ranging from 10^{-4} s, the phosphorescent quantum yield usually is quite small. An improvement in Φp is realized by decreasing the efficiency of external conversion. This is accomplished in several ways, including lowering the temperature, using a more viscous solvent, depositing the sample on a solid substrate, or trapping the molecule in solution. Figure 4 shows an example of phosphorescence.

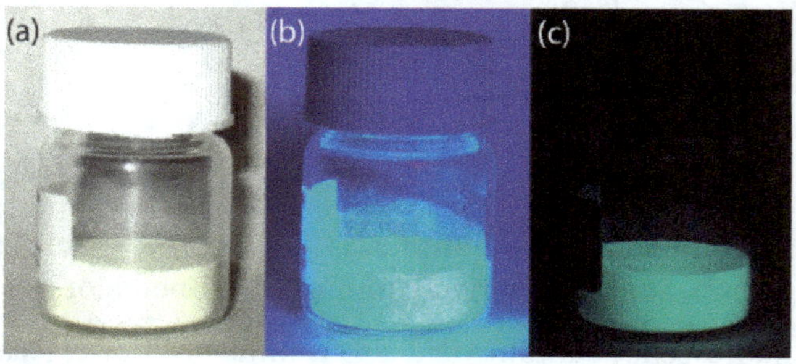

Fig. 4: An europium doped strontium silicate-aluminum oxide powder under (a) natural light, (b) a long-wave UV lamp, and (c) in total darkness. The photo taken in total darkness shows the phosphorescent emission. Source: modified from Splarka (commons.Wikipedia.org).

Excitation Versus Emission Spectra

Photoluminescence spectra are recorded by measuring the intensity of emitted radiation as a function of either the excitation wavelength or the emission wavelength. An *excitation spectrum* is obtained by monitoring emission at a fixed wavelength while varying the excitation wavelength. When corrected for variations in the source's intensity and the detector's response, a sample's excitation spectrum is nearly identical to its absorbance spectrum. The excitation spectrum provides a convenient means for selecting the best excitation wavelength for a quantitative or qualitative analysis.

In an emission spectrum a fixed wavelength is used to excite the sample and the intensity of emitted radiation is monitored as function of wavelength. Although a molecule has a single excitation spectrum, it has two emission spectra, one for fluorescence and one for phosphorescence. Figure 5 shows the UV absorption spectrum and the UV fluorescence emission spectrum for quinine.

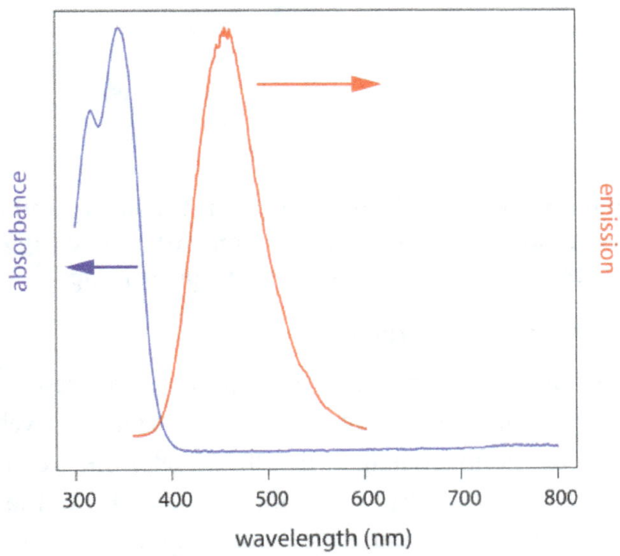

Fig. 5: Absorbance spectrum and fluorescence emission spectrum for quinine in 0.05 M H_2SO_4. The emission spectrum uses an excitation wavelength of 350 nm with a bandwidth of 20 nm. Both spectra are normalized so that the maximum absorbance is 1.00 and the maximum emission is 1.00. The actual maximum absorbance is 0.444 and the actual maximum emission is 126747. Source: data from Daniel Scott, Department of Chemistry & Biochemistry, DePauw University.

Instrumentation

The basic instrumentation for monitoring fluorescence and phosphorescence—a source of radiation, a means of selecting a narrow band of radiation, and a detector—are the same as those for absorption spectroscopy. The unique demands of fluorescence and phosphorescence, however, require some modifications to the instrument designs seen earlier in Figure 1 (filter photometer), Figure 2 (single-beam spectrophotometer),

Figure 3 (double-beam spectrophotometer), and Figure 4 (diode array spectrometer). The most important difference is that the detector cannot be placed directly across from the source. Figure 10.6.6 shows why this is the case. If we place the detector along the source's axis it receives both the transmitted source radiation, PT, and the fluorescent, If, or phosphorescent, Ip, radiation. Instead, we rotate the director and place it at 90o to the source.

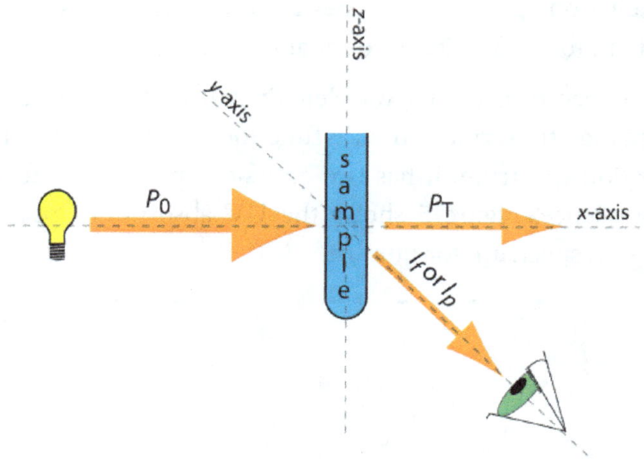

Fig. 6: Schematic diagram showing the orientation of the source and the detector when measuring fluorescence and phosphorescence. Contrast this to Figure7, which shows the orientation for absorption spectroscopy.

Instruments for Measuring Fluorescence

Figure 7 shows the basic design of an instrument for measuring fluorescence, which includes two wavelength selectors, one for selecting the source's excitation wavelength and one for selecting the analyte's emission wavelength. In a fluorimeter the excitation and emission wavelengths are selected using absorption or interference filters. The excitation source for a fluorimeter usually is a low-pressure Hg vapor lamp that provides intense emission lines distributed throughout the ultraviolet and visible region. When a monochromator is used to select the excitation and the emission wavelengths, the instrument is called a spectrofluorometer. With a monochromator the excitation source usually is a high-pressure Xe arc lamp, which has a continuous emission spectrum. Either instrumental design is appropriate for quantitative work, although only a spectrofluorometer can record an excitation or emission spectrum.

A Hg vapor lamp has emission lines at 254, 312, 365, 405, 436, 546, 577, 691, and 773 nm.

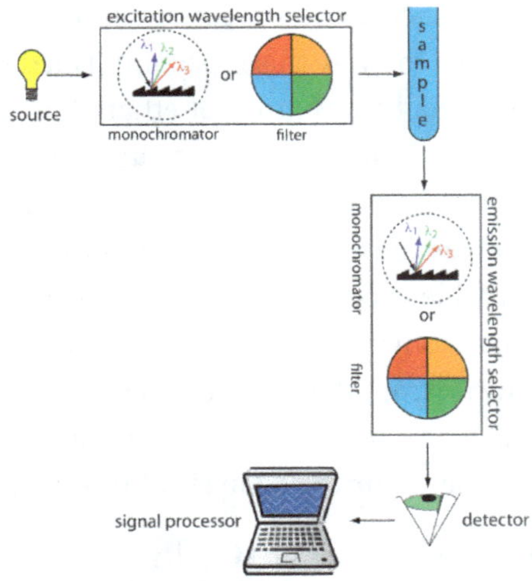

Fig. 7: Schematic diagram for measuring fluorescence showing the placement of the wavelength selectors for excitation and emission. When a filter is used the instrument is called a fluorimeter and when a monochromator is used the instrument is called a spectrofluorimeter.

The sample cells for molecular fluorescence are similar to those for molecular absorption (see Figure 6). Remote sensing using a fiber optic probe (see Figure 7) is possible using with either a fluorimeter or spectrofluorometer. An analyte that is fluorescent is monitored directly. For an analyte that is not fluorescent, a suitable fluorescent probe molecule is incorporated into the tip of the fiber optic probe. The analyte's reaction with the probe molecule leads to an increase or decrease in fluorescence.

Instruments for Measuring Phosphorescence

An instrument for molecular phosphorescence must discriminate between phosphorescence and fluorescence. Because the lifetime for fluorescence is shorter than that for phosphorescence, discrimination is achieved by incorporating a delay between exciting the sample and measuring the phosphorescent emission. Figure 10.6.8 shows how two out-of-phase choppers allow us to block fluorescent emission from reaching the detector when the sample is being excited and to prevent the source radiation from causing fluorescence when we are measuring the phosphorescent emission.

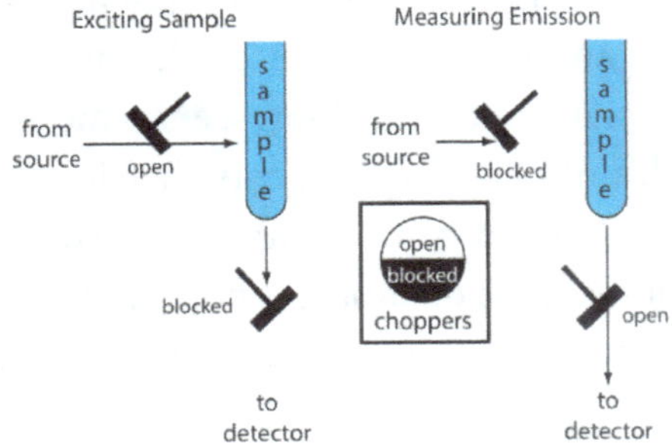

Fig. 8: Schematic diagram showing how choppers are used to prevent fluorescent emission from interfering with the measurement of phosphorescent emission.

Because phosphorescence is such a slow process, we must prevent the excited state from relaxing by external conversion. One way this is accomplished is by dissolving the sample in a suitable organic solvent, usually a mixture of ethanol, isopentane, and diethylether. The resulting solution is frozen at liquid-N_2 temperatures to form an optically clear solid. The solid matrix minimizes external conversion due to collisions between the analyte and the solvent. External conversion also is minimized by immobilizing the sample on a solid substrate, making possible room temperature measurements. One approach is to place a drop of a solution that contains the analyte on a small disc of filter paper. After drying the sample under a heat lamp, the sample is placed in the spectrofluorometer for analysis. Other solid substrates include silica gel, alumina, sodium acetate, and sucrose. This approach is particularly useful for the analysis of thin layer chromatography plates.

15.8 QUANTITATIVE APPLICATIONS

Molecular fluorescence and, to a lesser extent, phosphorescence are used for the direct or indirect quantitative analysis of analytes in a variety of matrices. A direct quantitative analysis is possible when the analyte's fluorescent or phosphorescent quantum yield is favorable. If the analyte is not fluorescent or phosphorescent, or if the quantum yield is unfavorable, then an indirect analysis may be feasible. One approach is to react the analyte with a reagent to form a product that is fluorescent or phosphorescent. Another approach is to measure a decrease in fluorescence or phosphores- cence when the analyte is added to a solution that contains a fluorescent or phosphorescent probe molecule. A decrease in emission is observed when the reaction between the analyte and the probe molecule enhances radiationless deactivation or results in a nonemitting product. The application of fluorescence and phosphorescence to inorganic and organic analytes are considered in this section.

Inorganic Analytes

Except for a few metal ions, most notably $UO^{+2}UO^{2+}$, most inorganic ions are not sufficiently fluorescent for a direct analysis. Many metal ions are determined indirectly by reacting with an organic ligand to form a fluorescent or, less commonly, a phosphorescent metal–ligand complex. One example is the reaction of Al^{3+} with the sodium salt of 2, 4, 3′-trihydroxyazobenzene-5′-sulfonic acid—also known as alizarin garnet R—which forms a fluorescent metal–ligand complex (Figure 9). The analysis is carried out using an excitation wavelength of 470 nm, with fluorescence monitored at 500 nm. Table 1 provides additional examples of chelating reagents that form fluorescent metal–ligand complexes with metal ions. A few inorganic nonmetals are determined by their ability to decrease, or quench, the fluorescence of another species. One example is the analysis for F– based on its ability to quench the fluorescence of the Al^{3+}–alizarin garnet R complex.

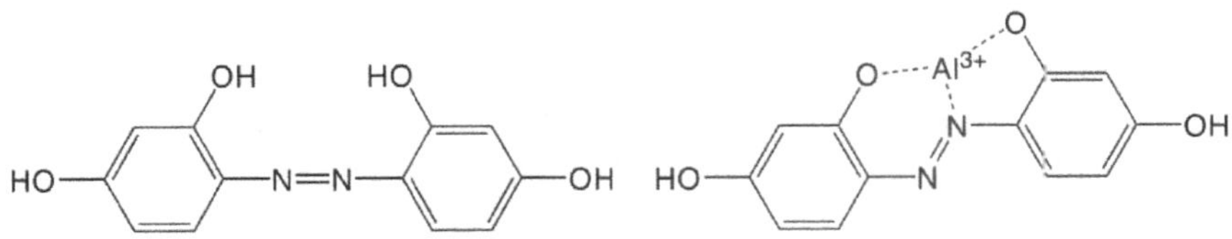

alizarin garnet R　　fluorescent complex

Fig. 9: Structure of alizarin garnet R and its metal–ligand complex with Al^{3+}.

Table 1 . Chelating Agents for the Fluorescent Analysis of Metal Ions

Chelating agent	Metal ions
8-hydroxyquinoline	Al^{3+}, Be^{2+}, Zn^{2+}, Li^{+}, Mg^{2+} (and others)
flavonal	Zr^{2+}, Sn^{4+}
benzoin	$B_4O_2^{-6}B_4O_6^{2-}$, Zn^{2+}
2,3′,4′,5,7−pentahydroxylflavone2′,3′,4′,5,7−pentahydroxylflavone	Be^{2+}
2-(*o*-hydroxyphenyl) benzoxazole	Cd^{2+}

Organic Analytes

As noted earlier, organic compounds that contain aromatic rings generally are fluorescent and aromatic heterocycles often are phosphorescent. Table 2 provides examples of several important biochemical, pharmaceutical, and environmental compounds that are analyzed quantitatively by fluorimetry or phosphorimetry. If an organic analyte is not naturally fluorescent or phosphorescent, it may be possible to incorporate it into a chemical reaction that produces a fluorescent or phosphorescent product. For example, the enzyme creatine phosphokinase is determined by using it to catalyze the formation of creatine from phosphocreatine. Reacting the creatine with ninhydrin produces a fluorescent product of unknown structure.

Table 2 . Examples of Naturally Photoluminescent Organic Analytes

Class	Compounds (f = fluorescence, p = phosphorescence)
Aromatic amino acids	Phenylalanine (f) Tyrosine (f) Tryptophan (f, p)
Vitamins	Vitamin a (f) Vitamin b2 (f) Vitamin b6 (f) Vitamin b12 (f) Vitamin e (f) Folic acid (f)

Catecholamines	Dopamine (f) Norepinephrine (f)
Pharmaceuticals and drugs	Quinine (f) Salicylic acid (f, p) Morphine (f) Barbiturates (f) Lsd (f) Codeine (p) Caffeine (p) Sulfanilamide (p)
Environmental pollutants	Pyrene (f) Benzo[a]pyrene (f) Organothiophosphorous pesticides (f) Carbamate insecticides (f) Ddt (p)

Standardizing the Method

From Equation 4 and Equation .6 we know that the intensity of fluorescence or phosphorescence is a linear function of the analyte's concentration provided that the sample's absorbance of source radiation (A=εbCA=εbC) is less than approximately 0.01. Calibration curves often are linear over four to six orders of magnitude for fluorescence and over two to four orders of magnitude for phosphorescence. For higher concentrations of analyte the calibration curve becomes nonlinear because the assumptions that led to Equation .4 and Equation 6 no longer apply. Nonlinearity may be observed for smaller concentrations of analyte fluorescent or phosphorescent contaminants are present. As discussed earlier, quantum efficiency is sensitive to temperature and sample matrix, both of which must be con- trolled when using external standards. In addition, emission intensity depends on the molar absorptivity of the photoluminescent species, which is sensitive to the sample matrix.

15.9 DETERMINATION OF QUININE IN URINE

The best way to appreciate the theoretical and the practical details discussed in this section is to carefully examine a typical analytical method. Although each method is unique, the following description of the determination of quinine in urine provides an instructive example of a typical procedure.

Description of the Method

Quinine is an alkaloid used to treat malaria. It is a strongly fluorescent compound in dilute solutions of H_2SO_4 (Φf=0.55Φf=0.55). Quinine's excitation spectrum has absorption bands at 250 nm and 350 nm and its emission spectrum has a single emission band at 450 nm. Quinine is excreted rapidly from the body in urine and is determined by measuring its fluorescence following its extraction from the urine sample.

Procedure

Transfer a 2.00-mL sample of urine to a 15-mL test tube and use 3.7 M NaOH to adjust its pH to between 9 and 10. Add 4 mL of a 3:1 (v/v) mixture of chloroform and isopropanol and shake the contents of the test tube for one minute. Allow the organic and the aqueous (urine) layers to separate and transfer the organic phase to a clean test tube. Add 2.00 mL of 0.05 M H_2SO_4 to the organic phase and shake the contents for one minute. Allow the organic and the aqueous layers to separate and transfer the aqueous phase to the sample cell. Measure the fluorescent emission at 450 nm using an excitation wavelength of 350 nm. Determine the concentration of quinine in the urine sample using a set of external standards in 0.05 M H_2SO_4, prepared from a 100.0 ppm solution of quinine in 0.05 M H_2SO_4. Use distilled water as a blank.

15.10 EVALUATION OF PHOTOLUMINESCENCE SPECTROSCOPY

Scale of Operation

Photoluminescence spectroscopy is used for the routine analysis of trace and ultratrace analytes in macro and meso samples. Detection limits for fluorescence spectroscopy are influenced by the analyte's quantum yield. For an analyte with Φf>0.5Φf>0.5, a picomolar detection limit is possible when using a high quality spectrofluorometer. For example, the detection limit for quinine sulfate, for which ΦΦ is 0.55, generally is between 1 part per billion and 1 part per trillion. Detection limits for phosphorescence are somewhat higher, with typical values in the nanomolar range for low-temperature phosphorimetry and in the micromolar range for room-temperature phosphorimetry using a solid substrate.

Accuracy

The accuracy of a fluorescence method generally is between 1–5% when spectral and chemical interferences are insignificant. Accuracy is limited by the same types of problems that affect other optical spectroscopic methods. In addition, accuracy is affected by interferences that affect the fluorescent quantum yield. The accuracy of phosphorescence is somewhat greater than that for fluorescence.

Precision

The relative standard deviation for fluorescence usually is between 0.5–2% when the analyte's concentration is well above its detection limit. Precision usually is limited by the stability of the excitation source. The precision for phosphorescence often is limited by reproducibility in preparing samples for analysis, with relative standard deviations of 5–10% being common.

Sensitivity

From Equation 4 and Equation 6 we know that the sensitivity of a fluorescent or a phosphorescent method is affected by a number of parameters. We already have considered the importance of quantum yield and the effect of temperature and solution composition on ΦfΦf and ΦpΦp. Besides quantum yield, sensitivity is improved by using an excitation source that has a greater emission intensity, P_0, at the desired wavelength, and by selecting an excitation wavelength for which the analyte has a greater molar absorptivity, εε. Another approach for improving sensitivity is to increase the volume from which emission is monitored. Figure 10.6.10 shows how rotating a monochromator's slits from their usual vertical orientation to a horizontal orientation increases the sampling volume. The result can in- crease the emission from the sample by 5–30×5–30×.

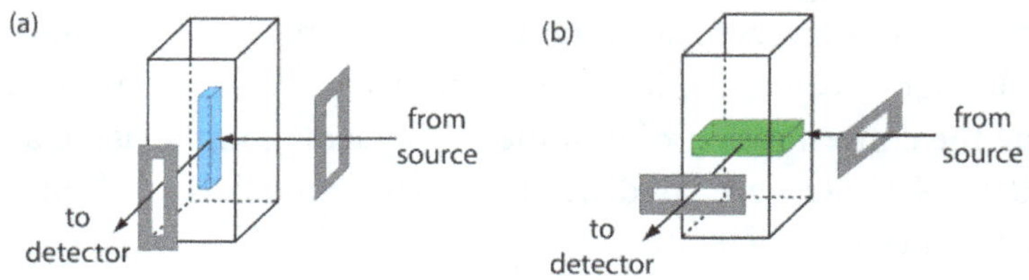

Fig. 10: Use of slit orientation to change the volume from which fluorescence is measured: (a) vertical slit orientation; (b) horizontal slit orientation. Suppose the slit's dimensions are 0.1 mm ×× 3 mm. In (a) the dimensions of the sampling volume are 0.1 mm × 0.1 mm ×× 3 mm, or 0.03 mm³. For (b) the dimensions of the sampling volume are 0.1 mm × 3 mm × 3 mm, or 0.9 mm³, a 30-fold increase in the sampling volume.

Selectivity

The selectivity of fluorescence and phosphorescence is superior to that of absorption spectrophotometry for two reasons: first, not every compound that absorbs radiation is fluorescent or phosphorescent; and, second, selectivity between an analyte and an interferent is possible if there is a difference in either their excitation or their emission spectra. The total emission intensity is a linear sum of that from each fluorescent or phosphorescent species. The analysis of a sample that contains n analytes, therefore, is accomplished by measuring the total emission intensity at n wavelengths.

Time, Cost, and Equipment

As with other optical spectroscopic methods, fluorescent and phosphorescent methods provide a rapid means for analyzing samples and are capable of automation. Fluorimeters are relatively inexpensive, ranging from several hundred to several thousand dollars, and often are satisfactory for quantitative work. Spectrofluorometers are more expensive, with models often exceeding $50,000.

15.11 ATOMIC EMISSION SPECTROSCOPY ATOMIC EMISSION SPECTRA

Atomic emission occurs when a valence electron in a higher energy atomic orbital returns to a lower energy atomic orbital. Figure 1 shows a portion of the energy level diagram for sodium, which consists of a series of discrete lines at wavelengths that correspond to the difference in energy between two atomic orbitals.

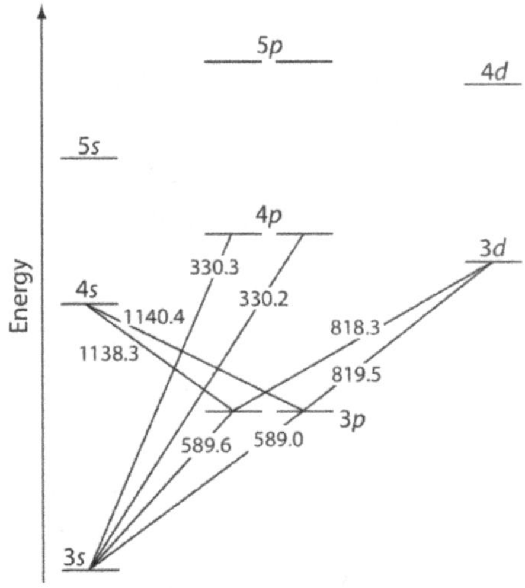

Fig. 1: Valence shell energy level diagram for sodium. The wavelengths corresponding to several transitions are shown.

The intensity of an atomic emission line, I_e, is proportional to the number of atoms, N∗N∗, that populate the excited state,

$$Ie = kN* \quad (1)$$

where *k* is a constant that accounts for the efficiency of the transition. If a system of atoms is in thermal equilibrium, the population of excited state *i* is related to the total concentration of atoms, *N*, by the Boltzmann distribution. For many elements at temperatures of less than 5000 K the Boltzmann distribution is approximated as

$$N*=N(gig0)e - Ei/kT \quad (2)$$

where gi and g0 are statistical factors that account for the number of equivalent energy levels for the excited state and the ground state, Ei is the energy of the excited state relative to a ground state energy, E0, k is Boltzmann's constant , and T is the temperature in Kelvin. From Equation 2 we expect that excited states with lower energies have larger populations and more intense emission lines. We also expect emission intensity to increase with temperature.

15.12 EQUIPMENT

An atomic emission spectrometer is similar in design to the instrumentation for atomic absorption. In fact, it is easy to adapt most flame atomic absorption spectrometers for atomic emission by turning off the hollow cathode lamp and monitoring the difference between the emission intensity when aspirating the sample and when aspirating a blank. Many atomic emission spectrometers, however, are dedicated instruments designed to take advantage of features unique to atomic emission, including the use of plasmas, arcs, sparks, and lasers as atomization and excitation sources, and an enhanced capability for multielemental analysis.

Atomization and Excitation

Atomic emission requires a means for converting into a free gaseous atom an analyte that is present in a solid, liquid, or solution sample. The same source of thermal energy used for atomization usually serves as the excitation source. The most common methods are flames and plasmas, both of which are useful for liquid or solution samples. Solid samples are analyzed by dissolving in a solvent and using a flame or plasma atomizer.

Flame Sources

Atomization and excitation in flame atomic emission is accomplished with the same nebulization and spray chamber assembly used in atomic absorption (Figure 1). The burner head consists of a single or multiple slots, or a Meker-style burner. Older atomic emission instruments often used a total consumption burner in which the sample is drawn through a capillary tube and injected directly into the flame.

A Meker burner is similar to the more common Bunsen burner found in most laboratories; it is designed to allow for higher temperatures and for a larger diameter flame.

Plasma Sources

A plasma is a hot, partially ionized gas that contains an abundant concentration of cations and electrons. The plasma used in atomic emission is formed by ionizing a flowing stream of argon gas, producing argon ions and electrons. A plasma's high temperature results from resistive heating as the electrons and argon ions move through the gas. Because a plasma operates at a much higher temperature than a flame, it provides for a better atomization efficiency and a higher population of excited states.

A schematic diagram of the inductively coupled plasma source (ICP) is shown in Figure 2 . The ICP torch consists of three concentric quartz tubes, surrounded at the top by a radio-frequency induction coil. The sample is mixed with a stream of Ar using a nebulizer, and is carried to the plasma through the torch's central capillary tube. Plasma formation is initiated by a spark from a Tesla coil. An alternating radio-frequency current in the induction coil creates a fluctuating magnetic field that induces the argon ions and the electrons to move in a circular path. The resulting collisions with the abundant unionized gas give rise to resistive heating, providing temperatures as high as 10000 K at the base of the plasma, and between 6000 and 8000 K at a height of 15–20 mm above the coil, where emission usually is measured. At these high temperatures the outer quartz tube must be thermally isolated from the plasma. This is accomplished by the tangential flow of argon shown in the schematic diagram.

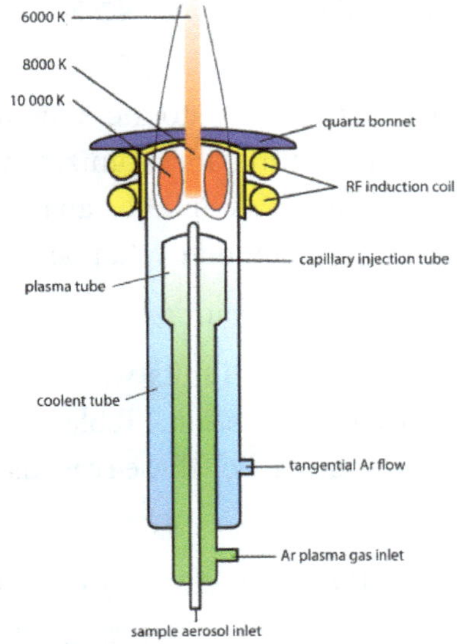

Fig. 2: Schematic diagram of an inductively coupled plasma torch.

Multielemental Analysis

Atomic emission spectroscopy is ideally suited for a multielemental analysis because all analytes in a sample are excited simultaneously. If the instrument includes a scanning monochromator, we can program it to move rapidly to an analyte's desired wavelength, pause to record its emission intensity, and then move to the next analyte's wavelength. This sequential analysis allows for a sampling rate of 3–4 analytes per minute.

Another approach to a multielemental analysis is to use a multichannel instrument that allows us to monitor simultaneously many analytes. A simple design for a multichannel spectrometer, shown in Figure 3, couples a monochromator with multiple detectors that are positioned in a semicircular array around the monochromator at positions that correspond to the wavelengths for the analytes.

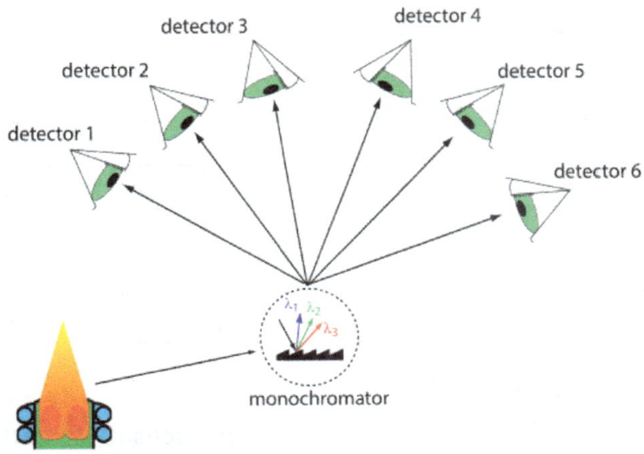

Fig. 3: Schematic diagram of a multichannel atomic emission spectrometer for the simultaneous analysis of several elements. Instruments may contain as many as 48–60 detectors.

15.13 QUANTITATIVE APPLICATIONS

Atomic emission is used widely for the analysis of trace metals in a variety of sample matrices. The development of a quantitative atomic emission method requires several considerations, including choosing a source for atomization and excitation, selecting a wavelength and slit width, preparing the sample for analysis, minimizing spectral and chemical interferences, and selecting a method of standardization.

Choice of Atomization and Excitation Source

Except for the alkali metals, detection limits when using an ICP are significantly better than those obtained with flame emission (Table 1). Plasmas also are subject to fewer spectral and chemical interferences. For these reasons a plasma emission source is usually the better choice.

Table . Detection Limits for Atomic Emission

Element	Detection limit (μg/ml): flame emission	Detection limit (μg/ml): icp
Ag	2	0.2
Al	3	0.2
As	2000	2
Ca	0.1	0.0001
Cd	300	0.07
Co	5	0.1
Cr	1	0.08
Fe	10	0.09
Hg	150	1
K	0.01	30
Li	0.001	0.02
Mg	1	0.02
Mn	1	0.01
Na	0.01	0.1
Ni	10	0.2
Pb	0.2	1
Pt	2000	0.9
Sn	100	3
Zn	1000	0.1

Source: Parsons, M. L.; Major, S.; Forster, A. R.; App. Spectrosc. 1983, 37, 411–418.

Selecting the Wavelength and Slit Width

The choice of wavelength is dictated by the need for sensitivity and the need to avoid interferences from the emission lines of other constituents in the sample. Because an analyte's atomic emission spectrum has an abundance of emission lines—particularly when using a high temperature plasma source—it is inevitable that there will be some overlap between emission lines. For example, an analysis for Ni using the atomic emission line at 349.30 nm is complicated by the atomic emission line for Fe at 349.06 nm.

A narrower slit width provides better resolution, but at the cost of less radiation reaching the detector. The easiest approach to selecting a wavelength is to record the sample's emission spectrum and look for an emission line that provides an intense signal and is resolved from other emission lines.

Preparing the Sample

Flame and plasma sources are best suited for samples in solution and in liquid form. Although a solid sample can be analyzed by directly inserting it into the flame or plasma, they usually are first brought into solution by digestion or extraction.

Minimizing Spectral Interferences

The most important spectral interference is broad, background emission from the flame or plasma and emission bands from molecular species. This background emission is particularly severe for flames because the temperature is insufficient to break down refractory compounds, such as oxides and hydroxides. Background corrections for flame emission are made by scanning over the emission line and drawing a baseline (Figure.4). Because a plasma's temperature is much higher, a background interference due to molecular emission is less of a problem. Although emission from the plasma's core is strong, it is insignificant at a height of 10–30 mm above the core where measurements normally are made.

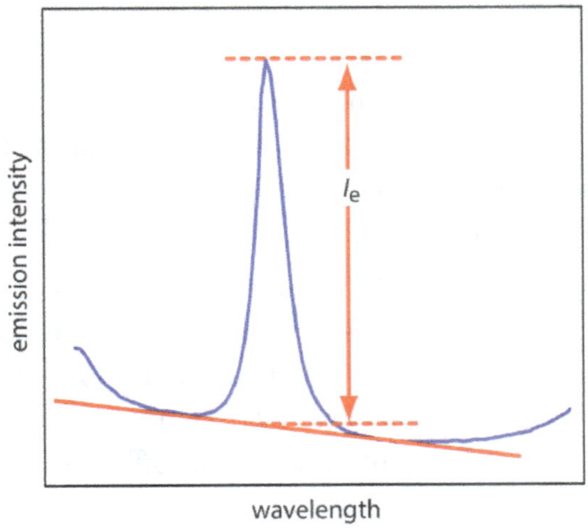

Fig. 4: Method for correcting an analyte's emission for the flame's background emission.

Minimizing Chemical Interferences

Flame emission is subject to the same types of chemical interferences as atomic absorption; they are minimized using the same methods: by adjusting the flame's composition and by adding protecting agents, releasing agents, or ionization suppressors. An additional chemical interference results from *self-absorption*. Because the flame's temperature is greatest at its center, the concentration of analyte atoms in an excited state is greater at the flame's center than at its outer edges. If an excited state atom in the flame's center emits a photon, then a ground state atom in the cooler, outer regions of the flame may absorb the photon, which decreases the emission intensity. For higher concentrations of analyte self-absorption may invert the center of the emission band (Figure 5).

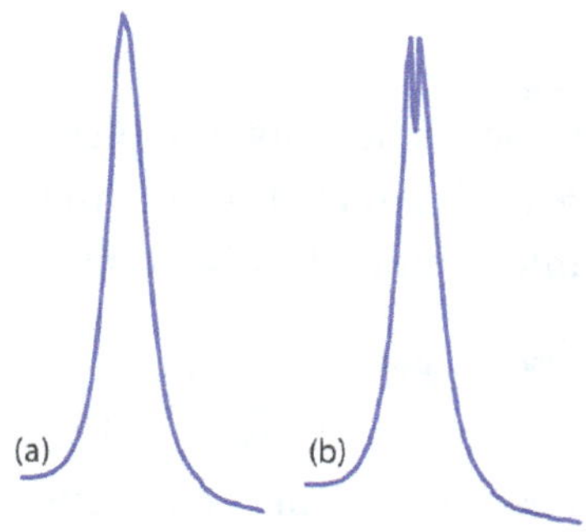

Fig. 5: Atomic emission lines for (a) a low concentration of analyte, and (b) a high concentration of analyte showing the effect of self-absorption.

Chemical interferences when using a plasma source generally are not significant because the plasma's higher temperature limits the formation of nonvolatile species For example, PO_3^{-4} is a significant interferent when analyzing samples for Ca^{2+} by flame emission, but has a negligible effect when using a plasma source. In addition, the high concentration of electrons from the ionization of argon minimizes ionization interferences.

Standardizing the Method

From Equation.1 we know that emission intensity is proportional to the population of the analyte's excited state, N∗N∗. If the flame or plasma is in thermal equilibrium, then the excited state population is proportional to the analyte's total population, *N*, through the Boltzmann distribution (Equation 2).

A calibration curve for flame emission usually is linear over two to three orders of magnitude, with ionization limiting linearity when the analyte's concentrations is small and self-absorption limiting linearity at higher concentrations of analyte. When using a plasma, which suffers from fewer chemical interferences, the calibration curve often is linear over four to five orders of magnitude and is not affected significantly by changes in the matrix of the standards.

Emission intensity is affected significantly by many parameters, including the temperature of the excitation source and the efficiency of atomization. An increase in temperature of 10 K, for example, produces a 4% increase in the fraction of Na atoms in the 3*p* excited state, an uncertainty in the signal that may limit the use of external standards. The method of internal standards is used when the variations in source parameters are difficult to control. To compensate for changes in the temperature of the excitation source, the internal standard is selected so that its emission line is close to the analyte's emission line. In addition, the internal standard should be subject to the same chemical interferences to compensate for changes in atomization efficiency. To accurately correct for these errors the analyte and internal standard emission lines are monitored simultaneously.

15.14 DETERMINATION OF SODIUM IN A SALT SUBSTITUTE

The best way to appreciate the theoretical and the practical details discussed in this section is to carefully examine a typical analytical method. Although each method is unique, the following description of the determination of sodium in salt substitutes provides an instructive example of a typical procedure.

Description of Method

Salt substitutes, which are used in place of table salt for individuals on low-sodium diets, replaces NaCl with KCl. Depending on the brand, fumaric acid, calcium hydrogen phosphate, or potassium tartrate also are present. Although intended to be sodium-free, salt substitutes contain small amounts of NaCl as an impurity. Typically, the concentration of sodium in a salt substitute is about 100 μg/g The exact concentration of sodium is determined by flame atomic emission. Because it is difficult to match the matrix of the standards to that of the sample, the analysis is accomplished by the method of standard additions.

Procedure

A sample is prepared by placing an approximately 10-g portion of the salt substitute in 10 mL of 3 M HCl and 100 mL of distilled water. After the sample has dissolved, it is transferred to a 250-mL volumetric flask and diluted to volume with distilled water. A series of standard additions is prepared by placing 25-mL portions of the diluted sample into separate 50-mL volumetric flasks, spiking each with a known amount of an approximately 10 mg/L standard solution of Na^+, and diluting to volume. After zeroing the instrument with an appropriate blank, the instrument is optimized at a wavelength of 589.0 nm while aspirating a standard solution of Na^+. The emission intensity is measured for each of the standard addition samples and the concentration of sodium in the salt substitute is reported in μg/g.

15.15 EVALUATION OF ATOMIC EMISSION SPECTROSCOPY

Scale of Operation

The scale of operations for atomic emission is ideal for the direct analysis of trace and ultratrace analytes in macro and meso samples. With appropriate dilutions, atomic emission can be applied to major and minor analytes.

Accuracy

When spectral and chemical interferences are insignificant, atomic emission can achieve quantitative results with accuracies of 1–5%. For flame emission, accuracy frequently is limited by chemical interferences. Because the higher temperature of a plasma source gives rise to more emission lines, accuracy when using plasma emission often is limited by stray radiation from overlapping emission lines.

Precision

For samples and standards in which the analyte's concentration exceeds the detection limit by at least a factor of 50, the relative standard deviation for both flame and plasma emission is about 1–5%. Perhaps the most important factor that affect precision is the stability of the flame's or the plasma's temperature. For example, in a 2500 K flame a temperature fluctuation of ±2.5±2.5 K gives a relative standard deviation of 1% in emission intensity. Significant improvements in precision are realized when using internal standards.

Sensitivity

Sensitivity is influenced by the temperature of the excitation source and the composition of the sample matrix. Sensitivity is optimized by aspirating a standard solution of analyte and maximizing the emission by adjusting the flame's composition and the height from which we monitor the emission. Chemical interferences, when present, decrease the sensitivity of the analysis. Because the sensitivity of plasma emission is less affected by the sample matrix, a calibration curve prepared using standards in a matrix of distilled water is possible even for samples that have more complex matrices.

Selectivity

The selectivity of atomic emission is similar to that of atomic absorption. Atomic emission has the further advantage of rapid sequential or simultaneous analysis of multiple analytes.

Time, Cost, and Equipment

Sample throughput with atomic emission is rapid when using an automated system that can analyze multiple analytes. For example, sampling rates of 3000 determinations per hour are possible using a multichannel ICP, and sampling rates of 300 determinations per hour when using a sequential ICP. Flame emission often is accomplished using an atomic absorption spectrometer, which typically costs between \$10,000–\$50,000. Sequential ICP's range in price from \$55,000–\$150,000, while an ICP capable of simultaneous multielemental analysis costs between \$80,000–\$200,000. Combination ICP's that are capable of both sequential and simultaneous analysis range in price from \$150,000–\$300,000. The cost of Ar, which is consumed in significant quantities, can not be overlooked when considering the expense of operating an ICP.

15.16 SPECTROSCOPY BASED ON SCATTERING

The blue color of the sky during the day and the red color of the sun at sunset are the result of light scattered by small particles of dust, molecules of water, and other gases in the atmosphere. The efficiency of a photon's scattering depends on its wavelength. We see the sky as blue during the day because violet and blue light scatter to a greater extent than other, longer wavelengths of light. For the same reason, the sun appears red at sunset because red light is less efficiently scattered and is more likely to pass through the atmosphere than other wavelengths of light. The scattering of radiation has been studied since the late 1800s, with applications beginning soon thereafter. The earliest quantitative applications of scattering, which date from the early 1900s, used the elastic scattering of light by colloidal suspensions to determine the concentration of colloidal particles.

Origin of Scattering

If we send a focused, monochromatic beam of radiation with a wavelength λλ through a medium of particles with dimensions <1.5λ<1.5λ, the radiation scatters in all directions. For example, visible radiation of 500 nm is scattered by particles as large as 750 nm in the longest dimension. Two general categories of scattering are recognized. In elastic scattering, radiation is first absorbed by the particles and then emitted without undergoing a change in the radiation's energy. When the radiation emerges with a change in energy, the scattering is inelastic. Only elastic scattering is considered in this text.

Elastic scattering is divided into two types: Rayleigh, or small-particle scattering, and large-particle scattering. Rayleigh scattering occurs when the scattering particle's largest dimension is less than 5% of the radiation's wavelength. The intensity of the scattered radiation is proportional to its frequency to the fourth power, ν4ν4—which accounts for the greater scattering of blue light than red light—and is distributed symmetrically (Figure1 a). For larger particles, scattering increases in the forward direction and decreases in the backward direction as the result of constructive and destructive interferences (Figure 1 b).

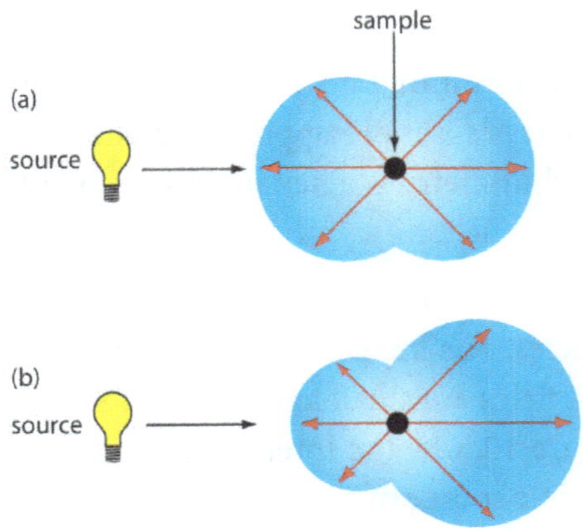

Fig. 1: Distribution of radiation for (a) Rayleigh, or small-particle scattering, and (b) large-particle scattering.

Turbidimetry and Nephelometry

Turbidimetry and nephelometry are two techniques that rely on the elastic scattering of radiation by a suspension of colloidal particles. In turbidimetry the detector is placed in line with the source and the decrease in the radiation's transmitted power is measured. In nephelometry the scattered radiation is measured at an angle of 90o to the source. The similarity of turbidimetry to absorbance spectroscopy and of nephelometry to fluorescence spectroscopy is evident in the instrumental designs . In fact, we can use a UV/Vis spectrophotometer for turbidimetry and we can use a spectrofluorometer for nephelometry.

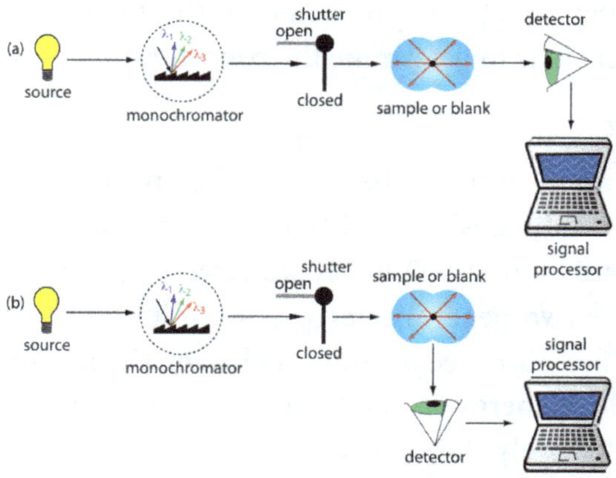

Fig. 2: Schematic diagrams for (a) a turbimeter, and (b) a nephelometer.

Turbidimetry or Nephelometry?

When developing a scattering method the choice between using turbidimetry or using nephelometry is determined by two factors. The most important consideration is the intensity of the scattered radiation relative to the intensity of the source's radiation. If the solution contains a small concentration of scattering particles, then the intensity of the transmitted radiation, IT, is approximately the same as the intensity of the source's radiation, I0. As we learned earlier in the section on molecular absorption, there is substantial uncertainty in determining a small difference between two intense signals. For this reason, nephelometry is a more appropriate choice for a sample that contains few scattering particles. Turbidimetry is a better choice when the sample contains a high concentration of scattering particles.

A second consideration in choosing between turbidimetry and nephelometry is the size of the scattering particles. For nephelometry, the intensity of scattered radiation at 90o increases when the particles are small and Rayleigh scattering is in effect. For larger particles, as shown in Figure 10.8.1 , the intensity of scattering decreases at 90o. When using an ultraviolet or a visible source of radiation, the optimum particle size is 0.1–1 μm. The size of the scattering particles is less important for turbidimetry where the signal is the relative decrease in transmitted radiation. In fact, turbidimetric measurements are feasible even when the size of the scattering particles results in an increase in reflection and refraction, although a linear relationship between the signal and the concentration of scattering particles may no longer hold.

Determining Concentration by Turbidimetry

For turbidimetry the measured transmittance, T, is the ratio of the intensity of source radiation transmitted by the sample, I_T, to the intensity of source radiation transmitted by a blank, I_0.

$$T = ITI0$$

The relationship between transmittance and the concentration of the scattering particles is similar to that given by Beer's law

$$-\log T = kbC \quad (1)$$

where C is the concentration of the scattering particles in mass per unit volume (w/v), b is the pathlength, and k is a constant that depends on several factors, including the size and shape of the scattering particles and the wavelength of the source radiation. The exact relationship is established by a calibration curve prepared using a series of standards that contain known concentrations of analyte. As with Beer's law, Equation 1 may show appreciable deviations from linearity.

Determining Concentration by Nephelometry

For nephelometry the relationship between the intensity of scattered radiation, I_S, and the concentration of scattering particles is

$$Is=kI0C \qquad (2)$$

where k is an empirical constant for the system and I_0 is the intensity of the source radiation. The value of k is determined from a calibration curve prepared using a series of standards that contain known concentrations of analyte.

Selecting a Wavelength for the Incident Radiation

The choice of wavelength is based primarily on the need to minimize potential interferences. For turbidimetry, where the incident radiation is transmitted through the sample, a monochromator or filter allow us to avoid wavelengths that are absorbed instead of scattered by the sample. For nephelometry, the absorption of incident radiation is not a problem unless it induces fluorescence from the sample. With a nonfluorescent sample there is no need for wavelength selection and a source of white light may be used as the incident radiation. For both techniques, other considerations in choosing a wavelength including the intensity of scattering, the trans- ducer's sensitivity (many common photon transducers are more sensitive to radiation at 400 nm than at 600 nm), and the source's intensity.

Preparing the Sample for Analysis

Although Equation 1 and Equation 2 relate scattering to the concentration of the scattering particles, the intensity of scattered radiation also is influenced by the size and the shape of the scattering particles. Samples that contain the same number of scattering particles may show significantly different values for $-\log T$ or I_S depending on the average diameter of the particles. For a quantitative analysis, therefore, it is necessary to maintain a uniform distribution of particle sizes throughout the sample and between samples and standards.

Most turbidimetric and nephelometric methods rely on precipitation reaction to form the scattering particles. Precipitate's properties, including particle size, are determined by the conditions under which it forms. To maintain a reproducible distribution of particle sizes between samples and standards, it is necessary to control parameters such as the concentration of reagents, the order of adding reagents, the pH and temperature, the agitation or stirring rate, the ionic strength, and the time between the precipitate's initial formation and the measurement of transmittance or scattering. In many cases a surface-active agent—such as glycerol, gelatin, or dextrin—is added to stabilize the precipitate in a colloidal state and to prevent the coagulation of the particles.

Applications

Turbidimetry and nephelometry are used to determine the clarity of water. The primary standard for measuring clarity is formazin, an easily prepared, stable polymer suspension

. A stock standard of formazin is prepared by combining a 1g/100mL solution of hydrazine sulfate, $N_2H_4H_2SO_4$, with a 10 g/100 mL solution of hexamethylenetetramine to produce a suspension of particles that is defined as 4000 nephelometric turbidity units (NTU). A set of external standards with NTUs between 0 and 40 is prepared by diluting the stock standard. This method is readily adapted to the analysis of the clarity of orange juice, beer, and maple syrup.

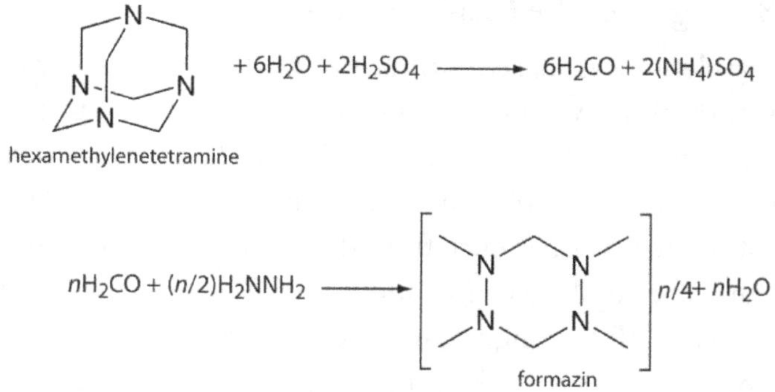

Fig. 3: Scheme for preparing formazin for use as a turbidity standard.

A number of inorganic cations and anions are determined by precipitating them under well-defined conditions. The transmittance or scattering of light, as defined by

Equation 1 or Equation 2, is proportional to the concentration of the scattering particles, which, in turn, is related by the stoichiometry of the precipitation reaction to the analyte's concentration. Several examples of analytes determined in this way are listed in Table 1 .

Table 1 . Examples of Analytes Determined by Turbidimetry or Nephelometry

Analyte	Precipitant	Precipitate
Ag^+	NaCl	AgCl
Ca^{2+}	$Na_2C_2O_4$	CaC_2O_4
Cl^-	$AgNO_3$	AgCl
CN^-	$AgNO_3$	AgCN
$CO_2{}^{-3}CO_3{}^{2-}$	$BaCl_2$	$BaCO_3$
F^-	$CaCl_2$	CaF_2
SO2−4SO42−	$BaCl_2$	$BaSO_4$

15.17 TURBIDIMETRIC DETERMINATION OF SULFATE IN WATER

The best way to appreciate the theoretical and the practical details discussed in this section is to carefully examine a typical analytical method. Although each method is unique, the following description of the determination of sulfate in water provides an instructive example of a typical procedure.

Description of Method

Adding BaCl2 to an acidified sample precipitates $SO_2{}^{-4}SO_4{}^{2-}$ as $BaSO_4$. The concentration of $SO_2{}^{-4}SO_4{}^{2-}$is determined either by turbidimetry or by nephelometry using an incident source of radiation of 420 nm. External standards that contain know concentrations of $SO_2{}^{-4}SO_4{}^{2-}$ are used to standardize the method.

Procedure

Transfer a 100-mL sample to a 250-mL Erlenmeyer flask along with 20.00 mL of an appropriate buffer. For a sample that contains more than 10 mg $SO_2{}^{-}4SO4^{2-}$/L, the buffer's composition is 30 g of $MgCl_26H_2O$, 5 g of $CH_3COONa3H_2O$, 1.0 g of KNO_3, and 20 mL of glacial CH_3COOH per liter. The buffer for a sample that contain less than 10 mg SO_2−4SO42−/L is the same except for the addition of 0.111 g of Na_2SO_4 per L.

Place the sample and the buffer on a magnetic stirrer operated at the same speed for all samples and standards. Add a spoonful of 20–30 mesh $BaCl_2$, using a measuring spoon with a capacity of 0.2–0.3 mL, to precipitate the $SO_2{}^{-4}SO_4{}^{2-}$ as $BaSO_4$. Begin timing when the $BaCl_2$ is added and stir the suspension for 2 ± 60 s. When the stirring is complete, allow the solution to sit without stirring for 0.5 ±5.0 min before measuring its transmittance or its scattering.

Prepare a calibration curve over the range 0–40 mg $SO_2^{-}4SO_4^{2-/}$L by diluting a stock standard that is 100-mg $SO_2^{-}4SO_4^{2-}$/L. Treat each standard using the procedure described above for the sample. Prepare a calibration curve and use it to determine the amount of sulfate in the sample.

16

STATISTICAL TREATMENT OF DATA

16.1 INTRODUCTION

Quantitative chemical analysis involves making experimental measurements that inherently contain errors. These errors lead to uncertainty in the results and it is necessary to quantify this uncertainty if a result is to be useful. Statistical methods allow us to accomplish this in a meaningful way. This topic introduces basic descriptive parameters used in statistics and the concept of probability distributions. This leads to the treatment of confidence intervals and statistical tests.

16.2 MEASUREMENT ERRORS

The final result of a quantitative analysis is typically one or more numbers that indicate the concentrations of analytics in the sample. There is always some uncertainty associated with each operation or measurement in an analysis, and so there is always some uncertainty in the final result. Knowing the uncertainty is as important as knowing the final result. Having data that are so uncertain as to be useless is no better than having no data at all. Thus there is a need to describe and reduce, if necessary and possible, this uncertainty.

Two terms that are very important in describing measurement errors are *accuracy* and *precision*, as defined in Topic 1. Accuracy relates to the closeness of an experimental result to the actual (correct) value, while precision describes the ability to reproduce the same result (not necessarily the correct one) when the experiment is repeated. Generally, unless we know the true answer, it is not possible to determine the accuracy. Precision can be determined, but a precise result does not necessarily mean an accurate one. However, it is normal to place more "faith" in the results of an analyst who has good precision, since good precision usually means greater attention to detail.

16.2.1 Types of Errors

Good accuracy should be the object of all chemical measurements. The question thus arises, how is accuracy measured? This topic is best introduced by discussing errors, since a measurement will be accurate if there are no errors. There are three general types of errors: blunders, determinate errors, and indeterminate errors.

16.2.1.1

Gross errors, sometimes called "blunders" are mistakes that are obvious to the analyst and that are unlikely to be repeated when another analysis is performed. An example of a gross error is forgetting to add a reagent that is required for analysis.

16.2.1.2

Determinate errors, often referred to as "bias", are also mistakes, but are less obvious to the analyst and are likely to be repeated when other analyses are performed. In theory many determinate errors, if recognized, could be eliminated by improving technique. An example of a determinate error would be the use of an analytical balance that has gone out of calibration to weigh a sample, producing results that are always low or high. Determinate errors are unidirectional in that they always produce an error in the same direction.

16.2.1.3

Indeterminate errors, often referred to as "random errors", are caused by the need to make estimates in the last figure of a measurement, by noise present in instruments, or other unbiased factors that lead to uncertainty in the estimated value. Such errors can be reduced, but never entirely eliminated. An example of indeterminate error is the variation in the values taken from a buret due to the need to estimate the last significant figure. The observer would certainly say, referring to Figure 1, that the volume delivered is between 40.2 and 40.3 mL. However, one can and should do better than this. Some observers might say that the volume was 40.24, some 40.25, and others 40.26 mL. This uncertainty in the last place is an example of indeterminate error. It is better to report the volume as 40.26 mL rather than to report its being between 40.2 and 40.3 mL, even though it is realized that the 0.06 is an intelligent estimate. Indeterminate errors are bidirectional; being random in nature they are as likely to produce a high result as a low result.

Note that determinate error may also be involved in estimating the last figure of a measurement. In the case being discussed, parallax error (looking at the meniscus at an angle) and number bias (selecting some numerals, e.g. 0 or 5, more often than others in the estimate) are two examples of determinate error. Since determinate errors are usually unidirectional, they will always cause an error in the final result, even if many measurements are averaged. Indeterminate errors, being bidirectional, will tend to cancel if enough measurements are averaged. These ideas are perhaps best illustrated by Figure 2.

In Case 1 shown in the figure, the precision is poor, but the accuracy will be good if enough data are taken and averaged, since the errors are approximately equally distributed about the correct answer. In Case 2, the precision is good and the accuracy is also good.

In Case 3, the precision is good, but the accuracy is poor due to a determinate error. If determinate error is present, averaging will not help to reduce its contribution.

The precision of the final result can always be improved by making replicate measurements and the extent of improvement is described in the discussion on confidence intervals. Even if many replicates are not required, it is standard practice in most analytical procedures to make two or three separate measurements so that precision can be assessed and gross errors identified.

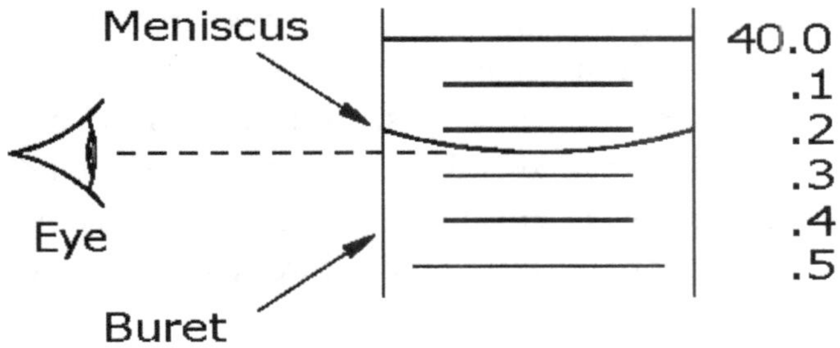

Fig. 1: Determination of the volume reading in a buret. The final digit is estimated from the position of the meniscus and gives rise to indeterminate errors.

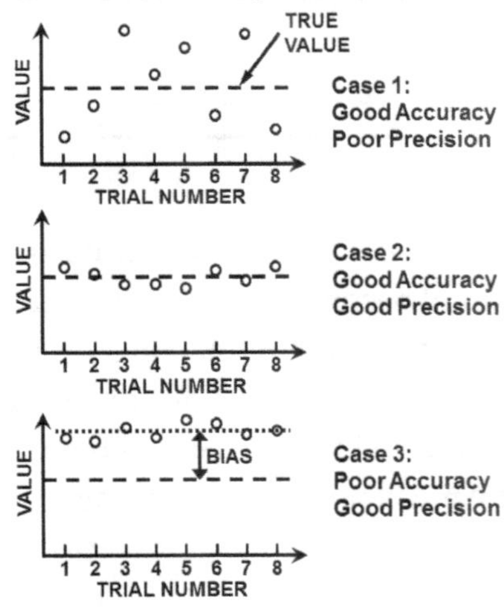

Fig. 2: Comparison of different situations of accuracy and precision.

To evaluate accuracy and assess indeterminate errors, the usual procedure is to analyze a standard (a sample of accurately known composition) and compare the analytical results with the known sample composition. If the values are the same, bias is assumed to be absent. Many organizations, *e.g.* the National Research Council of Canada and the National Institute of Standards and Technology in the U.S.A., provide such standards. A statistical test, described later, can be used to compare the results.

16.3 STATISTICAL PARAMETERS

In statistics we usually distinguish between parameters that pertain to a population and parameters that are estimated from a finite data set. A population would result if a very large number of measurements (strictly an infinite number) were made. A finite data set, or "sample", results when just a few measurements are made, as in the case of experimental measurements. The parameters calculated from a finite data set, usually represented by "English" characters, are only estimates of the population parameters, usually represented by Greek characters. It is the population parameters that one would like to obtain, if possible. Much of statistics is involved with trying to make inferences about the population parameters from the sample parameters.

Since there is always uncertainty involved in measurements, a data population or a data set will consist of a distribution of values rather than a single value. Such distributions have characteristics that are described by statistical parameters. The first such characteristic to be investigated is the "central tendency" of a distribution. Many distributions have some values that are more often observed than other values and these are usually toward the center of the distribution. Three methods of defining the "central tendency" are listed below.

16.3.1 The Mean

The mean, or average, is calculated by summing all the values, xi , in the set and dividing by the number of trials, N.

$$mean = \bar{x} = \frac{x_1 + x_2 + x_3 + \ldots + x_N}{N} = \frac{1}{N}\sum_{i=1}^{N} x_i \qquad 1$$

As N → ∞, x → μ, where μ is the population mean. In the absence of determinate error, μ is the true value, and as N becomes larger the estimate, x , of the population mean improves The median The median is the "central point" in a data set. Half of all the values in a set will lie above the median, half will lie below the median. If the set contains an odd number of data points, the median will be the central point of that set. If the set contains an even number of points, the median will be the average of the two central points.

16.3.2 The Mode

The mode is the value that occurs most often (the value with the highest probability). The mode is quite often not obvious in a finite data set where no two values are the same. In populations where errors are evenly distributed about the mean (the distribution is symmetric), the mean, the median, and the mode will all have the same value. In this text, the mean will be the most useful parameter for indicating the "central tendency" of a data set or population.

Accuracy and how one attempts to achieve and measure it have been discussed. The topic of precision comes next. How does one measure precision? Obviously one must have made more than one measurement to get an estimate of precision. The characteristic of a distribution that relates to precision is its "dispersion". The parameters used to describe the dispersion are listed below.

16.3.3 The Standard Deviation

The standard deviation (and those parameters related to it) is the only statistically acceptable measure of precision and is the only measure to be discussed in detail here. It is calculated from the mean and the values of the trials in the data set according to:

$$s=\sqrt{\frac{(x_1-\bar{x})^2+(x_2-\bar{x})^2+\ldots+(x_N-\bar{x})^2}{N-1}}=\sqrt{\frac{\sum(x_i-\bar{x})^2}{N-1}} \qquad 2$$

The values of $(x\ i\ -\ x)$ are the deviations from the mean and the quantity in the denominator is called "the degrees of freedom". When the standard deviation is estimated from the sample mean as opposed to the population mean, the degrees of freedom are one less than the number of measurements used in the set, or (*N*-1). The reason for this is that one degree of freedom is lost when calculating the mean of the data set. As with the mean, as $N \rightarrow \infty, s \rightarrow \sigma$, where σ is the population standard deviation. The sample standard deviation, *s*, is only an estimate of σ, but as *N* gets larger, that estimate gets better.

16.3.4 The Relative Standard Deviation

Often the relative standard deviation (RSD) is often more useful than the absolute standard deviation (given above), since it immediately provides an idea of the precision of the data set relative to its individual values. There are several ways of expressing the RSD.

x

$$\text{Percent RSD} = \%RSD = s_r \times 100\% = \frac{s}{\bar{x}} \times 100\% \qquad 3$$

$$\text{Percent RSD} = \%RSD = s_r \times 100\% = \frac{s}{\bar{x}} \times 100\% \qquad 4$$

x

$$\text{Percent RSD} = \%RSD = s_r \times 100\% = \frac{s}{\bar{x}} \times 100\% \qquad 5$$

The relative standard deviation in percent is often referred to as the "coefficient of variation". Note that the absolute standard deviation has the same units as the mean, whereas the relative standard deviation has no units. In some analytical determinations the relative standard deviation is reported in parts per thousand (ppt).

16.3.5 The Variance

The variance is another method of describing dispersion and is often preferred by statisticians because variances are directly additive (as will be seen later). The variance is simply the square of the standard deviation and is normally indicated by the symbol s^2, or σ^2 in the case of population variance.

16.3.6 The Range

The range or spread is simply the difference between the largest and smallest values in a data set.

16.4 THE NORMAL DISTRIBUTION

Many different data distributions are found in nature. Many types of data, including most chemical measurements subject to random errors, follow a Gaussian distribution. This is a symmetrical, bell-shaped distribution. Since so many data types follow this distribution it is usually referred to as the normal distribution and the mathematical equation that describes this distribution is called the normal error curve.

$$f(x) = \frac{1}{\sigma\sqrt{2\pi}} \exp\left\{-\frac{(x-\mu)^2}{2\sigma^2}\right\} \qquad 6$$

In this equation $f(x)$ is called the probability density function (PDF), x represents the trial values, μ is the population mean, and σ is the population standard deviation. The PDF gives the relative frequency that any value of x will be observed and is a function of the standard deviation and the position of x with respect to the mean. A graph of the normal distribution is shown in Figure 3. In this figure the mean is taken as zero and the standard deviation as unity. This is the standard representation of the normal error curve.

The ability to describe the normal distribution mathematically using Eqn. 6 is very useful. From Figure 3 it is obvious that the mean of a normally distributed data population is the value that occurs most often (has the highest frequency). Since the distribution is symmetrical, the mean, the median, and the mode are all the same. It is less obvious that inflection points occur at $\mu \pm \sigma$. The area under the curve between x^1 and x^2, as determined by integration, gives the probability that any trial value, x, will be found between x^1 and x^2.

The area between $+\infty$ and $-\infty$ is exactly 1, as it should be, since the probability of finding any value of x in this range must be unity. The area between μ and $\mu+\sigma$ is 0.3413, which means that 34.13% of all x values in a normally distributed population fall between the mean and the mean plus one standard deviation. Since the distribution is symmetrical, 34.13% of all values will fall between the mean and the mean minus one standard deviation. Thus 68.26% of all values of x will fall within $\pm 1\sigma$ of μ. Likewise it can be shown that 95.44% of all values of x will fall within $\pm 2\sigma$ of μ and 99.74% will fall within $\pm 3\sigma$ of μ. Thus, in a normally distributed population it will be very unlikely (the chances are less than 3 in 1000) that any value of x will be outside the $\pm 3\sigma$ limits.

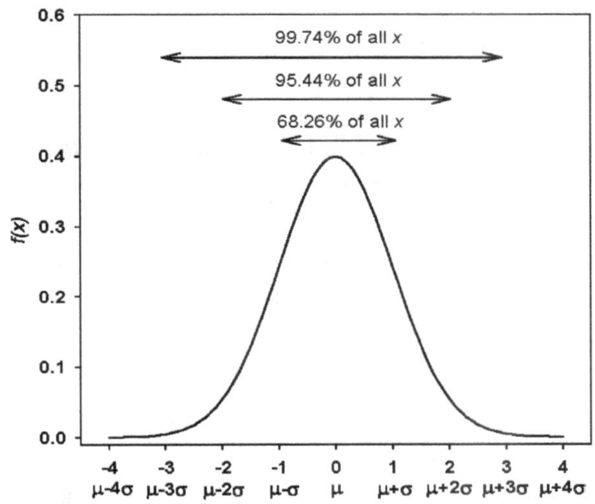

Fig. 3: Graph of the normal probability density function, which applies to many distributions in nature, including experimental measurement errors.

16.5 THE CONFIDENCE INTERVAL

One would often like to know how close the mean of a data set is likely to be to the true answer (the population mean), assuming that there is no determinate error. Although we can never know this exactly, we can use the normal distribution to establish a *confidence interval* (CI). This is a region in which we can assume that the true value exists with a certain probability. The bigger this region, the greater the probability that it contains the true value, but because of the nature of the normal distribution, we can never be 100% certain. The confidence interval is given by: Confidence interval, C.I.

$$\text{Confidence interval, C.I.} = \bar{x} \pm \frac{ts}{\sqrt{N}} \quad \text{with } P\% \text{ confidence} \qquad 7$$

As expected, the width of the confidence interval is dependent on the standard deviation (precision) of the measurements, since this defines the spread of the normal distribution. It also depends inversely on the square root of the number of measurements, *N*. This is because we are really interested in the standard deviation of the *mean* and not the individual measurements. The standard deviation of the mean of *N* values is given by:

$$sx = s\ \sqrt{N} \qquad 8$$

The inclusion of this term makes intuitive sense, since we expect the confidence interval to become narrower as we obtain more measurements. Finally, the quantity *t* is a constant that is related to the confidence interval we want to report (*e.g.* 95%). In an ideal situation, where we know the population standard deviation, σ, this would be the only factor defining *t*. However, since we use the sample standard deviation, *s*, the uncertainty in estimating σ must be taken into account, so *t* will become larger as *N* becomes smaller. The values of *t* (called the *critical values*) are taken from what is called a

Student's t-distribution, and these are shown in Table 1 as a function of degrees of freedom (*N*−1) and the confidence interval. An example will illustrate the calculation of some of the statistical quantities that have been introduced.

Table 1. Critical values for the Student's t-distribution at selected confidence levels.

Deg. of Freedom	Confidence Interval				
	80%	90%	95%	98%	99%
1	3.078	6.314	12.706	31.821	63.657
2	1.886	2.920	4.303	6.965	9.925
3	1.638	2.353	3.182	4.541	5.841
4	1.533	2.132	2.776	3.747	4.604
5	1.476	2.015	2.571	3.365	4.032
6	1.440	1.943	2.447	3.143	3.707
7	1.415	1.895	2.365	2.998	3.499
8	1.397	1.860	2.306	2.896	3.355
9	1.383	1.833	2.262	2.821	3.250
10	1.372	1.812	2.228	2.764	3.169
∞	1.282	1.645	1.960	2.326	2.576

Example 1: Statistical Parameters

Assume that the following values were obtained in the analysis of the weight of iron in 2.0000 g portions of a sample: 0.3791, 0.3784, 0.3793, 0.3779, and 0.3797 g. Calculate the mean, variance, standard deviation, relative standard deviation (RSD), the %RSD, the RSD in ppt, the range, the median, and the 80% and 95% confidence intervals.

The calculation of the statistical parameters is carried out as shown below, starting with the mean and degrees of freedom, since these are needed to calculate the variance and standard deviation.

$$\text{The mean} = \bar{x} = \frac{\sum x_i}{N} = \frac{0.3791 + 0.3784 + \ldots + 0.3797}{5} = \frac{1.8944}{5} = 0.37888\text{ g}$$

$$\text{The degrees of freedom} = \text{DF} = (N-1) = 4$$

$$\text{The variance} = s^2 = \frac{\sum (x_i - \bar{x})^2}{N-1} = \frac{(0.3791 - 0.37888)^2 + \ldots + (0.3797 - 0.37888)^2}{4}$$

$$= \frac{4.88\times10^{-8} + \ldots + 6.72\times10^{-7}}{4} = \frac{2.09\times10^{-6}\text{ g}^2}{4} = 5.2\times10^{-7}\text{ g}^2$$

$$\text{The standard deviation} = s = \sqrt{5.2\times10^{-7}\text{ g}^2} = 0.00072\text{ g}$$

$$\text{Relative standard deviation} = \text{RSD} = s_r = \frac{s}{\bar{x}} = \frac{0.00072\text{ g}}{0.37888\text{ g}} = 0.0019$$

$$\%RSD = r\,s \times \%0.19 = \%100 \times 0.0019 = \%100$$

$$RSD\ (ppt) = r\,s \times 1.9 = 1000 \times 0.0019 = 1000\ ppt$$

To easily see the range and median it is convenient to order the data in terms of increasing or decreasing values. This results in: 0.3779, 0.3784, 0.3791, 0.3793, and 0.3797 g. Since this data set has an odd number of trials, the median is simply the middle or 3rd datum, 0.3791 g. Note that for a finite data set the median and mean are not necessarily identical even if the distribution is symmetric. The range is 0.3797–0.3779 g or 0.0018 g.

From the critical *t* values listed in Table 1 (and in the Appendix, the value of *t* for 4 degrees of freedom and 80% confidence is 1.533, while that for 95% confidence is 2.776. With these values, we can calculate the corresponding confidence intervals.

$$80\%\ \text{C.I.} = \bar{x} \pm \frac{ts}{\sqrt{N}} = 0.37888\ \text{g} \pm \frac{(1.533)(0.00072\ \text{g})}{\sqrt{5}} = 0.3789 \pm 0.0005\ \text{g}$$

$$95\%\ \text{C.I.} = \bar{x} \pm \frac{ts}{\sqrt{N}} = 0.37888\ \text{g} \pm \frac{(2.776)(0.00072\ \text{g})}{\sqrt{5}} = 0.3789 \pm 0.0009\ \text{g}$$

Thus the population mean (and hopefully the true result) will lie between 0.3784 g and 0.3794 g 80% of the time and between 0.3780 g and 0.3798 g 95% of the time. Although this example illustrates the manual calculations necessary to calculate the statistical parameters directly, many calculators have the ability to evaluate these parameters through built-in functions. You are permitted (and advised) to use these to perform statistical calculations to save time and reduce errors, and *it is not necessary to show all of the calculations presented in the example if you use your calculator functions.*

However, you are responsible for knowing how to use these functions on your calculator. This involves knowing how to enter the data and extracting the results. Note that many calculators have two calculations for the standard deviation, one that uses (N–1) in the denominator (called the "sample standard deviation") and one that employs N in the denominator (called the "population standard deviation"). It is the former that should be used. In Figure 4, this is indicated by s*n*–1, but different designations are used on different calculators.

16.6 REPORTING RESULTS

An analytical result, or any experimental result, should be reported with an indication of its uncertainty or else it is not meaningful. In many circumstances, the uncertainty is implied by the last significant digit (±1 in the last digit is assumed), but when more reliable information, such as the standard deviation, is available, it should be used. Typically, the mean is reported as the determined value, since it should be the most precise estimate, and either the absolute or relative standard deviation is reported as well. For instance, the results in Example 1 could be reported in the following ways.

Fig. 4: Many calculators can compute statistical parameters automatically, as illustrated with the highlighted buttons shown here. You should learn how to use these.

0.3789 g ±0.0007 g, using the absolute standard deviation

0.3789 g with a RSD of 0.0019

0.3789 g ±0.19%

0.3789 g ±1.9 ppt

It is normally assumed that the number following a "±" will be the absolute standard deviation. The use of the various forms of the RSD can be confusing, particularly if a %RSD is given for a concentration that is also given as percentage, so if an RSD is given, this should be clearly indicated.

It is tempting to report the answer as 0.37888 g ± 0.00072 g, but this should not be done for two reasons: (1) the rules for significant figures do not allow this (it is assumed that you are familiar with these rules) and (2) the answer itself demonstrates how reporting should be done. The standard deviation indicates that there is an uncertainty of ±7 in the fourth place after the decimal point. Thus it makes no sense to list any figures past this point.

A disadvantage of using the standard deviation of the measurements in reporting is that it does not reflect the improvement in the precision of the result that comes from making more measurements. As the number of replicates increases, the estimate of the population standard deviation will improve ($s \rightarrow \sigma$) but it will not get smaller. Since the mean value is being reported, it is better to report the *standard deviation of the mean* ($sx = s\ N$) or the confidence interval (CI = $ts\ N$). Thus the result could also be reported as:

0.3789 g ±0.0003 g (using the standard deviation of the mean)

0.3789 g ±0.0009 g (using the 95% confidence interval)

Note that, although the confidence interval is wider than the standard deviation of the measurements, it is more statistically meaningful and will decrease as the number of measurements (*N*) increases.

Important points to remember in reporting analytical results are (1) whatever statistical measure is used to report the uncertainty of the result, it should be clearly indicated to avoid misinterpretation, and (2) when reporting absolute uncertainties, only one significant digit in of uncertainty should be given with the result and the number of decimal places shown in the result should be consistent with this uncertainty.

16.7 STATISTICAL TESTS FOR OUTLIERS

Statistical tests are often used to guide conclusions about experimental results in science. Such tests typically assume that some hypothesis, called the "null hypothesis" is true, and then test it against the "alternate hypothesis". While a statistical test cannot guarantee that the correct conclusion has been reached, it provides a certain level of confidence in the answer. Such tests involve the calculation of a *test statistic*, which is then compared to a tabulated *critical value* at a certain *confidence level*. The null hypothesis is rejected if the test statistic is greater than the critical value.

One of the most useful statistical tests involves testing for erroneous data. Sometimes a value in a set of replicate measurements appears so far from the other measurements that one suspects that the value must have been the result of some large unknown error not present in the other measurements. If this is the case, the suspect value is an *outlier* and, because it does not come from the same distribution as the other measurements, it should be excluded. On the other hand, the value may be just somewhat different from the other measurements. No statistical test can say with absolute certainty whether or not a value is an outlier, but they can provide an objective way to make a decision.

There are two widely used tests for outliers. The first is called **Dixon's Q-test** and uses as the test statistic the value calculated below.

$$Q_{exp} = \frac{|\text{suspect value} - \text{nearest value}|}{\text{range}} \qquad 9$$

Here, *Qexp* represents the experimental test statistic and the range is the maximum minus the minimum (including the suspect value). The test statistic is compared to the critical value, listed in Table 2.2, for the desired confidence level and number of measurements.

The second outlier test is the **Grubbs' test**, and the test statistic in this case is given by the following equation.

$$G_{exp} = \frac{|\text{suspect value} - \bar{x}|}{s} \qquad 10$$

Again, the test statistic is calculated and compared to the appropriate critical value, given in Table 3 (and the Appendix).

These two tests are essentially equivalent and will lead to the same conclusions in virtually all cases, so it does not matter which one is used. The Q-test has the feature that it does not require the calculation of a mean and standard deviation, which was a big advantage at a time when such calculations had to be done by hand, but this is not as important now.

Example 2: Outliers

The following values were reported for the concentration of lead (in parts per million, ppm) in replicate soil samples: 91.2, 80.3, 94.5, 89.6, 93.7, 90.5. Based on a statistical test, should any of the values be rejected as outliers? Carry out the test at the 95% confidence level.

Only one of the two tests needs to be applied, but for illustration, both will be applied in this example. For the Q-test, it is useful to sort the values first: 80.3, 89.6, 90.5, 91.2, 93.7, 94.5 Possible outliers could be at the high end or the low end, but we only test one of those. Since the difference between the bottom two values is larger than the top two, the lowest value is tested.

$$Q_{exp} = \frac{|80.3 - 89.6|}{94.5 - 80.3} = \frac{9.3}{14.2} = 0.655$$

Table 2. Critical values for Dixon's *Q* parameter at selected confidence levels.

No. of	Confidence Interval				
Observations	90%	95%	96%	98%	99%
3	0.941	0.970	0.976	0.988	0.994
4	0.765	0.829	0.846	0.889	0.926
5	0.642	0.710	0.729	0.780	0.821
6	0.560	0.625	0.644	0.698	0.740
7	0.507	0.568	0.586	0.637	0.680
8	0.468	0.526	0.543	0.590	0.634
9	0.437	0.493	0.510	0.555	0.598
10	0.412	0.466	0.483	0.527	0.568

Table 3. Critical values for Grubb's parameter at selected confidence levels.

Number of	Confidence Interval				
Observations	90%	95%	96%	98%	99%
3	1.153	1.154	1.154	1.155	1.155
4	1.463	1.481	1.485	1.493	1.496
5	1.671	1.715	1.725	1.749	1.764
6	1.822	1.887	1.904	1.944	1.973
7	1.938	2.020	2.042	2.097	2.139
8	2.032	2.127	2.152	2.221	2.274
9	2.110	2.215	2.244	2.323	2.387
10	2.176	2.290	2.322	2.410	2.482

The critical value for the *Q* statistic, $Qcrit$, at 95% for 6 observations is obtained from Table 2 as 0.625. Since $Qexp > Qcrit$, the null hypothesis (that the suspect value is not an outlier) is rejected, and we conclude that the value is an outlier.

For the Grubbs' test, we need the mean and the standard deviation.

$$\overline{x} = 89.97 \qquad s = 5.097 \qquad G_{exp} = \frac{|80.3 - 89.97|}{5.097} = \frac{9.67}{5.097} = 1.90$$

The critical value for Grubbs' statistic, *Gcrit*, from Table 3 is 1.887 ($N = 6$, 95%). Since *Gexp* > *Gcrit* , we again conclude that the suspect value should be rejected. Based on either test, the measured value of 80.7 should be excluded from calculations of the mean, standard deviation and other parameters. It is important to note that the outlier tests can be applied only once to a group of measurements.

What does it mean when a value is rejected in this manner? Since these tests were conducted at the 95% confidence level, it means that we will be correct in rejecting the value 95% of the time, or wrong in doing so 5% of the time. It is up to the analyst to decide the level of the test. If, for example, a 99% level were chosen, we would correctly reject values 99% of the time, but may mistakenly choose to retain outliers more often. In the above example, *Qcrit* at 99% is 0.740, and we would have retained the value. Statistics can never give absolute answers, only assign probabilities that we are correct. If a question does not specify a confidence level, a value of 95% will be assumed in this class.

Also note that, in any given situation, a suspect measurement can always be rejected on the basis of experimental observation rather than a statistical test. This is why it is important to note unusual circumstances (*e.g.* "added too much indicator") in your lab book.

16.8 STATISTICAL TEST TO COMPARE MEANS

It is quite common in analytical chemistry to want to compare two sets of measurements. For example, we may wish to determine if the amount of calcium in skim milk is the same as that in whole milk, or we may wish to compare the results of two laboratories on the same sample to determine if they are the same. There are several statistical tests that can be used for this purpose, but only one, called the *t*-test, will be discussed here.

It is impossible to determine if two values are the "same" without taking into account their uncertainty. For this reason, the *t*-test uses both the means and uncertainties in calculating the test statistic. The absolute difference in the means is divided by an "average" or pooled standard deviation that accounts for the number of measurements in each group. The test statistic is then compared to the critical value for the corresponding degrees of freedom, with the null hypothesis that the two means are the same. The important equations are given below.

$$t_{exp} = \frac{|\overline{x}_1 - \overline{x}_2|}{s_{pooled}\sqrt{1/N_1 + 1/N_2}}$$

$$s_{pooled} = \sqrt{\frac{(N_1 - 1)s_1^2 + (N_2 - 1)s_2^2}{N_1 + N_2 - 2}} \qquad 11$$

$$t_{exp} = \frac{|\bar{x}_1 - \bar{x}_2|}{s_{pooled}\sqrt{1/N_1 + 1/N_2}}$$

$$s_{pooled} = \sqrt{\frac{(N_1 - 1)s_1^2 + (N_2 - 1)s_2^2}{N_1 + N_2 - 2}} \qquad 12$$

Degrees of freedom = $v = N + N - 2$

Critical values for the *t*-statistic are found in Table 2.1 (given earlier) and the Appendix An example will illustrate the application of the *t*-test.

Example 3: Comparison of Means

Two analytical methods are tested to determine the total cholesterol in the same blood sample. The first method gives replicate values of 218, 226, 231, 231 and 211 mg/dL, while the second gives 229, 241, 235 and 248 mg/dL. Are the two sets of results significantly different at the 95% confidence level?

The statistical parameters for each set are first calculated.

$x1 = 223.40$ $s1 = 8.735$ $N1 = 5$

2 2 2 $x = 238.25$ $s = 8.139$ $N = 4$

These can be used to calculate the pooled standard deviation and the *t*-statistic.

$$s_{pooled} = \sqrt{\frac{(5-1)(8.735)^2 + (4-1)(8.139)^2}{5+4-2}} = 8.485$$

$$t_{exp} = \frac{|223.40 - 238.25|}{8.485\sqrt{1/5 + 1/4}} = 2.61 \qquad v = 5+4-2 = 7$$

Quantitative chemical analysis involves measurements with uncertainty arising from determinate and indeterminate errors that affect the precision and accuracy of results. Measurements can be described through statistical parameters such as the mean, standard deviation, and confidence intervals. Statistical tests can be used to determine the presence of outliers and also to test whether two sets of measurements provide the same result.

RELATED TERMINOLOGY

Absolute Zero - Absolute zero is 0k. It is the lowest possible temperature. Theoretically, at absolute zero, atoms stop moving.

Accuracy - Accuracy is a measure of how close a measured value is to its true value. For example, if an object is exactly a meter long and you measure it as 1.1 Meters long, that is more accurate than if you measured it at 1.5 Meters long.

Acid Anhydride - An acid anhydride is an oxide that forms an acid when it is reacted with water. For example, when so3- is added to water, it becomes sulfuric acid, h2so4.

Acid - There are several ways to define an acid, but they include any chemical that gives off protons or h+ in water. Acids have a ph less than 7. They turn the ph indicator phenolphthalein colorless and turn litmus paper red.

Actual Yield - The actual yield is the amount of product you actually obtain from a chemical reaction, as in the amount you can measure or weigh as opposed to a calculated value.

Addition Reaction - An addition reaction is a chemical reaction in which atoms add to a carbon-carbon multiple bond.

Alcohol - An alcohol is any organic molecule that has an -oh group.

Aldehyde - An aldehyde is any organic molecule that has a -coh group.

Alkali Metal - An alkali metal is a metal in group i of the periodic table. Examples of alkali metals include lithium, sodium, and potassium.

Alkali Metal: Any of the monovalent metals lithium, sodium, potassium, rubidium, caesium, and francium, belonging to group 1a of the periodic table. They are all very reactive and electropositive

Alkaline Earth Metal - An alkaline earth metal is an element belonging to group ii of the periodic table. Examples of alkaline earth metals are magnesium and calcium.

Alkaline Earth Metal: Any of the divalent electropositive metals beryllium, magnesium, calcium, strontium, barium, and radium, belonging to group 2a of the periodic table

Alkane: An alkane is an organic molecule that only contains single carbon-carbon bonds

Alkane: Any saturated aliphatic hydrocarbon with the general formula cnh2n+2

Alkene : An alkene is an organic molecule that contains at least one c=c or carbon-carbon double bond.

Alkyne:- An alkyne is an organic molecule that contains at least one carbon-carbon triple bond.

Allotrope:- Allotropes are different forms of a phase of an element. For example, diamond and graphite are allotropes of carbon.

Alloy: A metallic material, such as steel, brass, or bronze, consisting of a mixture of two or more metals or of metallic elements with nonmetallic elements. Alloys often have physical properties markedly different from those of the pure metals

Alpha particle - An alpha particle is another name for a helium nucleus, which contains two protons and two neutrons. It's called an alpha particle in reference to radioactive (alpha) decay.

Amine - An amine is an organic molecule in which one or more of the hydrogen atoms in ammonia have been replaced by an organic group. An example of an amine is methylamine.

Amino acid: Any of a group of organic compounds containing one or more amino groups, -nh2, and one or more carboxyl groups, -cooh. The alpha-amino acids rch(nh2)cooh (where r is either hydrogen or an organic group) are the component molecules of proteins; some can be synthesized in the body (nonessential amino acids) and others cannot and are thus essential components of the diet (essential amino acids)

Analysis: Analysisthe decomposition of a substance into its elements, radicals, or other constituents in order to determine the kinds of constituents present (qualitative analysis) or the amount of each constituent (quantitative analysis)

Anion: Aniona negatively charged ion; an ion that is attracted to the anode during electrolysis

Anode: Anodethe positive electrode in an electrolytic cell

Atomic Mass: Atomic massthe mass of an isotope of an element in atomic mass units

Atomic Numbe: Atomic numberthe number of protons in the nucleus of an atom of an elemen

Atomthe Smallest Quantity Of An Element: That can take part in a chemical reaction

Base - A base is a compound that produces oh- ions or electrons in water or that accepts protons. An example of a common base is sodium hydroxide, naoh.

Beta Particle - A beta particle is an electron, although the term is used when the electron is emitted in radioactive decay.

Binary Compound - A binary compound is one made up of two elements.

Binding Energy - Binding energy is the energy that holds protons and neutrons together in the atomic nucleus.

Boiling Point: The temperature at which a liquid boils at a given pressure, usually atmospheric pressure at sea level; the temperature at which the vapour pressure of a liquid equals the external pressure

Bond Energy - Bond energy is the amount of energy required to break one mole of chemical bonds.

Bond Length - Bond length is the average distance between the nuclei of two atoms that share a bond.

Bond: Something that binds, fastens, or holds together, such as a chain or rope

Brownian Motion: Random movement of microscopic particles suspended in a fluid, caused by bombardment of the particles by molecules of the fluid. First observed in 1827, it provided strong evidence in support of the kinetic theory of molecules

Buffer - A liquid that resists change in ph when an acid or base is added. A buffer consists of a weak acid and its conjugate base. An example of a buffer is acetic acid and sodium acetate.

Calorimetry - Calorimetry is the study of heat flow. Calorimetry may be used to find the heat of reaction of two compounds or the heat of combustion of a compound, for example.

Carbohydrate: any of a large group of organic compounds, including sugars, such as sucrose, and polysaccharides, such as cellulose, glycogen, and starch, that contain carbon, hydrogen, and oxygen, with the general formula cm(h2o)n: an important source of food and energy for animals

Carboxylic acid - A carboxylic acid is an organic molecule containing a -cooh group. An example of a carboxylic acid is acetic acid.

Catalyst - A catalyst is a substance that lowers the activation energy of a reaction or speeds it up without being consumed by the reaction. Enzymes are proteins that act as catalysts for biochemical reactions.

Catalyst: A substance that increases the rate of a chemical reaction without itself suffering any permanent chemical change

Cathode - A cathode is the electrode which gains electrons or is reduced. In other words, it is where reduction occurs in an electrochemical cell.

Cathodethe Negative electrode in an electrolytic cell; the electrode by which electrons enter a device from an external circuit

Cation: A positively charged ion; an ion that is attracted to the cathode during electrolysis

Chain reaction: A process in which a neutron colliding with an atomic nucleus causes fission and the ejection of one or more other neutrons, which induce other nuclei to split

Chain: Two or more atoms or groups bonded together so that the configuration of the resulting molecule, ion, or radical resembles a chain

Chemical Equation a representation of a chemical reaction using symbols of the elements to indicate the amount of substance, usually in moles, of each reactant and product

Chemical Property-A chemical property is a property that can only be observed when a chemical change occurs. Flammability is an example of a chemical property since you can't measure how flammable a substance is without igniting it (making/breaking chemical bonds).

Chromatography: The technique of separating and analysing the components of a mixture of liquids or gases by selective adsorption in, for example, a column of powder (column chromatography) or on a strip of paper (paper chromatography)

Combustion: A chemical process in which two compounds, such as sodium and chlorine, react together to produce heat and light

Compound: A substance that contains atoms of two or more chemical elements held together by chemical bonds

Condensation: A type of reaction in which two organic molecules combine to form a larger molecule as well as a simple molecule such as water, methanol, etc

Corrosiona Process in which a solid, esp a metal, is eaten away and changed by a chemical action, as in the oxidation of iron in the presence of water by an electrolytic process

Covalent bond-A covalent bond is a chemical bond formed when two atoms share two electrons.

Covalent Bond: A type of chemical bond involving the sharing of electrons between atoms in a molecule, esp the sharing of a pair of electrons by two adjacent atoms

Critical Mass- Critical mass is the minimum quantity of radioactive material needed to cause a nuclear chain reaction.

Critical Point -The critical point is the endpoint of the liquid-vapor line in a phase diagram, past which a supercritical liquid forms. At the critical point, the liquid and vapor phases become indistinguishable from one another.

Crystal - A crystal is an ordered, repeating three-dimensional pattern of ions, atoms, or molecules. Most crystals are ionic solids, although other forms of crystals exist.

Crystal: A piece of solid substance, such as quartz, with a regular shape in which plane faces intersect at definite angles, due to the regular internal structure of its atoms, ions, or molecules

Crystallizationthe Process in which crystals are formed either from something that has been melted or from a solution

Delocalization - Delocalization is when electrons become free to move all over a molecule, such as when double bonds occur on adjacent atoms in a molecule.

Denature - There are two common meanings for this in chemistry. First, it can refer to any process used to make ethanol unfit for consumption (denatured alcohol). Second, denaturing can mean breaking down the three-dimensional structure of a molecule, such as a protein is denatured when exposed to heat.

Diffusion - Diffusion is the movement of particles from an area of higher concentration to one of lower concentration.

Diffusion: The random thermal motion of atoms, molecules, clusters of atoms, etc, in gases, liquids, and some solids

Dilute(of a solution, suspension, mixture, etc) having a low concentration or a concentration that has been reduced by admixture; (of a substance) present in solution, esp a weak solution in water

Dilution - Dilution is when a solvent is added to a solution, making it less concentrated.

Dissociation -Dissociation is when a chemical reaction breaks a compound into two or more parts. For example, nacl dissociates into na+ and cl- in water.

Distillationthe Process of evaporating or boiling a liquid and condensing its vapour

Double Displacement reaction - a double displacement or double replacement reaction is when cations of two compounds switch places.

Effusion - Effusion is when a gas moves through an opening into a low-pressure container (e.G., Is drawn by a vacuum). Effusion occurs more quickly than diffusion because additional molecules aren›t in the way.

Electrode: An element in a semiconducting device that emits, collects, or controls the movement of electrons or holes

Electrolysis Electrolysis is using electricity to break the bonds in a compound to break it apart.

Electrolysis: The conduction of electricity by a solution or melt, esp the use of this process to induce chemical changes

Electrolyte - An electrolyte is an ionic compound that dissolves in water to produce ions, which can conduct electricity. Strong electrolytes completely dissociate in water, while weak electrolytes only partially dissociate or break apart in water.

Electron: A stable elementary particle present in all atoms, orbiting the nucleus in numbers equal to the atomic number of the element in the neutral atom; a lepton with a negative charge of 1.602 176 462 × 10–19 Coulomb, a rest mass of 9.109 381 88 × 10–31 Kilogram, a Sradius of 2.817 940 285 × 10–15 Metre, and a spin of

Electrovalency The valency of a substance in forming ions, equal to the number of electrons gained or los

Emulsiona Light-Sensitive coating on a base, such as paper or film, consisting of fine grains of silver bromide suspended in gelatin

Enantiomers - Enantiomers are molecules that are non superimposable mirror images of each other.

Endothermic -Endothermic describes a process that absorbs heat. Endothermic reactions feel cold.

Endpoint - The endpoint is when a titration is stopped, typically because an indicator has changed color. The endpoint need not be the same as the equivalence point of a titration.

Energy Level - An energy level is a possible value of energy that an electron can have in an atom.

Enthalpy - Enthalpy is a measure of the amount of energy in a system.

Entropy - Entropy is a measure of the disorder or randomness in a system.

Enzyme - An enzyme is a protein that acts as a catalyst in a biochemical reaction.

Equilibrium - Equilibrium occurs in reversible reactions when the forward rate of the reaction is the same as the reverse rate of the reaction.

Equivalence Point - The equivalence point is when the solution in a titration is completely neutralized. It is not the same as the endpoint of a titration because the indicator may not change colors precisely when the solution is neutral.

Ester - An ester is an organic molecule with a r-co-or› function group.

Ester: Any of a class of compounds produced by reaction between acids and alcohols with the elimination of water. Esters with low molecular weights, such as ethyl acetate, are usually volatile fragrant liquids; fats are solid esters

Ether: A colourless volatile highly flammable liquid with a characteristic sweetish odour, made by the reaction of sulphuric acid with ethanol: used as a solvent and anaesthetic. Formula: c2h5oc2h5

Evaporation: When a liquid changes to a vapor, caused by an increase in temperature and/or a decrease in pressure

Excess Reagent - Excess reagent is what you get when there is leftover reagent in a chemical reaction.

Excited state - An excited state is a higher energy state for an electron of an atom, ion, or molecule, compared with the energy of its ground state.

Exothermic - Exothermic describes a process that gives off heat.

Family - A family is a group of elements sharing similar properties. It is not necessarily the same thing as an element group. For example, the chalcogens or oxygen family consists of some different elements from the nonmetal group.

Fat: Any of a class of naturally occurring soft greasy solids that are esters of glycerol and certain fatty acids. They are present in some plants and in the adipose tissue of animals, forming a reserve energy source, and are used in making soap and paint and in the food industry

Fatty Acid: Any of a class of aliphatic carboxylic acids, such as palmitic acid, stearic acid, and oleic acid, that form part of a lipid molecule

Fermentation: A chemical reaction in which a ferment causes an organic molecule to split into simpler substances, esp the anaerobic conversion of sugar to ethyl alcohol by yeast

Fission: The act or process of splitting or breaking into parts

Foam: A mass of small bubbles of gas formed on the surface of a liquid, such as the froth produced by agitating a solution of soap or detergent in water

Formula: A representation of molecules, radicals, ions, etc, expressed in the symbols of the atoms of their constituent elements

Fuel: Any substance burned as a source of heat or power, such as coal or petrol

Gas: A substance in a physical state in which it does not resist change of shape and will expand indefinitely to fill any container. If very high pressure is applied a gas may become liquid or solid, otherwise its density tends towards that of the condensed phase

Halogen: Any of the chemical elements fluorine, chlorine, bromine, iodine, and astatine. They are all monovalent and readily form negative ions

Hydrocarbonany Organic compound containing only carbon and hydrogen, such as the alkanes, alkenes, alkynes, terpenes, and arenes

Hydrolysis: A chemical reaction in which a compound reacts with water to produce other compounds

Inerthaving only a limited ability to react chemically; unreactive

Inorganic: Relating to or denoting chemical compounds that do not contain carbon

Insoluble: Incapable of being dissolved; incapable of forming a solution, esp in waterionan electrically charged atom or group of atoms formed by the loss or gain of one or more electrons

Ionic bond: The chemical bond between two oppositely charged ions formed when one atom transfers electrons to another atom, as in the formation of sodium chloride; electrovalent bond

Ionization: The formation of ions as a result of a chemical reaction, high temperature, electrical discharge, particle collisions, or radiation

Isomer: A compound that exhibits isomerism with one or more other compounds

Isotope: One of two or more atoms with the same atomic number that contain different numbers of neutrons

Kelvin - Kelvin is a unit of temperature. A kelvin is equal in size to a degree celsius, although kelvin starts from absolute zero. Add 273.15 To a celsius temperature to get the kelvin value. Kelvin is not reported with a ° symbol. For example, you would simply write 300k not 300°k.

Ketone - A ketone is a molecule that contains a r-co-r' functional group. An example of a common ketone is acetone (dimethyl ketone).

Kinetic Energy -Kinetic energy is energy of motion. The more an object moves, the more kinetic energy it has.

Lanthanide Contraction- The lanthanide contraction refers to the trend in which lanthanide atoms become smaller as you move left to right across the periodic table even though they increase in atomic number.

Lanthanide or Rare-Earth Element: Any element of the lanthanide series

Lattice Energy-Lattice energy is the amount of energy released when one mole of a crystal forms from its gaseous ions.

Law of Conservation of Energy-The law of conservation of energy states the energy of the universe may change form, but its amount remains unchanged.

Ligand- A ligand is a molecule or ion stuck to the central atom in a complex. Examples of common ligands include water, carbon monoxide, and ammonia.

Liquida Substance in a physical state in which it does not resist change of shape but does resist change of size

Litmus Testa Test to establish the acidity or alkalinity of a mixture

Mass- Mass is the amount of matter in a substance. It is commonly reported in units of grams.

Melting Pointthe temperature at which a solid turns into a liquid. It is equal to the freezing point

Metalany of a number of chemical elements, such as iron or copper, that are often lustrous ductile solids, have basic oxides, form positive ions, and are good conductors of heat and electricity

Metalloida Nonmetallic element, such as arsenic or silicon, that has some of the properties of a metal

Mineralany of a class of naturally occurring solid inorganic substances with a characteristic crystalline form and a homogeneous chemical composition

Mixturea Substance consisting of two or more substances mixed together without any chemical bonding between them

Molarityanother Name (not in technical usage) for concentration

Mole - Avogadro›s number (6.02 X 1023) of anything.

Moleculethe Simplest unit of a chemical compound that can exist, consisting of two or more atoms held together by chemical bonds

Molethe Basic si unit of amount of substance; the amount that contains as many elementary entities as there are atoms in 0.012 Kilogram of carbon-12. The entity must be specified and may be an atom, a molecule, an ion, a radical, an electron, a photon, etc

Monomera Compound whose molecules can join together to form a polymer

Neutralneither Acidic nor alkaline

Neutrona Neutral elementary particle with a rest mass of 1.674 92716 × 10–27 Kilogram and spin ; classified as a baryon. In the nucleus of an atom it is stable, but when free it decays

Noble Gas or inert gasany of the unreactive gaseous elements helium, neon, argon, krypton, xenon, and radon

Node - A node is a location in an orbital with no probability of containing an electron.

Nonmetalany of a number of chemical elements that form negative ions, have acidic oxides, and are generally poor conductors of heat and electricity

Nucleon - A nucleon is a particle in the nucleus of an atom (proton or neutron).

Nucleusa Fundamental group of atoms in a molecule serving as the base structure for related compounds and remaining unchanged during most chemical reactions

Oilany of a number of viscous liquids with a smooth sticky feel. They are usually flammable, insoluble in water, soluble in organic solvents, and are obtained from plants and animals, from mineral deposits, and by synthesis. They are used as lubricants, fuels, perfumes, foodstuffs, and raw materials for chemicals

Oreany Naturally occurring mineral or aggregate of minerals from which economically important constituents, esp metals, can be extracted

Organicof, relating to, or belonging to the class of chemical compounds that are formed from carbon

Oxidation Number the oxidation number is the apparent charge on an atom. For example, the oxidation number of an oxygen atom is -2.

Oxidationany Process in which oxygen combines with an element or substance, either slowly, as in the rusting of iron, or rapidly, as in the burning of wood

Oxidation-Reductiona Reversible chemical process usually involving the transfer of electrons, in which one reaction is an oxidation and the reverse reaction is a reduction

Period - A period is a row (left to right) of the periodic table.

Periodic Tablea Table of the elements, arranged in order of increasing atomic number, based on the periodic law. Elements having similar chemical properties and electronic structures appear in vertical columns (groups)phpotential of hydrogen; a measure of the acidity or alkalinity of a solution equal to the common logarithm of the reciprocal of the concentration of hydrogen ions in moles per cubic decimetre of solution. Pure water has a ph of 7, acid solutions have a ph less than 7, and alkaline solutions a ph greater than 7

Plasticany One of a large number of synthetic usually organic materials that have a polymeric structure and can be moulded when soft and then set, esp such a material in a finished state containing plasticizer, stabilizer, filler, pigments, etc. Plastics are classified as thermosetting (such as bakelite) or thermoplastic (such as pvc) and are used in the manufacture of many articles and in coatings, artificial fibres, etc

Polymera Naturally occurring or synthetic compound, such as starch or perspex, that has large molecules made up of many relatively simple repeated units

Precipitatea Precipitated solid in its suspended form or after settling or filtering

Precision- Precision is how repeatable a measurement is. More precise measurements are reported with more significant figures.

Pressure- Pressure is force per area.

Product- A product is something made as a result of a chemical reaction.

Protona Stable, positively charged elementary particle, found in atomic nuclei in numbers equal to the atomic number of the element. It is a baryon with a charge of

1.602176462 × 10–19 Coulomb, a rest mass of 1.672 62159 × 10–27 Kilogram, and spin

Quantum Theory- Quantum theory is the description of energy levels and the predictions about the behavior of atoms at specific energy levels.

Radioactivity-Radioactivity occurs when the atomic nucleus is unstable and breaks apart, releasing energy or radiation.

Radioactivity The spontaneous emission of radiation from atomic nuclei. The radiation can consist of alpha, beta, and gamma radiation

Raoult's Law-Raoult's law states that the vapor pressure of a solution is directly proportional to the mole fraction of solvent.

Rate Determining Step- The rate determining step is the slowest step in any chemical reaction.

Rate Law- A rate law is a mathematical expression relating the speed of a chemical reaction as a function of concentration.

Reactiona Process that involves changes in the structure and energy content of atoms, molecules, or ions but not their nuclei

Reagenta Substance for use in a chemical reaction, esp for use in chemical synthesis and analysis

Redox Reaction- A redox reaction is a chemical reaction that involves oxidation and reduction.

Reduceto Undergo or cause to undergo a chemical reaction with hydrogen or formation of a hydride; to lose or cause to lose oxygen atoms; to undergo or cause to undergo an increase in the number of electrons.

Resonance Structure- Resonance structures are the set of lewis structures that can be drawn for a molecule when it has delocalized electrons.

Reversible Reaction- A reversible reaction is a chemical reaction which can go both ways: reactants make products and products make reactants.

Rms Velocity- The rms or root mean square velocity is the square root of the average of the squares of individual velocities of gas particles, which is a way of describing the average speed of gas particles.

Salt An ionic compound formed from reacting an acid and a base.

Salta White powder or colourless crystalline solid, consisting mainly of sodium chloride and used for seasoning and preserving food

Saponificationthe Conversion of an ester heated with an alkali into the corresponding alcohol and acid salt; specif., This process carried out with fats (glyceryl esters) to produce soap

Saturated(of a solution or solvent) containing the maximum amount of solute that can normally be dissolved at a given temperature and pressure

Soapa Cleaning or emulsifying agent made by reacting animal or vegetable fats or oils with potassium or sodium hydroxide. Soaps often contain colouring matter and perfume and act by emulsifying grease and lowering the surface tension of water so that it more readily penetrates open materials such as textiles

Solidof Concerned with, or being a substance in a physical state in which it resists changes in size and shape

Soluble(of a substance) capable of being dissolved, esp easily dissolved in some solvent, usually water

Solute- The solute is the substance that gets dissolved in a solvent. Usually, it refers to a solid that is dissolved in a liquid. If you are mixing two liquids, the solute is the one that is present in a smaller amount.

Solutiona Homogeneous mixture of two or more substances in which the molecules or atoms of the substances are completely dispersed. The constituents can be solids, liquids, or gasessolventa liquid capable of dissolving another substance

Solvent- This is the liquid that dissolves a solute in solution. Technically, you can dissolve gases into liquids or into other gases, too. When making a solution where both substances are in the same phase (e.G., Liquid-liquid), the solvent is the largest component of the solution.

Stp- **Stp** means standard temperature and pressure, which is 273k and 1 atmosphere.

Strong Acid - A strong acid is an acid that completely dissociates in water. An example of a strong acid is hydrochloric acid, hcl, which dissociates into h+ and cl- in water.

Strong Nuclear Force- The strong nuclear force is the force that holds the protons and neutrons in an atomic nucleus together.

Sublimation - Sublimation is when a solid changes directly into a gas. At atmospheric pressure, dry ice or solid carbon dioxide goes directly into carbon dioxide vapor, never becoming liquid carbon dioxide.

Sublimationthe Process or instance or sublimingsubstitution reaction

Sugara White crystalline sweet carbohydrate, a disaccharide, found in many plants and extracted from sugar cane and sugar beet: it is used esp as a sweetening agent in food and drinks. Formula: c12h22o11

Suspensiona Dispersion of fine solid or liquid particles in a fluid, the particles being supported by buoyancy

Synthesis-Synthesis is making a larger molecule from two or more atoms or smaller molecules.

Synthesisthe Process of producing a compound by a chemical reaction or series of reactions, usually from simpler or commonly available starting materialstransition metalany element belonging to one of three series of elements with atomic numbers between 21 and 30, 39 and 48, and 57 and 80. They have an incomplete penultimate electron shell and tend to exhibit more than one valency and to form complexes

System - A system includes everything you are evaluating in a situation.

Tbase The main ingredient of a mixture

Telementany of the 118 known substances (of which 93 occur naturally) that consist of atoms with the same number of protons in their nuclei

Temperature - Temperature is a measure of the average kinetic energy of particles.

Theoretical Yield-Theoretical yield is the amount of product which would result if a chemical reaction proceeded perfectly, to completion, with no loss.

Thermodynamics- Thermodynamics is the study of energy.

Titration- Titration is a procedure in which the concentration of an acid or base is determined by measuring how much base or acid is required to neutralize it.

Triple Point- the triple point is the temperature and pressure at which the solid, liquid, and vapor phases of a substance exist in equilibrium.

Unit Cell- A unit cell is the simplest repeating structure of a crystal.

Unsaturated- There are two common meanings for unsaturated in chemistry. The first refers to a chemical solution that does not contain all of the solute that can be dissolved in it. Unsaturated also refers to an organic compound which contains one or more double or triple carbon-carbon bonds.

Unsaturated(of a chemical compound, esp an organic compound) containing one or more double or triple bonds and thus capable of undergoing addition reactions

Unshared Electron Pair- An unshared electron pair or lone pair refers to two electrons that aren›t participating in chemical bonding.

Valence Electron- The valence electrons are the atom›s outermost electrons.

Valency-A property of atoms or groups, equal to the number of atoms of hydrogen that the atom or group could combine with or displace in forming compounds

Van der waals Forces-weak electrostatic forces between atoms and molecules caused by transient distortions in the distribution of electrons in the interacting atoms or molecules

Volatile Volatile refers to a substance that has a high vapor pressure.

Vsepr- Vsepr stands for valence shell electron pair repulsion. This is a theory used that predicts molecular shapes based on the assumption that electrons stay as far as possible from each other.

REFERENCES

A. A. Muzzarelli, "Chitins and chitosans for the repair of wounded skin, nerve, cartilage and bone," Carbohydrate Polymers, vol. 76, no. 2, pp. 167–182, 2009.

A. Botelho, N. A. P. Nogueira, G. M. Bastos et al., "Antimicrobial activity of the essential oil from Lippia sidoides, carvacrol and thymol against oral pathogens," Brazilian Journal of Medical and Biological Research, vol. 40, no. 3, pp. 349–356, 2007.

A. Botelho, V. S. Rao, C. B. M. Carvalho et al., "Lippia sidoides and Myracrodruon urundeuva gel prevents alveolar bone resorption in experimental periodontitis in rats," Journal of Ethnopharmacology, vol. 113, no. 3, pp. 471–478, 2007.

A. Cardoso, R. Cunha, and T. S. Pereira, "Germinac¸˜ao de sementes de Virola surinamensis (Rol.) Warb. (Myristicaceae) e Guarea guidonea (L.) Sleumer (Meliaceae)," Revista Brasileira de Sementes, vol. 16, no. 1, pp. 1–5, 1994.

A. Chen and G. W. Hwang, "Water-soluble self-acid-doped conducting polyaniline: structure and properties," Journal of the American Chemical Society, vol. 117, no. 40, pp. 10055– 10062, 1995.

A. Disegi, R. L. Kennedy, and R. Pillia, Cobalt-Base Alloys for Biomedical Applications, ASTM International Standards, 1999.

A. Rodrigues, "Revis˜ao taxonomica das esp´ecies de Virola Aublet (Myristicaceae) do Brasil," Acta Amazonica, vol. 10, supplement, pp. 1–127, 1980.

A. Schultz, T. N. Corso, S. J. Prosser, and S. Zhang, "A fully integrated monolithic microchip electrospray device for mass spectrometry," Analytical Chemistry, vol. 72, no. 17, pp. 4058– 4063, 2000.

A. Sumathykutty and J. M. Rao, "8-Hentriacontanol and other constituents from Piper attenuatum," Phytochemistry,

A. V. Carvalho, P. M. Galdino, M. V. M. Nascimento et al., "Antinociceptive and antiinflammatory activities of grandisin extracted from Virola surinamensis," Phytotherapy Research, vol. 24, no. 1, pp. 113–118, 2010.

A.A. Ansari: Effect of Vermicompost on the Productivity of Potato (Solanum tuberosum), Spinach (Spinacia oleracea) and Turnip (Brassica campestris); World Journal of Agricultural Sciences; 4 (3); 333-336. (2008).

A.E. Campos, Optimización de la digestión anaerobia de purines de cerdo mediante codigestión con residuos orgánicos de la industria agroalimentaria. Tesis doctoral. Departament de Medi Ambient i Ciencias del Sòl. Escola Tècnica Superior d'Enginyeria Agrària, Universitat de Lleida, Lleida, Spain (2001).

A.J. Biddlestone and K.R. Gray, Composting. In: Comprehensive Biotechnology. The principles, applications and regulations of biotechnology in industry, agriculture and medicine. Elsevier Science and Technology (1985).

A.Mirmohseni and A.Oladegaragoze, "Anti-corrosive properties of polyaniline coating on iron," Synthetic Metals, vol. 114, no. 2, pp. 105–108, 2000.

A.P. Francese, G. Aboagye-Mathiesen, T. Olesen, P.R. Cordoba and F. Siñeriz, World J. Microb. Biot. 16, 147 (2000).

Abad, K. Martinez, C. Planas, O. Palacios, J. Caixach and J. Rivera, Chemosphere 61, 1358 (2005).

Alvarenga, A.P. Gonçalves, R.M. Fernandes, A. de Varennes, G. Vallini, E. Duarte and A.C. Cunha-Queda, Sci. Total Environ. 406, 43 (2008).

Amir, M. Hafidi, G. Merlina, H. Hamdi, A. Jouraiphy, M.E. Gharous and J.C. Revel, Process. Biochem. 40, 2183 (2005).

Amlinger, M. Pollak and E. Favoino, Heavy metals and organic compounds from wastes used as organic fertilisers. ENV.A.2./ETU/2001/0024. Final Report to DG Environment, Brussels (2004).

Archives of Orthopaedic and Trauma Surgery, vol. 110, no. 2, pp. 61–74, 1991.

Arena, M.L. Mastellone and F. Perugini, Chem. Eng. J. 96 207 (2003).

Arvanitoyannis, I. S., Waste Management in Food Packaging Industries, in Book: Waste Management for the Food Industries, 2008, p 941-1045.

ASTM, American Society for Testing and Materials, 2000. Standard guide for riskbased corrective action. E2081-00, West Conshohocken, PA.

Azim, M.E., Verdegem, M.C.J., van Dam, A.A., Beveridge, M.C.M., Edit, 2005. Periphyton: Ecology, Exploitation and Management. CABI Publishing.

B. Blodgett, "Monomolecular films of fatty acids on glass," Journal of the American Chemical Society, vol. 56, no. 2, p. 495, 1934.

Bazzoffi, S. Pellegrini, A. Rocchini, M. Morandi and O. Grasselli, Soil Tillage Res. 48, 275 (1998).

Beck-Friis, S. Smars, H. Jönsson, H. Kirchmann, J. Agric. Eng. Res. 78, 423 (2001).

Bello-Mendoza and P.N. Shararratt, Water Sci. Technol. 40, 49 (1999).

Bendixen, Water Sci. Technol. 30, 171 (1994).

Benemann, J. R., 1979. Production of nitrogen fertilizer with nitrogen-fixing blue-green algae. Enzyme and Microbial Technology, 1, 83-90.

Bess, V.H. BioCycle 2000. 41(10), 71-72

Blatt, C.R., 1991. Comparison of several organic amendments with a chemical fertilizer for vegetable production. Scientia Horticulture, 47, 177-191.

Bombatkar, The Miracle Called Compost; The Other India Press, Pune, India (1996).

Bowen, G.D. and Rovira, A.D. 1999. The Rhizosphere and its management to improve plant growth. Advances in Agronomy, 66, 1-102.

Braz Filho, O. R. Gottlieb, A. A. De Moraes et al., "The chemistry of Brazilian myristicaceae. IX. Isoflavonoids from amazonian species," Lloydia, vol. 40, no. 3, pp. 236–238, 1977.

Brown L.D.; Lawton, K.; Carolus, R.L. Soil Sci. 1961. 92(6), 380-383.

C. Martinez and L. E. Cuca, "Flavonoids from Virola calophylloidea," Journal of Natural Products, vol. 50, no. 6, pp. 1045–1047, 1987.

C.A. Edwards and I. Burrows: The Potential of Earthworms Composts as Plant Growth Media. In Edward, C.A. and E.F. Neuhauser (Eds.), Earthworms in Waste and Environmental Management. SPB Academic Publishing, The Hague, The Netherlands; ISBN 90-5103-017-7, 21-32. (1988).

C.A. Edwards and I. Burrows: The Potential of Earthworms Composts as Plant Growth Media. In Edward, C.A. and E.F. Neuhauser (Eds.), Earthworms in Waste and Environmental Management. SPB Academic Publishing, The Hague, The Netherlands; ISBN 90-5103-017-7, 21-32. (1988).

C.A. Edwards and N.Q. Arancon: Vermicompost Supress Plant Pests and Diseases Attacks. In REDNOVA NEWS: http://www.rednova.com/ display/?id =55938 (2004).

Carrillo-González, J. Šimůnek, S. Sauvé and D. Adriano, Adv. Agron. 91, 111 (2006).

Casado-Vela, S. Sellés, C. Díaz-Crespo, J. Navarro-Pedreño, J. Mataix-Beneyto and I. Gómez, Waste Manage. 27, 1509 (2007).

Cayuela, M.L., Sinicco, T., Mondini, C., 2008. Mineralization dynamics and biochemical properties during initial decomposition of plant and animal residues in soil. Applied Soil Ecology, 41, 118-127.

Cebolla, V., Pera, J., 1983. Suppressive effects of certain soils and substrates against Fusarium wilt of carnation. Acta Horticulture, 150, 113-119.

Christian, Gary D., Analytical Chemistry, 5th ed., John Wiley & Sons, New York, 1994.

Ciavatta, M. Govi, A. Simoni and E. Sequi (1993) Bioresour. Technol. 43, 147 (1993).

Clift, A. Doig and G. Finnveden, Process Saf. Environ. Protect. 78 (4), 279 (2000).

Clore, W.J.; Stanberry, C.O. Proc. Amer. Soc. Hort. Sci. 1946. 46, 296-298.

Cloutier-Hurteau, S. Sauvé and F. Courchesne, Environ. Sci. Technol. 41, 8104 (2007).

Colón, J. Martínez-Blanco, X. Gabarrell, A. Artola, A. Sánchez, J. Rieradevall and X. Fonta, Resour. Conserv. Recycl. 54, 893 (2010).

Crews, T. E. and Peoples, M. B., 2004. Legume versus fertilizer sources of nitrogen: ecological tradeoffs and human needs. Agriculture, Ecosystems and Environment, 102,279-297.

Czub and M.S. McLachlan, Environ. Toxicol. Chem. 23, 2356 (2004).

D. Coley,M.Massa, C. E. Lovelock, and K. Winter, "Effects of elevated CO2 on foliar chemistry of saplings of nine species of tropical tree," Oecologia, vol. 133, no. 1, pp. 62–69, 2002.

D. Mackay and S. Paterson, Environ. Sci. Technol. 15, 1006 (1981).

D. Milton, D. Wypij, D. Kriebel, M. Walters, S.J. Hammond and J. Evans, Am. J. Ind. Med. 29, 3 (1996).

D.B. Archer, The microbiological basis of process control in methanogenic fermentation of soluble wastes. Enzyme Microbiology Technology 5, 170 (1983).

D.J. Batstone, J. Keller, I. Angelidaki, S. Kalyuzhny, S.G. Pavlostathis, A. Rozzi, W.Sanders, H. Siegrist, V.A. Vavilin, Anaerobic Digestion Model No. 1 (ADM1), Report No. 13, IWA Publishing; (2002).

Dai, M. Xu, J. Chen, X. Yang and Z. Ke, Chemosphere 66, 353 (2007).

Daskalopoulos, O. Badr and S.D. Probert, Appl. Energy 58, 209 (1997).

Day, R. A. and Arthur L. Underwood, Quantitative Analysis, 6th ed., Prentice Hall, Upper Saddle River, NJ, 1991.

Decision Maker's Guide to Solid Waste Management, Volume II, (EPA 530-R-95-023),

Diver, S. 2002. Notes on compost teas: a supplement to ATTRA Publication: compost teas for plant disease control. Appropriate Technology Transfer for Rural Areas. 15 July 2007. http://www.attra.ncat.org.

Dowd, B. M., Press, D., Huertos, M. L., 2008. Agricultural nonpoint source water pollution policy: The case of California's Central Coast. Agriculture, Ecosystems and Environment, 128, 151-161.

E. L. Maia and D. B. Rodrigues-Amaya, "Avaliac¸˜ao de um m´etodo simples e econˆomico para a metilac¸˜ao de ´acidos graxos com lip´ıdios de diversas esp´ecies de peixes," Revista do Instituto Adolfo Lutz, vol. 53, no. 1/2, pp. 27–35, 1993.

E. Lacoste, J. P. Chaumont, D. Mandin, M. M. Plumel, and F. J. A. Matos, "Antiseptic properties of the essential oil of Lippia sidoides Cham: application to the cutaneous microflora," Annales Pharmaceutiques Francaises, vol. 54, no. 5, pp. 228–230, 1996.

E. M. Crotti, J. L. C. Lopes, and N. P. Lopes, "Triple quadrupole tandem mass spectrometry of sesquiterpene lactones: a study of goyazensolide and its congeners," Journal of Mass Spectrometry, vol. 40, no. 8, pp. 1030–1034, 2005.

E. Marañón, Y. Fernández and L. Castrillón, Manual de estado del arte de la codigestion anaerobia de residuos ganaderos y agroindustriales - Handbook of state-ofart of co-digestion of agro-industrial wastes, ProBiogas PSE (2009).

E. Nelson and A. Rangarajan: Vermicompost: A Living Soil Amendment; Publication of Cornell University, Ithaca, U.S. (http://cwmi.css.cornell. edu/ vermicompost.htm) (2010).

E. R. Martins, D. M. Castro, D. C. Castelanni, and J. E. Dias, Plantas Medicinais, Universidade Federal de Vic¸osa, Vic¸osa, Brazil, 2000.

E. S. Barata, L. S. Santos, P. H. Ferri, J. D. Phillipson, A. Paine, and S. L. Croft, "Anti-leishmanial activity of neolignans from Virola species and synthetic analogues," Phytochemistry, vol. 55, no. 6, pp. 589–595, 2000.

E. S. Barata, P. M. Baker, O. R. Gottlieb, and E. A. R` uveda, "Neolignans of Virola surinamensis," Phytochemistry, vol. 17, no. 4, pp. 783–786, 1978.

E. S. Matveeva, C. F. Gimenez, and M. J. G. Tejera, "Charge transfer behaviour of the indium-tin oxide/polyaniline interface: dependence on pH and redox state of PANI," Synthetic Metals, vol. 123, no. 1, pp. 117–123, 2001.

E. Smet, H. Van Langenhove and I. De Bo, Atmos. Environ. 33, 1295 (1999).

E. Wildermuth, H. Stark, G. Friedrich et al., "Iron compounds," in Ullmann's Encyclopedia of Industrial Chemistry, Wiley-VCH, 2000.

E.M. Costa-Neto and M. V. M. Oliveira, "The use of medicinal plants in the county of Tanquinho, State of Bahia, Northeastern Brazil," Revista Brasileira de Plantas Medicinais, vol. 2, no. 2, pp. 1–8, 2000.

E.Z. Harrison, S.R. Oakes, M. Hysell, A. Hay, Sci. Total Environ. 367, 481 (2006).

Edelmann, A. Joss, H. Engeli. "Two step anaerobic digestion of organic solid wastes", Second International Symposium on Anaerobic Digestion of Solid Wastes, 15- 18, June 1999 (Barcelona, Spain) pp153–160.

Edelmann, K. Schleiss and A. Joss, Water Sci. Technol. 41, 263 (2000).

Edelmann, U. Baier, H. Engeli, Water Sci. Technol. 52, 203 (2005).

Eghball, J.F. Power, J.E. Gilley and J.W. Doran, J. Environ. Qual. 26 (1), 189 (1997).

Eiras, I. N. G. Passos, A. C. F.De Brito et al., "Electroactive nanocomposites made of poly(o-methoxyaniline) and natural polysaccharides," Qu ́ımica Nova, vol. 30, no. 5, pp. 1158–1162, 2007.

El-Hassan, S.A., Gowen, S.R., 2006. Formulation and delivery of the bacterial antagonist Bacillus subtilis for management of lentil vascular wilt caused by Fusarium oxysporum f.sp. lentis. Phytopathology 154, 148-155.

Environmental Protection Agency (EPA). 2000. Office of Transportation and Air Quality. National Water Quality Inventory, EPA, Washington, DC.

Environmental Protection Agency. Europe's environment. The fourth assessment. Copenhagen, Denmark. ISBN: 978-92-9167-932-4 (2007).

Epstein, W. 2003. The roles and regulation of potassium in bacteria. Progress in Nucleic Acid Research and Molecular Biology, 75, 293-320.

European Commission, Commission Regulation No. 208/2006 amending Annexes VI and VIII to Directive 1774/2002 as regards processing standards for biogas and composting plants and requirements for manure, (2006).

European Commission, Directive 1774/2002/EC laying down health rules concerning animal by-products not intended for human consumption, (2002).

European Commission, Final report on heavy metals and organic compounds from wastes used as organic fertilisers, ENV.A.2./ETU/2001/0024, July (2004).

Ewing, Galen Wood, Ed., Analytical Instrumentation Handbook, 2nd ed, Marcel Dekker, New York, 1997.

F. Cui, Y. H. Cui, Y. L. Sun, K. Zang, and W. D. Zhang, "Enhancement of dopamine sensing by layer-by-layer assembly of PVI-dmeOs and nafion on carbon nanotubes," Nanotechnology, vol. 21, pp. 215–601, 2010.

F. G. Martins, J. F. Andrade, A. C. Pimenta, L. M. Lourenco, J. R. M. Casto, and V. R. Balbo, "Spectrophotometric study of iron oxidation in the iron(II) thiocyanateacetone system and some analytical application," Electica Quimica, vol. 30, no. 3, pp. 63–71, 2005.

F. G. Rodrigues, H. D. M. Coutinho, A. R. Campos, S. G. de Lima, and J. G. M. da Costa, "Atividade antibacteriana e composic ̧ ̃ao qu ́ımica do ́ oleo essencial de Lippia microphylla cham," Acta Scientiarum, vol. 33, no. 2, pp. 141–144, 2011.

F. Madrid, R. López and F. Cabrera, Agric. Ecosyst. Environ. 119, 249 (2007).

F. N. Crespilho, V. Zucolotto, O. N. Oliveira Jr., and F. C.Nart, "Electrochemistry of layer-by-layer films: a review," International Journal of Electrochemical Science, vol. 1, pp. 194–214, 2006.

F. P. Oliveira, E. O. Lima, J. P. Siqueira-J′ unior, E. L. Souza, B. H. C. Santos, and H. M. Barreto, "Effectiveness of Lippia sidoides Cham. (Verbenaceae) essential oil in inhibiting the growth of Staphylococcus aureus strains isolated from clinical material," Brazilian Journal of Pharmacognosy, vol. 16, no. 4, pp. 510–516, 2006.

F. Pinamonti, G. Stringari, F. Gasperi and G. Zorzi, Resour. Conserv. Recycl. 21, 129 (1997).

F.R. Magdoff: Soil Organic Matter in Sustainable Agriculture' (www.compostforsoils.com.au) (2004).

F.R. McDougall, P.R. White, M. Franke and P. Hindle, Integrated Solid Waste Mangement: a Life

Fabre, I. Rustan, E. de Hoffmann, and J. Quetin-Leclercq, "Determination of flavone, flavonol, and flavanone aglycones by negative ion liquid chromatography electrospray ion trap mass spectrometry," Journal of the American Society for Mass Spectrometry, vol. 12, no. 6, pp. 707–715, 2001.

Fromme, T. Kuchler, T. Otto, K. Pilz, J. Muller and A. Wenzel, Water Res. 36, 1429 (2002).

Fujie, N. Matsutani, Y. Kinoshita, Y. Okamura, A. Saito, and S. Takeoka, "Adhesive, flexible, and robust polysaccharide nanosheets integrated for tissue-defect repair," Advanced Functional Materials, vol. 19, no. 16, pp. 2560–2568, 2009.

Fujimto, J. H. Kim, K. Ohmori, A. Ono, and S. Shiratori, "Flexible multilayer electrode films consisted of polyaniline

G. Decher, M. Eckle, J. Schmitt, and B. Struth, "Layer-by-layer assembled multicomposite films," Current Opinion in Colloid and Interface Science, vol. 3, no. 1, pp. 32–39, 1998.

G. Egiarte, G. Corti, M. Pinto, J. Arostegui, F. Macías, E. Ruíz-Romero and M. Camps Arbestain, Water Air Soil Pollut. 198, 133 (2009).

G. Finnveden, J. Johansson, P. Lind and Å. Moberg, J. Clean. Prod. 13, 213 (2005).

G. Fischer, T. Muller, R. Ostrowski and W. Dott, Chemosphere 38, 1745 (1999).

G. Hahn: 'Chitin Rich Vermicompost Repels Pests and Suppress Plant Diseases; (Personal Communication; Email: geohahn@gmail.com) (2011).

G. Lettinga, Antonie Leeuwenhoek 67, 3 (1995).

G. Martins, J. F. Andrade, A. C. Pimenta, L. M. Lourenco, J. R. M. Casto, and V. R. Balbo, "Spectrophotometric study of iron oxidation in the iron(II)/thiocyanate/acetone system and some analytical applications," Ecl´etica Qu´ımica, vol. 30, no. 3, pp. 63–71, 2005.

G.A. Benglini, Resour. Conserv. Recycl 52, 1373 (2008).

G.A.R. Hackett, C.A. Easton and S.J.B. Duff, Bioresour. Technol. 70, 217 (1999).

G.F. Fries, Sci. Total Environ. 185, 93 (1996).

G.G. Briggs, R.H. Bromilow and A.A. Evans, J. Pest. Sci. 13, 495 (1982).

G.M. Greenway and Q.J. Song, J. Environ. Monit. 4, 300 (2002).

G.O. Aspinall, The Polysaccharides, Academic Press,NewYork, NY, USA, 1982.

G.S. Senesi, G. Baldassarre, N. Senesi and B. Radina, Chemosphere 39, 343 (1999).

G.V. Schirato, F.M. F.Monteiro, F. O. Silva, J. L. L. Filho, A. M. A. C. Carneiro Le˜ao, and A. L. F. Porto, "The polysaccharide from Anacardium occidentale L. in the inflammatory phase of the cutaneous wound healing," Ciencia Rural, vol. 36, no. 1, pp. 149–154, 2006.

García, J.L. Moreno, T. Hernández, F. Costa and A. Polo, Bioresour. Technol. 53, 13 (1995)

Gill, Robin, Ed., Modern Analytical Geochemistry: An Introduction to Quantitative Chemical Analysis Techniques for Earth, Environmental and Materials Scientists, Addison Wesley, Harlow, U.K., 1997.

Gobbo-Neto, M. D. Santos, A. Kanashiro et al., "Evaluation of the anti-inflammatory and antioxidant activities of di-

Gong, M., Wang, J.D., Zhang, J., Yang, H., Lu, X.F., Pei, Y., Cheng, J.Q., 2006. Study of the antifungal ability of Bacillus subtilis strain PY-1 in vitro and identification of its antifungal substance (iturin A). Acta Biochim. Biophys. Sin. 38, 233-240.

Guaratini, P. J. Gates, E. Pinto, P. Colepicolo, and N. P. Lopes, "Differential ionisation of natural antioxidant polyenes in electrospray and nanospray mass spectrometry," Rapid Communications in Mass Spectrometry, vol. 21, no. 23, pp. 3842–3848, 2007.

Guaratini, P. J. Gates, K. H. M. Cardozo, P. M. B. G. M. Campos, P. Colepicolo, and N. P. Lopes, "Letter: radical ion and protonated molecule formation with retinal in electrospray and nanospray," European Journal of Mass Spectrometry, vol. 12, no. 1, pp. 71–74, 2006.

Guaratini, R. L. Vessecchi, F. C. Lavarda et al., "New chemical evidence for the ability to generate radical molecular ions of polyenes from ESI and HR-MALDI mass spectrometry," Analyst, vol. 129, no. 12, pp. 1223–1226, 2004.

Gujer and A.J.B. Zehnder, Water Sci. Technol. 15, 127 (1983).

H. Hoitink: Compost Use for Disease Suppression; In 'On farm Composting Handbook' (www.plantpath.osu.edu) (2008).

H. Hyun, A.C. Chang, D.R. Parker and A.L. Page, J. Environ. Qual. 27, 329 (1998). R. Millares, E.M. Beltrán, M.A. Porcel, M.L. Beringola, J.V. Martín, R. Calvo and M.M. Delgado, Foro del Oliva y el Medio Ambiente OLI-2007 (in Spanish). Available at: www.expoliva.com/expoliva2003. Accessed 27 January 2010. (2003).

H. S. Silva, N. P. Lopes, M. J. Kato, and M. Yoshida, "Fatty acids from Myristicaceous seeds of myristic acid-rich species," Anais da Associac¸˜ao Brasileira de Qu´ımica, vol. 46, pp. 232–235, 1997.

Haoyi, Z. Guoxiu, and Y. Gaohua, "Determination of cobalt in terephthalic acid by picramazochrom spectrophotometry," Chemical Analysis and Meterage, vol. 1, 2009.

Hargreaves, J.C.; Adl, M.S.; Warman, P.R. Agric. Eco. Environ. 2008a. 123(1-3), 1-14.

Hargreaves, J.C.; Adl, M.S.; Warman, P.R. J. Sci. Fd. Agric. 2009. 89, 390-397.

Hargreaves, J.C.; Adl, M.S.; Warman, P.R.; Rupasinghe, H.P.V. Plant Soil 2008b. 308, 213-226.

Harris, Daniel C., Quantitative Chemical Analysis, 5th ed., W. H. Freeman & Co., New York, 1998.

Hartmann, B.K. Ahring, Water Sci. Technol. 53, 7 (2006).

Hladik and S. Miquel, "Seedling types and plant establishment in an African rain forest," in Reproductive Ecology

Hobson, and A. Wheatley, Anaerobic digestion–modern theory and practice, Elsevier Applied Science, London (1992).

Hoitink, H.A.J., Boehm, M.J., 1999. Biocontrol within the context of soil microbial communities: a substrate-dependent phenomenon. Annual Review of Phytopathology, 37,427-446.

Hryhorczuk, L. Curtis, P. Scheff, J. Chung, M. Rizzo, C. Lewis, N. Keys and M. Moomey, Ann. Agric. Environ. Med. 8, 177 (2001).

IPCC (Intergovernmental Panel on Climate Change), IPCC Guidelines for National Greenhouse Gas Inventories. Volume 5, chapter 4 (2006).

Irisarri, P., Gonnet, S., Monza, J., 2001. Cyanobacteria in Uruguayan rice fields: diversity, nitrogen fixing ability and tolerance to herbicides and combined nitrogen. Journal of Biotechnology, 91, 95-103.

Iwata, F.A. Gunther and W.E. Westlake, Bull. Environ. Contam. Toxicol. 11, 523 (1974).

J. Bard and L. R. Faulkner, ElectrochemicalMethods, Fundamentals and Applications, John Wiley & Sons, New York, NY, USA, 2nd edition, 2001.

J. E.Huheey, E. A. Keiter, and R. L. Keiter, Inorganic Chemistry, Harper Collins, New York, NY, USA, 4th edition, 1993.

J. G. S. Maia and E. H. A. Andrade, "Database of the amazon aromatic plants and their essential oils," Quimica Nova, vol. 32, no. 3, pp. 595–622, 2009.

J. J. Langer, M. Filipiak, J. Kecinska, J. Jasnowska, J. Włodarczak, and B. Buładowski, "Polyaniline biosensor for

J. Laska and J. Widlarz, "One-step polymerization leading to conducting polyaniline," Synthetic Metals, vol. 135, pp. 261– 262, 2003.

J. Pichtel and M. Anderson, Bioresour. Technol. 60, 223 (1997).

J. T. Ellis, I. Schulman, and C. H. Smith, "Generalized siderosis with fibrosis of liver AND pancreas in cooley's (Mediterranean) anemia with observations on the pathogenesis of the siderosis AND fibrosis," American Journal of Pathology, vol. 30, no. 2, pp. 287–309, 1954.

J.B. Holm-Nielsen, T. Al Seadi and P. Oleskowicz-Popiel, Bioresource Technol. 100, 5478 (2009)

J.C. Buckerfield and K.A. Webster: Worm-Worked Waste Boost Grape Yield: Prospects for Vermicompost Use in Vineyards; The Australian and New Zealand Wine Industry Journal, 13: 73-76. (1998).

J.C. Buckerfield, T.C. Flavel, K.E. Lee and K.A. Webster: Vermicompost in Solid and Liquid Forms as a Plant-Growth Promoter; Pedobiologia, 43: 753-759. (1999).

J.C. Hargreaves, M.S. Adl, P.R. Warman, Agr. Ecosyst. Environ. 123, 1 (2008).

J.D. Murphy and E. McKeogh, Renew. Energ. 29, 1043 (2004).

J.G. Speight and S. Lee, Environmental Technology Handbook, 2nd Edition. Taylor and Francis (2000).

J.J. Mortvedt, F.R. Cox, L.M. Shuman and R.M. Welch, Micronutrients in Agriculture, 2nd Ed. The Soil Science Society of America, Madison, WI (1991).

J.O Azeez, I.O. Adekunle, O.O. Atiku, K.B. Akande and S.O. Jamiu-Azeez, Waste Manage. 29, 2582 (2009).

J.R. Barton and D. Dalley, V.S. Patel, Waste Manage. 16 (1-3), 35 (1996).

J.W. Gillett, Cornell Composting – Resources: MSW composting fact sheets. Municipal Solid Waste Composting: Issues in Risk Assessment and Management/Worker Health and Safety. Institute for Comparative and Environmental Toxicology at Cornell University (1996).

J.W.C. Wong, K.K. Ma, K.M. Fang and C. Cheung, Bioresour. Technol. 67, 43 (1999).

J.Y. Kim, J.K. Park, B. Emmons, D.E. Armstrong, Water Environ. Res. 67, 1044 (1995).

Jack: Suppression of plant pathogens with vermicomposts; In CA Edwards, NQ Arancon, and RL Sherman (Eds.) 'Vermiculture Technology: Earthworms, Organic Wastes and Environmental Management'; Boca Raton, FL, CRC Press (US); p.623.

Justesen, "Collision-induced fragmentation of deprotonated methoxylated flavonoids, obtained by electrospray

K. Malik and A. L. J. Rao, "Spectrophotometric determination of iron(III) dimethyldithiocarbamate (ferbam)," Talanta, vol. 44, no. 2, pp. 177–183, 1997.

K. Malik, K. N. Kaul, B. S. Lark, W. Faubel, and A. L. J. Rao, "Spectrophotometric determination of cobalt, nickel palladium, copper, ruthenium and molybdenum using sodium isoamylxanthate in presence of surfactants," Turkish Journal of Chemistry, vol. 25, no. 1, pp. 99–105, 2001.

K. S. Bawa, P. S. Ashton, and S. M. Nor, "Reproductive ecology of tropical forest Plants: management issues," in Reproductive Ecology of Tropical Forest Plants, K. S. Bawa and M. Hadley, Eds., p. 3, UNESCO & Parthenon Publishing Group, Paris, France, 1990.

K. Sharma and I. Singh, "Spectrophotometric trace determination of iron in food, milk, and tea samples using a new

K. Tarafder and R. Thakur, "Surfactant-mediated extraction of iron and its spectrophotometric determination in rocks, minerals, soils, stream sediments and water samples," Microchemical Journal, vol. 80, no. 1, pp. 39–43, 2005.

K.A Webster: Vermicompost Increases Yield of Cherries for Three Years after a Single Application; EcoResearch, South Australia, (www.ecoresearch.com.au). (2005).

K.C Pun, R.Y.H Cheung and M.H Wong, Mar. Pollut. Bull. 31 (4–12), 394 (1995).

Karmegam and T. Daniel: Effect of Vermicompost and Chemical Fertilizer on Growth and Yield of Hyacinth Bean (Lablab purpureas); Dynamic Soil, Dynamic Plant; Global Science Books; 2 (2): 77-81. (2008).

Kaschl, V. Römheld and Y. Chen, Sci. Total Environ. 291, 45 (2002).

Keeney, D. R., Nitrogen management for maximum efficiency and minimum pollution. Severnson , F. J., Nitrogen in Agricultural Soils. Am. Soc. of Agron. Maidson Wis: Agronomy, 1982, 605-649.

Khan, R.A., Saxena, S.K., 1997. Integrated management of root knot nematode Meloidogyne javanica infecting tomato using organic materials and Paecilomyces lilacininus. Bioresource Technology, 61, 247-250.

knema furfuracea," Phytochemistry, vol. 29, no. 6, pp. 1985– 1988, 1990.

Korboulewsky, S. Dupouyet and G. Bonin, J. Environ. Qual. 31, 1522 (2002).

Korcak, R.F. Hort Rev. 1988. 10, 183-227.

L. Hua, W.X. Wu, Y.X. Liu, C.M. Tientchen and Y.X. Chen, Biomed. Environ. Sci. 21, 345 (2008).

L. I. Cheng-hong, G. E. Chang-hua, L. Hua-ding, and P. Fuyou, "Spectrophotometric determination of iron with 2-(5-

L. M. A. Macambira, C. H. S. Andrade, F. J. A. Matos, A. A. Craveiro, and R. Braz Filho, "Naphthoquinoids from Lippia sidoides," Journal of Natural Products, vol. 49, no. 2, pp. 310– 312, 1986.

L. M. Wang, C. Song, and J. Jin, "Spectrophotometric determination of iron by extraction of its ternary complex with

L. R.Da Cunha, R. C.M. de Paula, and J. P. A. Feitosa, "Polissacar ′ıdeos da biodiversidade brasileira: UMA oPortunidade de transformar conhecimento em valor econˆomico," Qu′ımica Nova, vol. 32, no. 3, pp. 649–660, 2009.

L. Roca-Pérez, C. Martínez, P. Marcilla and R. Boluda, Chemosphere 75, 781 (2009).

L. Sahlström, Bioresour. Technol. 87, 161 (2003).

L. Walker, W. Charles and R. Cord-Ruwisch, Bioresour. Technol. 100, 3799 (2009).

L.C. Martins das Neves, A. Converti, T.C. Vessoni Penna, Chem. Eng. Technol. 32 (8), 1147 (2009).

Ladha, J. K., Pathak, H., Krupnik, T. J., Six, J., van Kessel, C. 2005. Efficiency of fertilizer nitrogen in cereal production: retrospects and prospects. Advances in Agronomy, 87, 85-156.

Lagowski, J. J. Introduction to Semimicro Qualitative Analysis, 5th ed. Prentice-Hall: Englewood Cliffs, NJ, 1977.; (b) Shriner, R. L.; Fuson, R. C.; Curtin, D. Y. The Systematic Identification of Organic Compounds, 5th ed. John Wiley and Sons: New York, 1964.

Larena, I., Sabuquillo, P., Melgarejo, P., De Cal, A., 2003. Biocontrol of Fusarium and Verticillium wilt of tomato by Penicillium oxalicum under greenhouse and field conditions. Phytopathology, 151, 507-512.

Larkin, R.P., Fravel, D.R., 1998. Efficacy of various fungal and bacterial biocontrol organisms for control of Fusarium wilt of tomato. Plant Disease, 82, 1022-1028.

Lee, J., Effect of application methods of organic fertilizer on growth, soil chemical properties and microbial densities in organic bulb onion production. Scientia Horticulture, 2010, 124, 299-305.

Lee, J.J., Park, R.D., Kim, Y.W., Shim, J.H., Chae, D.H., Rim, Y.S., Sohn, B.K., Kim, T.H., Kim, K.Y., 2004. Effect of food waste compost on microbial population, soil enzyme activity and lettuce growth. Bioresource Technology,93, 21-28.

Li, J., Zhao, B. Q., Li, X. Y., Jiang, R. B., So, H. B., 2008. Effects of long-term combined application of organic and mineral fertilizers on microbial biomass, soil enzyme activities and soil fertility. Agricultural Sciences in China, 7, 336-343.

Lippia sidoides (Cham.) Verbenaceae," Brazilian Journal of Pharmacognosy, vol. 12, supplement 1, pp. 66–67, 2002.

Lopes, C., Herva, M., Franco, A., Roca, E., Organic waste valorization in agriculture and its environmental risk implications. In: Pollution Control: Management, Technology and Regulations. Air, Water and Soil Science and Technology Series. Pages 1-45. Editorial: Nova Science Publishers, Inc., New York (2011).

M. Dohányos, J. Zábranská and P. Jenícek, Water Sci. Technol. 36, 145 (1997).

M. F. Zampa, A. C. F. de Brito, I. L. Kitagawa et al., "Natural gum-assisted phthalocyanine immobilization in electroactive nanocomposites: physicochemical characterization and sensing applications," Biomacromolecules, vol. 8, no. 11, pp. 3408–3413, 2007.

M. Ferreira, L. R. Dinelli, K. Wohnrath, A. A. Batista, and O. N. Oliveira Jr., "Langmuir-Blodgett films from polyaniline/ ruthenium complexes as modified electrodes for detection of dopamine," Thin Solid Films, vol. 446, no. 2, pp. 301–306, 2004.

M. López Alonso, J.L. Benedito, M. Miranda, C. Castillo, J. Hernández and R.F. Shore, Sci. Total Environ. 246, 237 (2000b)

M. M. M. Pinto, A. Kijjoa, I. O. Mondranondra, A. B. Guti´errez, and W. Herz, "Lignans and other constituents of

M. Nagabhushana, G. T. Chandrappa, B. Nagappa, and N. H. Nagaraj, "Diformylhydrazine as analytical reagent for

M. O. Simo˜es and V. Spitzer, "O´ leos vola´teis," in Farmacognosia: Da Planta Ao Medicamento, C. M. O. Sim˜oes, E. P. Schenkel, G. Gosmann, J. C. P. Mello, L. A. Mentz, and P. R. Petrovick, Eds., pp. 467–496, UFRGS, Porto Alegre, Brazil, 5th edition, 2004.

M. Pivatto, A. E. M. Crotti, N. P. Lopes et al., "Electrospray ionization mass spectrometry screening of piperidine

M. Raposo and O. N. Oliveira Jr., "Energies of adsorption of poly(o-methoxyaniline) layer-by-layer films," Langmuir, vol. 16, no. 6, pp. 2839–2844, 2000.

M. Schlegelmilch, J. Streese, W. Biedermann, T. Herold and R. Stegmann, Waste Manage. 25, 917 (2005).

M. Soldano, C. Fabbri and S. Piccinini, "Co-digestion plant in dairy cattle farm in Emilia Romagna Region (Italy)", International Conference: Progress in Biogas, 19-21, Sept. 2007 (Sttutgart, Germany) pp95-99.

M. Soliva and S. Paulet, In: Boixadeira J, Teira MR (eds) Agricultural aplication of organic wastes. University of Lleida, Spain, pp 63 (in Spanish) (2001).

M. Tarantini, A. Dominici Loprieno, E. Cucchi and F. Frenquellucci, Energy 34, 613 (2009).

M. Tejada, A.M. García-Martínez and J. Parrado, Catena 77, 238 (2009).

M. van Tongeren, L. van Amelsvoort and D. Heederik, Int. J. Occup. Environ. Health 3, 30 (1997).

M. W. Hentze and L. C. K¨uhn, "Molecular control of vertebrate iron metabolism: mRNA-based regulatory circuits

M. Wilm and M. Mann, "Analytical properties of the nanoelectrospray ion source," Analytical Chemistry, vol. 68, no. 1,

M.A. Bustamante, C. Paredes, F.C. Marhuenda-Egea, A. Pérez-Espinosa, M.P. Bernal and R. Moral, Chemosphere 72, 551 (2008).

M.D. Kim, M. Song, M. Jo, S.G. Shin, J.H. Khim and S. Hwang, Appl. Microbiol. Biotechnol. 85, 1611 (2010).

M.D. Perez-Murcia, R. Moral, J. Moreno-Caselles, A. Perez-Espinosa and C. Paredes,

M.D. Webber, H.R. Rogers, C.D. Watts, A.B.A. Boxall, R.D. Davis, R. Scoffin, Sci. Total Environ. 185, 27 (1996).

M.P. Bernal, C. Paredes, M.A. Sánchez-Monedero and J. Cegarra, Bioresour. Technol.63, 91 (1998).

Methodology. National Academy Press (1993).

microgram quantities with 2-hydroxy-3-methoxy benzaldehyde thiosemicarbazone," Journal of the Korean Chemical Society, vol. 51, no. 4, pp. 331–338, 2007.

Mishra, K., Kumar, A., Pandey, K., 2010.RAPD based genetic diversity among different isolates of Fusarium oxysporum f. sp. lycopersici and their comparative biocontrol. World Journal of Microbial Biotechnology, 26, 1079-1085.

N. P. Lopes, P. A. Dos Santos, M. J. Kato, and M. Yoshida, "New butenolides in plantlets of Virola surinamensis (Myristicaceae)," Chemical and Pharmaceutical Bulletin, vol. 52, no. 10, pp. 1255–1257, 2004.

N. P. Lopes, P. Chicaro, M. J. Kato, S. Albuquerque, and M. Yoshida, "Flavonoids and lignans from Virola surinamensis

N. P. Lopes, S. C. Franca de, A. M. S. Pereira et al., "A butanolide from seedlings and micropropagated leaves of

N. P. Lopes, T. Fonseca, J. P. G. Wilkins, J. Staunton, and P.J. Gates, "Novel gas-phase ion-molecule aromatic nucleophilic substitution in β-carbolines," Chemical Communications, vol. 9, no. 1, pp. 72–73, 2003.

N. Sadhale: Recommendation to incorporate earthworms in soil of pomegranate to obtain high quality fruits. In Surpala's Vrikshayurveda, Verse 131. The Science of Plant Life by Surpala, 10th Century A.D. Asian Agri-History Bulletin; No. 1. Secunderabad, India. (1996).

N. Sridhara Cari, C.T. Kamala and D.S. Suman Raj, Ecotoxicol. Environ. Saf. 69, 513 (2008).

N. Veerachalee, P. Taweema, and A. Songsasen, "Complexation and spectrophotometric determination of cobalt(II) ion with 3-(2_-thiazolylazo)-2,6-diaminopyridine," Kasetsart Journal— Natural Science, vol. 41, no. 4, pp. 675–680, 2007.

N.A.E. Kader, P. Robin, J.-M. Paillat and P. Leterme, Bioresour. Technol. 98, 2619 (2007).

N.Q. Arancon and C. A. Edwards): Effects of Vermicompost on Plant Growth; In: Guerrero, R.D. and Guerrero-del Castillo, M.R. (Eds): Vermitechnologies for Developing Countries; Philippine Fisheries Association, Laguna, Philippines; .32 - 65. (2006).

N.Q. Arancon, C.A. Edwards, P. Bierman, C. Welch, and J.D. Metzger : Influences of Vermicomposts on Field Strawberries: 1. Effects on Growth and Yields; Bioresource Technology, 93, No. 2; 145-153. (2004).

N.Q. Arancon, P.A. Galvis, and C.A. Edwards: Suppression of insect pest populations and damage to plants by vermicomposts; Bioresource Technology, 96(10): 1137-1142. (2005).

N.S. Bolan: Enhancing Soil Carbon Sequestration Utilizing Compost; Paper presented at Int. Symposium on 'Organic Matter Management and Compost Use in Horticulture'; Adelaide, Australia, April 4-7, 2011; University of South Australia (Nanthi.bolan@unisa.edu.au).

National Institute of Standards and Technology, PC Version of the NIST/EPA/NIH Mass Spectral Database, U.S. Department of Commerce, Gaithersburg, Md, USA, 1998.

Nussinovitch, Plant Gum Exudates of the World: Sources, Distribution, Properties, and Applications, CRC Press, New York, NY, USA, 2010.

Oren, A., 2004. A proposal for further integration of the cyanobacteria under the bacteriological Code. Int. J. Syst. Evol. Microbiol. 54, 1895-1902.

Osunlaja, S.O., 1989. Effect of organic soil amendments on the incidence of brown spot disease in maize caused by Physoderma maydis. Journal of Basic Microbiology, 29, 501-505.

P. Danelutte, A. J. Cavalheiro, and M. J. Kato, "Lignoids in seedlings of Virola sebifera," Phytochemical Analysis, vol. 11, no. 6, pp. 383–386, 2000.

P. Pramanik, G.K. Ghosh, P.K. Ghosal and P. Banik: Changes in Organic-C, N, P and K and Enzyme Activities in Vermicompost of Biodegradable Organic Wastes Under Liming and Microbial Inoculants. J. of Bioresource Technology, 98, 2485-2494. (2007).

P. Roberts, G.E. Jones and D.L. Jones: Yield Responses of Wheat (Triticum aestivum) to Vermicompost; Compost Science and Utilization; 15: . 6-15. (2007).

P. Sørensen and H.B. Møller. Fate of nitrogen in pig and cattle slurries applied to the soil-crop system. In: Adani, F., Schievano, A., Boccasile, G. (Eds.), Anaerobic Digestion: Opportunities for Agriculture and Environment. DiProVe University of Milan, Milan, pp 27–37 (2009).

P. Sosnowski, A. Wieczorek and S. Ledakowicz, Adv. Environ. Res. 7, 609 (2003).

P. Sykes, K. Jones and J.D. Wildsmith, Resour. Conserv. Recycl. 52, 410 (2007).

P. Vandevivere, L. De Baere, W. Verstraete, Types of anaerobic digesters for solid wastes. In: Biomethanization of the organic fraction of municipal solid wastes, IWA Publishing; (2003).

P. Villar, M. Callejon, E. Alonso, J.C. Jimenez and A. Guiraum, Chemosphere 64, 535 (2006).

P. Zhuang, M.B. McBride, H. Xia, N. Li and Z. Li, Sci. Total Environ. 407, 1551 (2009).

P.G. Stroot, K.D. McMahoni, R.I. Mackie and L. Raskini, Water Res. 3, 1804 (2001).

P.L. Giusquiani, M. Pagliai, G. Gigliotti, D. Businelli and A. Benetti, J. Environ. Qual. 24 175 (1995).

Q. Lin and L. Zhang, "Simultaneous determination of dopamine and ascorbic acid at glutamic acid modified graphite electrode," Analytical Letters, vol. 34, no. 10, pp 1585–1589, 2001.

R. Duarte-Davidson and K.C. Jones, Sci. Total Environ. 185, 59 (1996).

R. Follet,: Soil Management Concepts and Carbon Sequestration in Croplands Soils'; Soil and Tillage Research; 61; pp. 77-92. (2001).

R. Gibson, M.J. Wang, E. Padgett and A.J. Beck, Chemosphere 61, 1336 (2005).

R. K. Lima, M. G. Cardoso, J. C. Moraes, S. M. Carvalho, V. G. Rodrigues, and L. G. L. Guimar˜aes, "composic¸˜ao qu´ımica e efeito fumigante do ´ oleo essencial de Lippia sidoides cham. e monoterpenos sobre Tenebrio molitor (L.) (coleoptera: tenebrionidae)," Ciencia e Agrotecnologia, vol. 35, no. 4, pp. 664– 671, 2011.

R. K. Patil and D. G. Dhuley, "Solvent extraction and spectrophotometric determination of Fe(II) with 1,3-diphenyl-4-

R. K. Sinha, Sunil Herat, Dalsukh Valani and Krunal Chauhan: Vermiculture and Sustainable Agriculture'; American-Eurasian J. of Agricultural and Environmental Sciences; ISSN 1818: 5 (S); 01- 55; IDOSI Publication (Special Issue) (2009).

R. K. Sinha: Vermiculture Revolution: The Technological Revival of Charles Darwin's Unheralded Soldier's of Mankind; NOVA Science Publication, U.S.A. (Under Publication) (2011).

R. Lăcătuşu, C. Răuţă, S. Cărstea and I. Ghelase, Appl. Geochem. 11, 105 (1996).

R. Lal and J. Bruce: The Potential of World's Croplands Soils to Sequester Carbon (C) and Mitigate the Greenhouse Effect; Environmental Science and Policy; 2; 177-185. (1999).

R. Michel, M. Nolte, M. Reich, and F. Loer, "Systemic effects of implanted prostheses made of cobalt-chromium alloys,"

R. Oliver, E. May and J. Williams, Water Res. 39, 4436 (2005).

R. P. Adams, Identification of Essential Oil Components by Gas Chromatography/Mass Spectroscopy, Allured, Carol Stream, Ill, USA, 4th edition, 2007.

R. Pipatti and M.S. Manso, 2006 IPPC guidelines for national greenhouse gas inventories, Volume 5 Waste, Hayama, Kanagawa (2006).

R. Reddy, P. Radhika, J. R. Kumar, D. N. Priya, and K. Rajgopal, "Extractive spectrophotometric determination

R. Rynk, On-farm composting handbook. Cooperative extension, Northeast Regional Agricultural Engineering Service, Ithaca, New York (1992).

R. S.Nunes, H. S. Xavier, P. J. RolimNeto,D. P. Santana, and U. P. Albuquerque, "Botanical standardization of Lippia sidoides Cham. (Verbenaceae)," Acta Farmaceutica Bonaerense, vol. 19, no. 2, pp. 115–118, 2000.

R. Sharma: Vermiculture for Sustainable Agriculture: Study of the Agronomic Impact of Earthworms and their Vermicompost on Growth and Production of Wheat Crops; Ph.D. Thesis; Awarded by University of Rajasthan, Jaipur, India (Supervisor: Dr. Rajiv K. Sinha). (2001).

R. Singh, R.R. Sharma, S. Kumar, R.K. Gupta and R.T.Patil: Vermicompost Substitution Influences Growth, Physiological Disorders, Fruit Yield and Quality of Strawberry (Fragaria ananassa Duch.); J. of Bioresoursce Technology; 99: 8507-8511. (2008).

R. Takeara, S. Albuquerque, N. P. Lopes, and J. L. C. Lopes, "Trypanocidal activity of Lychnophora staavioides Mart.

R. W. Doskotch and M. S. Flom, "Acuminatin, a new bisphenylpropide from Magnolia acuminata," Tetrahedron, vol.

R.C. Brändli, T. Kupper, T.D. Bucheli, M. Zennegg, S. Huber, D. Ortelli, J. Müller, C. Schaffner, S. Iozza, P. Schmid, U. Berger, P. Edder, M. Oehme, F.X. Stadelmann and J. Tarradellas, J. Environ. Monit. 9, 465 (2007b).

R.C. Brändli, T.D. Bucheli, T. Kupper, M. Zennegg, U. Berger, P. Edder, M. Oehme, J. Müller, C. Schaffner, R. Furrer, P. Schmid, S. Huber, D. Ortelli, S. Iozza, F.X. Stadelmann and J. Tarradellas, Organohalogen Compd. 68, 863 (2006).

R.C. Brändli, T.D. Bucheli, T. Kupper, R. Furrer, W.A. Stahel, F.X. Stadelmann and J. Tarradellas, J. Environ. Monit. 9, 456 (2007a).

R.D. Guerrero: Vermicompost Production and its Use in Crop Production in The Philippines; Int. J. of Global Environmental Issues ; In Rajiv K. Sinha et. al., (Eds.) Special Issue on 'Vermiculture Technology'; 10; 378-383; Inderscience Pub. (2010).

R.D. Kale, B.C. Mallesh, B. Kubra, and D.J. Bagyaraj : Influence of Vermicompost Application on the Available Macronutrients and Selected Microbial Populations in a Paddy Field; Soil Biology and Biochemistry; 24 (12): 1317-1320. (1992).

R.D. Kale. and K. Bano: Field Trials With Vermicompost. An Organic Fertilizer; In Proc. of National Seminar on 'Organic Waste Utilization by Vermicomposting'. GKVK Agricultural University, Bangalore, India. (1986) (dr.rdkale@gmail.com).

R.D. Kale: Vermicompost – crown jewel of organic farming (A Memoir); N.D. Kale, Malleswaram, Banglore, India; (dr.rdkale@gmail.com) (2006).

R.E. Alcock, J. Bacon, R.D. Bardget, A.J. Beck, P.M. Haygarth, R.G.M. Lee, C.A. Parker and K.C. Jones, Environ. Pollut. 93, 83 (1996).

R.E. Speece, Anaerobic biotechnology for industrial wastewaters. Arche Press, Nashville, TN (1996).

R.K. Rosenbaum, T.E. McKone and O. Jolliet, Environ. Sci. Technol. 43, 8191 (2009).

R.K. Sinha and Dalsukh Valani: Earthworms: Charles Darwin's 'Unheralded Soldiers of Mankind and Farmer's Friend' Working Day and Night Under the Soil: Their Role as Soil Managers; In Justin A Daniels (Ed.) Advances in Environmental Research – 9 (Chapter 8); NOVA Science Publishers, N.Y., USA; ISBN: 978 – 1 – 61728 – 999 - 6. (2010).

R.K. Sinha, D. Valani, B. K. Soni and V. Chandran: Earthworms Vermicompost: A Sustainable Alternative to Chemical Fertilizers for Organic Farming; Agricultural Issues and Policies; NOVA Science Publishers, N.Y., USA; ISBN 978-1-61122-580-8. (2011).

R.K. Sinha, D. Valani, K. Chauhan and S. Agarwal: Embarking on a Second Green Revolution for Sustainable Agriculture by Vermiculture Biotechnology Using Earthworms: Reviving the Dreams of Sir Charles Darwin; J. of Agricultural Biotechnology and Sustainable Development; 2(7): 113-128; Academic Journals, USA; (2010).

R.K. Sinha, Dalsukh Valani, Vinod Chandran and Brijal K. Soni : Earthworms – The Soil Managers: Their Role in Restoration and Improvement of Soil Fertility; Agricultural Issues and Policies; NOVA Science Publishers, N.Y., USA; ISBN 978-1- 61122-514-3; (2011).

R.K. Sinha, P.K. Singh, D. Valani and S. Agarwal: Earthworms Vermicompost: An Economically Viable and Environmentally Sustainable Alternative to Destructive Chemical Fertilizers: Some Experimental Studies on Potted and Farmed Cereal and

Vegetable Crops; In Justin A Daniels (Ed.) Advances in Environmental Research – 9 (Chapter 2); NOVA Science Publishers, N.Y., USA; ISBN: 978 – 1 – 61728 – 999 - 6. (2009).

R.K. Sinha, S. Agarwal, K.Chauhan and D. Valani : The Wonders of Earthworms and its Vermicompost in Farm Production: Charles Darwin's 'Friends of Farmers', With Potential to Replace Destructive Chemical Fertilizers from Agriculture; J. Of Agricultural Sciences, 1 (2): 76-94; Scientific Research Publication, USA (2010).

R.K. Suhane: Vermicompost (In Hindi); Pub. Of Rajendra Agriculture University, Pusa, Bihar; pp: 88 (www.kvksmp.org) (Email: info@kvksmp.org). (2007).

R.L. Neilson: Presence of Plant Growth Substances in Earthworms, Demonstrated by the Paper Chromatography and Went Pea Test; Nature, (London); 208: 1113-1114. (1965).

R.M. Atiyeh, C.A. Lee Edward, N.Q. Arancon and J.D. Metzger: The Influence of Humic Acids Derived from Earthworm-Processed Organic Wastes on Plant Growth; J. of Bioresource Technology; 84: 7-14. (2002).

R.M. Atiyeh, S. Subler, C.A. Edwards, G. Bachman, J.D. Metzger and W. Shuster : Effects of Vermicomposts and Composts on Plant Growth in Horticultural Container Media and Soil; Pedobiologia; 44; 579-590.(2000).

R.P. Singh and M. Agrawal, Waste Manage. 28, 347 (2008).

R.T. Haug, Lewis Publishers, Boca Raton, FL (1993)

R.V. Krishnamoorthy and S.N. Vajranabhaiah: Biological Activity of Earthworm Casts: An Assessment of Plant Growth Promoter Levels in the Casts; Proc. of Indian Academy of Sciences (Animal Science): 95: 341-351. (1986).

Radin, A.M.; Warman, P.R. Interntl. J.Veg. Sci. 2010. 16(4), 374-391.

Ravey, M. Spectroscopy 1990, 5(7), 11. 2. de Haseth, J. Spectroscopy 1990, 5(7), 11.

Raviv, M., Reuveni, R., Zaidman, B.Z., 1998. Improved medium for organic transplant. Biology and Agriculture Horticulture, 16, 53-64.

Recycled Organics Units, 2nd Edition, Life cycle inventory and life cycle assessment for windrow composting systems. Department of Environment and Conservation, The University of New South Wales, Sydney, Australia (2007).

Reddy and K. Ohkura: Vermicomposting of Rice Straw and its Effect on Sorghum Growth; Tropical Ecology; 45 (2): 327-331. (2004).

Relwani, L. L., Manna, G. B. 1964. Effect of blue-green algae in combination with urea on paddy yield. Ibid, 33, 687.

Relwani, L. L., Subrahmanyan, R., 1963. Role of blue-green chemical nutrients and partial soil sterilization on paddy yield. Ibid, 32, 441-43.

Relwani, L. L.,1963. Role of blue-green algae on paddy yield. Current Science, 32, 417-418.

Relwani, L. L.,1965. Response of paddy varieties to blue-green algae and methods of propagation. Ibid, 5, 34.

Resch, M. Grasmug, W. Smeets, R. Braun and R. Kirchmay, Water Sci. Technol. 53, 213 (2006).

Response of Faba bean (Vicia faba L.) to combined inoculation with Azotobacter and Rhizobium leguminosarum bv. Viceae. Applied Soil Ecology, 12, 51-59.

Review, vol. 52, no. 3, pp. 260–299, 1986.

Rodelas, B., González-López, J., Pozo, C., Salmerón, V., Martínez-Toledo, M. V. 1999.

Rojo, F.G., Reynoso, M.M., Ferez, M., Chulze, S.N., Torres, A.M., 2007. Biological control by Trichoderma species of Fusarium solani causing peanut crown root rot under field conditions. Crop Protection, 26, 549-555.

Rubinson, Kenneth A. and Judith Faye Rubinson, Contemporary Instrumental Analysis, Prentice Hall, Upper Saddle River, NJ, 2000.

S. Alencar, F. N. Crespilho, M. R. M. C. Santos, V. Zucolotto, O. N. Oliveira Jr., and W. C. J. Silva, "Influence of film architecture on the charge-transfer reactions of metallophthalocyanine layer-by-layer films," The Journal of Physical Chemistry C, vol. 111, pp. 12817–12821, 2007.

S. Bhadra, D. Khastgir, N. K. Singha, and J. H. Lee, "Progress in preparation, processing and applications of polyaniline," Progress in Polymer Science, vol. 34, no. 8, pp. 783–810, 2009.

S. Bhatia, Sonu, R. K. Sinha and R. Sharma : Seeking Alternatives to Chemical Fertilisers for Sustainable Agriculture: A Study on the Impact of Vermiculture on the Growth and Yield of Potted Wheat Crops (Triticum aestivum Linn); International J. Of Environmental Education and Information; University of Salford, UK, 19 (4): 295-304.(2000).

S. Bhatia: Earthworm and Sustainable Agriculture: Study of the Role of Earthworm in Production of Wheat Crop; Ph.D Thesis Awarded by University of Rajasthan, Jaipur, India. (Supervisor: Rajiv K. Sinha). (2000).

S. Bose and A.K. Bhattacharyya, Chemosphere 70, 1264 (2008).

S. Bose, S. Chandrayan, V. Rai, A.K. Bhattacharyya and A.L. Ramanathan, Bioresour Technol. 99, 4467 (2008).

S. Brown, C. Kruger and S. Subler, J. Environ. Qual. 37, 1396 (2008).

S. Bruun, T.L., Hansen, T.H. Christensen, J. Magid and L.S. Jensen, Environ. Model. Assess. 11, 251 (2006).

S. C. Davis and L. G. Ovington, "Electrical stimulation and ultrasound in wound healing," Dermatologic Clinics, vol. 11, no. 4, pp. 775–781, 1993.

S. Compant, B. Duffy, J. Nowak, C. Clement and E.A. Barka : Use of plant growthpromoting bacteria (from soil) for bio-control of plant diseases; Principles, mechanisms of action and future prospects; Appl. Environ. Microbiol; 71 (9): 4951 – 4959. (2005).

S. Dagnall, Bioresource. Technol. 52, 275 (1995).

S. Froelich, M. P. Gupta, K. Siems, and K. Jenett-Siems, "Phenylethanoid glycosides from Stachytarpheta cayennensis

S. G. Prabhulkar and R.M. Patil, "2-Hydroxy-1-naphthalidine salicylohydrazone as an analytical reagent for extractive spectrophotometric determination of a biologically and industrially important metal Cobalt(II)," International Journal of Chemical Sciences, vol. 6, no. 3, pp. 1480–1485, 2008.

S. Huang, B. D. Humphrey, and A. G. Macdiarmid, "Polyaniline, a novel conducting polymer. Morphology and chemistry of its oxidation and reduction in aqueous electrolytes," Journal of the Chemical Society, Faraday Transactions.1, vol. 82, no. 8, pp. 2385–2400, 1986.

S. Khan, Q. Cao, Y.M. Zheng, Y.Z. Huang and Y.G. Zhu, Environ. Pollut. 152, 686 (2008).

S. L. Experimental Design: A Chemometric Approach, Elsevier: Amsterdam, 1987.

S. L. Goulart and C. R. Marcati, "Anatomia comparada do lenho em raiz e caule de Lippia salviifolia Cham. (Verbenaceae)," Revista Brasileira de Botanica, vol. 31, no. 2, pp. 263– 275, 2008.

S. Lafitte-Trouqué and C.F. Forster, Bioresour. Technol. 71, 77 (2000).

S. Nomeda, P. Valdas, S.-Y Chen and J.-G. Lin, Waste Manage. 28, 1637 (2008).

S. Pajon: 'The Worms Turn-Argentina'; Intermediate Technology Development Group, Case Study Series 4, (Quoted in Munroe, 2007).

S. Palaniswamy: Earthworm and Plant Interactions; Paper presented in ICAR Training Program; Tamil Nadu Agricultural University, Coimbatore. (1996).

S. S. Patil and A. D. Sawant, "Pyridine-2-acetaldehyde salicyloylhydrazone as reagent for extractive and spectrophotometric determination of cobalt(II) at trace level," Indian Journal of Chemical Technology, vol. 8, no. 2, pp. 88–91, 2001.

S. Scheu: Microbial Activity and Nutrient Dynamics in Earthworms Casts. J. Of Biological Fertility Soils, 5: 230-234. (1987).

S. Subler, Edwards Clive and Metzger James: Comparing Vermicomposts and Composts; Biocycle, 39: 63-66. (1998).

S. Subler, Edwards Clive and Metzger James: Comparing Vermicomposts and Composts; Biocycle, 39: 63-66. (1998).

S. Suthar: Impact of vermicompost and composted farmyard manure on growth and yield of garlic (Allium stivum L.); International Journal of Plant Production, 3(1), 1-12. (2009).

S.A. Vigil and G. Tchobanoglous, Comparison of the Environmental Effects of Aerobic and Anaerobic Composting Technologies, Report 94-WPIOO.05, Air and Waste management association (1994).

S.C. Tiwari, B.K. Tiwari and R.R. Mishra: Microbial Populations, Enzyme Activities and Nitrogen-Phosphorus-Potassium Enrichment in Earthworm Casts and in Surrounding Soil of a Pineapple Plantation. J. of Biology and Fertility of Soils; 8, 178- 182. (1989).

S.D. Nighawan and J.S. Kanwar: Physico-chemical Properties of Earthworm Castings; Indian J. of Agricultural Sciences; 22: 357-375. (1952).

S.G. Pavlostathis and E. Giraldo-Gómez, Kinetic of anaerobic treatment: a critical review. Crit. Rev. Environ. Contr. 21, 411 (1991).

S.M. Al-Salem and P. Lettieri, Eur. J. Sci. Res. 34 (3), 395 (2009).

S.N. Jordan, G.J. Mullen and M.C. Murphy, Bioresour. Technol. 99, 411 (2008).

S.N. Misi and C.F. Forster, Bioresour. Technol. 80, 19 (2001).

S.R. Wild and K.C. Jones, Sci. Total Environ. 119, 85 (1992).

S.V. Krupa, Environ. Pollut. 124, 179 (2003).

Sabuquillo, P., De Cal, A., Melgarejo, P., 2006. Biocontrol of tomato wilt by Penicillium oxalicum formulations in different crop conditions. Biological Control, 37, 256-265.

Sarjolta: Use of Vermicompost in Apple Orchards in Himachal Pradesh, India; Personal Communication (Email: vineet787@gmail. com). (2009).

Schroeder, G.L., Wohlfarth, G., Alkon, A., Halevy, A., Krueger, H., 1990. The dominance of algal-based food webs in fish ponds receiving chemical fertilizers plus organic manures. Aquaculture, 86, 219-229.

Settle, Frank A., Ed., Handbook of Instrumental Techniques for Analytical Chemistry, Prentice Hall, Upper Saddle River, NJ, 1997.

Shen, Y.W., Liu, Y. D., Wu, G.Q., Ao, H. Y., Qiu, C. Q., 2005. Efficiency test on organic and inorganic fertilizers with cyanobacteria (Microcysits) in several crops. Acta Hydrobiologica Sinica, 29, 400-405.

Skoog, Douglas A., Donald M. West, and F. James Holler, Fundamentals of Analytical Chemistry, 7th ed., Saunders College Publishing Co., Philadelphia, 1997.

Skoog, Douglas A., F. James Holler, and Timothy A. Nieman, Principles of Instrumental Analysis, 5th ed., Saunders College Publishing Co., Philadelphia, 1998.

Smith, S.R. Environ. Interntl. 2009. 35(1), 142-156.

Sohn, S.M., Han, D.H., Kim, Y.H., 1996. Chemical characteristics of soils cultivated by the conventional farming, greenhouse cultivation and organic farming and accumulation ofNO3- in Chinese cabbage and lettuce. Journal of Korean Organic Agriculture, 5, 149-165.

Spolaore, P., Joannis-Cassan, C., Duran, E., Isambert, A., 2006. Commercial applications of microalgae. J. Biosci. Bioeng. 101, 87-96.

Stewart, I., Falconer, I. R., 2008. Cyanobacteria and cyanobacterial toxins. Walsh, P.J., Smith, S.L., Fleming, L.E., Eds, Oceans and human health: risks and remedies from the seas, Academic Press.

Suárez-Estrella, F., Vargas-Garcia, C., Lopez, C.M.J., Capel, J.M., 2007. Antagonistic activity of bacteria and fungi from horticultural compost against Fusarium oxysporum f. sp. melonis. Crop Protection, 26, 46-53.

Subrahmanyan, R., 1964a. Observations on the role of blue-green algae on rice yield compared with that of conventional fertilizers. Ibid, 33, 485-486.

Subramanian, G., Ed., Quality Assurance in Environmental Monitoring: Instrumental Methods, VCH, Weinheim, Germany, 1995.

T. Guaratini, R. Vessecchi, E. Pinto, P. Colepicolo, and N. P. Lopes, "Balance of xanthophylls molecular and protonated molecular ions in electrospray ionization," Journal of Mass Spectrometry, vol. 40, no. 7, pp. 963–968, 2005.

T. Katami, T. Hayakawa, M. Furukawa, and S. Shibata, "Extraction—spectrophotometric determination of iron with

T. Łuczak, "Preparation and characterization of the dopamine film electrochemically deposited on a gold template and its applications for dopamine sensing in aqueous solution," Electrochimica Acta, vol. 53, no. 19, pp. 5725–5731, 2008.

T. Manios, E.I. Stentiford and P.A. Millner, Ecol. Eng. 20, 65 (2003).

T. Müller, R. Thiflen, S. Braun, W. Dott and G. Fischer, Environ. Sci. Pollut. Res. 11 (2), 91 (2004).

T. Paré, H. Dinel, M. Schnitzer and S. Dumontet, Biol. Fertil. Soils 26, 173 (1998).

T. S. Fiuza, S. M. T. Sab ʹoia-Morais, J. R. Paula et al., "Composition and chemical variability in the essential oils of Hyptidendron canum (Pohl ex Benth.) Harley," Journal of Essential Oil Research, vol. 22, no. 2, pp. 159–163, 2010.

T. Wu, X. Wang, D. Li and Z. Yi, Atmos. Environ. 44, 5065 (2010).

T.B. Chen, J.W.C. Wong, H.Y. Zhou and M.H. Wong, Environ. Pollut. 96, 61 (1997).

Talapatra, P. K. Chaudhuri, and S. K. Talapatra, "(-)- Maglifloenone, a novel spirocyclohexadienone neolignan and the determination of iron in titanium base alloys using 4-(2-Pyridylazo) resorcinol and a long chain quaternary ammonium salt," Journal of the Indian Chemical Society, vol. 66, pp. 437–439, 1989.

thiocyanate-phenanthroline-OP," Fenxi Shiyanshi (Analytical Laboratory), vol. 24, no. 1, pp. 77–79, 2005.

Thomas and F. McDougall, J. Clean. Prod. 13, 321 (2005).

Tomati, A. Grappelli and E. Galli: The Presence of Growth Regulators in Earthworm-Worked Wastes; In Proceeding of International Symposium on 'Earthworms'; Italy; 31 March-5 April, 1985; 423-436. (1987).

Tropical Forest Plants, K. S. Bawa and M. Hadley, Eds., pp. 191–218, UNESCO & The Parthenon Publising Group, Paris,

twigs and their in vitro activity against Trypanosoma cruzi," Planta Medica, vol. 64, no. 7, pp. 667–669, 1998.

UNEP/GEMS: The Contamination of Food; Publication of United Nation Environment Program; Environment Library No. 5, Nairobi, Kenya. (1992).

V.K. Sharma, M. Canditelli, F. Fortuna and G. Cornacchia, Energy Convers Manag 38, 453 (1997).

Virola surinamensis," Phytochemistry, vol. 35, no. 6, pp. 1469–1470, 1994.

W.H. Elmer: Influence of earthworm activity on soil microbes and soil-borne diseases of vegetables; Plant Diseases; 93 (2): 175-179. (2009).

Walter, F. Martínez and V. Cala, Environ. Pollut. 139, 507 (2006).

Wang, Z. Sun, Y. Liu, D. Zheng, X. Liu, and S. Li : Earthworm polysaccharide and its anti-bacterial function on plant-pathogen microbes in vitro; European J. of Soil Biol.; 43: 135-142. (2007).

Warman, P.R. Dynamic Soil. Dynamic Plant, Theme: Compost. Global Science Books. submitted 12/2010.

Warman, P.R.; Hargreaves, J.C.; Radin, A.M. In: Proceedings 7th International Conference, Orbit 2010. Eds: Lasardi, K., Manios, T., Bidlingmaier, W., Abeliotis, K., de Bertoldi, M., Diaz, L., Stentiford, I. Heraklion, Crete, Greece, 2010. CD ROM. pp. 516-521.

Weller, D.M., Raaijmakers, J.M., Mc Spadden Gardener, B.B., Thomashow, L.S., 2002. Microbial populations responsible for specific soil suppressiveness to plant pathogens. Annual Reviwe of Phytopathology, 40, 309-348.

Whipps, J.M., 1997. Developments in the biological control of soilborne pathogens. Advance in Botany Research, 26, 1-134. .

Wu, Li-Xiang, Guo, and J. Cun, Metallurgical Analysis, vol. 24, no. 3, pp. 66–68, 2004.

X., Kuiwu and D. Xingting: 'Culture and Utilization of Earthworms", Nanjing Publisher, China. (1988).

X.F. Lou and J. Nair, Bioresour. Technol. 100, 3792 (2009).

X.Z. Li, Q.L. Zhao and X.D. Hao, Waste Manage. 196, 409 (1999).

Yan, S. Z., Yang, Q. Y., Chen, Y. Y., 2004. Antagonism of complex microbial fertilizer and functional actinomycetes against soil-borne plant pathogenic fungi. Chinese Journal of Biological Control, 20, 49-52.

Ying, J.F., Herridge, D.F., Peoples, M.B., Rerkasem, B., 1992. Effects of N fertilization on N2 fixation and N balances of soybean grown after lowland rice. Plant Soil, 147, 235-242.

Z. Zhai, "Catalytic kinetic spectrophotometric determination of trace copper with copper(II)-p-acetylchlorophosphonazo- hydrogen peroxide system," Bulletin of the Chemical Society of Ethiopia, vol. 23, no. 3, pp. 327–335, 2009.

Z.J. Wang and C.J. Banks, Process Biochemistry 38, 1267 (2003).

Zhang, Q., Qin, T., Zhang, H.Y., Luan, X. W., 2005. The progress of microbial fertilizer application. Xinjiang Agricultural Science (in Chinese), 42, 159-180.

Zhao, Q., Dong, C., Yang, X., Mei, X., Ran, W., Shen, Q., Xu, Y., 2010. Biocontrol of Fusarium wilt disease for Cucumis melo melon using bio-organic fertilizer. Applied Soil Ecology, 47, 67-75.

Zheljazkov, V.D.; Warman, P.R. Compost Sci. Util. 2002 10(3),197–203.

Zhuang, S.D., 2003. The status, problem and strategy on the development and application of microbial fertilizer. Fujiang Agricultural Science and Technology, 1, 34-45.

INDEX

www.ingramcontent.com/pod-product-compliance
Lightning Source LLC
LaVergne TN
LVHW062213200926
848484LV00006B/1387
9781787150386